ENDANGERED AND THREATENED PLANTS OF THE UNITED STATES

EDWARD S. AYENSU

and

ROBERT A. DeFILIPPS

With the Assistance of

SAM E. FOWLER, MARY G. MANGONE,

CAROL MATTI-NATELLA, *and* WILLIAM E. RICE

Published jointly by

the SMITHSONIAN INSTITUTION *and the* WORLD WILDLIFE FUND, INC.

Washington, D.C.

1978

Distributed by the Smithsonian Institution Press

© 1978 Smithsonian Institution. All rights reserved.

ISBN 0-87474-222-6

Library of Congress Cataloging in Publication Data

Ayensu, Edward S.
  Endangered and threatened plants of the United
States.

  Ed. of 1975, by the Smithsonian Institution, published under title: Report on endangered and threatened plant species of the United States.
  Includes bibliographies.
  1. Rare plants—United States. 2. Plant conservation—United States. I. DeFilipps, Robert A., joint
author. II. Smithsonian Institution. III. Smithsonian Institution. Report on endangered and threatened plant species of the United States. IV. Title.
QK86.U6A93   1978   333.9'5   77-25138
ISBN 0-87474-222-6

Cover: *Pedicularis furbishiae,* the Furbish lousewort
belonging to the snapdragon family, is one of the
proposed endangered plant species of the United
States. This plant, which was presumed extinct in
Maine since 1917, was rediscovered along the St. John
River in 1976. The plant was also recently found in
New Brunswick, Canada, after an absence of
thirty-five years. Illustration by Alice R. Tangerini.

FOREWORD

The Endangered Species Act of 1973 (Public Law 93-205, enacted
December 28, 1973) directed the Secretary of the Smithsonian Institution to
review species of plants that are or may become endangered or threatened
and methods of adequately conserving such species.

As a result of that request, the *Report on Endangered and Threatened
Plant Species of the United States* was prepared and transmitted to the
United States Congress on January 9, 1975. The Report was subsequently
published by the U. S. Government Printing Office (1975) for the use of the
House Committee on Merchant Marine and Fisheries, of which the Subcommittee
on Fisheries and Wildlife Conservation and the Environment oversees the
Endangered Species Act of 1973.

Having accepted the Report as a petition within the context of the
Endangered Species Act, the Director of the U.S. Fish and Wildlife Service
published on July 1, 1975, a notice of review of the status of the plants
recommended in the Report. A further step towards official listing was
achieved on June 16, 1976, when the Fish and Wildlife Service proposed
endangered status for many of the recommended plants. The endangered
list in the present publication is essentially the same as the June 16
proposal. In August of 1977, the Director of the Fish and Wildlife
Service officially listed four California plants as endangered species. The
aforementioned steps, however, indicate the Interior Department's intent to
pursue a course of action that will lead to official determination of the
status of endangered and threatened plants listed in this publication.

i

A number of significant events have continued to highlight the concern of international and national groups over the situation of endangered and threatened plants. Plant conservation was a dominant concern of the 12th International Botanical Congress held in Leningrad, U.S.S.R., in July 1975. Among the resolutions adopted at the closing plenary session of that Congress was an appeal to all governments to make their efforts toward food production consistent with the principles of conservation and the need to preserve the genetic diversity of the Earth's flora and fauna.

Recognizing that man's activities have caused the extinction or severe reduction in numbers of numerous plant species in many parts of the world, and that such destruction seems to be increasing, the 12th Botanical Congress urged the governments of all nations to conserve wild plants in every way possible by the establishment and maintenance of ecosystem reserves and national parks.

The Function of Living Plant Collections in Conservation and in Conservation-oriented Research and Public Education was the subject of an international conference held at the Royal Botanic Gardens, Kew, England, during September 1975 under the sponsorship of NATO. The conference strongly asserted that the basic requirement for the preservation of the threatened floras of the world was conservation of their natural habitats. The resolutions adopted at that conference called for early world-wide action through the extension of nature-protection policies and the establishment of an adequate network of ecosystem reserves in all the major floristic regions.

The 13th Technical Meeting of the International Union for the Conservation of Nature and Natural Resources (IUCN) was held at Kinshasha, Zaire, in September 1975. Major emphasis was given to tropical rain forests and the need for conserving plants and plant genetic resources threatened by advancing technological, land, and population pressures.

An international symposium on threatened and endangered plant species of the Americas was sponsored by the New York Botanical Garden in May 1976, at which the present and future significance of those plants in their ecosystems was considered in detail.

A meeting of the Conference of Parties to the Convention on International Trade in Endangered Species of Wild Fauna and Flora was held in Berne, Switzerland, from November 2 to 6, 1976. The effects of international commerce in endangered plants upon the ultimate extinction of the species from the wild were probed, as was the regulation of such trade by the issuance of export and import permits.

In 1975 and 1976 endangered species conferences were held in various states including Tennessee, Alabama, Maryland, New Mexico, North Carolina, and Oregon. These conferences often resulted in new or revised state lists of endangered plants. Rare, endangered, and threatened biota of the southeastern states was the subject of a symposium co-sponsored by the Association of Southeastern Biologists and Ecological Society of America at New Orleans in April 1976.

The Smithsonian Report of 1975, which regarded about 10 percent of the native flora of the continental United States as endangered and threatened, was given prominent exposition in the Living Resources section of the Sixth Annual Report of the President's Council on Environmental Quality (1975, pp. 408-412).

Oversight hearings on the administration of the Endangered Species Act of 1973, at which the Smithsonian report was a conspicuous feature of the discussions and testimony, were held before the House Subcommittee on Fisheries and Wildlife Conservation and the Environment in Washington, D.C., on October 1, 2, and 6, 1975, and the Senate Subcommittee on Environment on May 6, 1976.

Since the publication of the lists of endangered, threatened, extinct and commercially exploited plants in the 1975 Smithsonian report, additional data on the taxonomy, geographical range, and endangered status of the species, subspecies, and varieties, as well as suggestions of taxa to be included or deleted from the earlier listing, have been received by the Endangered Flora Project of the Smithsonian Institution. This new and vital data based on field and specialist knowledge has been generously provided by a network of botanical colleagues and correspondents across the United States, including the curatorial staff of the Smithsonian's Department of Botany.

The revised lists reflect this new data, and we have also included a section on endangered and threatened plants of Puerto Rico and the Virgin Islands, extensive bibliographies on endangered plants and the conservation of plant habitats, and examples of computerized information sheets and locality maps.

The report to Congress in 1975 and this publication were prepared under the supervision of Dr. Edward S. Ayensu, Director of the Endangered Species Program. He was assisted by Dr. Robert A. DeFilipps.

The revised lists in this review, as compared with the 1975 Report, still comprise approximately 10 percent of the native vascular flora of the continental United States, 2140 species, subspecies, and varieties, and about 50 percent of the Hawaiian flora.

We hope that this publication will be of practical use to all involved
with the conservation of America's endangered and threatened plants as part
of this nation's biological heritage.

                        S. Dillon Ripley
                        Secretary
                        Smithsonian Institution

ACKNOWLEDGMENTS

Since the publication of the *Report on Endangered and Threatened Plant Species of the United States,* which Secretary S. Dillon Ripley presented to the 94th Congress in 1975, the Endangered Flora Project has received numerous positive responses from many people who are concerned with the state of the endangered flora of this nation. The Project has been much encouraged by the active participation of many of our botanical colleagues (see page 235 for a complete list of contributors), who have supplied us with specialized taxonomic and ecological information on various plant taxa, as well as phytogeographical information on many regions of this country. I am most grateful to all of them for their sustained interest in this endeavor.

I am particularly pleased with the effort and encouragement of the many members of the Garden Club of America in their appeals to government agencies to lend support for this Project.

An undertaking of this scope and magnitude cannot progress without the sustained effort and dedication of the Project staff. I am especially grateful for the services and support of Dr. Robert DeFilipps, Ms. Carol Matti-Natella, Mr. Sam Fowler, Dr. William Rice, and Ms. Mary Mangone. In addition, staff members of the Department of Botany and Dr. Porter Kier, Director of the National Museum of Natural History, and his staff have offered freely their time and services to the Project. I also want to thank Dr. David Challinor, Assistant Secretary for Science and his staff for their support.

Dr. T. Gary Gautier, Mr. David Bridge and other members of the Smithsonian's Automatic Data Processing Program have rendered invaluable aid in the computerization of the plant lists, locality maps and data sheets presented in this review.

The entire project could not have achieved its stated objectives without the personal support and encouragement of Secretary S. Dillon Ripley. We owe the success of this Project to him.

Edward S. Ayensu
Director
Endangered Species Program
Smithsonian Institution

CONTENTS

TABLE 1

Statistical Comparison of Endangered, Threatened, Extinct* Native Plants of the United States

	Continental United States (including Alaska)				Hawaii			
	Species, subspecies and varieties		Percent of Continental flora		Species, subspecies and varieties		Percent of Hawaiian flora	
	1975	1977	1975	1977	1975	1977	1975	1977
Endangered	761	839	3.8	4.2	639	646	29.0	29.4
Threatened	1238	1211	6.1	6.1	194	197	8.8	8.9
Extinct	100	90	0.5	0.4	255	270	11.6	12.3
Total	2099	2140	10.4	10.7	1088	1113	49.4	50.6

*Total native flora of continental United States (including Alaska) = approximately 20,000 species.
Total native flora of Hawaii = approximately 2200 species.
1975 statistics are from Report on Endangered and Threatened Plant Species of the United States (1975).

TABLE 2

STATE TOTALS OF RECOMMENDED ENDANGERED, THREATENED, AND EXTINCT NATIVE PLANT SPECIES

STATE	ENDANGERED	THREATENED	EXTINCT	TOTAL
Alabama	33	57	0	90
Alaska	6	13	0	19
Arizona	62	104	2	168
Arkansas	10	18	0	28
California	275	372	33	680
Colorado	41	23	4	68
Connecticut	3	9	0	12
Delaware	2	9	0	11
District of Columbia	0	2	0	2
Florida	77	121	4	202
Georgia	26	68	4	98
Hawaii	646	197	270	1113
Idaho	23	38	0	61
Illinois	7	16	1	24
Indiana	2	13	1	16
Iowa	3	7	0	10
Kansas	1	2	0	3
Kentucky	7	23	0	30
Louisiana	1	8	3	12
Maine	6	11	0	17
Maryland	1	11	0	12
Massachusetts	1	10	1	12
Michigan	2	14	0	16
Minnesota	3	10	0	13
Mississippi	2	17	0	19

STATE	ENDANGERED	THREATENED	EXTINCT	TOTAL
Missouri	9	19	0	28
Montana	6	12	0	18
Nebraska	0	4	0	4
Nevada	49	82	4	135
New Hampshire	6	8	0	14
New Jersey	4	13	0	17
New Mexico	26	30	1	57
New York	6	20	0	26
North Carolina	20	55	2	77
North Dakota	0	3	0	3
Ohio	3	9	0	12
Oklahoma	11	10	1	22
Oregon	57	122	3	182
Pennsylvania	4	13	1	18
Rhode Island	1	5	0	6
South Carolina	13	40	0	53
South Dakota	0	2	0	2
Tennessee	23	46	1	70
Texas	97	127	15	239
Utah	56	133	5	194
Vermont	4	7	1	12
Virginia	13	30	1	44
Washington	17	67	2	86
West Virginia	0	11	0	11
Wisconsin	3	12	0	15
Wyoming	3	21	2	26

INTRODUCTION

Throughout the world, most of the efforts to conserve living organisms have been concentrated on endangered animals, particularly mammals and birds. *Red Data Books* containing detailed information on endangered animals have been prepared for a number of years by the Survival Service Commission of the International Union for the Conservation of Nature and Natural Resources (IUCN). The volume on birds in the IUCN series is prepared by experts from the International Council for Bird Preservation. Only comparatively recently have the individual species comprising the world's endangered flora (as opposed to conservation of vegetation in general) been given serious consideration; a *Red Data Book* on angiosperms (flowering plants) was started in 1970.

The number of species of flowering plants in the world is variously estimated as approximately 226,000 (R. Melville, 1970 b), 235,000 (Raven, Evert, and Curtis, 1976), and 250,000 (G. Lucas, 1975). Dr. Ronald Melville of the Royal Botanic Gardens at Kew has estimated that of the world's flora, approximately 10 percent of the species is endangered or vulnerable and threatened with extinction by the year 2000 AD. The Smithsonian Institution's Endangered Flora Project is involved in gathering data on the endangered status of native flora of the United States, which numbers approximately 22,200 species.

This report presents the second national lists of recommended endangered, threatened, recently extinct, and commercially exploited, species of native vascular plants of the United States. The presentation has been considerably expanded beyond that of the 1975 Report, and includes a treatment on Puerto

Rico and the Virgin Islands; chronology of events pertaining to the federal government's response to the Endangered Species Act of 1973; bibliographies of state lists, endangered flora of the United States and the world, the role of botanical gardens in preservation, and the need for a diversity of plants; computerized data sheets; examples of computer-drawn maps of exploited species and of data sheets on some of the plants listed in Appendix I to the Convention on International Trade.

The bibliographies attempt to bring together a wide range of pertinent articles from the growing literature on endangered plant species, altered habitats, and broken ecosystems. They may also serve as a guide to the botanical and conservation-oriented journals in which more articles are likely to appear in the future. Occasionally a cited article is concerned with more than one of the primary subject headings (United States, World, Diversity, Botanical Gardens), but such articles have been listed under the single heading thought to be the most generally applicable.

The recommendations for conservation of endangered plants which were presented to Congress in the 1975 Report are reiterated with additional relevant comments, and the value of retaining diverse plant species, the causes of rarity, and plant conservation are discussed. The methods and criteria used in preparing the lists, the present status of the lists, and future plans are also presented.

Status of the Lists

Historical Perspective

The efforts of the Department of the Interior towards official listing of endangered and threatened United States plants may be discussed in relation to its response to three entities: Convention on International Trade in Endangered Species of Wild Fauna and Flora (1973); Endangered Species Act of 1973; and the Smithsonian Institution *Report on Endangered and Threatened Plant Species of the United States* (1975).

The importance of the present lists is that they will be utilized as basic sources by the Department of the Interior in its progress towards an official listing of endangered and threatened plant species.

A chronology of progress towards the listing of endangered and threatened plants of the United States by the federal government is given below.

1973

<u>Convention</u> <u>on</u> <u>International</u> <u>Trade</u> <u>in</u> <u>Endangered</u> <u>Species</u> <u>of</u> <u>Wild</u> <u>Fauna</u> <u>and</u> <u>Flora</u> was signed by the United States on March 3. This Convention undertakes to regulate the export, import, and transit of specimens and products of plant species threatened or not yet threatened with world-wide extinction. The Convention was prepared and adopted by the Plenipotentiary Conference to Conclude an International Convention on Trade in Certain Species of Wildlife, held during February and March in Washington, D.C.

<u>Endangered</u> <u>Species</u> <u>Act</u> <u>of</u> <u>1973</u> became law on December 28. (Public Law 93-205, 93rd Congress). Unlike the Endangered Species Acts of 1966 and 1969, provisions for plant species are included in the Act of 1973.

1975

Smithsonian Institution <u>Report</u> <u>on</u> <u>Endangered</u> <u>and</u> <u>Threatened</u> <u>Plant</u> <u>Species</u> <u>of</u> <u>the</u> <u>United</u> <u>States</u> was submitted to the U.S. Congress on January 9. (House Document No. 94-51, Serial No. 94-A, 94th Congress, 1st Session, Government Printing Office, 200 pages). This Report was printed for the use of the Committee on Merchant Marine and Fisheries.

<u>Notice</u> <u>of</u> <u>Review</u> of status of <u>Aconitum</u> novaeboracense, <u>Sullivantia</u> <u>renifolia</u>, <u>Primula</u> <u>mistassinica</u>, and <u>Saxifraga</u> <u>forbesii</u> was published by the U.S. Fish and Wildlife Service on April 21. <u>Federal</u> <u>Register</u> 40(77): 17612.

Notices on the concept and determination of <u>Critical</u> <u>Habitat</u>, as it relates to Section 7 of the Endangered Species Act of 1973, were published by the U.S. Fish and Wildlife Service on April 22 and May 16. <u>Federal</u> <u>Register</u> 40(78): 17764-17765; 40(96): 21499-21501.

<u>Notice</u> <u>of</u> <u>Review</u> of status of all plants listed in the Smithsonian Report was published by the U.S. Fish and Wildlife Service on July 1. <u>Federal</u> <u>Register</u> 40(127): 27823-27924.

<u>Convention</u> <u>on</u> <u>International</u> <u>Trade</u> <u>in</u> <u>Endangered</u> <u>Species</u> <u>of</u> <u>Wild</u> <u>Fauna</u> <u>and</u> <u>Flora</u> was ratified by the United States on July 1, and the United States became a party to the Convention as of that date.

<u>Proposed</u> <u>endangered</u> <u>status</u> for plants in <u>Appendix</u> <u>I</u> to the Convention on International Trade was published by the U.S. Fish and Wildlife Service on September 26. <u>Federal</u> <u>Register</u> 40(188): 44329-44333. At present, no native plants of the United States are in Appendix I.

<u>Oversight</u> <u>Hearings</u> on implementation and administration of the Endangered Species Act of 1973 were held in Washington, D.C., on October

1, 2, and 6. Published as Hearings before the Subcommittee on Fisheries and Wildlife Conservation and the Environment of the Committee on Merchant Marine and Fisheries (House of Representatives, Ninety-Fourth Congress, First Session. Serial No. 94-17, Government Printing Office, 367 pages). This publication was printed for the use of the Committee on Merchant Marine and Fisheries, 1976.

1976

Executive Order Number 11911, designating the Management Authority and Scientific Authority for the United States, under the Convention on International Trade, was signed by the President on April 13.

Hearing to amend the Endangered Species Act of 1973 to authorize appropriations and to review progress in administration of the Act was held in Washington, D.C., on May 6. Published as Hearing before the Subcommittee on Environment of the Committee on Commerce (United States Senate, Ninety-Fourth Congress, Second Session. Serial No. 94-82, Government Printing Office, 190 pages), this report was printed for the use of the Committee on Commerce.

Proposed Regulations for endangered and threatened plants under the Endangered Species Act of 1973 were published by the U.S. Fish and Wildlife Service on June 7. Federal Register 41(110): 22915-22922.

Proposed Implementation of the Convention on International Trade was published by the U.S. Fish and Wildlife Service on June 16. Federal Register 41(117): 24367-24378.

Proposed endangered status for approximately 1700 United States plants (many of which are recommended as endangered in the present publication) was

published by the U.S. Fish and Wildlife Service on June 16. Federal
Register 41(117): 24523-24572.

Public hearings regarding the Fish and Wildlife Service's proposals to
list some 1700 United States plants as endangered, to list plants in Appendix
I to the Convention on International Trade as endangered, and the proposed
regulations for plants, were held in Honolulu, Hawaii (July 14); El Segundo,
California (July 22); Kansas City, Missouri (July 28); Washington, D.C.
(August 4). Excerpts of testimony appear in the Endangered Species Technical
Bulletin 1(3), September 1976; comments on the hearings have been published
in National Parks and Conservation Magazine 50(10): 29-30 (October 1976).

First Conference of the Parties to the Convention on International
Trade in Endangered Species of Wild Fauna and Flora was held on November
2-6 in Berne, Switzerland. Among the items adopted at the Conference were
a recommendation that inventories be made of endangered fauna and flora now
in museums and herbaria to reduce demand for specimens from the wild; a
procedure for marking shipments of specimens exchanged between herbaria,
using standard customs forms; and a resolution urging governments to protect
island fauna and flora and their habitats. A printed Report of the U.S.
Delegation was issued December 10 at a consultative meeting held at the
Department of the Interior.

1977

Proposed Determination of Critical Habitat for two coastal California
plants, Erysimum capitatum var. angustatum (Contra Costa Wallflower) and
Oenothera deltoides ssp. howellii (Antioch Dunes Evening Primrose), was
published by the U.S. Fish and Wildlife Service on February 8. Federal
Register 42(26):7972-7975.

Implementation of the Convention on International Trade in Endangered Species of Wild Fauna and Flora was published by the U.S. Fish and Wildlife Service on February 22. Federal Register 42(35): 10461-10488. Effective May 23, 1977.

The plants listed in the present publication are considered to be qualified for endangered or threatened status under the Endangered Species Act of 1973 by a consensus of the best botanical expertise available. It must be noted, however, that the authority to determine officially the status of plants, and to list plants as endangered or threatened pursuant to the Act, resides in the Secretary of the Interior.

The Endangered Species Act of 1973 charges the Secretary of the Interior with maintaining lists of those plants and animals throughout the world that either are in danger of extinction (endangered species) or are likely to become so (threatened species). Once a plant species or subspecies has been determined to be endangered or threatened, as indicated by its placement upon the official List of Endangered or Threatened Plants published in the Federal Register, certain restrictions can ensue with respect to interstate and international commerce in that plant (or parts derived from it).

The U.S. Fish and Wildlife Service of the Department of the Interior is developing (through publication in the Federal Register), regulations for endangered and threatened plants. These regulations will provide for the issuance of permits for scientific purposes or for activities that would enhance the propagation or survival of the species. Similar regulations for animals were published in the Federal Register on September 26, 1975, and are a portion of Part 17 of Title 50 of the Code of Federal Regulations.

Any threatened or endangered species also receives the protection

provided by Section 7 of the Endangered Species Act of 1973. That Section,

entitled "Interagency Cooperation", reads as follows:

Sec. 7. The Secretary [of the Interior] shall review other programs administered by him and utilize such programs in furtherance of the purposes of this Act. All other Federal departments and agencies shall, in consultation with and with the assistance of the Secretary, utilize their authorities in furtherance of the purposes of this Act by carrying out programs for the conservation of endangered species and threatened species listed pursuant to section 4 of this Act and by taking such action necessary to insure that actions authorized, funded, or carried out by them do not jeopardize the continued existence of such endangered species and threatened species or result in the destruction or modification of habitat of such species which is determined by the Secretary, after consultation as appropriate with the affected States, to be critical.

An official list of endangered wildlife already had been developed

when the Endangered Species Act was signed on December 28, 1973. No

such list of plants had been developed and Section 12 of the Act authorized

the Secretary of the Smithsonian Institution to review the problem of

jeopardized flora and to make recommendations concerning their conservation.

The present publication and its January 9, 1975, predecessor constitute a

part of the Smithsonian Institution's response.

The lists contained herein have been forwarded to the U.S. Fish and

Wildlife Service with the recommendation that these plant taxa be placed

on the appropriate official list as provided for by the Endangered Species

Act. Such action requires two steps:

1. The Service must first publish the lists in the Federal Register

in the form of a Proposed Rulemaking and must allow at least sixty days

for any interested parties to submit any comments they care to offer on the proposal. Any information received during the comment period must be analyzed and summarized. In some cases public hearings may be necessary. Final decisions made by the Service are based upon the original proposal and an evaluation of information received from this review.

2. The final decision on those plant taxa that have been determined to be endangered or threatened species in the proposal is announced via a Final Rulemaking in the Federal Register. At that time the species are eligible for the benefits provided for by the Act.

"Critical Habitat" for any endangered or threatened species can be determined as necessary through similar proposed and final rulemakings in the Federal Register, again with provisions for comments by any interested parties. Under certain circumstances, land can be purchased for plants on the official list, using funds made available from the Land and Water Conservation Fund Act of 1965.

The 1975 Report of the Smithsonian Institution was accepted as a petition under Section 4(c) (2) of the Endangered Species Act of 1973, by the U.S. Fish and Wildlife Service publishing a Notice of Review of the plant taxa named therein in the July 1, 1975, Federal Register. The Service published, in the Federal Register of June 16, 1976, a Proposed Rulemaking that some 1700 of the plant taxa named therein are Endangered Species within the authority of the Endangered Species Act of 1973. The plants recommended for threatened status in this publication will be taken under consideration by the Service.

The Endangered Flora Project of the Smithsonian Institution compiles and maintains data on the plants independently of the Department of the Interior, and certain data resulting from its activites are transmitted to the Office of Endangered Species of the U.S. Fish and Wildlife Service (Department of the Interior).

Scope of the Lists

These are the second published lists of recommended endangered and
threatened, commercially exploited, and recently extinct species of native
plants of the United States. The lists should be considered as a picture
of the present status of higher plants. This status is ever-changing as
knowledge increases, and species may be removed or added to future lists
during the process of updating and improvement.

This publication concerns plants which are thought to be endangered
or threatened on a *national basis* in the United States. A species that is
on the edge of its range and rare in one particular state is not included
if that species is common in the neighboring states or elsewhere in the
country. A species may be rare at the edge of its range, but not endangered
or threatened as a whole. In determining endangered, threatened, and
extinction status, the *total* range and abundance of the species or other
recognized taxonomic unit has been considered, and the lists are thus
emphatically *not* a collection of *individual* state lists. In keeping with
this concept, therefore, a species which may have become extirpated in
one state but remains abundant in another would not be placed on the
national list.

The lists comprise only the vascular plants: angiosperms (flowering
plants), gymnosperms (pines and relatives), and pteridophytes (ferns and
fern-allies). The lower plant groups such as algae, fungi, lichens, mosses,
and liverworts are not included. Only species that are indigenous to the
United States (including Alaska and Hawaii) are included. No introduced,
adventive, naturalized, or domesticated species have been listed.

Forms, races, apomicts, and hybrids are not included. In some genera
such as *Crataegus* (hawthorns) and *Rubus* (blackberries), hybridization

between species has produced a plethora of forms that have been given various taxonomic ranks and frequently have been designated as species by some botanists. Because of taxonomic disagreement in these genera, they are not dealt with in the lists. The differences between normal species and hybrids have been summarized by W. H. Wagner, Jr., BioScience 19(9): 785 (1969). At present, hybrids cannot be subject to protection under the Endangered Species Act of 1973.

Plants that are very rare in the United States but also are native in Canada, Europe, Mexico, the West Indies, South America, or other regions, are not included unless their rarity or endangered status and exact distribution are known from throughout their range. Most of such species with ranges outside of the country have been excluded since data on their statuses and ranges generally are not available.

Well-recognized subspecies and varieties are included. Although the Endangered Species Act of 1973 literally provides for plant species (including subspecies), but not varieties, varieties are listed in this publication because of their biological significance, and because the term "subspecies" may be interpreted to include or be the equivalent to "variety".

The U.S. Fish and Wildlife Service has proposed in the Federal Register 41(117): 24525 (June 16, 1976) that "those plants named as 'varieties' in the Smithsonian Institution report and its revision are here [in the proposal of endangered status for some 1700 plant taxa] considered to be subspecies and, therefore, 'species' as defined in section 3(11) of the Act."

The following explanation of the subspecies-variety situation is provided by Dr. F. R. Fosberg, Curator, Department of Botany, Smithsonian Institution.

The species is the basic unit of classification of both animals and plants. Basically a species is a population of individuals more similar among themselves than to other individuals and separated from other such populations by marked morphological discontinuities (gaps in form and structure). By common agreement (and accepted codes of nomenclatural rules) each such species is designated by a binomial--a name made up of two words.

The individuals in such species are usually variable in one or more sets of characteristics. Often there exist subpopulations of individuals characterized by one or more of these variations and usually, by a separate geographic range. Usually, such populations are not sharply set off from others within the same species, but are connected by individuals of intermediate character. The tendency to differentiate into such subpopulations is equally shared by plants and animals and the phenomenon does not differ essentially between the two.

By custom that is built into the zoological code, zoologists have come to call such subpopulations "subspecies". Botanists have traditionally called these subpopulations "varieties". Certain botanists, in imitation of zoologists, have come to use the term "subspecies" for what others call "varieties". Certain other botanists have reserved the term "subspecies" for groups of closely related varieties within species. Botanists also sometimes recognize another category of usually sporadic, very minor variations which they term "forms". The International Code of Botanical Nomenclature, unlike the zoological code, does not, except by implication, specify which of these terms shall be used for the common sort of subpopulations recognizable within species, but only prescribes in which hierarchical order the terms are to be used.

Hence there is no real difference between the subspecies of the zoologist and the variety of the botanist. For all practical purposes

they are equivalent and merely designate subdivisions of species. In an
evolutionary sense they each may represent a species in process of formation.
This leads us to the fact that there is no defined, universal criterion
for determining how far two populations must have diverged before they
cease being subspecies or varieties of one species and become two separate
species. Botanists and zoologists frequently disagree among themselves
on this question.

The fact is that there is no fundamental difference in kind between
these categories. What is one man's subspecies or variety is another man's
species. The terms are principally useful in indicating degree of divergence
and of genetic isolation.

Thus, for legal purposes no distinction should, indeed no distinction
can, be maintained between these categories. The terms are useful in
communication between scientists and other informed people to indicate
degree of difference and closeness of relationship.

For the above reasons, the lists of recommended endangered and threatened plants compiled by the Smithsonian Institution include populations
designated by all three of these terms ("species", "subspecies", "variety")
on an equal basis. Thus, in order to be included, the populations must
merely be distinguishable and endangered, threatened, or so rare that they
could easily be destroyed by act of man.

Definitions of Plant Status

Since the factors that determine endangered status listed in Section
4(a) of the Endangered Species Act of 1973 apply primarily to animals, the
factors were modified to apply to plants, and the following terms and
meanings were agreed upon by the contributors to the 1975 Report.

Endangered species. Those species of plants in danger of extinction throughout all or a significant portion of their range. Existence may be endangered because of the destruction, drastic modification, or severe curtailment of their habitat, or because of overexploitation, disease, or even unknown reasons. Plant taxa that occur in very limited areas, such as those known from the type locality only, or those which grow in restricted fragile habitats, usually are considered endangered.

Threatened species. Those species of plants that are likely to become endangered within the foreseeable future throughout all or a significant portion of their range.

Recently extinct or possibly extinct species. Those species of plants no longer known to exist after repeated search of the type localities and other known or likely habitats. A number of species are extinct in the wild but preserved in cultivation.

The question of rarity presents itself in considerations of plant status. A *rare* species of plant is one that has a small population in its range. It may be found in a restricted geographic region or it may occur sparsely over a wider area. When a rare species, subspecies, or variety appears on a list as endangered, threatened, or possibly extinct, it becomes particularly subject to two critical questions:

1. Has the taxon been correctly categorized as a natural entity? Perhaps it is a taxonomically untenable segregate of a more common or widespread species; it may be a rarely produced hybrid, polyploid, aberrant, or mutant; it could be an unusual phenotype owing to some unusual edaphic condition or environmental stress.

2. Is the "extinct" species truly extinct? Perhaps it actually does occur in small numbers or is more widespread, but has not been

observed or collected because it is small and inconspicuous, or it flowers rarely, briefly, or at unusual times, germinates infrequently or has long dormant periods; it may occur only in difficult or inaccessible terrain in hidden habitats rarely visited by collectors.

Development of the Lists

<u>Report</u> <u>on</u> <u>Endangered</u> <u>and</u> <u>Threatened</u> <u>Plant</u> <u>Species</u> <u>of</u> <u>the</u> <u>United</u> <u>States</u> (1975). The distribution and abundance of flowering plants in the United States are fairly well known, but there are gaps in our knowledge, unexplored geographic areas, and undescribed taxa awaiting the investigator. The northeastern states are quite well known botanically; much work remains to be done, however, on the flora of the southwestern states, and there are no recent accurate state floras for any of the states from Florida through Louisiana. Fortunately, Texas, California, and the Pacific Northwest recently have been studied, and the flora of the intermountain basin states of the West is being published. Several floras of Alaska have appeared, but exploration into the more remote regions of the state is needed.

The first step in preparing the lists was to review the best available published floras of regions, states, and more localized areas of the United States. Species with very limited distribution or rare status were listed, and all available data was compiled.

In addition to the floras, taxonomic revisions of various families, genera, and species-groups were consulted, since they are of importance in determining the correct scientific name, range, and status of the species. The monographs and revisions often contain distribution maps of the species treated.

Collections in herbaria were checked, and taxonomic specialists were consulted, yielding further information.

The lists then were compared with all available individual state lists of rare and endangered plants. Some states have official committees that assign categories and publish lists. Lists from other states are compiled by concerned individuals or informal groups from universities or private organizations. About 30 states have developed lists, some of which are highly documented. Others are merely lists of wildflowers that should not be collected, but are not actually rare or endangered. The several states have different criteria for determining which plants are endangered and threatened.

For species endemic to a single state, the status given in the state lists proved to be valuable; however, for the more wide-ranging species, the local designations of abundance often had only limited value for the national lists.

Special data cards used in preparing the lists permitted compilation of all available information on taxonomy, habitat, former and present distribution, abundance, threats, and present status, with the authorities and references cited. Examples of these data cards are shown on pages 342-346. The family, species, range by state, and status data from the cards were then computerized and printed out in lists.

These preliminary computer lists were compiled with the assistance of Dr. Dale W. Jenkins, consultant to the Endangered Flora Project and former Director of the Smithsonian's Ecology Program, in conjunction with advice from the staff of the Department of Botany, National Museum of Natural History, particularly Dr. John J. Wurdack. The Hawaiian list was compiled by Dr. F. Raymond Fosberg with the assistance of other Smithsonian and Hawaiian botanists. The entire program was under the supervision of Dr. Edward S. Ayensu, Director of the Endangered Species Program.

The preliminary lists were next submitted to a Workshop held in September 1974 by the Smithsonian Institution, with staff support from the Office of Endangered Species and International Activities of the Department of the Interior. The participants, reflecting a broad spectrum of eminent botanists and various institutions, organizations, herbaria, and arboreta, evaluated and improved the preliminary lists based on their specialist knowledge, as well as helped prepare recommendations for conserving plant species.

After a number of species had been referred back to specialists for verification, the revised list was recirculated for final evaluation to the Workshop participants, other specialists, and to compilers of some state lists. The invited participants in the Workshop are listed on pages 239-241.

Various botanists made special contributions that were particularly valuable in specific plant groups: Dr. E. S. Ayensu (Orchidaceae); Dr. R. C. Barneby (Astragalus); Dr. L. Benson (Cactaceae); Dr. L. Constance (Apiaceae, Hydrophyllaceae); Dr. A. Cronquist (Asteraceae); Lauramay Dempster (Galium); Dr. D. E. Fairbrothers; Dr. L. R. Heckard (Scrophulariaceae); Dr. D. Isely (Fabaceae); Dr. J. T. Mickel (ferns); Dr. P. H. Raven (Onagraceae); Dr. R. W. Read (Orchidaceae); Dr. J. L. Reveal (Eriogonum); Dr. R. C. Rollins (Brassicaceae); Dr. A. R. Smith (ferns); Dr. J. L. Strother (Asteraceae); Dr. R. L. Stuckey (Rorippa); Dr. R. Tryon (ferns); Dr. W. H. Wagner (ferns).

Endangered and Threatened Plants of the United States. Publication of the 1975 Report precipitated the arrival of a wealth of additional information to the Endangered Flora Project, resulting in significant revision, up-dating and further documentation of the lists. When research and mapping of species uncovered questions regarding their taxonomy, distribution, or rarity, the recognized authorities on the taxa were consulted. New contacts were made with interested botanists and state rare plant committees throughout the country. Information continued to come from the original contributors to the lists.

The influx of newly published and unpublished data and expert opinion has substantially augmented the assessment of the status of our endangered, threatened, exploited, and extinct flora. A few species thought to be extinct have been rediscovered, some formerly thought to be rare have been found to be more abundant or even taxonomically untenable, and a number of previously overlooked species have been commended to our attention. A list of individuals who made contributions of new data for this publication is on pages 235-238.

Value of the Lists

These lists should indicate to federal and state agencies, and to interested conservation groups, those species which are in need of special protection. The lists should be of value to land-use planners and to all agencies involved in preparing environmental impact statments in conformance with the National Environmental Policy Act of 1969. An awareness of endangered and threatened plants should be urged upon all people involved in the exploitation, development, management, and preservation of the land.

A major value of the lists should be to assist in the selection of areas for preservation and in the designation of natural areas to be protected. Without the benefit of such lists in past years, these decisions often have been based more on endangered animal species, interesting or scenic geological areas, or the preservation of unique habitats and forest types. The lists should prove to be of much greater value when the geographic distribution of each species has been accurately plotted on maps.

The lists should also stimulate further detailed reviewing and checking of the status of the species, and more extensive field study. The discovery of more populations will result in the shifting, adding, or removing of some species from the various categories.

Progress and Future Plans

A computer program to map the locality of the habitats of endangered and threatened species, by latitudinal and longitudinal coordinates, has been initiated. Distribution mapping will show eventually which species occur in already protected areas such as national and state parks and forests, preserves and natural areas, and will also indicate which species need protection anew. Mapping will also permit the designation of centers of endemism and aggregations of botanically diverse species growing in the same place, and will assist in conserving critical habitats for their protection. Computerized pilot maps are shown on pages 311-314.

The versatile SELGEM computer system of the National Museum of Natural History, which generated the printout lists included in this publication, is now being used to develop computerized information sheets on endangered and threatened plants based on the data cards in the Project's files. Preliminary examples of the sheets are shown on pages 315-341.

The project continues to collect additional relevant information on the status, abundance, and range of the recommended taxa, and is always interested in receiving such information, including data on the exploitation of species.

At the generous invitation of the Plant Sciences Data Center of the American Horticultural Society, a computer-tape of the listings of endangered and threatened species in this publication will be matched with its accession records, in order to determine which of the taxa are now in cultivation in botanical gardens.

In the Endangered Species Act of 1973, the term "United States", when used in a geographical context, includes all 50 states, the District of Columbia, the Commonwealth of Puerto Rico, American Samoa, the Virgin Islands, Guam, and the Trust Territory of the Pacific Islands. To the present the Project has considered the 50 states, the District of Columbia, the Commonwealth of Puerto Rico, and the Virgin Islands. In the future the geographic coverage of the Project's information-gathering efforts will be enlarged to include American Samoa, Guam, the Commonwealth of the Marianas, and the Caroline Islands. Data on the plants is filed at the Smithsonian Institution.

The advisability of providing a list of common names for all the recommended taxa will be investigated. Common names can cause confusion because they frequently refer to an entire genus, or even to unrelated genera. Also, a species may have different local names in different geographical regions. Very few of the plants in this report are common enough to have acquired standard, generally accepted common names.

The Project intends to study endangered and threatened species among the lower plant groups, which include algae, fungi, lichens, mosses, and liverworts.

In the future, it may be possible to give detailed consideration to interesting disjunct populations and species, which are classified as rare in the United States, but which also occur in other countries.

Value of Maintaining a Diversity of Plant Species

Plants are essential to the maintenance of life on this planet. They are a basic component of man's environment. The diverse green plant cover of the Earth helps to make our environment delightful and beautiful, and full of variety. But more significantly, plants fix solar energy in the form of carbohydrates and are the vital, ultimate sources of food, clothing, shelter, and fuel required for man's existence. Thus, human beings, as well as domestic and all other species of animals, are dependent on plants for their survival. The Earth's vegetation is essential for the maintenance of the environment in a livable condition, by preventing wind and water erosion and aiding in the development of fertile soil, in storing water, and in maintaining or providing subsurface water. A wide diversity of plant species and populations, therefore, is required to stock the many different habitats and ecosystems of the Earth and is necessary to maintain ecological stability. Man's largely monocultural system of agriculture has taught him much about the problems of ecological stability and the need for protection from diseases, insect pests, weeds, depletion of soil nutrients, and similar ecological problems.

Each species, subspecies, and variety of plant represents a unique type of life, a biological germ plasm or gene pool with special characteristics and values. These unique genetic resources are specific and different. When one of these types becomes extinct, the gene pool cannot be duplicated or reestablished and is lost forever. The extinction of a species limits man's options in the potential uses of the known or unknown values of each gene pool. Imagine the loss to mankind if the bread mold *Penicillium* had been eliminated before the discovery of the first antibiotic, penicillin, or if *Cinchona* had been destroyed before quinine was discovered as a cure for malaria.

A large untapped potential of enormous value to man exists if he preserves the diversity of plant species. It has been estimated that thousands of new alkaloids can be discovered in plants, including possible cancer cures. New plant medicinals and drugs presently are being discovered. Using yams (*Dioscorea*) as a natural intermediate in the search for cortisone precursors illustrates a valuable shortcut in a normally lengthy and expensive chemical synthesis. "Endod," an effective molluscicide from pokeweed (*Phytolacca*), recently was discovered to control snails which transmit schistosomiasis, a parasitic worm disease that afflicts at least 200 million people in Asia, Africa, Latin America, and the Caribbean. Plants not only produce natural biologically active chemicals, such as insecticides, but also provide man with knowledge of the chemical structures to synthesize even more effective chemicals and pesticides. They even indirectly stimulate research in other fields of study: the attempts to synthesize quinine gave a new impetus to organic chemistry.

The value of maintaining a diversity of plant species for potential uses that the present generation may not foresee, may be further illustrated by the jojoba bush (*Simmondsia chinensis*, Buxaceae). Recently there have been renewed investigations of the potential use of this southwestern desert plant as a commerically cultivated crop, since the liquid wax contained in its seeds is virtually identical to sperm whale oil, a strategic material used as a lubricant in industry. The use of jojoba oil as a replacement for whale oil could well reduce the whaling pressure on the remaining populations of these severely endangered mammals.

Development of new and improved food, fiber, and timber crop species from the existing diversity of under-utilized plants is a very important potential. With improved technology, man is able to utilize new species, and to improve plant production and yield by genetic selection and manipulation and the use of biologically active chemicals. Man will need all of the different types of plant germ plasm possible to increase future food production, a major and growing world problem.

In addition to developing untapped potentials, we also must preserve the ancestors of today's crop plants to provide protection against plant diseases and insect pests and to ensure vigorous and basic stocks for future breeding. The rare progenitors and relatives of corn, wheat, rice, and similar crops must be preserved in their present genetic state, and they must be protected from cross-breeding and hybridization with newer stocks in order to preserve their original values. Then, if "miracle" strains grown in monoculture should fail, the ancient types will still exist for use in breeding. Some ancestral stocks such as certain progenitors of corn are very rare and some apparently are extinct. The ancestors of fiber, timber, and horticultural plant species also should be preserved.

It should be pointed out that many species of endangered and threatened plants have special value to man since they are able to grow in unique and difficult places. Thus, these species have special characteristics that give them value in stocking such habitats with a "pre-adapted" ground cover. Some are bio-indicators of needed minerals and metal ores such as uranium. Many endangered and threatened species, such as cacti and lilies, possess known or potential horticultural value. Others which are rapidly becoming rarer in the wild, such as ginseng and golden seal, are of medicinal value. In fact, the horticultural values of some plants have been the cause of intensive exploitation, leading to depletion or extermination in the wild.

Causes of Rarity

The distribution and abundance or rarity of plant species are the result of a large number of factors, not only naturally caused but often man-caused. Species have been developing, spreading, retreating, and becoming extinct for at least 135 million years of geological history. Development and extinction of species is usually a long, slow process. Man, however, has accelerated the process in the twentieth century because he has greatly modified the surface of the Earth with his highly advanced technology. In the past (and certainly at present) he has shown an almost total lack of concern for other forms of life that share the planet with him. As a result, many species of plants recently have become extinct, and a large number are highly vulnerable to extinction unless they are soon provided with special protection or the removal of threats.

The long, slow geological and climatic changes that have taken place on Earth have resulted in the development of new species, the extinction of species, and changes in their distribution and abundance. Some of the major natural occurrences that affect plants are the uplift and sinking of land, the development of mountain ranges and islands, volcanic eruptions, flooding, erosion, glaciation, xerothermic periods, droughts, expansion of deserts, filling of lakes and ponds, and fire.

Natural biotic factors also affect the distribution and abundance of plants. Among the most important are newly evolved competitive species (with increased vigor), plant diseases, animal damage from overgrazing, insect damage, and the destruction of seeds and fruit. When species become very old or senescent, such as small populations of relicts from floras of older geological periods, they may suffer genetic depletion or loss of genetic variability and become inbred and unadaptive, or de-

velop narrow specialization that results in their rarity and perhaps extinction.

Natural factors act over long periods of time with the result that
the development and loss of species probably has been fairly well
balanced. Stress factors, mentioned above, not only result in extinction,
but they can also result in the development of new species.

Man has changed drastically the surface of the Earth as a result
of his enormously increased population and his technological advances.
The building of more dams, power plants, and mines, increased irrigation
and agriculture, and the development of cities with the resulting
dumps and pollution are further threatening to destroy or modify more
of his natural environment.

Plant populations often are reduced severely by the activities
of man, particularly in the destruction of habitats as well as in
the destruction of plants and plant parts themselves. Plant habitats
are destroyed outright by bulldozing, flooding, and the lowering of
water tables by wells and drainage. Timber removal, particularly
clear-cutting, with the resultant erosion, is a direct method of
destruction of habitats.

Plant populations are also destroyed as a result of commercial or
private exploitation, with the collection of entire plants, their
seeds, fruits, or flowers. Despite the high frequency of unsuccessful
transplants from the wild to cultivated gardens, collectors of rarities
persist in causing serious reduction of critical species.

Into the continental United States, man has accidentally or purpose-
fully introduced over 1800 species of foreign vascular plants that are
adventives or escapes, some of which have become naturalized or have
been cultivated. Nearly 2000 species have been introduced into Hawaii

in this manner. Many of these species are vigorous, aggressive weed
pests that choke agricultural crop fields, damage lawns and pastures,
and overgrow lakes and waterways. These introduced species, free of
the native diseases and pests that formerly held them in biological
balance, often reproduce prolifically and can win out in competition with
the native species. By taking over the habitat of the naturally rare
Sarracenis alabamensis (Alabama canebrake pitcher plant), the introduced Japanese honeysuckle contributes one of the several factors
(including collecting, drainage, herbicides, agriculture, and absence of
bog-burning) responsible for the decline of this carnivorous species
(Case & Case, 1974).

Plant species and populations are destroyed indirectly or changed
as a result of various other activities set in motion by man: overgrazing;
introduction of plant diseases, insects, and other animal pests; use of
chemicals, fertilizers, herbicides, and other biocides, which include
air, water, and soil pollutants; and the destruction of pollinators such
as insects, birds, and bats. The local introduction of diseases, insects,
birds, or other organisms may result in their spread over the entire
continent, with serious effects unknown until after the damage is done.

There are instances where a variety of threats are now operating in
the same region. For example, among the environmental problems associated
with the interior wetlands and saline tidal areas of South Florida,
comprising the Everglades, Big Cypress Swamp, and coastal mangrove forests
(from Lake Okeechobee south to the Florida Keys), Patrick Gleason (1974)
has indicated the following which concern exotic plants and changing plant
communities:

1. Ecological effects of naturalized introduced plants (numbering

over 50 species), e.g. the cajeput tree, *Melaleuca quinquenervia*; Brazilian pepper tree, *Schinus terebinthifolius*.

2. Controlled burning on pinelands and sawgrass (*Cladium jamaicense*) marshes.

3. Invasion of *Melaleuca* into northern Everglades.

4. Tropical hardwood hammock and bayhead hammock destruction by tracked vehicles, fire, high water levels, and extended hydroperiods.

5. Replacement of sawgrass by weed species.

6. Ecological effects of mangrove removal and preservation of mangrove zone.

7. Role of periphyton in the Everglades food chain.

8. Lumbering of cypress (*Taxodium distichum*) stands.

9. Protection of rare plants, mainly ferns and orchids, from theft.

Direct and indirect human threats to plants and plant habitats in the United States may be summarized as follows.

Off-road vehicles. Dune buggies; motorcycles; trail bikes; snowmobiles; swamp buggies; air boats.

Mining. Strip mining; shale oil recovery; subsurface mining.

Forestry practices. Clear-cutting; replacing native trees with exotic timber trees.

Biocide spraying. Insecticides; herbicides.

Construction and real estate development. Roads; factories; golf courses; power plants; shopping centers; housing tracts; condominiums; land-clearing; landscaping.

Introduction of competitive weeds. Chokers of native vegetation.

Over-grazing. By domesticated or feral goats, sheep, cattle, deer, pigs, rabbits, burros, horses, with associated trampling in some cases.

<u>Fire</u>. Destructive fires; preventing natural fires.

<u>Agriculture</u>. Fields cleared for monoculture crops.

<u>Water management</u>. Flooding; stream channelization; irrigation; dams; drainage of swamps.

<u>Illegal poaching</u>. From federal, state-owned and private land.

<u>Commercial exploitation</u>. Cacti and carnivorous plants, among many others.

<u>Collecting by private individuals</u>. For transplanting to gardens.

<u>Trampling of vegetation by humans</u>. Inviting accelerated soil erosion.

Habitats of Endangered and Threatened Plants

Certain of the recommended endangered and threatened plants occur in seemingly rather homogeneous or undifferentiated habitats within the vast expanses of forest, grassland, and desert in the country. By and large, however, many endangered and threatened plants occupy niches in locally unique, unusual, or isolated habitats. Those habitats are ecologically and geographically restricted, fragile, or otherwise specialized due to variously balanced combinations of climatic, geological, and biological factors; they provide an environmental anchor for those plants whose genetically determined tolerances and requirements have become of adaptive advantage for survival under those particular sets of conditions.

One such restricted species is *Dodecatheon frenchii* (French's shooting star, Primulaceae), which occurs in unglaciated areas of southern Illinois, Arkansas, and Kentucky. It grows in small colonies in stream canyons, shaded cave-like sandstone rock shelters at the base of wet overhanging cliffs, and beneath overhanging ledges on wooded hillsides. Incapable of leaving the habitat, it thrives in conditions of lower temperature, higher humidity, lower evaporation rate, and indirect sunlight, as compared to the surrounding habitat.

Specialized habitats in which endangered and threatened plants are found, with an example of one plant from each, may be categorized as follows. The locations of the types of habitats are not confined to the states from which the plant examples are taken.

Serpentine rock. *Allium hoffmanii* (Liliaceae) in California.

Cedar barrens and glades. *Lesquerella perforata* (Brassicaceae) in Tennessee.

Sandy pinelands. *Polygala lewtonii* (Polygalaceae) in Florida.

Shale barrens. *Clematis viticaulis* (Ranunculaceae) in Virginia.

Shorelines. *Ptilimnium fluviatile* (Apiaceae) in several southeastern states.

Sand dunes. *Swallenia alexandrae* (Poaceae) cn Eureka Dunes, California.

Rocky cliffsides. *Polygala maravillasensis* (Polygalaceae) in Trans-Pecos Texas.

Talus slopes. *Eriogonum cronquistii* (Polygalaceae) in Utah.

Mountain tops. *Paronychia monticola* (Caryophyllaceae) in Davis Mountains, Texas.

Sphagnum bogs. *Sarracenia oreophila* (Sarraceniaceae) in several southeastern states.

Islands. *Pritchardia remota* (Arecaceae) on Nihoa Island, Hawaii.

Peninsulas. *Iris lacustris* (Iridaceae) on Door Peninsula, Wisconsin.

Hot, alkaline, or salt springs. *Eriogonum argophyllum* (Polygonaceae) at Sulphur Hot Springs, Nevada.

Canyons. *Acer grandidentatum* var. *sinuosum* (Aceraceae) on Edwards Plateau, Texas.

Vernal pools. *Neostapfia colusana* (Poaceae) in California.

Swamps. *Roystonea elata* (Arecaceae) in Florida.

Tidal estuaries. *Cardamine longii* (Brassicaceae) in Maine, Maryland, and Virginia.

The Endangered Species Act of 1973 is intended to prevent the further decline, and to bring about the restoration, of endangered and threatened species and of the habitats upon which such species depend. The administration and managemenet of areas known as "critical habitats" will provide an important means for protecting plant species once they have been officially given endangered or threatened status under the Act. Notices published in the Federal Register of April 22 and May 16, 1975, which describe the concept of "critical habitat", indicate that for any given endangered species or threatened species, habitat is considered *critical* if the destruction, disturbance, modification, or subjection to human activity of any constituent element of the habitat might be expected to result in a reduction in the numbers or distribution of that species, or in a restriction of the potential and reasonable expansion or recovery of that species. Among the vital needs which are relevent in determining critical habitat for a given species are space for normal growth of the species and sites for reproduction.

The area of a critical habitat could encompass the entire habitat of a species, or any portion thereof if any constituent element is necessary to the normal needs or survival of that species. Designation of critical habitat for a species is done by the Department of the Interior, and since the designation applies only to federal agencies, it is essentially an official notification to these agencies that their conservation responsibilities pursuant to the endangered Species Act of 1973 are applicable in a certain area.

The term "habitat", itself, could be considered to consist of a spatial environment in which a species lives and all elements of that environment including (but not limited to) land and water area, physical structure and topography, flora, fauna, climate, human activity, and the quality and chemical content of soil, water, and air.

An interesting phenomenon concerning the plants recommended as endangered and threatened is that aggregations of botanically diverse species of them often occur in the same habitat. The region of the Apalachicola River in northwestern Florida, for example, harbors a concentration of more than a dozen endemic rare and endangered species found only in that location. The various endemic species associated with California's unique vernal pools are the subject of a special issue of *Fremontia* (October 1976), a journal of the California Native Plant Society.

Rare and endangered species are usually narrow endemics; they occur only in restricted ranges. In the United States there are several major centers of endemism or areas with concentrations of endemic species known only from specific regions: Florida, the southern Appalachian Mountains, Texas, California, and Hawaii. There is also considerable endemism in the southwestern region comprising Nevada, Arizona, and Utah, and in the Pacific Northwest of Oregon and Washington. Endemism occurs to a comparatively limited degree in the northeastern states, with some in the Ohio River Valley, but it is very limited in the glaciated area of the Great Lakes region, and it is lowest in the extensive Great Plains region.

Conservation

Conservation of rare species of plants requires the preservation and protection of habitats upon which they depend for growth and reproduction. *In situ* perpetuation of sufficient populations of endangered and threatened plants is required to ensure their survival.

Three key elements which are involved in this *in situ* perpetuation are prevention of the destruction of populations and their habitats; monitoring and research on population levels and viability; and prevention of collection and commercial exploitation.

Prevention of the Destruction of Plant Habitats

It is essential that habitat preservation be recognized as the critical factor in all conservation activities and in legislative and executive acts, amendments, orders, and regulations.

Since preservation and protection of endangered and threatened plants depends primarily on the preservation of their habitats-- with about 3000 endangered and threatened species--it is not usually practical to attempt to preserve an individual or species by building a fence around it. A more practical method for preservation involves mapping the ranges of all the species to determine where they are aggregated. The habitats in which they cluster could then be preserved.

Knowledge about endemic centers and aggregations of species is of considerable value in determining and establishing priorities for preserving land to protect these plants.

A variety of methods can and should be used for protecting and preserving habitats and populations, including landmark designations, conservation easements, tax breaks, acquisition, and penalty procedures. These methods should be fully explored and the findings made available for executive and legislative action in appropriate cases.

Much protective activity can be accomplished in the existing
conservation programs. The National Park Service should incorporate
the list of recommended endangered and threatened species as criteria
in its Natural Landmark Program (by being included in the regional
natural landmark surveys, endangered and threatened species populations
could be made eligible for designation as Natural Landmarks); and the
Federal Committee on Ecological Reserves should encourage natural area
programs in the states to do the same.

Endangered and threatened plants should be recognized as basic
elements in land-use plans and inventories when the federal government
is involved either in a direct capacity or in the role of a guiding
or advisory party. Federal agencies involved in land management,
including the Bureau of Land Management, Fish and Wildlife Service,
National Park Service, Forest Service, Energy Research and Development
Administration and Department of Defense, should recognize endangered
and threatened species as natural resources and include their populations
and supporting habitats as elements in those agencies' natural resource
surveys or inventories. State and local land-planning agencies and bodies
should be encouraged to include endangered and threatened species in
their planning, with particular attention to the plants' critical
supporting habitats.

Locating the habitats of endangered plant species is one of the
major objectives of the Heritage programs conducted by The Nature
Conservancy, an organization which preserves natural areas through land
acquisition for scientific and educational purposes.

Monitoring and Research

Protection of plant habitats alone may be insufficient for the permanent preservation of some populations of endangered and threatened species. Consequently, it is essential that monitoring of populations be undertaken to first determine, and then to maintain or increase, the trends in their abundance and viability.

In developing plans for the preservation and protection of endangered plant species, it will be necessary to understand the reasons for the rarity of each species and the specific causes of endangerment. Species may be local endemics occuring in small geographical areas or in highly specialized habitats. They may be limited to refugia, or protected places, where, owing to special factors, they can persist.

The habitats of endangered plants often are unstable or geologically young areas with unique characteristics that discourage the growth of highly competitive, aggressive species. Many species occur in the pioneer or early stages of ecological succession in a given region. It is an interesting fact that many species of rare plants often are found congregated in these habitats. The rare species growing on specialized substrates, such as serpentine, are not restricted to the substrate, but are preserved from extinction caused by competition with other plant species when the latter are unable to grow on serpentine.

Endemic species are those with restricted or limited geographic distribution. These may be relict species that are ancient, that show little variability, and that are highly restricted to specific habitats, for example, the Giant Sequoia (*Sequoiadendron giganteum*), the Florida Torreya (*Torreya taxifolia*), the Florida Yew (*Taxus floridana*), the Lost Franklinia (*Franklinia alatamaha*), and the Georgia Plume (*Elliottia racemosa*) with its low fertility, poor germination, and difficult regeneration from cuttings. Relict rare species that occur in small populations can lose their store of genetic variability and become depauperized. Extinction thus can result from a narrow specialization to a restricted habitat. Many rare species, however, that once were highly restricted in range grow well when introduced into other areas. The Lost Franklinia was known only from a habitat of about two acres in Georgia; now it grows well in cultivation in the eastern United States and overseas.

Other rare species are newly evolved and have not yet had time to occur in a wider range. These often are aggressive and capable of withstanding various threats. If the biology of the rare species, therefore, is understood, effective action can be taken to help preserve them.

Prevention of Commercial Exploitation

The longer the period between the appearance of these lists and the official listing of the taxa pursuant to the Endangered Species Act of 1973, the greater becomes the danger of "last-minute" and extensive collecting of those which are at present commercially exploitable on an interstate basis without a permit.

Since a relatively small number of the endangered plant species are in commercial trade, effort in enforcing the Act should be concentrated on close monitoring of interstate traffic in those species. This subset of species that appear in commercial trade is identified in a separate list of commercially exploited species (pages 52-59).

Another potential danger involved in public listing of endangered species is the pressure that might be put on the endangered or threatened plants by well-meaning but misguided persons, who think that they should

dig up the last specimens from their native environment to "grow"
them in their own gardens--usually with fatal results. This problem
must be clearly stated, and appropriate warnings of its futility should
be issued.

Preservation in Botanical Gardens

While cultivation or artificial propagation is not an acceptable
alternative to *in situ* perpetuation of species, it indeed should
be done when destruction of the habitat appears imminent. Research
must begin on endangered and threatened species far in advance of
theoretically imminent habitat destruction, in order to ensure that
potential genetic resources will not be completely lost due to unforeseen
events.

In order to successfully perpetuate the majority of endangered and
threatened plants in botanical gardens, much research is needed on
propagation techniques, breeding systems, pollinators and soil preferences
of the species. Even when these factors have been investigated and
accommodated there are risks to be safeguarded in bringing such plants
into cultivation. Rare species can be exposed to hybridization with
related taxa; records can be lost over long periods of time; a gardener
may not be able to resist selecting from the progeny the best, the color
variant, the bizarre mutant, and the species will thus not stay the same,
but will become a cultivar.

Programs for cultivating endangered and threatened plants are
underway, among other places, at North Carolina Botanical Garden
(Chapel Hill, North Carolina); Waimea Arboretum (Oahu, Hawaii);
Tennessee Valley Authority (Norris, Tennessee); and Pacific Tropical
Botanical Garden (Kauai, Hawaii).

A computer-tape of the endangered and threatened plants listed
in the 1975 Report has been cross-referenced with the master data
file maintained by the Plant Sciences Data Center of the American
Horticultural Society. This was done in order to determine which of those
plants are currently being cultivated within 39 protected collections
(botanical gardens) in the United States whose accession records are
on file at the Center. The records on file are generally of plants
grown for scientific or ornamental value. The accessions of Hawaiian
botanical gardens were not available for consideration.

At the Data Center, the 3187 taxa listed in the Report were matched
by computer against the Center's file of 139,161 garden records.
A listing of *exact* matches between the taxa in the Report and the
Center's accession files, shows that 132 taxa, or 4.14 percent of those
listed in the 1975 Report, are in cultivation among the gardens.

Basic information on the computer processing (matching) techniques
used in making the comparisons is available from Mr. Richard A. Brown,
Director, Plant Sciences Data Center, American Horticultural Society,
Mount Vernon, Virginia 22121.

Suggested Priorities for Governmental Action

There are eight immediately identifiable arenas in which action
is urgently required.

1. The Secretary of the Interior should officially determine the
   status of the recommended endangered and threatened plants,
   pursuant to Section 4 of the Endangered Species Act of 1973.

2. The Endangered Species Act of 1973 should be amended to
   prohibit the "taking" (picking, collecting) of endangered

and threatened plants. At present, Section 9(a)(2) of the Act basically only prohibits the export and import of endangered plants, and prohibits the interstate moving or sale of such plants in the course of a commercial activity, without a permit.

3. These lists of recommended endangered and threatened plants should be submitted by the Secretary of the Interior, in his capacity as United States Management Authority, for inclusion in Appendix III to the Convention on International Trade in Endangered Species of Wild Fauna and Flora. The Secretary of the Interior could then exercise the authority given to him in Section 5 of the Endangered Species Act, to acquire by purchase, donation or otherwise, land on which endangered and threatened plants listed in the Appendices to the Convention occur.

4. Section 5 of the Endangered Species Act of 1973 should be amended so as to authorize the Secretary of the Interior to acquire land on which occur endangered and threatened species of plants listed under the Act.

5. The U.S. Fish and Wildlife Service should make final its proposal, in the Federal Register of June 16, 1976, that those plants named as "varieties" in this publication are to be considered as subspecies and, therefore, "species" as defined in Section 3(11) of the Endangered Species Act of 1973.

6. Individual states should pass laws prohibiting the *taking* of officially endangered and threatened plants. Very few states have laws protecting specific plants, and fewer have laws protecting endangered or threatened plants.

7. Consideration of the plants recommended for endangered or

threatened status in this publication should be made a requisite in the preparation of Environmental Impact Statements under the National Environmental Policy Act of 1969.

8. When considering the designation of habitats on federal land as "Natural Areas," the agencies should take into account the presence of plants recommended for endangered and threatened status and designate the habitats in which they occur.

RECOMMENDATIONS FOR THE CONSERVATION OF ENDANGERED AND THREATENED PLANTS (1975)

Some of the recommendations made in the 1975 Report have, to one extent or another, received attention from concerned individuals and organizations. Much work, however, remains to be accomplished.

The National Parks and Conservation Association, Washington, D.C., and the Garden Club of America, New York City, have expressed their support of the recommendations. Resolutions of support and endorsement for all of the recommendations have been passed by the Board of the Cactus and Succulent Society of America (July 1975); the Central Arizona Cactus and Succulent Society (September 1975); the Long Beach (California) Cactus Club (October 1975); the Michigan Cactus and Succulent Society (November 1975); the South Coast (California) Cactus and Succulent Society (November 1975); the National Capital Cactus and Succulent Society (December 1975); the Imperial Valley (California) Cactus and Succulent Society (January 1976); the Midwest (Ohio) Cactus and Succulent Society (January 1976); the Post Garden Club of Fort Meade, Maryland (February 1976); and the Solano (California) Cactus and Succulent Society (September 1976).

1. *Preservation of endangered and threatened species of plants in their native habitat should be adopted as the best method of ensuring their survival. Cultivation or artificial propagation of these species is an unsatisfactory alternative to* in situ *perpetuation and should be used only as a last resort, when extinction appears certain, with the purpose of re-establishing the species in its natural habitat.*

Habitat preservation must be given the highest priority in all conservation activities, particularly when dealing with the critical habitats of endangered species. Modification or destruction of critical habitats by human activities could result not only in a further reduction in population and distribution, but also in restriction of population expansion and recovery.

Transplantation and artificial cultivation should be a last resort, always with the ultimate objective of re-establishing the species in its natural habitat. Attempts to protect individual plants by fencing, for example, without preservation of the habitat or ecosystem upon which they depend will not provide successful perpetuation.

Protection and preservation of critical habitats and populations can be given high priority by landmark designations, conservation easements, acquisition, the institution of firm penalty procedures, and the habitats' designation as Natural Landmarks and Research Natural Areas.

2. *The species of endangered and threatened plants that occur on federal and state lands should be mapped and given continued protection. More specific attention should be given by federal*

departments and agencies to the prevention of destruction or
modification of critical habitats of endangered and threatened
flora in accordance with the Endangered Species Act of 1973 and
the National Environmental Policy Act of 1969.

The species of endangered and threatened plants on federal and state lands should be determined and exact locations should be mapped and made known to appropriate authorities.

It would be advisable for the executive branch of the federal government, through the Council on Environmental Quality, to require consideration of plant species recommended for endangered and threatened status in the review of environmental impact analyses and in the statements issued under the National Environmental Policy Act of 1969.

Federal agencies that are involved in land management (including the Bureau of Land Management, the Fish and Wildlife Service, the Bureau of Outdoor Recreation, the National Park Service, the Forest Service, the Soil Conservation Service, the Energy Research and Development Administration, the Department of Defense) should be reminded that endangered and threatened plant species and their supporting habitats are basic natural resources in the agencies' land-use plans and in their natural resource surveys or inventories. State and local land-use planning agencies and similar bodies should be encouraged always to include in their guidelines on planning legislation a list of endangered and threatened plant species and the supporting habitats.

3. *In accordance with Section 4 of the Endangered Species Act*

of 1973, the Secretary of the Interior should review the lists in this report and publish proposed lists of endangered and threatened plants in the Federal Register.

The Secretary of the Interior is required by Section 4 to determine, after consultation, the endangered or threatened status of plant species and to publish the resulting lists in the Federal Register.

A Notice of Review of the status of the taxa listed in the 1975 Report was published by the U.S. Fish and Wildlife Service in the Federal Register of July 1, 1975. The Service has also published a Proposed Rule-making of Endangered Status for many of the plants recommended as endangered in this publication, in the Federal Register of June 16, 1976.

4. *The Secretary of the Interior is advised to ensure that the commercially exploited species of plants in this report are given urgent protection. Appropriate government agencies should be alerted and existing laws should be fully enforced.*

The Convention on Internation Trade in Endangered Species of Wild Fauna and Flora (1973) was implemented by the United States on February 22, 1977. The Convention is designed to regulate trade in species threatened with extinction through the issuance of export and import permits by the management authorities of the ratifying countries. The Secretary of the Interior is the United States Management Authority for the Convention.

The U.S. Fish and Wildlife Service published, in the Federal

<u>Register</u> of September 26, 1975, a proposal of endangered status for the plants listed in Appendix I to the Convention; the species will be reviewed on an individual basis to determine if they are qualified under the Endangered Species Act of 1973. Export and import permits are required for trade in Appendix I plants. Although none of the plants currently listed on Appendix I are native to the United States, it is likely that a number of cacti are qualified for Appendix I protection.

Plants in Appendix II to the Convention presently include all species of Cactaceae in the Americas and all of the world's Orchidaceae, as well as American ginseng (*Panax quinquefolius*). Import permits are required for trade in Appendix II plants.

Although the U.S. government issues permits and thus helps to regulate trade as a Party to the Convention, for practical purposes, federal interstate protection of the commercially exploited species in this publication must await their final listing pursuant to the Endangered Species Act of 1973, and revision of the Act to prohibit the picking ("taking") of endangered and threatened plants.

5. *It is recommended that the list of the species of endangered and threatened plants in this report should be submitted by the Secretary of the Interior to the Convention on International Trade in Endangered Species of Wild Fauna and Flora for inclusion in Appendix III. This listing will enable the Secretary of the Interior, acting as the United States Management Authority to the Convention, to acquire lands for the preservation of endangered species of plants.*

6. *Since protection alone may not be sufficient for the survival of some populations of endangered and threatened species, monitoring of*

population levels is needed. For declining populations, research is necessary to determine the causes of rarity and to ascertain what can be done to save the species.

Protection of plant habitats may be insufficient for the preservation of some populations of threatened and endangered species. Consequently, it is essential that monitoring of populations be undertaken to observe population size (decline or increase), condition of habitats, and ability of reproduction.

Federal agencies involved in land management should monitor population levels in areas reserved, protected, or otherwise identified as refugia for one or more threatened or endangered species.

State agencies are being encouraged to enter into cooperative agreements with the Fish and Wildlife Service for conservation of endangered species. Such agencies and departments with research and/ or land management programs already underway should be encouraged to conduct or sponsor expanded research on the biology of endangered and threatened plant species, investigation that is necessary to appraise the survival status of these species and to provide guidance for management in order to maintain, perpetuate, or restore the populations.

7. *A "Registry of Endangered and Threatened Plants" should be established on a permanent basis to continue to collect, evaluate, and update all pertinent information which would be available to interested national and international organizations.*

A national registry office would be required to maintain such a register on a permanent basis, and to collect, evaluate, synthesize, and publish information on all endangered and threatened plant species,

commencing with the vascular plants of the United States. The registry
and coordination should include central card files and maps, a specialized
library, use of a computer, and a small staff of experts. The register
would require continual updating of information on the location, status,
habitat requirements, reproductive behavior, population size, and commercial
and private exploitation of endangered and threatened plant species. The
register should be available to government agencies and qualified investigators
from the general public.

8. *The lists of endangered and threatened plants should be given wide
exposure and publicity. Colored illustrations should be displayed in
public places, in publications, and on postage stamps. Interested
organizations should be encouraged to assist in publicizing the need for
protection and preservation of endangered and threatened species of plants.*

Publication of the 1975 Report caused a surge of interest in the
condition of the native flora, and thousands of copies were distributed
at the request of members of the general public, professional botanists,
numerous state and federal agencies, garden clubs, members of Congress,
conservation organizations, libraries, parks, museums, botanical gardens,
journalists, power and utility companies, ecological and environmental
consulting firms of all sorts, schools and universities.

Illustrations of endangered plants should be widely disseminated
and publicized, and copies should be made available to the public at
large and to appropriate organizations. Colored illustrations of
endangered plant species of the United States should be prepared for general
distribution and for prominent display at parks, nature reserves, museums,
and tourist centers. They should be sent to botanical gardens; to

horticultural, gardening, and conservation groups; and to educational establishments from elementary schools to universities.

Posters of this kind have been prepared in New South Wales (Australia), Holland, France, Switzerland, and Great Britain. For a number of years the Swiss have posted in public places colored pictures of rare and endangered species that are protected by law. Since the rich alpine flora is a great attraction and temptation to tourists, the intent of such displays is to make the public aware of the tenuous existence maintained by some populations of the rare plants that they encounter.

In 1974 the Botanical Society of the British Isles published a color poster depicting 21 of the rarest endangered species in Great Britain, which was distributed to the public at minimal charge. In 1975 the 21 plants became totally protected by an Act of Parliament, and a revised edition of the color poster was published and made available to the public.

The large area of the United States and the number of endangered and threatened species affords an opportunity for the preparation of regional posters that would emphasize the endangered species of particular states and regions where endemism is high, and those of selected natural areas. In addition, garden and horticultural groups should be asked to encourage and cooperate with local and state authorities in efforts to publicize these species. The Nantucket Island Garden Club in Massachusetts has produced a color poster (1975) entitled "Protect Nantucket's Wildflowers," which depicts fourteen species, with the message that, if picked or uprooted, these beautiful flowers will disappear.

The U.S. Postal Service should consider issuing a series of postage stamps portraying endangered and threatened plant species, as done by several other countries. Of the plants listed in this report, only one, the Lost Franklinia (*Franklinia alatamaha*), has been shown on a United States postage stamp, issued in 1969.

9. *No new federal legislation is required at this time. However, after a reasonable period, a review of the effectiveness of the Endangered Species Act of 1973 may be required to provide better protection to the endangered and threatened plant species.*

The Endangered Species Act of 1973 provides unequal protection to listed plant taxa as compared to listed animal taxa. It differs for plants as follows: (1) The Act does not prohibit the "taking" of endangered and threatened plant species in the United States; it seeks to regulate commercial interstate traffic in such plants; (2) the term "species" as applied to plants includes subspecies only and does not literally include varieties; (3) The Secretary of the Interior does not have the authority to acquire land for the purposes of conserving endangered and threatened plants that are not listed in the Appendices to the Convention on International Trade in Endangered Species of Wild Fauna and Flora.

Oversight hearings on implementation and administration of the Endangered Species Act of 1973 and it amendments, and to review the problems and issues encountered, were held in Washington D.C., on October 1, 2, and 6, 1975, before the Subcommittee on Fisheries and Wildlife Conservation and the Environment of the Committee on Merchant Marine and Fisheries, House of Representatives (Endangered Species Oversight, Serial No. 94-17, U.S. Government Printing Office (1976), 367 pages). Comments on the testimony appear in Science News 10: 230 (October 11, 1975), National Wildlife Federation

Conservation Report No. 34: 448-449 (October 10, 1975), and National Parks and Conservation Magazine 49: 18-19 (December, 1975).

A hearing to amend the Act to authorize appropriations and to review progress in administration of the Act was held in Washington, D.C., on May 6, 1976, before the Subcommittee on Environment of the Committee on Commerce, United States Senate (Serial No. 94-82, U.S. Government Printing Office (1976), 190 pages).

LISTS OF ENDANGERED, THREATENED, AND EXTINCT SPECIES

Commercially Exploited Endangered and
Threatened Species of the Continental United States

The 1975 Report's list of endangered and threatened plants that are commercially exploited and privately collected from nature has been augmented for this publication with additional data supplied by people knowledgeable of the current trade, and by use of current catalogues in the collections of the National Agricultural Library (Beltsville, Maryland) and the Library of the U.S. National Arboretum (Washington, D.C.), as well as catalogues sent to the Endangered Flora Project by private donors.

Commercial plant catalogues usually quote the highest prices for the naturally rare plants, with accompanying comments on their rarity and desirability. There is a very large trade in the sale of exotic-looking carnivorous plants, such as the Venus fly trap and pitcher plants. Commercial cactus dealers in some cases will collect specimens by the truckload from the wild and sell them immediately or shortly after they are potted.

Efforts should be made to end the current fad for rarity, which is encouraged by dealers who advertise the rarity of their offerings. Instead, advertising emphasis should be directed towards species grown from seed or cuttings and not collected from the wild. There is a growing trend of response to this suggestion by cactus fanciers around the country. Dealers and the public should be encouraged to leave endangered flowers undisturbed or to photograph them carefully *in situ*.

Many species cannot be successfully transplanted; if any roots are broken or damaged, the plants will die from fungus attack. This is especially true for certain rare orchids. Since the occurrence of many

species of rare and endangered plants is frequently confined to specialized
habitats, these species are particularly difficult to cultivate in the
absence of their delicately balanced natural conditions. An informed
public could decrease significantly the commercial and private pressure
under which these exploited species survive.

The following reliable accounts of the nature of commercial activity
and other exploitation of vulnerable plants are given for the purpose
of indicating the scope and the growing need for control of such activities.

1. More than one-half million individual rattlesnake-orchids
(*Goodyera pubescens*) have been collected in Tennessee for sale in terrariums.

2. About one-half million individual cactus plants per year are
removed from the area near Terlingua and LaJitas, Texas, and sold to
private dealers, who in turn sell them.

3. At Terlingua, Texas, a shed containing 25,000 to 30,000 cacti
has been observed. Several dealers are known to specialize in this type
of mass harvesting and sale. Illegal aliens are paid one penny per plant
to harvest all the small globular cacti they can find. Dealers sell
the plants for about two cents to one dollar per plant. One and a half
ton trucks full of cacti and piles of dried-out cacti that collectors
neglected to pick up in their trucks often have been observed.

4. Up to $350 is asked for large Arizona barrel cacti (*Ferocactus*,
Echinocactus) in New York City, picked from nature. Smaller, but
sizable, specimens for sale at $20 to $30 each are commonly observed
across the country.

Mr. Ernest Douglas' article in the *Christian Science Monitor* of
November 17, 1975, states that, despite the stringent Arizona Native Plant
Law, more than 90 percent of the desert plants stolen in Arizona are

cacti, with barrel cacti and young giant saguaros being the favorites of the thieves. In response to the Arizona Cactus Law, poachers are changing their sites of predation to public land in California's Mojave Desert, as detailed in Mr. Robert Jones' article of December 19, 1976, in the *Los Angeles Times*.

5. *Epithelantha bokei*, recommended for threatened status in this publication, has been poached from even the remotest parts of Big Bend National Park, Texas.

6. An individual in a western state has destroyed certain cacti on his property in order to avoid cooperating in their protection under any possible future laws.

7. Carnivorous plants are being decimated. A commercial dealer from a northern state made a wholesale raid on DeSoto State Park in Alabama, completely wiping out that locality of *Sarracenia oreophila* and thereby effectively reducing the known good stands of this species by 25 percent.

8. The Green Swamp area of North Carolina is being overcollected of Venus flytraps *(Dionaea)* and butterworts *(Pinguicula)*, even though *Dionaea* is protected by state law in North Carolina.

9. The endangered Dehesa beargrass *(Nolina interrata)*, desirable for its attractive caudex, has been commercially collected from its type locality in San Diego County, California. Six of the ten colonies at the locality were removed for sale.

10. In 1975, a nursery was said to have received a total of 16,000 compass barrel cacti *(Ferocactus acanthodes)*, all thought to be field-collected. There are some indications that they came out of the Clark Mountains in California, possibly on Bureau of Land Management land.

Species and Varieties	Status	Exploitation Status*	Locality
Araliaceae			
Panax quinquefolius	Threatened	Expl	32 states
Arecaceae			
Rhapidophyllum hystrix	Threatened	Expl	Alabama Florida Georgia Mississippi
Cactaceae			
Ancistrocactus tobuschii	Endangered	Priv	Texas
Cereus eriophorus var. fragrans	Endangered	Priv	Florida
Cereus gracilis var. aboriginum	Endangered	Priv	Florida
Cereus gracilis var. simpsonii	Endangered	Priv	Florida
Cereus greggii	Threatened	Expl Priv	Arizona New Mexico Texas
Cereus robinii var. deeringii	Endangered	Priv	Florida
Cereus robinii var. robinii	Endangered	Priv	Florida
Coryphantha dasyacantha var. varicolor	Threatened	Expl Priv	Texas
Coryphantha duncanii	Threatened	Expl Priv	Texas
Coryphantha hesteri	Threatened	Expl Priv	Texas
Coryphantha minima	Endangered	Expl Priv	Texas
Coryphantha ramillosa	Endangered	Expl	Texas
Coryphantha recurvata	Threatened	Expl	Arizona

*Expl = commercially exploited; Priv = privately collected.

Species and Varieties	Status	Exploitation Status	Locality
Coryphantha scheeri var. robustispina	Threatened	Priv	Arizona
Coryphantha sneedii var. leei	Endangered	Expl Priv	New Mexico
Coryphantha sneedii var. sneedii	Endangered	Expl Priv	New Mexico Texas
Coryphantha strobiliformis var. durispina	Endangered	Expl	Texas
Coryphantha sulcata var. nickelsiae	Threatened	Expl	Texas
Coryphantha vivipara var. alversonii	Threatened	Expl Priv	Arizona California
Coryphantha vivipara var. rosea	Threatened	Expl Priv	Arizona California Nevada
Echinocactus horizonthalonius var. nicholii	Endangered	Expl	Arizona
Echinocereus chloranthus var. neocapillus	Endangered	Expl	Texas
Echinocereus fendleri var. kuenzleri	Endangered	Priv	New Mexico
Echinocereus ledingii	Threatened	Expl Priv	Arizona
Echinocereus lloydii	Endangered	Priv	New Mexico Texas
Echinocereus reichenbachii var. albertii	Endangered	Priv	Texas
Echinocereus reichenbachii var. chisosensis	Threatened	Priv	Texas
Echinocereus reichenbachii var. fitchii	Threatened	Expl Priv	Texas

Species and Varieties	Status	Exploitation Status	Locality
Echinocereus triglochidiatus var. arizonicus	Endangered	Priv	Arizona
Echinocereus viridiflorus var. correllii	Threatened	Expl	Texas
Echinocereus viridiflorus var. davisii	Endangered	Expl Priv	Texas
Epithelantha bokei	Threatened	Expl	Texas
Ferocactus acanthodes var. eastwoodiae	Threatened	Expl	Arizona
Ferocactus viridescens	Endangered	Priv	California
Mammillaria orestera	Threatened	Expl	Arizona New Mexico
Mammillaria thornberi	Threatened	Expl	Arizona
Neolloydia erectocentra var. acunensis	Threatened	Expl Priv	Arizona
Neolloydia erectocentra var. erectocentra	Threatened	Expl Priv	Arizona
Neolloydia mariposensis	Endangered	Expl Priv	Texas
Neolloydia warnockii	Threatened	Priv	Texas
Opuntia basilaris var. brachyclada	Threatened	Expl	California
Pediocactus bradyi	Endangered	Expl Priv	Arizona
Pediocactus knowltonii	Endangered	Expl Priv?	Colorado New Mexico
Pediocactus papyracanthus	Threatened	Expl Priv?	Arizona New Mexico

Species and Varieties	Status	Exploi-tation Status	Locality
Pediocactus paradinei	Threatened	Expl Priv	Arizona
Pediocactus peeblesianus var. fickeiseniae	Threatened	Expl Priv	Arizona
Pediocactus peeblesianus var. peeblesianus	Endangered	Expl Priv	Arizona
Pediocactus sileri	Endangered	Expl Priv	Arizona Utah
Sclerocactus glaucus	Endangered	Expl Priv	Colorado Utah
Sclerocactus mesae-verdae	Endangered	Expl Priv	Colorado New Mexico
Sclerocactus polyancistrus	Threatened	Expl	California Nevada
Sclerocactus spinosior	Threatened	Expl Priv	Arizona Colorado Utah
Sclerocactus wrightiae	Endangered	Expl Priv	Utah
Thelocactus bicolor var. flavidispinus	Threatened	Priv	Texas
Cochlospermaceae			
Amoreuxia wrightii	Threatened	Expl	Texas
Crassulaceae			
Dudleya spp.	Endangered Threatened	Priv	California
Dudleya collomae	Endangered	Priv	Arizona
Graptopetalum bartramii	Endangered	Priv	Arizona
Graptopetalum rusbyi	Threatened	Expl Priv	Arizona New Mexico

Species and Varieties	Status	Exploi-tation Status	Locality
Parvisedum leiocarpum	Endangered	Priv	California
Sedum moranii	Endangered	Expl	Oregon
Sedum nevii	Endangered	Expl	Alabama Tennessee
Cycadaceae			
Zamia integrifolia	Endangered	Expl Priv	Florida Georgia
Droseraceae			
Dionaea muscipula	Threatened	Expl Priv	North Carolina South Carolina
Ericaceae			
Rhododendron minus var. chapmanii	Endangered	Expl	Florida
Rhododendron vaseyi	Threatened	Priv	North Carolina
Fabaceae			
Cladrastis lutea	Threatened	Expl	Alabama Arkansas Georgia Illinois Indiana Kentucky Missouri North Carolina Oklahoma Tennessee
Iridaceae			
Iris tenuis	Endangered	Priv	Oregon
Liliaceae			
Agave utahensis var. eborispina	Threatened	Expl	Nevada
Agave utahensis var. nevadensis	Threatened	Expl Priv	California

Species and Varieties	Status	Exploitation Status	Locality
Calochortus greenei	Endangered	Expl? Priv	California Oregon
Erythronium oregonum	Threatened	Priv	Oregon Washington
Fritillaria gentneri	Threatened	Priv	Oregon
Lilium iridollae	Endangered	Expl	Alabama Florida
Lilium occidentale	Endangered	Priv	California Oregon
Lilium washingtonianum var. minus	Threatened	Priv	California
Nolina interrata	Endangered	Expl	California
Zephyranthes simpsonii	Threatened	Priv	Florida
Zephyranthes treatiae	Threatened	Priv	Florida
Orchidaceae			
Cypripedium californicum	Threatened	Expl	California Oregon
Encyclia boothiana var. erythronioides	Threatened	Priv	Florida
Papaveraceae			
Arctomecon merriamii	Endangered	Expl Priv	California Nevada
Polemoniaceae			
Gilia ripleyi	Threatened	Priv	California Nevada
Phlox missoulensis	Endangered	Expl	Montana
Polygonaceae			
Eriogonum kennedyi var. austromontanum	Threatened	Expl	California

Species and Varieties	Status	Exploi-tation Status	Locality
Polypodiaceae			
Polystichum kruckebergii	Threatened	Expl Priv	Oregon Washington
Portulacaceae			
Lewisia columbiana var. wallowensis	Threatened	Expl	Idaho Oregon
Lewisia tweedyi	Threatened	Expl Priv	Washington
Primulaceae			
Primula cusickiana	Endangered	Priv	Idaho Oregon
Primula nevadensis	Endangered	Priv	Nevada
Ranunculaceae			
Aquilegia jonesii	Threatened	Expl	Montana Wyoming
Aquilegia saximontana	Endangered	Priv	Colorado
Hydrastis canadensis	Threatened	Expl	23 states
Sarraceniaceae			
Sarracenia alabamensis ssp. alabamensis	Endangered	Expl Priv	Alabama
Sarracenia jonesii	Endangered	Expl Priv	North Carolina South Carolina
Sarracenia oreophila	Threatened	Expl Priv	Alabama
Scrophulariaceae			
Penstemon barrettiae	Endangered	Priv	Oregon Washington
Synthyris schizantha	Threatened	Priv	Oregon Washington

Species and Varieties	Status	Exploi-tation Status	Locality
Taxaceae			
Taxus floridana	Endangered	Expl	Florida

Extinct Species of Higher
Plants of the Continental United States

"Extinct" is a term that probably is too strong and definitive to use for plant species that, on currently available evidence, have "disappeared" during the last two hundred years. Knowledge of the flora usually is insufficient to state with certitude that a species has recently become extinct. Accordingly, the majority of the suspected examples of extinct species have been labeled "probably extinct," "possibly extinct," or "extinct in nature but cultivated."

For plants that occurred during past geological periods and are known only from fossils (with which this review is not concerned), it is safe to use the term "extinct." Even some species known at first only from fossils, however, have been discovered later to be living in isolated refugia. In the present list, the designation "extinct" means that, after a species was described and known, it disappeared and has not been rediscovered even with repeated searches over a number of years in the type locality and similar or likely place. There is always the possibility that species categorized as "extinct" will be rediscovered, especially those occurring in difficult terrain seldom visited by collectors.

The species included in the "probably extinct" and "possibly extinct" lists were selected from the botanical literature, herbarium specimens, and the advice of various specialists. Considering the large area of knowledge involved, however, and the numerous botanists and other specialists who may have in hand unpublished data on various species, it has not been possible to contact all possible sources of information. It is hoped that publication of this list will stimulate such persons to publish or otherwise make available their own valuable data. In addition, the list should stimulate intensified field studies and research in an effort to rediscover the existence, substantiate the extinction, or clarify the taxonomy of these species.

Recent field work by botanists around the country has led to the rediscovery of a number of taxa listed as extinct in the 1975 Report; those which occur in small numbers are now recommended as endangered. Some examples of rediscovered taxa are:

Pedicularis furbishiae (Furbish Lousewort, Scrophulariaceae) of the St. John River Valley was last collected in Maine in 1917 and in adjacent New Brunswick, Canada, in 1943. It was considered extinct until a colony was found in Maine in 1976 and hailed by the press as a "botanical Lazarus."

The existence of living *Eriogonum ericifolium* var. *ericifolium* (Heath-leaf Buckwheat, Polygonaceae) of Arizona was in doubt, since the variety was known only from 1865 and 1887 collections. A herbarium search later revealed specimens collected in the late 1930s, and plants were again found in nature in 1976.

Schoenolirion texanum (Sunnybell, Liliaceae) formerly was thought to be endemic and extinct in Texas. Although still either extirpated or very nearly so in Texas, populations have recently been located in Arkansas and Alabama.

Astragalus perianus (Rydberg Milk-vetch, Fabaceae) is a Utah endemic which was found in 1975 for the first time since the type collection of 1905. Formerly its habitat preference was unknown, since data had not been recorded with the original collection.

Physaria grahamii (Graham's Twinpod, Brassicaceae) of Utah (and probably adjacent Colorado) has the largest leaves in the genus, but

the plant itself is only six inches high. After the original collection
in 1935 it evaded field collectors until it was rediscovered in abundance
in 1976.

The last recorded sighting of *Dudleya traskiae* (Crassulaceae), an
endemic of Santa Barbara Island, California, was made in 1968. These
rare plants had been browsed to the stumps by the large wild rabbit
population of the island and had virtually disappeared. After the
rabbit population was reduced, dormant stems of a very few plants were
seen in 1975, indicating that the species has survived.

Despite intensive searches over the years, *Betula uber* (Virginia
Round-leaf Birch, Betulaceae) was unknown as a living plant since its
original discovery in 1914. However, in 1975, after much tracking through
Virginia forests, small populations were rediscovered and documented
by two independent investigators. Seedlings are being grown at the National
Arboretum.

Taxa newly added to the extinct list include *Clarkia mosquinii* subsp.
xerophila, the type locality and only known habitat of which was covered
by Lake Oroville in Butte County, California, in 1968, and *Arabis gunnisoniana*
from Gunnison County, Colorado, of which the known localities are now,
likewise, under water.

Species, subspecies, or varieties that have been extirpated or are
thought to be extirpated in one state, but are extant in other states, are
not included in this list. For example, the endangered orchid *Isotria
medeoloides* has recently been extirpated in Missouri, where its former
site in Bollinger County was destroyed by a clear-cut. It is, however,
extant in other parts of its range and thus not listed as extinct.

Species that are extinct in the wild but living in cultivation, e.g.,
Franklinia alatamaha, are included with proper notation.

The 90 listed taxa of extinct and probably or possibly extinct plants represent approximately 0.4 percent of the native vascular flora of the continental United States.

Species and Varieties	Locality	Date Last Collected	Status*
Aizoaceae			
Sesuvium trianthemoides	Kennedy Co., Tex.	1947	PrEx
Apiaceae			
Lilaeopsis masonii	Napa and Solano Co., Calif.		PoEx
Apocynaceae			
Apocynum jonesii	Coconino Co., Ariz.	1884	PoEx
Asclepiadaceae			
Matelea radiata	Brooks Co., Tex.	1909	PoEx
Asteraceae			
Calycadenia fremontii	Butte and Tehama Co., Calif.	1950	PoEx
Eriophyllum nubigenum var. nubigenum	Mariposa Co., Calif.	1938	PoEx
Greenella discoidea	Cochise Co., Ariz.	c.1883	PrEx
Helianthus nuttallii ssp. parishii	Los Angeles, Orange and San Bernardino Co., Calif.	1932	PoEx
Helianthus praetermissus	Valencia Co., New Mex.	1851	PrEx, Tax?
Hymenoxys texana	Harris Co., Tex.	1900	PrEx

*Ex = Recently extinct or extinct in wild. Not collected during extensive search over a number of years; type locality or sites may be destroyed.

PrEx = Probably extinct. Not recollected after one or a few visits to the type locality and/or similar sites.

PoEx = Possibly extinct. Not recollected or known only from type locality for a number of years; may not have been searched for.

Tax? = Taxonomic status in doubt, perhaps owing to scarcity of material; possible hybrid.

Cult = May be extinct in nature, but the plants are in cultivation or have been transplanted.

Species and Varieties	Locality	Date Last Collected	Status
Perityle rotundata	Presidio Co., Tex.	1852	PrEx
Solidago porteri	Jasper Co., Ga.; Buncombe and Jackson Co., N.C.	1899	PoEx, Tax?
Boraginaceae			
Cryptantha aperta	Mesa Co., Colo.	1892	PrEx
Cryptantha insolita	Clark Co., Nev.	1905	PrEx
Mertensia toiyabensis	Lander Co., Nev.	1882	PoEx
Brassicaceae			
Arabis fructicosa	Yellowstone National Park Co., Wyo.	1899	PrEx
Arabis gunnisoniana	Gunnison Co., Colo.	1938	PrEx
Lesquerella macrocarpa	Sweetwater Co., Wyo.	1901	PrEx
Smelowskia holmgrenii	Nye Co., Nev.	1947	PoEx
Thelypodium tenue	Presidio Co., Tex.	1942	PoEx
Bromeliaceae			
Hechtia texensis	Brewster Co., Tex.	1885	PoEx
Burmanniaceae			
Thismia americana	Cook Co., Ill.	1913	Ex
Cactaceae			
Coryphantha scheeri var. uncinata	El Paso Co., Tex.	1912	PoEx
Echinocereus blanckii var. angusticeps	Hidalgo Co., Tex.	1934	PoEx
Opuntia strigil var. flexospina	Webb Co., Tex.	1911	PoEx
Campanulaceae			
Campanula robinsiae	Hernando Co., Fla.	1958	PoEx

Species and Varieties	Locality	Date Last Collected	Status
Githopsis filicaulis	San Diego Co., Calif.	1884	PoEx
Githopsis latifolia	Plumas Co., Calif.	1912	PoEx
Caryophyllaceae			
Arenaria livermorensis	Jeff Davis Co., Tex.	1934	PoEx
Ceratophyllaceae			
Ceratophyllum floridanum	Big Pine Key, Fla.	1953	PoEx
Chenopodiaceae			
Atriplex tularensis	Kern Co., Calif.	1921	PoEx
Cuscutaceae			
Cuscuta warneri	Millard Co., Utah	1957	PoEx
Ericaceae			
Arctostaphylos hookeri ssp. franciscana	San Francisco Co., Calif.	1968	PoEx, Cult
Fabaceae			
Astragalus columbianus	Yakima Co., Wash.	1883	PoEx
Astragalus desereticus	Sanpete Co., Utah	1909	PoEx
Astragalus humillimus	Montezuma Co., Colo.	1875	PoEx
Astragalus lentiginosus var. ursinus	Iron Co., Utah	1877	PoEx, Tax?
Astragalus linifolius	Mesa Co., Colo.	1926	PoEx
Astragalus pycnostachyus ssp. lanosissimus	Ventura and Los Angeles Co., Calif.	1967	PoEx
Astragalus robbinsii var. robbinsii	Chittenden Co., Vt.	1893	Ex
Lotus argophyllus var. adsurgens	San Clemente Is., Calif.	1939	PrEx

Species and Varieties	Locality	Date Last Collected	Status
Psoralea macrophylla	Polk Co., N.C.	1897	Ex
Psoralea stipulata	Clark and Floyd Co., Ind.	1839	PoEx
Vicia reverchonii	Dallas Co., Tex.; Seminole Co., Okla.	1879	PrEx
Hydrocharitaceae			
Elodea brandegeae	Nevada Co., Calif.	1908	PoEx
Elodea linearis	Davidson Co., Tenn.	1875	PoEx
Elodea nevadensis	Washoe Co., Nev.	1887	PoEx
Elodea schweinitzii	Northampton Co., Pa.	1832	PrEx
Hydrophyllaceae			
Phacelia cinerea	San Nicolas Is., Calif.	1901	PoEx
Phacelia lenta	Walla Walla Co., Wash.	1883	PoEx
Isoetaceae			
Isoetes louisianensis	Washington Parish, La.	1974	PoEx, Cult
Juncaceae			
Juncus leiospermus	Tehama Co., Calif.	1916	PoEx
Juncus pervetus	Barnstable Co., Mass.	1916	PrEx
Lamiaceae			
Hedeoma pilosum	Brewster Co., Tex.	1940	PrEx
Monardella leucocephala	Merced and Stanislaus Co., Calif.	1941	PoEx
Scutellaria ocmulgee	Bibb Co., Ga.	1895	PoEx
Liliaceae			
Calochortus monanthus	Siskiyou Co., Calif.	1876	PoEx
Fritillaria adamantina	Douglas Co., Ore.	1936	PoEx

Species and Varieties	Locality	Date Last Collected	Status
Malvaceae			
Malacothamnus abbottii	Monterey Co., Calif.	1899	PoEx
Malacothamnus mendocinensis	Mendocino Co., Calif.	1939	PoEx
Sidalcea covillei	Inyo Co., Calif.	c.1976	PrEx
Onagraceae			
Clarkia mosquinii ssp. xerophila	Butte Co., Calif.	1968	PoEx
Orchidaceae			
Spiranthes parksii	Brazos Co., Tex.	1945	PrEx, Tax?
Orobanchaceae			
Orobanche valida	Ventura and Los Angeles Co., Calif.	1929	PoEx
Poaceae			
Dissanthelium californicum	Monterey Co., Santa Catalina and San Clemente Is., Calif.	1912	PoEx
Pleuropogon oregonus	Union and Lake Co., Ore.	1901	PoEx
Polygonaceae			
Chorizanthe valida	Sonoma and Marin Co., Calif.	1962	PoEx
Polygonum montereyense	Monterey Co., Calif.	1917	PoEx
Ranunculaceae			
Ranunculus acriformis var. aestivalis	Garfield Co., Utah	1950	PrEx

Species and Varieties	Locality	Date Last Collected	Status
Rosaceae			
Potentilla multijuga	Los Angeles Co., Calif.	1890	PrEx
Saxifragaceae			
Lithophragma maximum	San Clemente Is., Calif.	1936	PoEx
Scrophulariaceae			
Agalinis caddoensis	Caddo Parish, La.	1913	PoEx
Agalinis stenophylla	Hillsboro Co., Fla.	1897	PoEx
Bacopa simulans	Charles City Co., Va.	1941	PrEx
Castilleja leschkeana	Marin Co., Calif.	1960	PoEx
Castilleja ludoviciana	Jeff Davis Parish, La.	1915	PoEx
Cordylanthus mollis ssp. mollis	San Francisco Bay, Calif.	1966	PoEx
Cordylanthus palmatus	San Joaquin and Sacramento Valleys, Calif.	1970	PoEx, Cult
Mimulus brandegei	Santa Cruz Is., Calif.	1932	PoEx
Mimulus traskiae	Santa Catalina Is., Calif.	1904	PoEx
Orthocarpus pachystachyus	Siskiyou Co., Calif.	1913	PoEx
Penstemon garrettii	Duchesne and Wasatch Co., Utah	1905	PoEx
Seymeria harvardii	Maverick Co., Tex.	1882	PoEx
Synthyris missurica ssp. hirsuta	Douglas Co., Ore.	1881	PoEx
Solanaceae			
Lycium hassei	Santa Catalina and San Clemente Is., Calif.	1936	PoEx
Lycium verrucosum	San Nicolas Is., Calif.	1901	PoEx

Species and Varieties	Locality	Date Last Collected	Status
Solanum bahamense var. rugelii	Key West, Fla.	1860	PrEx
Solanum carolinense var. hirsutum	Baldwin and Muscogee Co., Ga.	1834	PrEx

Sterculiaceae

Nephropetalum pringlei	Hidalgo Co., Tex.	1888	PoEx

Theaceae

Franklinia alatamaha	McIntosh Co., Ga.	1803	Ex, Cult

Endangered Species of the
Continental United States

Endangered species of plants are in danger of extinction throughout
all or a significant portion of their range. These plants usually are
rare, with limited geographical distribution, and often occur in fragile,
restricted habitats.

Endangered species are more immediately in danger than threatened
species and may become extinct unless they are protected. Some endangered
species have little reproductive capability and are so narrowly endemic
or specialized to a certain habitat or specific condition that they are
unable to withstand threats or dangers normally withstood by most plants.
Many also are endangered by human or human-induced activities that are
destroying their habitats. At the present time, 839 kinds or 4.2 percent
of the native higher plants are considered to be endangered in the continental
United States.

Also included in the endangered plant list are 90 kinds of plants
considered as "recently extinct," "probably extinct," or "possibly extinct."
The extinct taxa are marked with an asterisk (*) in front of the genus name.
These are included in the list in order to help provide them with protection
if they are rediscovered as the result of more intensive searches. If they
are rediscovered, they will be included automatically in the endangered
category unless another status is determined.

As the flora is better studied, species may be removed from the
"endangered" list and placed on the "threatened" list or deleted as a
result of the discovery of more populations or wider distribution of a
species. Also, some endangered species can be expected to become extinct.

ENDANGERED SPECIES IN THE CONTINENTAL UNITED STATES

FAMILY	SPECIES	STATE
ACANTHACEAE	JUSTICIA COOLEYI	FLORIDA
ACANTHACEAE	JUSTICIA CRASSIFOLIA	FLORIDA
AIZOACEAE	*SESUVIUM TRIANTHEMOIDES	TEXAS
ALISMATACEAE	SAGITTARIA FASCICULATA	NORTH CAROLINA, SOUTH CAROLINA
ANACARDIACEAE	RHUS KEARNEYI	ARIZONA
ANNONACEAE	ASIMINA TETRAMERA	FLORIDA
APIACEAE	ANGELICA CALLII	CALIFORNIA
APIACEAE	CYMOPTERUS MINIMUS	UTAH
APIACEAE	CYMOPTERUS NIVALIS	NEVADA
APIACEAE	ERYNGIUM ARISTULATUM VAR. PARISHII	CALIFORNIA
APIACEAE	ERYNGIUM RACEMOSUM	CALIFORNIA
APIACEAE	*LILAEOPSIS MASONII	CALIFORNIA
APIACEAE	LOMATIUM BRADSHAWII	OREGON
APIACEAE	LOMATIUM GREENMANII	OREGON
APIACEAE	LOMATIUM LAEVIGATUM	WASHINGTON, OREGON
APIACEAE	LOMATIUM RAVENII	CALIFORNIA
APIACEAE	LOMATIUM SUKSDORFII	WASHINGTON, OREGON
APIACEAE	LOMATIUM TUBEROSUM	WASHINGTON
APIACEAE	OXYPOLIS GREENMANII	FLORIDA
APIACEAE	SANICULA MARITIMA	CALIFORNIA
APIACEAE	SANICULA SAXATILIS	CALIFORNIA
APIACEAE	SANICULA TRACYI	CALIFORNIA, OREGON
APIACEAE	SIUM FLORIDANUM	FLORIDA
APIACEAE	TAUSCHIA HOOVERI	WASHINGTON
APOCYNACEAE	*APOCYNUM JONESII	ARIZONA
APOCYNACEAE	CYCLADENIA HUMILIS VAR. JONESII	UTAH
ARECACEAE	ROYSTONEA ELATA	FLORIDA
ARISTOLOCHIACEAE	HEXASTYLIS NANIFLORA	NORTH CAROLINA, SOUTH CAROLINA, VIRGINIA
ARISTOLOCHIACEAE	HEXASTYLIS SPECIOSA	ALABAMA
ASCLEPIADACEAE	ASCLEPIAS EASTWOODIANA	NEVADA
ASCLEPIADACEAE	ASCLEPIAS MEADII	INDIANA, ILLINOIS, IOWA, KANSAS, MISSOURI
ASCLEPIADACEAE	MATELEA ALABAMENSIS	ALABAMA
ASCLEPIADACEAE	MATELEA EDWARDSENSIS	TEXAS
ASCLEPIADACEAE	*MATELEA RADIATA	TEXAS
ASCLEPIADACEAE	MATELEA TEXENSIS	TEXAS
ASTERACEAE	AMBROSIA CHEIRANTHIFOLIA	TEXAS
ASTERACEAE	ANTENNARIA ARCUATA	IDAHO, WYOMING
ASTERACEAE	ARTEMISIA ARGILOSA	COLORADO
ASTERACEAE	ASTER BLEPHAROPHYLLUS	NEW MEXICO
ASTERACEAE	ASTER CHILENSIS VAR. LENTUS	CALIFORNIA
ASTERACEAE	ASTER GORMANII	OREGON
ASTERACEAE	ASTER PINIFOLIUS	FLORIDA

ENDANGERED SPECIES IN THE CONTINENTAL UNITED STATES

FAMILY	SPECIES	STATE
ASTERACEAE	ASTER VIALIS	OREGON
ASTERACEAE	BALDUINA ATROPURPUREA	GEORGIA, FLORIDA
ASTERACEAE	BLENNOSPERMA BAKERI	CALIFORNIA
ASTERACEAE	BLENNOSPERMA NANUM VAR. ROBUSTUM	CALIFORNIA
ASTERACEAE	BRICKELLIA VIEJENSIS	TEXAS
ASTERACEAE	*CALYCADENIA FREMONTII	CALIFORNIA
ASTERACEAE	CIRSIUM CLOKEYI	NEVADA
ASTERACEAE	CIRSIUM CRASSICAULE	CALIFORNIA
ASTERACEAE	CIRSIUM FONTINALE VAR. FONTINALE	CALIFORNIA
ASTERACEAE	CIRSIUM FONTINALE SSP. OBISPOENSE	CALIFORNIA
ASTERACEAE	CIRSIUM HYDROPHILUM VAR. HYDROPHILUM	CALIFORNIA
ASTERACEAE	CIRSIUM LONCHOLEPIS	CALIFORNIA
ASTERACEAE	CIRSIUM RHOTHOPHILUM	CALIFORNIA
ASTERACEAE	COREOPSIS INTERMEDIA	LOUISIANA, TEXAS
ASTERACEAE	DYSSODIA TEPHROLEUCA	TEXAS
ASTERACEAE	ECHINACEA TENNESSEENSIS	TENNESSEE
ASTERACEAE	ENCELIOPSIS COVILLEI	CALIFORNIA
ASTERACEAE	ERIGERON BASALTICUS	WASHINGTON
ASTERACEAE	ERIGERON CALVUS	CALIFORNIA
ASTERACEAE	ERIGERON DELICATUS	OREGON, CALIFORNIA
ASTERACEAE	ERIGERON ERIOPHYLLUS	ARIZONA
ASTERACEAE	ERIGERON FLAGELLARIS VAR. TRILOBATUS	UTAH
ASTERACEAE	ERIGERON FOLIOSUS VAR. BLOCHMANAE	CALIFORNIA
ASTERACEAE	ERIGERON GEISERI VAR. CALCICOLA	TEXAS
ASTERACEAE	ERIGERON KACHINENSIS	UTAH
ASTERACEAE	ERIGERON KUSCHEI	ARIZONA
ASTERACEAE	ERIGERON LATUS	IDAHO
ASTERACEAE	ERIGERON MAGUIREI	UTAH
ASTERACEAE	ERIGERON RELIGIOSUS	UTAH
ASTERACEAE	ERIGERON RHIZOMATUS	NEW MEXICO
ASTERACEAE	ERIGERON SIONIS	UTAH
ASTERACEAE	ERIOPHYLLUM LANATUM VAR. HALLII	CALIFORNIA
ASTERACEAE	ERIOPHYLLUM MOHAVENSE	CALIFORNIA
ASTERACEAE	*ERIOPHYLLUM NUBIGENUM VAR. NUBIGENUM	CALIFORNIA
ASTERACEAE	EUPATORIUM RESINOSUM VAR. KENTUCKIENSE	KENTUCKY
ASTERACEAE	GAILLARDIA FLAVA	UTAH
ASTERACEAE	GALINSOGA SEMICALVA VAR. PERCALVA	ARIZONA
ASTERACEAE	GNAPHALIUM OBTUSIFOLIUM VAR. SAXICOLA	WISCONSIN
ASTERACEAE	*GREENELLA DISCOIDEA	ARIZONA
ASTERACEAE	GRINDELIA FRAXINO-PRATENSIS	NEVADA, CALIFORNIA
ASTERACEAE	GRINDELIA HOWELLII	IDAHO, MONTANA
ASTERACEAE	GRINDELIA OOLEPIS	TEXAS
ASTERACEAE	HAPLOPAPPUS CANUS	CALIFORNIA

ENDANGERED SPECIES IN THE CONTINENTAL UNITED STATES

FAMILY	SPECIES	STATE
ASTERACEAE	HAPLOPAPPUS EASTWOODIAE	CALIFORNIA
ASTERACEAE	HAPLOPAPPUS FREMONTII SSP. MONOCEPHALUS	COLORADO
ASTERACEAE	HAPLOPAPPUS SALICINUS	ARIZONA
ASTERACEAE	HAPLOPAPPUS SPINULOSUS SSP. LAEVIS	NEW MEXICO
ASTERACEAE	HELIANTHUS EXILIS	CALIFORNIA
ASTERACEAE	HELIANTHUS NIVEUS SSP. TEPHRODES	CALIFORNIA
ASTERACEAE	*HELIANTHUS NUTTALLII VAR. PARISHII	CALIFORNIA
ASTERACEAE	HELIANTHUS PARADOXUS	TEXAS, NEW MEXICO
ASTERACEAE	*HELIANTHUS PRAETERMISSUS	NEW MEXICO
ASTERACEAE	HELIANTHUS SMITHII	ALABAMA, GEORGIA
ASTERACEAE	HEMIZONIA CONJUGENS	CALIFORNIA
ASTERACEAE	HEMIZONIA FLORIBUNDA	CALIFORNIA
ASTERACEAE	HEMIZONIA MINTHORNII	CALIFORNIA
ASTERACEAE	HEMIZONIA MOHAVENSIS	CALIFORNIA
ASTERACEAE	HETEROTHECA JONESII	UTAH
ASTERACEAE	HETEROTHECA RUTHII	TENNESSEE
ASTERACEAE	HOLOCARPHA MACRADENIA	CALIFORNIA
ASTERACEAE	*HYMENOXYS TEXANA	TEXAS
ASTERACEAE	JAMESIANTHUS ALABAMENSIS	ALABAMA
ASTERACEAE	LASTHENIA BURKEI	CALIFORNIA
ASTERACEAE	LASTHENIA CONJUGENS	CALIFORNIA
ASTERACEAE	LAYIA DISCOIDEA	CALIFORNIA
ASTERACEAE	LIATRIS OHLINGERAE	FLORIDA
ASTERACEAE	LYGODESMIA GRANDIFLORA VAR. STRICTA	UTAH
ASTERACEAE	MACHAERANTHERA AUREA	TEXAS
ASTERACEAE	MACHAERANTHERA LEUCANTHEMIFOLIA	NEVADA
ASTERACEAE	MARSHALLIA MOHRI	ALABAMA, GEORGIA
ASTERACEAE	MICROSERIS DECIPIENS	CALIFORNIA
ASTERACEAE	PALAFOXIA ARIDA VAR. GIGANTEA	CALIFORNIA
ASTERACEAE	PARTHENIUM LIGULATUM	COLORADO, UTAH
ASTERACEAE	PARTHENIUM TETRANEURIS	COLORADO
ASTERACEAE	PECTIS RUSBYI	ARIZONA
ASTERACEAE	PENTACHAETA LYONII	CALIFORNIA
ASTERACEAE	PERITYLE BISETOSA VAR. BISETOSA	TEXAS
ASTERACEAE	PERITYLE BISETOSA VAR. SCALARIS	TEXAS
ASTERACEAE	PERITYLE CINEREA	TEXAS
ASTERACEAE	PERITYLE GILENSIS VAR. SALENSIS	ARIZONA
ASTERACEAE	PERITYLE LINDHEIMERI VAR. HALIMIFOLIA	TEXAS
ASTERACEAE	*PERITYLE ROTUNDATA	TEXAS
ASTERACEAE	PERITYLE VITREOMONTANA	TEXAS
ASTERACEAE	PSEUDOBAHIA BAHIAEFOLIA	CALIFORNIA
ASTERACEAE	PSEUDOBAHIA PEIRSONII	CALIFORNIA
ASTERACEAE	PYRROCOMA RADIATUS	IDAHO, OREGON

ENDANGERED SPECIES IN THE CONTINENTAL UNITED STATES

FAMILY	SPECIES	STATE
ASTERACEAE	SAUSSUREA WEBERI	COLORADO, MONTANA
ASTERACEAE	SENECIO FRANCISCANUS	ARIZONA
ASTERACEAE	SENECIO HALLII VAR. DISCOIDEA	COLORADO
ASTERACEAE	SENECIO LAYNEAE	CALIFORNIA
ASTERACEAE	SENECIO PORTERI	COLORADO, OREGON
ASTERACEAE	SILPHIUM BRACHIATUM	TENNESSEE
ASTERACEAE	SILPHIUM INTEGRIFOLIUM VAR. GATTINGERI	TENNESSEE
ASTERACEAE	SOLIDAGO ALBOPILOSA	KENTUCKY
ASTERACEAE	*SOLIDAGO PORTERI	NORTH CAROLINA, GEORGIA
ASTERACEAE	SOLIDAGO SHORTII	KENTUCKY
ASTERACEAE	SPHAEROMERIA RUTHIAE	UTAH
ASTERACEAE	STEPHANOMERIA MALHEURENSIS	OREGON
ASTERACEAE	STEPHANOMERIA SCHOTTII	ARIZONA
ASTERACEAE	TANACETUM COMPACTUM	NEVADA
ASTERACEAE	TOWNSENDIA APRICA	UTAH
ASTERACEAE	TRACYINA ROSTRATA	CALIFORNIA
ASTERACEAE	VIGUIERA LUDENS	TEXAS
BERBERIDACEAE	BERBERIS HARRISONIANA	ARIZONA
BERBERIDACEAE	BERBERIS NEVINII	CALIFORNIA
BERBERIDACEAE	BERBERIS SONNEI	CALIFORNIA
BETULACEAE	BETULA UBER	VIRGINIA
BORAGINACEAE	AMSINCKIA GRANDIFLORA	CALIFORNIA
BORAGINACEAE	*CRYPTANTHA APERTA	COLORADO
BORAGINACEAE	CRYPTANTHA ATWOODII	ARIZONA
BORAGINACEAE	*CRYPTANTHA INSOLITA	NEVADA
BORAGINACEAE	CRYPTANTHA OCHROLEUCA	UTAH
BORAGINACEAE	CRYPTANTHA ROOSIORUM	CALIFORNIA
BORAGINACEAE	CRYPTANTHA SHACKLETTEANA	ALASKA
BORAGINACEAE	CRYPTANTHA WEBERI	COLORADO
BORAGINACEAE	DASYNOTUS DAUBENMIREI	IDAHO
BORAGINACEAE	HACKELIA CRONQUISTII	OREGON
BORAGINACEAE	HACKELIA DAVISII	IDAHO
BORAGINACEAE	HACKELIA OPHIOBIA	OREGON
BORAGINACEAE	HACKELIA VENUSTA	WASHINGTON
BORAGINACEAE	*MERTENSIA TOIYABENSIS	NEVADA
BORAGINACEAE	PLAGIOBOTHRYS DIFFUSUS	CALIFORNIA
BORAGINACEAE	PLAGIOBOTHRYS HIRTUS SSP. HIRTUS	OREGON
BORAGINACEAE	PLAGIOBOTHRYS LAMPROCARPUS	OREGON
BRASSICACEAE	ARABIS BREWERI VAR. PECUNIARIA	CALIFORNIA
BRASSICACEAE	ARABIS CRANDALLII	COLORADO
BRASSICACEAE	*ARABIS FRUCTICOSA	WYOMING
BRASSICACEAE	*ARABIS GUNNISONIANA	COLORADO
BRASSICACEAE	ARABIS KOEHLERI VAR. KOEHLERI	OREGON

ENDANGERED SPECIES IN THE CONTINENTAL UNITED STATES

FAMILY	SPECIES	STATE
BRASSICACEAE	ARABIS MCDONALDIANA	CALIFORNIA
BRASSICACEAE	ARABIS OXYLOBULA	COLORADO
BRASSICACEAE	ARABIS PERSTELLATA VAR. AMPLA	TENNESSEE
BRASSICACEAE	ARABIS PERSTELLATA VAR. PERSTELLATA	ALABAMA, KENTUCKY
BRASSICACEAE	BRAYA HUMILIS SSP. VENTOSA	COLORADO
BRASSICACEAE	CARDAMINE CONSTANCEI	IDAHO
BRASSICACEAE	CARDAMINE MICRANTHERA	NORTH CAROLINA
BRASSICACEAE	CARDAMINE PATTERSONII	OREGON
BRASSICACEAE	DENTARIA INCISA	TENNESSEE
BRASSICACEAE	DRABA APRICA	ARKANSAS, GEORGIA, MISSOURI, OKLAHOMA, SOUTH CAROLINA
BRASSICACEAE	DRABA ARIDA	NEVADA
BRASSICACEAE	DRABA ASPRELLA VAR. ASPRELLA	ARIZONA
BRASSICACEAE	DRABA ASPRELLA VAR. KAIBABENSIS	ARIZONA
BRASSICACEAE	DRABA PAUCIFRUCTA	NEVADA
BRASSICACEAE	ERYSIMUM CAPITATUM VAR. ANGUSTATUM	CALIFORNIA
BRASSICACEAE	ERYSIMUM FRANCISCANUM VAR. FRANCISCANUM	CALIFORNIA
BRASSICACEAE	ERYSIMUM TERETIFOLIUM	CALIFORNIA
BRASSICACEAE	EUTREMA PENLANDII	COLORADO
BRASSICACEAE	GLAUCOCARPUM SUFFRUTESCENS	UTAH
BRASSICACEAE	LEAVENWORTHIA ALABAMICA VAR. BRACHYSTYLA	ALABAMA
BRASSICACEAE	LEAVENWORTHIA AUREA	OKLAHOMA, TEXAS
BRASSICACEAE	LEAVENWORTHIA CRASSA VAR. CRASSA	ALABAMA
BRASSICACEAE	LEAVENWORTHIA CRASSA VAR. ELONGATA	ALABAMA
BRASSICACEAE	LEAVENWORTHIA EXIGUA VAR. LACINIATA	KENTUCKY
BRASSICACEAE	LEAVENWORTHIA EXIGUA VAR. LUTEA	ALABAMA, TENNESSEE
BRASSICACEAE	LEPIDIUM BARNEBYANUM	UTAH
BRASSICACEAE	LEPIDIUM DAVISII	IDAHO
BRASSICACEAE	LESQUERELLA AUREA	NEW MEXICO
BRASSICACEAE	LESQUERELLA DENSIPILA	TENNESSEE, ALABAMA
BRASSICACEAE	LESQUERELLA FILIFORMIS	MISSOURI
BRASSICACEAE	LESQUERELLA FREMONTII	WYOMING
BRASSICACEAE	LESQUERELLA LATA	NEW MEXICO
BRASSICACEAE	LESQUERELLA LYRATA	ALABAMA
BRASSICACEAE	*LESQUERELLA MACROCARPA	WYOMING
BRASSICACEAE	LESQUERELLA PERFORATA	TENNESSEE
BRASSICACEAE	LESQUERELLA PRUINOSA	COLORADO
BRASSICACEAE	LESQUERELLA STONENSIS	TENNESSEE
BRASSICACEAE	LESQUERELLA TUMULOSA	UTAH
BRASSICACEAE	LESQUERELLA VALIDA	TEXAS, NEW MEXICO
BRASSICACEAE	SELENIA JONESII	TEXAS
BRASSICACEAE	SIBARA FILIFOLIA	CALIFORNIA
BRASSICACEAE	SISYMBRIUM KEARNEYI	ARIZONA

ENDANGERED SPECIES IN THE CONTINENTAL UNITED STATES

FAMILY	SPECIES	STATE
BRASSICACEAE	SMELOWSKIA BOREALIS VAR. VILLOSA	ALASKA
BRASSICACEAE	*SMELOWSKIA HOLMGRENII	NEVADA
BRASSICACEAE	SMELOWSKIA OVALIS VAR. CONGESTA	CALIFORNIA
BRASSICACEAE	STREPTANTHUS ALBIDUS SSP. ALBIDUS	CALIFORNIA
BRASSICACEAE	STREPTANTHUS CALLISTUS	CALIFORNIA
BRASSICACEAE	STREPTANTHUS FARNSWORTHIANUS	CALIFORNIA
BRASSICACEAE	STREPTANTHUS LEMMONII	ARIZONA
BRASSICACEAE	STREPTANTHUS MORRISONII SSP. HIRTIFLORUS	CALIFORNIA
BRASSICACEAE	STREPTANTHUS NIGER	CALIFORNIA
BRASSICACEAE	STREPTANTHUS SPARSIFLORUS	TEXAS
BRASSICACEAE	STREPTANTHUS SQUAMIFORMIS	OKLAHOMA, ARKANSAS
BRASSICACEAE	THELYPODIUM REPANDUM	IDAHO
BRASSICACEAE	*THELYPODIUM TENUE	TEXAS
BRASSICACEAE	THELYPODIUM TEXANUM	TEXAS
BRASSICACEAE	WAREA AMPLEXIFOLIA	FLORIDA
BRASSICACEAE	WAREA CARTERI	FLORIDA
BROMELIACEAE	*HECHTIA TEXENSIS	TEXAS
BURMANNIACEAE	*THISMIA AMERICANA	ILLINOIS
CACTACEAE	ANCISTROCACTUS TOBUSCHII	TEXAS
CACTACEAE	CEREUS ERIOPHORUS VAR. FRAGRANS	FLORIDA
CACTACEAE	CEREUS GRACILIS VAR. ABORIGINUM	FLORIDA
CACTACEAE	CEREUS GRACILIS VAR. SIMPSONII	FLORIDA
CACTACEAE	CEREUS ROBINII VAR. ROBINII	FLORIDA
CACTACEAE	CEREUS ROBINII VAR. DEERINGII	FLORIDA
CACTACEAE	CORYPHANTHA MINIMA	TEXAS
CACTACEAE	CORYPHANTHA RAMILLOSA	TEXAS
CACTACEAE	*CORYPHANTHA SCHEERI VAR. UNCINATA	TEXAS
CACTACEAE	CORYPHANTHA SNEEDII VAR. LEEI	NEW MEXICO
CACTACEAE	CORYPHANTHA SNEEDII VAR. SNEEDII	TEXAS, NEW MEXICO
CACTACEAE	CORYPHANTHA STROBILIFORMIS VAR. DURISPINA	TEXAS
CACTACEAE	ECHINOCACTUS HORIZONTHALONIUS VAR. NICHOLII	ARIZONA
CACTACEAE	*ECHINOCEREUS BLANCKII VAR. ANGUSTICEPS	TEXAS
CACTACEAE	ECHINOCEREUS CHLORANTHUS VAR. NEOCAPILLUS	TEXAS
CACTACEAE	ECHINOCEREUS ENGELMANNII VAR. HOWEI	CALIFORNIA
CACTACEAE	ECHINOCEREUS ENGELMANNII VAR. PURPUREUS	UTAH
CACTACEAE	ECHINOCEREUS FENDLERI VAR. KUENZLERI	NEW MEXICO
CACTACEAE	ECHINOCEREUS LLOYDII	TEXAS, NEW MEXICO
CACTACEAE	ECHINOCEREUS REICHENBACHII VAR. ALBERTII	TEXAS
CACTACEAE	ECHINOCEREUS RUSSANTHUS	TEXAS
CACTACEAE	ECHINOCEREUS TRIGLOCHIDIATUS VAR. ARIZONICUS	ARIZONA
CACTACEAE	ECHINOCEREUS VIRIDIFLORUS VAR. DAVISII	TEXAS
CACTACEAE	FEROCACTUS VIRIDESCENS	CALIFORNIA
CACTACEAE	NEOLLOYDIA GAUTII	TEXAS

ENDANGERED SPECIES IN THE CONTINENTAL UNITED STATES

FAMILY	SPECIES	STATE
CACTACEAE	NEOLLOYDIA MARIPOSENSIS	TEXAS
CACTACEAE	OPUNTIA BASILARIS VAR. TRELEASEI	CALIFORNIA, ARIZONA
CACTACEAE	*OPUNTIA STRIGIL VAR. FLEXOSPINA	TEXAS
CACTACEAE	PEDIOCACTUS BRADYI	ARIZONA
CACTACEAE	PEDIOCACTUS KNOWLTONII	NEW MEXICO, COLORADO
CACTACEAE	PEDIOCACTUS PEEBLESIANUS VAR. PEEBLESIANUS	ARIZONA
CACTACEAE	PEDIOCACTUS SILERI	ARIZONA, UTAH
CACTACEAE	SCLEROCACTUS GLAUCUS	UTAH, COLORADO
CACTACEAE	SCLEROCACTUS MESAE-VERDAE	COLORADO, NEW MEXICO
CACTACEAE	SCLEROCACTUS WRIGHTIAE	UTAH
CAMPANULACEAE	CAMPANULA CALIFORNICA	CALIFORNIA
CAMPANULACEAE	*CAMPANULA ROBINSIAE	FLORIDA
CAMPANULACEAE	*GITHOPSIS FILICAULIS	CALIFORNIA
CAMPANULACEAE	*GITHOPSIS LATIFOLIA	CALIFORNIA
CAMPANULACEAE	LEGENERE LIMOSA	CALIFORNIA
CAPPARIDACEAE	CLEOME MULTICAULIS	ARIZONA, COLORADO, NEW MEXICO, TEXAS
CAPRIFOLIACEAE	VIBURNUM BRACTEATUM	ALABAMA, GEORGIA
CARYOPHYLLACEAE	ARENARIA ALABAMENSIS	ALABAMA
CARYOPHYLLACEAE	*ARENARIA LIVERMORENSIS	TEXAS
CARYOPHYLLACEAE	ARENARIA URSINA	CALIFORNIA
CARYOPHYLLACEAE	CERASTIUM ARVENSE VAR. VILLOSISSIMUM	PENNSYLVANIA
CARYOPHYLLACEAE	CERASTIUM CLAWSONII	TEXAS
CARYOPHYLLACEAE	GEOCARPON MINIMUM	MISSOURI, ARKANSAS
CARYOPHYLLACEAE	PARONYCHIA CHARTACEA	FLORIDA
CARYOPHYLLACEAE	PARONYCHIA CONGESTA	TEXAS
CARYOPHYLLACEAE	PARONYCHIA MACCARTII	TEXAS
CARYOPHYLLACEAE	PARONYCHIA RUGELII VAR. INTERIOR	FLORIDA, GEORGIA
CARYOPHYLLACEAE	SILENE DOUGLASII VAR. ORARIA	OREGON
CARYOPHYLLACEAE	SILENE INVISA	CALIFORNIA
CARYOPHYLLACEAE	SILENE MARMORENSIS	CALIFORNIA
CARYOPHYLLACEAE	SILENE PLANKII	TEXAS, NEW MEXICO
CARYOPHYLLACEAE	SILENE POLYPETALA	FLORIDA, GEORGIA
CARYOPHYLLACEAE	SILENE RECTIRAMEA	ARIZONA
CARYOPHYLLACEAE	SILENE SPALDINGII	WASHINGTON, OREGON, IDAHO, MONTANA
CARYOPHYLLACEAE	STELLARIA IRRIGUA	COLORADO
CELASTRACEAE	FORSELLESIA PUNGENS VAR. GLABRA	CALIFORNIA
CERATOPHYLLACEAE	*CERATOPHYLLUM FLORIDANUM	FLORIDA
CHENOPODIACEAE	ATRIPLEX GRIFFITHSII	ARIZONA, NEW MEXICO
CHENOPODIACEAE	ATRIPLEX KLEBERGORUM	TEXAS
CHENOPODIACEAE	ATRIPLEX PLEIANTHA	COLORADO
CHENOPODIACEAE	*ATRIPLEX TULARENSIS	CALIFORNIA
CHENOPODIACEAE	NITROPHILA MOHAVENSIS	CALIFORNIA
CHENOPODIACEAE	SUAEDA DURIPES	TEXAS

ENDANGERED SPECIES IN THE CONTINENTAL UNITED STATES

FAMILY	SPECIES	STATE
CISTACEAE	HUDSONIA ERICOIDES SSP. MONTANA	NORTH CAROLINA
CISTACEAE	LECHEA MARITIMA VAR. VIRGINICA	VIRGINIA
CISTACEAE	LECHEA MENSALIS	TEXAS
CONVOLVULACEAE	CALYSTEGIA STEBBINSII	CALIFORNIA
CONVOLVULACEAE	DICHONDRA OCCIDENTALIS	CALIFORNIA
CONVOLVULACEAE	IPOMOEA EGREGIA	ARIZONA
CONVOLVULACEAE	IPOMOEA LEMMONI	ARIZONA
CRASSULACEAE	DUDLEYA BETTINAE	CALIFORNIA
CRASSULACEAE	DUDLEYA CANDELABRUM	CALIFORNIA
CRASSULACEAE	DUDLEYA COLLOMAE	ARIZONA
CRASSULACEAE	DUDLEYA CYMOSA SSP. MARCESCENS	CALIFORNIA
CRASSULACEAE	DUDLEYA MULTICAULIS	CALIFORNIA
CRASSULACEAE	DUDLEYA NESIOTICA	CALIFORNIA
CRASSULACEAE	DUDLEYA STOLONIFERA	CALIFORNIA
CRASSULACEAE	DUDLEYA TRASKIAE	CALIFORNIA
CRASSULACEAE	GRAPTOPETALUM BARTRAMII	ARIZONA
CRASSULACEAE	LENOPHYLLUM TEXANUM	TEXAS
CRASSULACEAE	PARVISEDUM LEIOCARPUM	CALIFORNIA
CRASSULACEAE	SEDUM MORANII	OREGON
CRASSULACEAE	SEDUM NEVII	ALABAMA, TENNESSEE
CRASSULACEAE	SEDUM RADIATUM SSP. DEPAUPERATUM	CALIFORNIA, OREGON
CUPRESSACEAE	CUPRESSUS GOVENIANA VAR. ABRAMSIANA	CALIFORNIA
CUSCUTACEAE	CUSCUTA ATTENUATA	OKLAHOMA
CUSCUTACEAE	CUSCUTA HOWELLIANA	CALIFORNIA
CUSCUTACEAE	*CUSCUTA WARNERI	UTAH
CYCADACEAE	ZAMIA INTEGRIFOLIA	FLORIDA, GEORGIA
CYPERACEAE	CAREX ABORIGINUM	IDAHO
CYPERACEAE	CAREX ALBIDA	CALIFORNIA
CYPERACEAE	CAREX BILTMOREANA	VIRGINIA, NORTH CAROLINA, GEORGIA
CYPERACEAE	CAREX ELACHYCARPA	MAINE
CYPERACEAE	CAREX JACOBI-PETERI	ALASKA
CYPERACEAE	CAREX LATEBRACTEATA	OKLAHOMA
CYPERACEAE	CAREX SPECUICOLA	ARIZONA
CYPERACEAE	CAREX TOMPKINSII	CALIFORNIA
CYPERACEAE	CYPERUS GRAYIOIDES	ILLINOIS
CYPERACEAE	ELEOCHARIS CYLINDRICA	TEXAS
CYPERACEAE	FIMBRISTYLIS PERPUSILLA	GEORGIA
CYPERACEAE	RHYNCHOSPORA CALIFORNICA	CALIFORNIA
CYPERACEAE	RHYNCHOSPORA CRINIPES	ALABAMA
CYPERACEAE	RHYNCHOSPORA KNIESKERNII	NEW JERSEY, DELAWARE
CYPERACEAE	SCIRPUS ANCISTROCHAETUS	VERMONT, NEW YORK, PENNSYLVANIA, VIRGINIA
DIAPENSIACEAE	PYXIDANTHERA BARBULATA VAR. BREVIFOLIA	NORTH CAROLINA, SOUTH CAROLINA

ENDANGERED SPECIES IN THE CONTINENTAL UNITED STATES

FAMILY	SPECIES	STATE
DIAPENSIACEAE	SHORTIA GALACIFOLIA VAR. BREVISTYLA	SOUTH CAROLINA, NORTH CAROLINA
DIAPENSIACEAE	SHORTIA GALACIFOLIA VAR. GALACIFOLIA	NORTH CAROLINA, SOUTH CAROLINA, GEORGIA
ERICACEAE	ARCTOSTAPHYLOS AURICULATA	CALIFORNIA
ERICACEAE	ARCTOSTAPHYLOS BAKERI	CALIFORNIA
ERICACEAE	ARCTOSTAPHYLOS DENSIFLORA	CALIFORNIA
ERICACEAE	ARCTOSTAPHYLOS EDMUNDSII VAR. PARVIFOLIA	CALIFORNIA
ERICACEAE	ARCTOSTAPHYLOS GLANDULOSA SSP. CRASSIFOLIA	CALIFORNIA
ERICACEAE	ARCTOSTAPHYLOS GLUTINOSA	CALIFORNIA
ERICACEAE	*ARCTOSTAPHYLOS HOOKERI SSP. FRANCISCANA	CALIFORNIA
ERICACEAE	ARCTOSTAPHYLOS HOOKERI SSP. HEARSTIORUM	CALIFORNIA
ERICACEAE	ARCTOSTAPHYLOS HOOKERI SSP. RAVENII	CALIFORNIA
ERICACEAE	ARCTOSTAPHYLOS IMBRICATA	CALIFORNIA
ERICACEAE	ARCTOSTAPHYLOS MYRTIFOLIA	CALIFORNIA
ERICACEAE	ARCTOSTAPHYLOS PACIFICA	CALIFORNIA
ERICACEAE	ARCTOSTAPHYLOS PALLIDA	CALIFORNIA
ERICACEAE	ARCTOSTAPHYLOS PUMILA	CALIFORNIA
ERICACEAE	ARCTOSTAPHYLOS REFUGIOENSIS	CALIFORNIA
ERICACEAE	ELLIOTTIA RACEMOSA	GEORGIA
ERICACEAE	KALMIA CUNEATA	NORTH CAROLINA, SOUTH CAROLINA
ERICACEAE	KALMIOPSIS LEACHIANA	OREGON
ERICACEAE	MONOTROPSIS REYNOLDSIAE	FLORIDA
ERICACEAE	RHODODENDRON MINUS VAR. CHAPMANII	FLORIDA
ERIOCAULACEAE	ERIOCAULON KORNICKIANUM	TEXAS, ARKANSAS, OKLAHOMA
EUPHORBIACEAE	ANDRACHNE ARIDA	TEXAS
EUPHORBIACEAE	ARGYTHAMNIA APHOROIDES	TEXAS
EUPHORBIACEAE	ARGYTHAMNIA ARGYRAEA	TEXAS
EUPHORBIACEAE	CHAMAESYCE DELTOIDEA SSP. SERPYLLUM	FLORIDA
EUPHORBIACEAE	CHAMAESYCE GARBERI	FLORIDA
EUPHORBIACEAE	CHAMAESYCE PORTERIANA VAR. KEYENSIS	FLORIDA
EUPHORBIACEAE	CHAMAESYCE PORTERIANA VAR. SCOPARIA	FLORIDA
EUPHORBIACEAE	CROTON ALABAMENSIS	ALABAMA, TENNESSEE
EUPHORBIACEAE	CROTON ELLIOTTII	FLORIDA, GEORGIA, ALABAMA
EUPHORBIACEAE	CROTON GLANDULOSUS VAR. SIMPSONII	FLORIDA
EUPHORBIACEAE	CROTON WIGGINSII	CALIFORNIA, NEVADA
EUPHORBIACEAE	DITAXIS DIVERSIFLORA	NEVADA
EUPHORBIACEAE	EUPHORBIA FENDLERI VAR. TRILIGULATA	TEXAS
EUPHORBIACEAE	EUPHORBIA GOLONDRINA	TEXAS
EUPHORBIACEAE	MANIHOT WALKERAE	TEXAS
EUPHORBIACEAE	PHYLLANTHUS ERICOIDES	TEXAS
FABACEAE	ACACIA EMORYANA	TEXAS
FABACEAE	AMORPHA OUACHITENSIS	ARKANSAS, OKLAHOMA
FABACEAE	APIOS PRICEANA	KENTUCKY, TENNESSEE, ILLINOIS, MISSISSIPPI

FAMILY	SPECIES	STATE
FABACEAE	ASTRAGALUS AGNICIDUS	CALIFORNIA
FABACEAE	ASTRAGALUS AMNIS-AMISSI	.IDAHO
FABACEAE	ASTRAGALUS ATRATUS VAR. INSEPTUS	IDAHO
FABACEAE	ASTRAGALUS BEATHII	ARIZONA
FABACEAE	ASTRAGALUS BEATLEYAE	NEVADA
FABACEAE	ASTRAGALUS CASTETTERI	NEW MEXICO
FABACEAE	ASTRAGALUS CLARIANUS	CALIFORNIA
FABACEAE	*ASTRAGALUS COLUMBIANUS	WASHINGTON
FABACEAE	ASTRAGALUS CREMNOPHYLAX	ARIZONA
FABACEAE	ASTRAGALUS CRONQUISTII	UTAH
FABACEAE	*ASTRAGALUS DESERETICUS	UTAH
FABACEAE	ASTRAGALUS DETERIOR	COLORADO
FABACEAE	ASTRAGALUS FUNEREUS	CALIFORNIA, NEVADA
FABACEAE	ASTRAGALUS HAMILTONII	UTAH
FABACEAE	ASTRAGALUS HARRISONII	UTAH
FABACEAE	*ASTRAGALUS HUMILLIMUS	COLORADO
FABACEAE	ASTRAGALUS ISELYI	UTAH
FABACEAE	ASTRAGALUS JAEGERIANUS	CALIFORNIA
FABACEAE	ASTRAGALUS JOHANNIS-HOWELLII	CALIFORNIA
FABACEAE	ASTRAGALUS KENTROPHYTA VAR. DOUGLASII	WASHINGTON, OREGON
FABACEAE	ASTRAGALUS LENTIGINOSUS VAR. MARICOPAE	ARIZONA
FABACEAE	ASTRAGALUS LENTIGINOSUS VAR. MICANS	CALIFORNIA
FABACEAE	ASTRAGALUS LENTIGINOSUS VAR. SESQUIMETRALIS	CALIFORNIA, NEVADA
FABACEAE	*ASTRAGALUS LENTIGINOSUS VAR. URSINUS	UTAH
FABACEAE	*ASTRAGALUS LINIFOLIUS	COLORADO
FABACEAE	ASTRAGALUS LUTOSUS	UTAH, COLORADO
FABACEAE	ASTRAGALUS MICROCYMBUS	COLORADO
FABACEAE	ASTRAGALUS MISELLUS VAR. PAUPER	WASHINGTON
FABACEAE	ASTRAGALUS MONOENSIS	CALIFORNIA
FABACEAE	ASTRAGALUS NATURITENSIS	COLORADO
FABACEAE	ASTRAGALUS NYENSIS	NEVADA
FABACEAE	ASTRAGALUS OOCALYCIS	COLORADO, NEW MEXICO
FABACEAE	ASTRAGALUS OSTERHOUTII	COLORADO
FABACEAE	ASTRAGALUS PERIANUS	UTAH
FABACEAE	ASTRAGALUS PHOENIX	NEVADA
FABACEAE	ASTRAGALUS PORRECTUS	NEVADA
FABACEAE	ASTRAGALUS PROIMANTHUS	WYOMING
FABACEAE	ASTRAGALUS PURSHII VAR. OPHIOGENES	IDAHO, OREGON
FABACEAE	*ASTRAGALUS PYCNOSTACHYUS SSP. LANOSISSIMUS	CALIFORNIA
FABACEAE	ASTRAGALUS RAVENII	CALIFORNIA
FABACEAE	ASTRAGALUS ROBBINSII VAR. ALPINIFORMIS	OREGON
FABACEAE	ASTRAGALUS ROBBINSII VAR. JESUPI	NEW HAMPSHIRE, VERMONT
FABACEAE	ASTRAGALUS ROBBINSII VAR. OCCIDENTALIS	NEVADA

ENDANGERED SPECIES IN THE CONTINENTAL UNITED STATES

FAMILY	SPECIES	STATE
FABACEAE	*ASTRAGALUS ROBBINSII VAR. ROBBINSII	VERMONT
FABACEAE	ASTRAGALUS SCHMOLLAE	COLORADO
FABACEAE	ASTRAGALUS SERENOI VAR. SORDESCENS	NEVADA
FABACEAE	ASTRAGALUS SILICEUS	NEW MEXICO
FABACEAE	ASTRAGALUS SINUATUS	WASHINGTON
FABACEAE	ASTRAGALUS STERILIS	OREGON, IDAHO
FABACEAE	ASTRAGALUS TENER VAR. TITI	CALIFORNIA
FABACEAE	ASTRAGALUS UNCIALIS	NEVADA
FABACEAE	ASTRAGALUS XIPHOIDES	ARIZONA
FABACEAE	BAPTISIA ARACHNIFERA	GEORGIA
FABACEAE	BRONGNIARTIA MINUTIFOLIA	TEXAS
FABACEAE	CAESALPINIA DRUMMONDII	TEXAS
FABACEAE	CALLIANDRA BIFLORA	TEXAS
FABACEAE	CASSIA KEYENSIS	FLORIDA
FABACEAE	CENTROSEMA ARENICOLA	FLORIDA
FABACEAE	GALACTIA PINETORUM	FLORIDA
FABACEAE	GENISTIDIUM DUMOSUM	TEXAS
FABACEAE	HOFFMANNSEGGIA TENELLA	TEXAS
FABACEAE	LATHYRUS HITCHCOCKIANUS	CALIFORNIA, NEVADA
FABACEAE	LATHYRUS JEPSONII SSP. JEPSONII	CALIFORNIA
FABACEAE	LESPEDEZA LEPTOSTACHYA	ILLINOIS, WISCONSIN, IOWA, MINNESOTA
FABACEAE	*LOTUS ARGOPHYLLUS VAR. ADSURGENS	CALIFORNIA
FABACEAE	LOTUS SCOPARIUS SSP. TRASKIAE	CALIFORNIA
FABACEAE	LUPINUS BURKEI SSP. CAERULEOMONTANUS	OREGON
FABACEAE	LUPINUS GUADALUPENSIS	CALIFORNIA
FABACEAE	LUPINUS LUDOVICIANUS	CALIFORNIA
FABACEAE	LUPINUS MILO-BAKERI	CALIFORNIA
FABACEAE	LUPINUS TIDESTROMII VAR. LAYNEAE	CALIFORNIA
FABACEAE	LUPINUS TIDESTROMII VAR. TIDESTROMII	CALIFORNIA
FABACEAE	LUPINUS TRACYI	CALIFORNIA
FABACEAE	OXYTROPIS KOBUKENSIS	ALASKA
FABACEAE	PETALOSTEMUM FOLIOSUM	ILLINOIS, TENNESSEE, ALABAMA
FABACEAE	PETALOSTEMUM REVERCHONII	TEXAS
FABACEAE	PETALOSTEMUM SABINALE	TEXAS
FABACEAE	PSORALEA EPIPSILA	UTAH, ARIZONA
FABACEAE	*PSORALEA MACROPHYLLA	NORTH CAROLINA
FABACEAE	*PSORALEA STIPULATA	INDIANA
FABACEAE	TRIFOLIUM AMOENUM	CALIFORNIA
FABACEAE	TRIFOLIUM ANDERSONII SSP. BEATLEYAE	NEVADA
FABACEAE	TRIFOLIUM LEMMONII	CALIFORNIA, NEVADA
FABACEAE	TRIFOLIUM POLYODON	CALIFORNIA
FABACEAE	TRIFOLIUM THOMPSONII	WASHINGTON
FABACEAE	TRIFOLIUM TRICHOCALYX	CALIFORNIA

ENDANGERED SPECIES IN THE CONTINENTAL UNITED STATES

FAMILY	SPECIES	STATE
FABACEAE	VICIA OCALENSIS	FLORIDA
FABACEAE	*VICIA REVERCHONII	TEXAS, OKLAHOMA
FAGACEAE	CASTANEA PUMILA VAR. OZARKENSIS	MISSOURI, ARKANSAS, OKLAHOMA
FAGACEAE	QUERCUS GRACILIFORMIS	TEXAS
FAGACEAE	QUERCUS HINCKLEYI	TEXAS
FAGACEAE	QUERCUS TARDIFOLIA	TEXAS
FRANKENIACEAE	FRANKENIA JOHNSTONII	TEXAS
FUMARIACEAE	CORYDALIS AQUAE-GELIDAE	OREGON
FUMARIACEAE	DICENTRA OCHROLEUCA	CALIFORNIA
GENTIANACEAE	BARTONIA TEXANA	TEXAS
GENTIANACEAE	CENTAURIUM NAMOPHILUM	CALIFORNIA, NEVADA
GENTIANACEAE	FRASERA GYPSICOLA	NEVADA
GENTIANACEAE	FRASERA PAHUTENSIS	NEVADA
GENTIANACEAE	GENTIANA BISETAEA	OREGON
GENTIANACEAE	GENTIANA PENNELLIANA	FLORIDA
GERANIACEAE	GERANIUM TOQUIMENSE	NEVADA
HYDROCHARITACEAE	*ELODEA BRANDEGEAE	CALIFORNIA
HYDROCHARITACEAE	*ELODEA LINEARIS	TENNESSEE
HYDROCHARITACEAE	*ELODEA NEVADENSIS	NEVADA
HYDROCHARITACEAE	*ELODEA SCHWEINITZII	PENNSYLVANIA
HYDROPHYLLACEAE	ERIODICTYON ALTISSIMUM	CALIFORNIA
HYDROPHYLLACEAE	ERIODICTYON CAPITATUM	CALIFORNIA
HYDROPHYLLACEAE	PHACELIA ARGILLACEA	UTAH
HYDROPHYLLACEAE	PHACELIA BEATLEYAE	NEVADA
HYDROPHYLLACEAE	PHACELIA CAPITATA	OREGON
HYDROPHYLLACEAE	*PHACELIA CINEREA	CALIFORNIA
HYDROPHYLLACEAE	PHACELIA COOKEI	CALIFORNIA
HYDROPHYLLACEAE	PHACELIA FILIFORMIS	ARIZONA
HYDROPHYLLACEAE	PHACELIA FORMOSULA	COLORADO
HYDROPHYLLACEAE	PHACELIA INDECORA	UTAH
HYDROPHYLLACEAE	*PHACELIA LENTA	WASHINGTON
HYDROPHYLLACEAE	PHACELIA MAMMILLARENSIS	UTAH
HYDROPHYLLACEAE	PHACELIA PALLIDA	TEXAS
HYDROPHYLLACEAE	PHACELIA SUBMUTICA	COLORADO
HYDROPHYLLACEAE	PHACELIA WELSHII	ARIZONA
HYPERICACEAE	HYPERICUM CUMULICOLA	FLORIDA
IRIDACEAE	IRIS TENAX SSP. KLAMATHENSIS	CALIFORNIA
IRIDACEAE	IRIS TENUIS	OREGON
ISOETACEAE	ISOETES LITHOPHYLLA	TEXAS
ISOETACEAE	*ISOETES LOUISIANENSIS	LOUISIANA
JUGLANDACEAE	JUGLANS HINDSII	CALIFORNIA
JUNCACEAE	*JUNCUS LEIOSPERMUS	CALIFORNIA
JUNCACEAE	*JUNCUS PERVETUS	MASSACHUSETTS

ENDANGERED SPECIES IN THE CONTINENTAL UNITED STATES

FAMILY	SPECIES	STATE
LAMIACEAE	ACANTHOMINTHA ILICIFOLIA	CALIFORNIA
LAMIACEAE	ACANTHOMINTHA OBOVATA SSP. DUTTONII	CALIFORNIA
LAMIACEAE	BRAZORIA PULCHERRIMA	TEXAS
LAMIACEAE	CONRADINA BREVIFOLIA	FLORIDA
LAMIACEAE	CONRADINA GLABRA	FLORIDA
LAMIACEAE	CONRADINA VERTICILLATA	KENTUCKY, TENNESSEE
LAMIACEAE	DICERANDRA FRUTESCENS	FLORIDA
LAMIACEAE	DICERANDRA IMMACULATA	FLORIDA
LAMIACEAE	HEDEOMA GRAVEOLENS	FLORIDA
LAMIACEAE	*HEDEOMA PILOSUM	TEXAS
LAMIACEAE	MACBRIDEA ALBA	FLORIDA
LAMIACEAE	*MONARDELLA LEUCOCEPHALA	CALIFORNIA
LAMIACEAE	MONARDELLA LINOIDES SSP. VIMINEA	CALIFORNIA
LAMIACEAE	MONARDELLA MACRANTHA VAR. HALLII	CALIFORNIA
LAMIACEAE	MONARDELLA PRINGLEI	CALIFORNIA
LAMIACEAE	MONARDELLA UNDULATA VAR. FRUTESCFNS	CALIFORNIA
LAMIACEAE	PHYSOSTEGIA CORRELLII	TEXAS
LAMIACEAE	POGOGYNE ABRAMSII	CALIFORNIA
LAMIACEAE	POGOGYNE CLAREANA	CALIFORNIA
LAMIACEAE	POGOGYNE DOUGLASII SSP. PARVIFLORA	CALIFORNIA
LAMIACEAE	POGOGYNE NUDIUSCULA	CALIFORNIA
LAMIACEAE	PYCNANTHFMUM CURVIPES	GEORGIA, ALABAMA, TENNESSEE
LAMIACEAE	SALVIA BLODGETTII	FLORIDA
LAMIACEAE	SALVIA COLUMBARIAE VAR. ZIEGLERI	CALIFORNIA
LAMIACEAE	*SCUTELLARIA OCMULGEE	GEORGIA
LAMIACEAE	TRICHOSTFMA AUSTROMONTANUM SSP. COMPACTUM	CALIFORNIA
LAURACEAE	LINDERA MELISSIFOLIA	ALABAMA, ARKANSAS, FLORIDA, MISSISSIPPI, MISSOURI, NORTH CAROLINA, SOUTH CAROLINA, GEORGIA
LENNOACEAE	AMMOBROMA SONORAE	CALIFORNIA
LILIACEAE	AGAVE ARIZONICA	ARIZONA
LILIACEAE	AGAVE MCKELVEYANA	ARIZONA
LILIACEAE	AGAVF SCHOTTII VAR. TRELEASEI	ARIZONA
LILIACEAE	AGAVE TOUMFYANA VAR. BELLA	ARIZONA
LILIACEAE	ALLIUM AASEAE	IDAHO
LILIACEAE	ALLIUM DICTUON	WASHINGTON
LILIACEAE	ALLIUM HICKMANII	CALIFORNIA
LILIACEAE	ALLIUM PASSEYI	UTAH
LILIACEAE	BLOOMERIA HUMILIS	CALIFORNIA
LILIACEAE	BRODIAEA CORONARIA VAR. ROSEA	CALIFORNIA
LILIACEAE	BRODIAEA FILIFOLIA	CALIFORNIA
LILIACEAE	BRODIAEA ORCUTTII	CALIFORNIA
LILIACEAE	BRODIAEA PALLIDA	CALIFORNIA
LILIACEAE	CALOCHORTUS CLAVATUS SSP. RECURVIFOLIUS	CALIFORNIA

ENDANGERED SPECIES IN THE CONTINENTAL UNITED STATES

FAMILY	SPECIES	STATE
LILIACEAE	CALOCHORTUS COERULEUS VAR. WESTONII	CALIFORNIA
LILIACEAE	CALOCHORTUS GREENEI	OREGON, CALIFORNIA
LILIACEAE	CALOCHORTUS INDECORUS	OREGON
LILIACEAE	CALOCHORTUS LONGEBARBATUS VAR. PECKII	OREGON
LILIACEAE	*CALOCHORTUS MONANTHUS	CALIFORNIA
LILIACEAE	CALOCHORTUS PERSISTENS	CALIFORNIA
LILIACEAE	CALOCHORTUS TIBURONENSIS	CALIFORNIA
LILIACEAE	CHLOROGALUM GRANDIFLORUM	CALIFORNIA
LILIACEAE	CHLOROGALUM PURPUREUM VAR. PURPUREUM	CALIFORNIA
LILIACEAE	CHLOROGALUM PURPUREUM VAR. REDUCTUM	CALIFORNIA
LILIACEAE	ERYTHRONIUM GRANDIFLORUM SSP. PUSATERII	CALIFORNIA
LILIACEAE	ERYTHRONIUM PROPULLANS	MINNESOTA
LILIACEAE	*FRITILLARIA ADAMANTINA	OREGON
LILIACEAE	FRITILLARIA PHAEANTHERA	CALIFORNIA
LILIACEAE	FRITILLARIA RODERICKII	CALIFORNIA
LILIACEAE	HARPEROCALLIS FLAVA	FLORIDA
LILIACEAE	HYMENOCALLIS CORONARIA	ALABAMA, GEORGIA, FLORIDA, SOUTH CAROLINA
LILIACEAE	HYPOXIS LONGII	OKLAHOMA, VIRGINIA
LILIACEAE	LILIUM IRIDOLLAE	ALABAMA, FLORIDA
LILIACEAE	LILIUM OCCIDENTALE	OREGON, CALIFORNIA
LILIACEAE	LILIUM PITKINENSE	CALIFORNIA
LILIACEAE	NOLINA ATOPOCARPA	FLORIDA
LILIACEAE	NOLINA BRITTONIANA	FLORIDA
LILIACEAE	NOLINA INTERRATA	CALIFORNIA
LILIACEAE	POLIANTHES RUNYONII	TEXAS
LILIACEAE	SCHOENOLIRION TEXANUM	ALABAMA, ARKANSAS, TEXAS
LILIACEAE	TOFIELDIA GLUTINOSA SSP. ABSONA	IDAHO
LILIACEAE	TRILLIUM PERSISTENS	GEORGIA, SOUTH CAROLINA
LILIACEAE	TRILLIUM PUSILLUM VAR. VIRGINIANUM	VIRGINIA, MARYLAND, NORTH CAROLINA
LILIACEAE	ZIGADENUS VAGINATUS	UTAH
LIMNANTHACEAE	LIMNANTHES BAKERI	CALIFORNIA
LIMNANTHACEAE	LIMNANTHES FLOCCOSA SSP. GRANDIFLORA	OREGON
LIMNANTHACEAE	LIMNANTHES FLOCCOSA SSP. PUMILA	OREGON
LIMNANTHACEAE	LIMNANTHES GRACILIS VAR. PARISHII	CALIFORNIA
LIMNANTHACEAE	LIMNANTHES VINCULANS	CALIFORNIA
LINACEAE	HESPEROLINON CONGESTUM	CALIFORNIA
LINACEAE	HESPEROLINON DIDYMOCARPUM	CALIFORNIA
LINACEAE	LINUM ARENICOLA	FLORIDA
LINACEAE	LINUM CARTERI VAR. CARTERI	FLORIDA
LINACEAE	LINUM CARTERI VAR. SMALLII	FLORIDA
LINACEAE	LINUM WESTII	FLORIDA
LOASACEAE	MENTZELIA LEUCOPHYLLA	NEVADA, CALIFORNIA

ENDANGERED SPECIES IN THE CONTINENTAL UNITED STATES

FAMILY	SPECIES	STATE
LOASACEAE	MENTZELIA NITENS VAR. LEPTOCAULIS	ARIZONA
LOASACEAE	MENTZELIA PACKARDIAE	OREGON
LOGANIACEAE	SPIGELIA GENTIANOIDES	FLORIDA
LOGANIACEAE	SPIGELIA LOGANIOIDES	FLORIDA
LYTHRACEAE	CUPHEA ASPERA	FLORIDA
MALVACEAE	CALLIRHOF SCABRIUSCULA	TEXAS
MALVACEAE	GAYA VIOLACEA	TEXAS
MALVACEAE	HIBISCUS CALIFORNICUS	CALIFORNIA
MALVACEAE	HIBISCUS DASYCALYX	TEXAS
MALVACEAE	ILIAMNA REMOTA	ILLINOIS, INDIANA, VIRGINIA
MALVACEAE	LAVATERA ASSURGENTIFLORA	CALIFORNIA
MALVACEAE	*MALACOTHAMNUS ABBOTTII	CALIFORNIA
MALVACEAE	MALACOTHAMNUS CLEMENTINUS	CALIFORNIA
MALVACEAE	*MALACOTHAMNUS MENDOCINENSIS	CALIFORNIA
MALVACEAE	MALACOTHAMNUS PALMERI VAR. INVOLUCRATUS	CALIFORNIA
MALVACEAE	SIDALCEA CAMPESTRIS	OREGON
MALVACEAE	*SIDALCEA COVILLEI	CALIFORNIA
MALVACEAE	SIDALCEA NELSONIANA	OREGON
MALVACEAE	SIDALCEA OREGANA SSP. VALIDA	CALIFORNIA
MALVACEAE	SPHAERALCEA FENDLERI VAR. ALBESCENS	ARIZONA
MELASTOMATACEAE	RHEXIA PARVIFLORA	FLORIDA, GEORGIA
NAJADACEAE	NAJAS CAESPITOSA	UTAH
NYCTAGINACEAE	ABRONIA ALPINA	CALIFORNIA
NYCTAGINACEAE	ABRONIA BIGELOVII	NEW MEXICO
NYCTAGINACEAE	HERMIDIUM ALIPES VAR. PALLIDIUM	UTAH
NYCTAGINACEAE	MIRABILIS MACFARLANEI	OREGON, IDAHO
OLEACEAE	FORESTIERA SEGREGATA VAR. PINETORUM	FLORIDA
OLEACEAE	FRAXINUS GOODDINGII	ARIZONA
ONAGRACEAE	CAMISSONIA MEGALANTHA	ARIZONA, NEVADA, UTAH
ONAGRACEAE	CAMISSONIA NEVADENSIS	NEVADA
ONAGRACEAE	CAMISSONIA SPECUICOLA SSP. SPECUICOLA	ARIZONA
ONAGRACEAE	CLARKIA BOREALIS SSP. ARIDA	CALIFORNIA
ONAGRACEAE	CLARKIA FRANCISCANA	CALIFORNIA
ONAGRACEAE	CLARKIA IMBRICATA	CALIFORNIA
ONAGRACEAE	CLARKIA LINGULATA	CALIFORNIA
ONAGRACEAE	CLARKIA MOSQUINII SSP. MOSQUINII	CALIFORNIA
ONAGRACEAE	*CLARKIA MOSQUINII SSP. XEROPHILA	CALIFORNIA
ONAGRACEAE	CLARKIA SPECIOSA SSP. IMMACULATA	CALIFORNIA
ONAGRACEAE	GAURA NEOMEXICANA SSP. COLORADENSIS	COLORADO
ONAGRACEAE	OENOTHERA AVITA SSP. EUREKENSIS	CALIFORNIA
ONAGRACEAE	OENOTHERA DELTOIDES VAR. HOWELLII	CALIFORNIA
ONAGRACEAE	OENOTHERA PSAMMOPHILA	IDAHO
ONAGRACEAE	OENOTHERA SESSILIS	ARKANSAS

ENDANGERED SPECIES IN THE CONTINENTAL UNITED STATES

FAMILY	SPECIES	STATE
ORCHIDACEAE	ISOTRIA MEDEOLOIDES	RHODE ISLAND, NEW HAMPSHIRE, VERMONT, MASSACHUSETTS, NEW YORK, CONNECTICUT, PENNSYLVANIA, VIRGINIA, ILLINOIS, MAINE, NEW JERSEY, NORTH CAROLINA, MICHIGAN
ORCHIDACEAE	SPIRANTHES LANCEOLATA VAR. PALUDICOLA	FLORIDA
ORCHIDACEAE	*SPIRANTHES PARKSII	TEXAS
ORCHIDACEAE	TRIPHORA CRAIGHEADII	FLORIDA
ORCHIDACEAE	TRIPHORA LATIFOLIA	FLORIDA
OROBANCHACEAE	*OROBANCHE VALIDA	CALIFORNIA
PAPAVERACEAE	ARCTOMECON HUMILIS	ARIZONA, UTAH
PAPAVERACEAE	ARCTOMECON MERRIAMII	CALIFORNIA, NEVADA
PAPAVERACEAE	ARGEMONE PLEIACANTHA SSP. PINNATISECTA	NEW MEXICO
PLUMBAGINACEAE	LIMONIUM CAROLINIANUM VAR. ANGUSTATUM	FLORIDA
POACEAE	AGROSTIS BLASDALEI VAR. MARINENSIS	CALIFORNIA
POACEAE	AGROSTIS HENDERSONII	CALIFORNIA, OREGON
POACEAE	ANDROPOGON ARCTATUS	FLORIDA, ALABAMA
POACEAE	ARISTIDA FLORIDANA	FLORIDA
POACEAE	CALAMAGROSTIS INEXPANSA VAR. NOVAE-ANGLIAE	MAINE, VERMONT, NEW HAMPSHIRE
POACEAE	CALAMAGROSTIS INSPERATA	OHIO, MISSOURI
POACEAE	CALAMAGROSTIS PERPLEXA	NEW YORK
POACEAE	CALAMOVILFA ARCUATA	TENNESSEE, OKLAHOMA
POACEAE	CALAMOVILFA CURTISSII	FLORIDA
POACEAE	*DISSANTHELIUM CALIFORNICUM	CALIFORNIA
POACEAE	FESTUCA DASYCLADA	UTAH, COLORADO
POACEAE	GLYCERIA NUBIGENA	NORTH CAROLINA, TENNESSEE
POACEAE	MUHLENBERGIA VILLOSA	TEXAS
POACEAE	NEOSTAPFIA COLUSANA	CALIFORNIA
POACEAE	ORCUTTIA CALIFORNICA VAR. CALIFORNICA	CALIFORNIA
POACEAE	ORCUTTIA CALIFORNICA VAR. INAEQUALIS	CALIFORNIA
POACEAE	ORCUTTIA CALIFORNICA VAR. VISCIDA	CALIFORNIA
POACEAE	ORCUTTIA GREENEI	CALIFORNIA
POACEAE	ORCUTTIA MUCRONATA	CALIFORNIA
POACEAE	ORCUTTIA PILOSA	CALIFORNIA
POACEAE	ORCUTTIA TENUIS	CALIFORNIA
POACEAE	PANICUM HIRSTII	GEORGIA, NEW JERSEY
POACEAE	PANICUM MUNDUM	VIRGINIA, NORTH CAROLINA
POACEAE	PANICUM THERMALE	CALIFORNIA
POACEAE	PLEUROPOGON HOOVERIANUS	CALIFORNIA
POACEAE	*PLEUROPOGON OREGONUS	OREGON
POACEAE	POA ATROPURPUREA	CALIFORNIA
POACEAE	POA FIBRATA	CALIFORNIA
POACEAE	POA INVOLUTA	TEXAS
POACEAE	POA NAPENSIS	CALIFORNIA
POACEAE	POA PACHYPHOLIS	WASHINGTON

ENDANGERED SPECIES IN THE CONTINENTAL UNITED STATES

FAMILY	SPECIES	STATE
POACEAE	SCHIZACHYRIUM RHIZOMATUM	FLORIDA
POACEAE	SPOROBOLUS PATENS	ARIZONA
POACEAE	STIPA LEMMONII VAR. PUBESCENS	CALIFORNIA
POACEAE	SWALLENIA ALEXANDRAE	CALIFORNIA
POACEAE	TRIPSACUM FLORIDANUM	FLORIDA
POACEAE	TRISETUM ORTHOCHAETUM	MONTANA
POACEAE	ZIZANIA TEXANA	TEXAS
POLEMONIACEAE	COLLOMIA MACROCALYX	OREGON
POLEMONIACEAE	ERIASTRUM TRACYI	CALIFORNIA
POLEMONIACEAE	GILIA CAESPITOSA	UTAH
POLEMONIACEAE	GILIA PENSTEMONOIDES	COLORADO
POLEMONIACEAE	NAVARRETIA PAUCIFLORA	CALIFORNIA
POLEMONIACEAE	NAVARRETIA PLIEANTHA	CALIFORNIA
POLEMONIACEAE	NAVARRETIA SETILOBA	CALIFORNIA
POLEMONIACEAE	PHLOX HIRSUTA	CALIFORNIA
POLEMONIACEAE	PHLOX IDAHONIS	IDAHO
POLEMONIACEAE	PHLOX LONGIPILOSA	OKLAHOMA
POLEMONIACEAE	PHLOX MISSOULENSIS	MONTANA
POLEMONIACEAE	PHLOX NIVALIS SSP. TEXENSIS	TEXAS
POLEMONIACEAE	PHLOX PULCHRA	ALABAMA
POLEMONIACEAE	POLEMONIUM OCCIDENTALE VAR. LACUSTRE	MINNESOTA
POLEMONIACEAE	POLEMONIUM PAUCIFLORUM SSP. HINCKLEYI	TEXAS
POLYGALACEAE	POLYGALA LEWTONII	FLORIDA
POLYGALACEAE	POLYGALA MARAVILLASENSIS	TEXAS
POLYGALACEAE	POLYGALA RIMULICOLA	TEXAS, NEW MEXICO
POLYGONACEAE	CHORIZANTHE LEPTOCERAS	CALIFORNIA
POLYGONACEAE	CHORIZANTHE ORCUTTIANA	CALIFORNIA
POLYGONACEAE	CHORIZANTHE PARRYI VAR. FERNANDINA	CALIFORNIA
POLYGONACEAE	CHORIZANTHE SPINOSA	CALIFORNIA
POLYGONACEAE	*CHORIZANTHE VALIDA	CALIFORNIA
POLYGONACEAE	DEDECKERA EUREKENSIS	CALIFORNIA
POLYGONACEAE	ERIOGONUM ALPINUM	CALIFORNIA
POLYGONACEAE	ERIOGONUM AMMOPHILUM	UTAH
POLYGONACEAE	ERIOGONUM ANEMOPHILUM	NEVADA
POLYGONACEAE	ERIOGONUM APRICUM VAR. APRICUM	CALIFORNIA
POLYGONACEAE	ERIOGONUM APRICUM VAR. PROSTRATUM	CALIFORNIA
POLYGONACEAE	ERIOGONUM ARETIOIDES	UTAH
POLYGONACEAE	ERIOGONUM ARGOPHYLLUM	NEVADA
POLYGONACEAE	ERIOGONUM BREEDLOVEI	CALIFORNIA
POLYGONACEAE	ERIOGONUM BUTTERWORTHIANUM	CALIFORNIA
POLYGONACEAE	ERIOGONUM CANINUM	CALIFORNIA
POLYGONACEAE	ERIOGONUM CAPILLARE	ARIZONA
POLYGONACEAE	ERIOGONUM CHRYSOPS	OREGON

ENDANGERED SPECIES IN THE CONTINENTAL UNITED STATES

FAMILY	SPECIES	STATE
POLYGONACEAE	ERIOGONUM CORYMBOSUM VAR. DAVIDSEI	UTAH
POLYGONACEAE	ERIOGONUM CORYMBOSUM VAR. REVEALIANUM	UTAH
POLYGONACEAE	ERIOGONUM CROCATUM	CALIFORNIA
POLYGONACEAE	ERIOGONUM CRONQUISTII	UTAH
POLYGONACEAE	ERIOGONUM DARROVII	ARIZONA, NEVADA
POLYGONACEAE	ERIOGONUM DICLINUM	CALIFORNIA, OREGON
POLYGONACEAE	ERIOGONUM EPHEDROIDES	UTAH, COLORADO
POLYGONACEAE	ERIOGONUM ERICIFOLIUM VAR. ERICIFOLIUM	ARIZONA
POLYGONACEAE	ERIOGONUM ERICIFOLIUM VAR. THORNEI	CALIFORNIA
POLYGONACEAE	ERIOGONUM FLAVUM VAR. AQUILINUM	ALASKA
POLYGONACEAE	ERIOGONUM GIGANTEUM VAR. COMPACTUM	CALIFORNIA
POLYGONACEAE	ERIOGONUM GILMANII	CALIFORNIA
POLYGONACEAE	ERIOGONUM GRANDE VAR. TIMORUM	CALIFORNIA
POLYGONACEAE	ERIOGONUM GYPSOPHILUM	NEW MEXICO
POLYGONACEAE	ERIOGONUM HIRTELLUM	CALIFORNIA
POLYGONACEAE	ERIOGONUM HUMIVAGANS	UTAH
POLYGONACEAE	ERIOGONUM HYLOPHILUM	UTAH
POLYGONACEAE	ERIOGONUM INTERMONTANUM	UTAH
POLYGONACEAE	ERIOGONUM INTRAFRACTUM	CALIFORNIA
POLYGONACEAE	ERIOGONUM KENNEDYI VAR. PINICOLA	CALIFORNIA
POLYGONACEAE	ERIOGONUM LEMMONII	NEVADA
POLYGONACEAE	ERIOGONUM LOGANUM	UTAH
POLYGONACEAE	ERIOGONUM LONGIFOLIUM VAR. HARPERI	ALABAMA, TENNESSEE
POLYGONACEAE	ERIOGONUM MICROTHECUM VAR. JOHNSTONII	CALIFORNIA
POLYGONACEAE	ERIOGONUM MORTONIANUM	ARIZONA
POLYGONACEAE	ERIOGONUM NEALLEYI	TEXAS
POLYGONACEAE	ERIOGONUM NUDUM VAR. MURINUM	CALIFORNIA
POLYGONACEAE	ERIOGONUM OVALIFOLIUM VAR. VINEUM	CALIFORNIA
POLYGONACEAE	ERIOGONUM PARVIFOLIUM VAR. LUCIDUM	CALIFORNIA
POLYGONACEAE	ERIOGONUM PELINOPHILUM	COLORADO
POLYGONACEAE	ERIOGONUM SMITHII	UTAH
POLYGONACEAE	ERIOGONUM SUFFRUTICOSUM	TEXAS
POLYGONACEAE	ERIOGONUM THOMPSONAE VAR. ATWOODII	ARIZONA
POLYGONACEAE	ERIOGONUM TRUNCATUM	CALIFORNIA
POLYGONACEAE	ERIOGONUM UMBELLATUM VAR. MINUS	CALIFORNIA
POLYGONACEAE	ERIOGONUM UMBELLATUM VAR. TORREYANUM	CALIFORNIA
POLYGONACEAE	ERIOGONUM VISCIDULUM	NEVADA
POLYGONACEAE	ERIOGONUM WRIGHTII VAR. OLANCHENSE	CALIFORNIA
POLYGONACEAE	ERIOGONUM ZIONIS VAR. COCCINEUM	ARIZONA
POLYGONACEAE	ERIOGONUM ZIONIS VAR. ZIONIS	UTAH
POLYGONACEAE	POLYGONELLA CILIATA VAR. BASIRAMIA	FLORIDA
POLYGONACEAE	POLYGONELLA MYRIOPHYLLA	FLORIDA
POLYGONACEAE	POLYGONELLA PARKSII	TEXAS

ENDANGERED SPECIES IN THE CONTINENTAL UNITED STATES

FAMILY	SPECIES	STATE
POLYGONACEAE	*POLYGONUM MONTEREYENSE	.CALIFORNIA
POLYGONACEAE	POLYGONUM PENSYLVANICUM VAR. EGLANDULOSUM	OHIO
POLYGONACEAE	POLYGONUM TEXENSE	TEXAS
POLYGONACEAE	RUMEX ORTHONEURUS	ARIZONA
POLYPODIACEAE	ASPLENIUM ANDREWSII	ARIZONA, COLORADO, UTAH
POLYPODIACEAE	CHEILANTHES FIBRILLOSA	CALIFORNIA
POLYPODIACEAE	LEPTOGRAMMA PILOSA VAR. ALABAMENSIS	ALABAMA
POLYPODIACEAE	PHYLLITIS SCOLOPENDRIUM VAR. AMERICANUM	MICHIGAN, NEW YORK, TENNESSEE
POLYPODIACEAE	POLYSTICHUM ALEUTICUM	ALASKA
PORTULACACEAE	CALYPTRIDIUM PULCHELLUM	CALIFORNIA
PORTULACACEAE	LEWISIA MAGUIREI	NEVADA
PORTULACACEAE	PORTULACA SMALLII	NORTH CAROLINA, GEORGIA
PORTULACACEAE	TALINUM APPALACHIANUM	ALABAMA
POTAMOGETONACEAF	POTAMOGETON CLYSTOCARPUS	TEXAS
PRIMULACEAE	PRIMULA CAPILLARIS	NEVADA
PRIMULACEAE	PRIMULA CUSICKIANA	IDAHO, OREGON
PRIMULACEAE	PRIMULA NEVADENSIS	NEVADA
RANUNCULACEAE	ACONITUM NOVEBORACENSE	WISCONSIN, IOWA, NEW YORK
RANUNCULACEAE	ANEMONE OREGANA VAR. FELIX	OREGON, WASHINGTON
RANUNCULACEAE	AQUILEGIA BARNEBYI	COLORADO
RANUNCULACEAE	AQUILEGIA CAERULEA VAR. DAILEYAE	COLORADO
RANUNCULACEAE	AQUILEGIA CANADENSIS VAR. AUSTRALIS	FLORIDA
RANUNCULACEAE	AQUILEGIA CHAPLINEI	TEXAS, NEW MEXICO
RANUNCULACEAE	AQUILEGIA HINCKLEYANA	TEXAS
RANUNCULACEAE	AQUILEGIA MICRANTHA VAR. MANCOSANA	COLORADO
RANUNCULACEAE	AQUILEGIA SAXIMONTANA	COLORADO
RANUNCULACEAE	CLEMATIS ADDISONII	VIRGINIA
RANUNCULACEAE	CLEMATIS GATTINGERI	ALABAMA, TFNNESSEE
RANUNCULACEAE	CLEMATIS MICRANTHA	FLORIDA
RANUNCULACEAE	CLEMATIS VITICAULIS	VIRGINIA
RANUNCULACEAE	DELPHINIUM ALABAMICUM	ALABAMA
RANUNCULACEAE	DELPHINIUM BAKERI	CALIFORNIA
RANUNCULACEAE	DELPHINIUM KINKIENSE	CALIFORNIA
RANUNCULACEAE	DELPHINIUM LEUCOPHAEUM	OREGON
RANUNCULACEAE	DELPHINIUM LUTEUM	CALIFORNIA
RANUNCULACEAE	DELPHINIUM PAVONACEUM	OREGON
RANUNCULACEAE	*RANUNCULUS ACRIFORMIS VAR. AESTIVALIS	UTAH
RANUNCULACEAE	RANUNCULUS FASCICULARIS VAR. CUNEIFORMIS	TEXAS
RANUNCULACEAE	RANUNCULUS INAMOENUS VAR. SUBAFFINIS	ARIZONA
RANUNCULACEAE	THALICTRUM COOLEYI	NORTH CAROLINA
RANUNCULACEAE	TROLLIUS LAXUS	CONNECTICUT, PENNSYLVANIA, DELAWARE, OHIO, NEW HAMPSHIRE, MAINE, NEW JERSEY, NEW YORK
RHAMNACEAE	CEANOTHUS FERRISAE	CALIFORNIA

ENDANGERED SPECIES IN THE CONTINENTAL UNITED STATES

FAMILY	SPECIES	STATE
RHAMNACEAE	CEANOTHUS HEARSTIORUM	CALIFORNIA
RHAMNACEAE	CEANOTHUS MARITIMUS	CALIFORNIA
RHAMNACEAE	CEANOTHUS MASONII	CALIFORNIA
RHAMNACEAE	COLUBRINA STRICTA	TEXAS
RHAMNACEAE	CONDALIA HOOKERI VAR. EDWARDSIANA	TEXAS
ROSACEAE	CERCOCARPUS TRASKIAE	CALIFORNIA
ROSACEAE	COWANIA SUBINTEGRA	ARIZONA
ROSACEAE	FILIPENDULA OCCIDENTALIS	OREGON
ROSACEAE	GEUM GENICULATUM	NORTH CAROLINA, TENNESSEE
ROSACEAE	GEUM PECKII	NEW HAMPSHIRE
ROSACEAE	GEUM RADIATUM	NORTH CAROLINA, TENNESSEE
ROSACEAE	HORKELIA WILDERAE	CALIFORNIA
ROSACEAE	IVESIA CALLIDA	CALIFORNIA
ROSACEAE	IVESIA CRYPTOCAULIS	NEVADA
ROSACEAE	IVESIA EREMICA	NEVADA
ROSACEAE	NEVIUSIA ALABAMENSIS	ALABAMA, MISSOURI, ARKANSAS
ROSACEAE	PETROPHYTUM CINERASCENS	WASHINGTON
ROSACEAE	POTENTILLA HICKMANII	CALIFORNIA
ROSACEAE	*POTENTILLA MULTIJUGA	CALIFORNIA
ROSACEAE	POTENTILLA ROBBINSIANA	NEW HAMPSHIRE
ROSACEAE	POTENTILLA RUPINCOLA	COLORADO
ROSACEAE	POTENTILLA SIERRA-BLANCAE	NEW MEXICO
ROSACEAE	PRUNUS GENICULATA	FLORIDA
ROSACEAE	PRUNUS GRAVESII	CONNECTICUT
RUBIACEAE	GALIUM ANGUSTIFOLIUM SSP. BORREGOENSE	CALIFORNIA
RUBIACEAE	GALIUM CALIFORNICUM SSP. LUCIENSE	CALIFORNIA
RUBIACEAE	GALIUM CALIFORNICUM SSP. PRIMUM	CALIFORNIA
RUBIACEAE	GALIUM CALIFORNICUM SSP. SIERRAE	CALIFORNIA
RUBIACEAE	GALIUM CATALINENSE SSP. ACRISPUM	CALIFORNIA
RUBIACEAE	GALIUM COLLOMAE	ARIZONA
RUBIACEAE	GALIUM GLABRESCENS SSP. MODOCENSE	CALIFORNIA
RUBIACEAE	GALIUM GRANDE	CALIFORNIA
RUBIACEAE	GALIUM HARDHAMAE	CALIFORNIA
RUBIACEAE	GALIUM HILENDIAE SSP. KINGSTONENSE	CALIFORNIA, NEVADA
RUBIACEAE	GALIUM SERPENTICUM SSP. SCOTTICUM	CALIFORNIA
RUTACEAE	ZANTHOXYLUM PARVUM	TEXAS
SALICACEAE	POPULUS HINCKLEYANA	TEXAS
SALICACEAE	SALIX FLORIDANA	FLORIDA, GEORGIA
SAPOTACEAE	BUMELIA THORNEI	GEORGIA
SARRACENIACEAE	SARRACENIA ALABAMENSIS SSP. ALABAMENSIS	ALABAMA
SARRACENIACEAE	SARRACENIA JONESII	NORTH CAROLINA, SOUTH CAROLINA
SAXIFRAGACEAE	BENSONIELLA OREGANA	CALIFORNIA, OREGON
SAXIFRAGACEAE	HEUCHERA MISSOURIENSIS	MISSOURI

ENDANGERED SPECIES IN THE CONTINENTAL UNITED STATES

FAMILY	SPECIES	STATE
SAXIFRAGACEAE	*LITHOPHRAGMA MAXIMUM	CALIFORNIA
SAXIFRAGACEAE	PARNASSIA KOTZEBUEI VAR. PUMILA	WASHINGTON
SAXIFRAGACEAE	RIBES ECHINELLUM	FLORIDA, SOUTH CAROLINA
SCHIZAEACEAE	SCHIZAEA GERMANII	FLORIDA
SCROPHULARIACEAE	*AGALINIS CADDOENSIS	LOUISIANA
SCROPHULARIACEAE	*AGALINIS STENOPHYLLA	FLORIDA
SCROPHULARIACEAE	AMPHIANTHUS PUSILLUS	SOUTH CAROLINA, GEORGIA
SCROPHULARIACEAE	*BACOPA SIMULANS	VIRGINIA
SCROPHULARIACEAE	BACOPA STRAGULA	VIRGINIA
SCROPHULARIACEAE	CASTILLEJA AQUARIENSIS	UTAH
SCROPHULARIACEAE	CASTILLEJA CHLOROTICA	OREGON
SCROPHULARIACEAE	CASTILLEJA CHRISTII	IDAHO
SCROPHULARIACEAE	CASTILLEJA CILIATA	TEXAS
SCROPHULARIACEAE	CASTILLEJA CRUENTA	ARIZONA
SCROPHULARIACEAE	CASTILLEJA GRISEA	CALIFORNIA
SCROPHULARIACEAE	*CASTILLEJA LESCHKEANA	CALIFORNIA
SCROPHULARIACEAE	*CASTILLEJA LUDOVICIANA	LOUISIANA
SCROPHULARIACEAE	CASTILLEJA OWNBEYANA	OREGON
SCROPHULARIACEAE	CASTILLEJA REVEALII	UTAH
SCROPHULARIACEAE	CASTILLEJA SALSUGINOSA	NEVADA
SCROPHULARIACEAE	CASTILLEJA ULIGINOSA	CALIFORNIA
SCROPHULARIACEAE	CORDYLANTHUS BRUNNEUS SSP. CAPILLARIS	CALIFORNIA
SCROPHULARIACEAE	CORDYLANTHUS EREMICUS SSP. BERNARDINUS	CALIFORNIA
SCROPHULARIACEAE	CORDYLANTHUS MARITIMUS SSP. MARITIMUS	CALIFORNIA, OREGON
SCROPHULARIACEAE	*CORDYLANTHUS MOLLIS SSP. MOLLIS	CALIFORNIA
SCROPHULARIACEAE	CORDYLANTHUS NIDULARIUS	CALIFORNIA
SCROPHULARIACEAE	*CORDYLANTHUS PALMATUS	CALIFORNIA
SCROPHULARIACEAE	CORDYLANTHUS RIGIDUS SSP. LITTORALIS	CALIFORNIA
SCROPHULARIACEAE	CORDYLANTHUS TENUIS SSP. PALLESCENS	CALIFORNIA
SCROPHULARIACEAE	GRATIOLA HETEROSEPALA	CALIFORNIA
SCROPHULARIACEAE	LIMOSELLA PUBIFLORA	ARIZONA
SCROPHULARIACEAE	LINDERNIA SAXICOLA	GEORGIA, NORTH CAROLINA
SCROPHULARIACEAE	*MIMULUS BRANDEGEI	CALIFORNIA
SCROPHULARIACEAE	MIMULUS GEMMIPARUS	COLORADO
SCROPHULARIACEAE	MIMULUS GUTTATUS SSP. ARENICOLA	CALIFORNIA
SCROPHULARIACEAE	MIMULUS PYGMAEUS	CALIFORNIA
SCROPHULARIACEAE	MIMULUS RINGENS VAR. COLPOPHILUS	MAINE
SCROPHULARIACEAE	*MIMULUS TRASKIAE	CALIFORNIA
SCROPHULARIACEAE	MIMULUS WHIPPLEI	CALIFORNIA
SCROPHULARIACEAE	ORTHOCARPUS CASTILLEJOIDES VAR. HUMBOLDTIENSIS	CALIFORNIA
SCROPHULARIACEAE	*ORTHOCARPUS PACHYSTACHYUS	CALIFORNIA
SCROPHULARIACEAE	ORTHOCARPUS SUCCULENTIS	CALIFORNIA
SCROPHULARIACEAE	PEDICULARIS DUDLEYI	CALIFORNIA

ENDANGERED SPECIES IN THE CONTINENTAL UNITED STATES

FAMILY	SPECIES	STATE
SCROPHULARIACEAE	PEDICULARIS FURBISHIAE	MAINE
SCROPHULARIACEAE	PENSTEMON BARRETTIAE	OREGON, WASHINGTON
SCROPHULARIACEAE	PENSTEMON CLUTEI	ARIZONA
SCROPHULARIACEAE	PENSTEMON CONCINNUS	UTAH
SCROPHULARIACEAE	PENSTEMON DECURVUS	NEVADA
SCROPHULARIACEAE	PENSTEMON DISCOLOR	ARIZONA
SCROPHULARIACEAE	*PENSTEMON GARRETTII	UTAH
SCROPHULARIACEAE	PENSTEMON GLAUCINUS	OREGON
SCROPHULARIACEAE	PENSTEMON GRAHAMII	UTAH
SCROPHULARIACEAE	PENSTEMON KECKII	NEVADA
SCROPHULARIACEAE	PENSTEMON LEMHIENSIS	IDAHO, MONTANA
SCROPHULARIACEAE	PENSTEMON NYEENSIS	NEVADA
SCROPHULARIACEAE	PENSTEMON PAHUTENSIS	NEVADA
SCROPHULARIACEAE	PENSTEMON PERSONATUS	CALIFORNIA
SCROPHULARIACEAE	PENSTEMON RETRORSUS	COLORADO
SCROPHULARIACEAE	PENSTEMON RUBICUNDUS	NEVADA
SCROPHULARIACEAE	PENSTEMON SPATULATUS	OREGON
SCROPHULARIACEAE	SCROPHULARIA MACRANTHA	NEW MEXICO
SCROPHULARIACEAE	*SEYMERIA HAVARDII	TEXAS
SCROPHULARIACEAE	*SYNTHYRIS MISSURICA SSP. HIRSUTA	OREGON
SCROPHULARIACEAE	SYNTHYRIS RANUNCULINA	NEVADA
SOLANACEAE	*LYCIUM HASSEI	CALIFORNIA
SOLANACEAE	*LYCIUM VERRUCOSUM	CALIFORNIA
SOLANACEAE	*SOLANUM BAHAMENSE VAR. RUGELII	FLORIDA
SOLANACEAE	*SOLANUM CAROLINENSE VAR. HIRSUTUM	GEORGIA
STERCULIACEAE	FREMONTODENDRON DECUMBENS	CALIFORNIA
STERCULIACEAE	*NEPHROPETALUM PRINGLEI	TEXAS
STYRACACEAE	STYRAX PLATANIFOLIA VAR. STELLATA	TEXAS
STYRACACEAE	STYRAX TEXANA	TEXAS
TAXACEAE	TAXUS FLORIDANA	FLORIDA
TAXACEAE	TORREYA TAXIFOLIA	FLORIDA, GEORGIA
THEACEAE	*FRANKLINIA ALATAMAHA	GEORGIA
URTICACEAE	URTICA CHAMAEDRYOIDES VAR. RUNYONII	TEXAS
VALERIANACEAE	VALERIANELLA TEXANA	TEXAS
VERBENACEAE	VERBENA TAMPENSIS	FLORIDA

Threatened Species of the
Continental United States

Threatened species of plants are those that presently are not
endangered but are likely to become so within the foreseeable future
throughout all or a significant portion of their range.

Threatened plants are usually rare, with a restricted range, or
they occur in specialized habitats. They are less likely to occur in
highly fragile habitats; if they do, they usually are not subjected
to severe threats by man or other agents. The threatened category is
somewhat difficult to differentiate from the endangered, but usually it
includes species that occur in larger populations or that occur over a
wider geographic range. The difference between "threatened immediately"
versus "threatened in the foreseeable future" is also difficult to differentiate because conditions often can change rapidly. If large areas
suddenly become threatened by wide-scale earth moving operations, for
example, any threatened species which occur there would have to be reclass-
ified into the endangered category.

At present, 1211 kinds, or 6.1 percent, of the native higher plants
of the continental United States are in the threatened category.

THREATENED SPECIES IN THE CONTINENTAL UNITED STATES

FAMILY	SPECIES	STATE
ACANTHACEAE	DYSCHORISTE CRENULATA	TEXAS
ACANTHACEAE	ELYTRARIA CAROLINIENSIS VAR. ANGUSTIFOLIA	FLORIDA
ACANTHACEAE	JUSTICIA MORTUIFLUMINIS	VIRGINIA
ACANTHACEAE	JUSTICIA RUNYONII	TEXAS
ACANTHACEAE	JUSTICIA WARNOCKII	TEXAS
ACANTHACEAE	JUSTICIA WRIGHTII	TEXAS
ACANTHACEAE	STENANDRIUM FASCICULARIS	TEXAS
ACERACEAE	ACER GRANDIDENTATUM VAR. SINUOSUM	TEXAS
ANACARDIACEAE	RHUS MICHAUXII	GEORGIA, NORTH CAROLINA, SOUTH CAROLINA
ANNONACEAE	ASIMINA PULCHELLA	FLORIDA
ANNONACEAE	ASIMINA RUGELII	FLORIDA
APIACEAE	ALETES FILIFOLIUS	TEXAS, NEW MEXICO
APIACEAE	ANGELICA SCABRIDA	NEVADA
APIACEAE	ANGELICA WHEELERI	NEVADA, UTAH
APIACEAE	CICUTA BOLANDERI	CALIFORNIA
APIACEAE	CYMOPTERUS CORRUGATUS	OREGON, NEVADA
APIACEAE	CYMOPTERUS COULTERI	UTAH
APIACEAE	CYMOPTERUS DESERTICOLA	CALIFORNIA
APIACEAE	CYMOPTERUS DUCHESNENSIS	UTAH
APIACEAE	CYMOPTERUS HIGGINSII	UTAH
APIACEAE	CYMOPTERUS ROSEI	UTAH
APIACEAE	ERYNGIUM CUNEIFOLIUM	FLORIDA
APIACEAE	ERYNGIUM PETIOLATUM	OREGON, WASHINGTON
APIACEAE	ERYNGIUM PINNATISECTUM	CALIFORNIA
APIACEAE	EURYTAENIA HINCKLEYI	TEXAS
APIACEAE	LIGUSTICUM PORTERI VAR. BREVILOBUM	UTAH
APIACEAE	LILAEOPSIS CAROLINENSIS	LOUISIANA, NORTH CAROLINA, SOUTH CAROLINA, VIRGINIA
APIACEAE	LOMATIUM CONGDONII	CALIFORNIA
APIACEAE	LOMATIUM CUSPIDATUM	WASHINGTON
APIACEAE	LOMATIUM FOENICULACEUM SSP. INYOENSE	CALIFORNIA, IDAHO, NEVADA
APIACEAE	LOMATIUM HENDERSONII	OREGON, IDAHO
APIACEAE	LOMATIUM HOWELLII	CALIFORNIA, OREGON
APIACEAE	LOMATIUM LATILOBUM	UTAH
APIACEAE	LOMATIUM MINIMUM	UTAH
APIACEAE	LOMATIUM MINUS	OREGON
APIACEAE	LOMATIUM OREGANUM	OREGON
APIACEAE	LOMATIUM PECKIANUM	CALIFORNIA, OREGON
APIACEAE	LOMATIUM RIGIDUM	CALIFORNIA
APIACEAE	LOMATIUM ROLLINSII	OREGON, IDAHO
APIACEAE	LOMATIUM SERPENTINUM	OREGON, WASHINGTON, IDAHO
APIACEAE	LOMATIUM THOMPSONII	WASHINGTON
APIACEAE	MUSINEON LINEARE	UTAH

THREATENED SPECIES IN THE CONTINENTAL UNITED STATES

FAMILY	SPECIES	STATE
APIACEAE	NEOPARRYA LITHOPHILA	COLORADO
APIACEAE	OXYPOLIS CANBYI	DELAWARE, GEORGIA
APIACEAE	PERIDERIDIA BACIGALUPII	CALIFORNIA
APIACEAE	PERIDERIDIA ERYTHRORHIZA	OREGON
APIACEAE	PERIDERIDIA GAIRDNERI SSP. GAIRDNERI	CALIFORNIA
APIACEAE	PERIDERIDIA LEPTOCARPA	CALIFORNIA
APIACEAE	PERIDERIDIA PRINGLEI	CALIFORNIA
APIACEAE	PTILIMNIUM FLUVIATILE	ALABAMA, MARYLAND, NORTH CAROLINA, WEST VIRGINIA, SOUTH CAROLINA
APIACEAE	PTILIMNIUM NODOSUM	SOUTH CAROLINA, GEORGIA, ALABAMA
APIACEAE	RHYSOPTERUS PLURIJUGUS	OREGON
APIACEAE	SANICULA HOFFMANNII	CALIFORNIA
APIACEAE	SANICULA PECKIANA	CALIFORNIA, OREGON
APIACEAE	TAUSCHIA GLAUCA	CALIFORNIA, OREGON
APIACEAE	TAUSCHIA HOWELLII	CALIFORNIA, OREGON
APIACEAE	TAUSCHIA STRICKLANDII	WASHINGTON
APIACEAE	ZIZIA LATIFOLIA	FLORIDA
APOCYNACEAE	AMSONIA GLABERRIMA	LOUISIANA, TEXAS
APOCYNACEAE	AMSONIA PALMERI	ARIZONA
APOCYNACEAE	AMSONIA PEEBLESII	ARIZONA
APOCYNACEAE	AMSONIA REPENS	TEXAS
APOCYNACEAE	AMSONIA THARPII	TEXAS
AQUIFOLIACEAE	ILEX OPACA VAR. ARENICOLA	FLORIDA
ARALIACEAE	PANAX QUINQUEFOLIUS	ALABAMA, ARKANSAS, CONNECTICUT, DELAWARE, GEORGIA, ILLINOIS, INDIANA, IOWA, KENTUCKY, LOUISIANA, MAINE, MARYLAND, MASSACHUSETTS, MICHIGAN, MINNESOTA, MISSISSIPPI, MISSOURI, NEBRASKA, NEW HAMPSHIRE, NEW JERSEY, NEW YORK, NORTH CAROLINA, OHIO, OKLAHOMA, PENNSYLVANIA, SOUTH CAROLINA, TENNESSEE, VERMONT, VIRGINIA, WEST VIRGINIA, WISCONSIN
ARECACEAE	RHAPIDOPHYLLUM HYSTRIX	FLORIDA, MISSISSIPPI, GEORGIA, ALABAMA
ARISTOLOCHIACEAE	HEXASTYLIS CONTRACTA	NORTH CAROLINA, TENNESSEE
ARISTOLOCHIACEAE	HEXASTYLIS LEWISII	VIRGINIA, NORTH CAROLINA
ASCLEPIADACEAE	ASCLEPIAS CUTLERI	UTAH, ARIZONA
ASCLEPIADACEAE	ASCLEPIAS RUTHIAE	UTAH
ASCLEPIADACEAE	ASCLEPIAS VIRIDULA	FLORIDA
ASCLEPIADACEAE	MATELEA BREVICORONATA	TEXAS
ASCLEPIADACEAE	MATELEA FLORIDANA	FLORIDA
ASCLEPIADACEAE	MATELEA PARVIFLORA	TEXAS
ASTERACEAE	AMBROSIA PUMILA	CALIFORNIA
ASTERACEAE	ANTENNARIA SOLICEPS	NEVADA
ASTERACEAE	ANTENNARIA SUFFRUTESCENS	OREGON, CALIFORNIA
ASTERACEAE	ARNICA AMPLEXICAULIS VAR. PIPERI	WASHINGTON, OREGON
ASTERACEAE	ARNICA VENOSA	CALIFORNIA

THREATENED SPECIES IN THE CONTINENTAL UNITED STATES

FAMILY	SPECIES	STATE
ASTERACEAE	ARNICA VISCOSA	CALIFORNIA, OREGON
ASTERACEAE	ARTEMISIA ALEUTICA	ALASKA
ASTERACEAE	ARTEMISIA PAPPOSA	IDAHO
ASTERACEAE	ARTEMISIA PORTERI	WYOMING
ASTERACEAE	ASTER BRACHYPHOLIS	FLORIDA
ASTERACEAE	ASTER BRICKELLIOIDES	CALIFORNIA, OREGON
ASTERACEAE	ASTER CHILENSIS SSP. HALLII	OREGON, WASHINGTON
ASTERACEAE	ASTER CURTUS	WASHINGTON, OREGON
ASTERACEAE	ASTER GLAUCESCENS	WASHINGTON
ASTERACEAE	ASTER JESSICAE	IDAHO, WASHINGTON
ASTERACEAE	ASTER LEMMONII	ARIZONA
ASTERACEAE	ASTER PEIRSONII	CALIFORNIA
ASTERACEAE	ASTER PLUMOSUS	FLORIDA
ASTERACEAE	ASTER SCABRICAULIS	TEXAS
ASTERACEAE	ASTER SPINULOSUS	FLORIDA
ASTERACEAE	ASTER VERUTIFOLIUS	MISSISSIPPI
ASTERACEAE	ASTRANTHIUM ROBUSTUM	TEXAS
ASTERACEAE	BAHIA BIGELOVII	TEXAS
ASTERACEAE	BALSAMORHIZA ROSEA	WASHINGTON
ASTERACEAE	BENITOA OCCIDENTALIS	CALIFORNIA
ASTERACEAE	BOLTONIA ASTEROIDES VAR. DECURRENS	ILLINOIS, MISSOURI
ASTERACEAE	BRICKELLIA BRACHYPHYLLA VAR. HINCKLEYI	TEXAS
ASTERACEAE	BRICKELLIA BRACHYPHYLLA VAR. TERLINGUENSIS	TEXAS
ASTERACEAE	BRICKELLIA CORDIFOLIA	FLORIDA, GEORGIA, ALABAMA
ASTERACEAE	BRICKELLIA DENTATA	TEXAS
ASTERACEAE	BRICKELLIA EUPATORIOIDES VAR. FLORIDANA	FLORIDA
ASTERACEAE	BRICKELLIA KNAPPIANA	CALIFORNIA
ASTERACEAE	BRICKELLIA SHINERI	TEXAS
ASTERACEAE	CACALIA DIVERSIFOLIA	FLORIDA, GEORGIA, ALABAMA
ASTERACEAE	CACALIA RUGELIA	NORTH CAROLINA, TENNESSEE
ASTERACEAE	CHAENACTIS EVERMANNII	IDAHO
ASTERACEAE	CHAENACTIS NEVII	OREGON
ASTERACEAE	CHAENACTIS PARISHII	CALIFORNIA
ASTERACEAE	CHAENACTIS RAMOSA	WASHINGTON
ASTERACEAE	CHAENACTIS THOMPSONII	WASHINGTON
ASTERACEAE	CHAETOPAPPA HERSHEYI	TEXAS, NEW MEXICO
ASTERACEAE	CHAMAECHAENACTIS SCAPOSA	COLORADO, UTAH
ASTERACEAE	CIRSIUM CAMPYLON	CALIFORNIA
ASTERACEAE	CIRSIUM CILIOLATUM	OREGON
ASTERACEAE	CIRSIUM DAVISII	IDAHO
ASTERACEAE	CIRSIUM HYDROPHILUM VAR. VASEYI	CALIFORNIA
ASTERACEAE	CIRSIUM PITCHERI	INDIANA, ILLINOIS, MICHIGAN, WISCONSIN

THREATENED SPECIES IN THE CONTINENTAL UNITED STATES

FAMILY	SPECIES	STATE
ASTERACEAE	CIRSIUM RYDBERGII	UTAH, ARIZONA
ASTERACEAE	CIRSIUM TURNERI	TEXAS
ASTERACEAE	CIRSIUM VINACEUM	NEW MEXICO
ASTERACEAE	COREOPSIS HAMILTONII	CALIFORNIA
ASTERACEAE	COREOPSIS LATIFOLIA	GEORGIA, NORTH CAROLINA, SOUTH CAROLINA
ASTERACEAE	COREOPSIS PULCHRA	ALABAMA
ASTERACEAE	ECHINACEA LAEVIGATA	PENNSYLVANIA, SOUTH CAROLINA, GEORGIA, NORTH CAROLINA, VIRGINIA, ALABAMA
ASTERACEAE	ENCELIA FRUTESCENS VAR. RESINOSA	ARIZONA
ASTERACEAE	ENCELIOPSIS NUDICAULIS VAR. CORRUGATA	NEVADA
ASTERACEAE	ERIGERON ABAJOENSIS	UTAH
ASTERACEAE	ERIGERON AEQUIFOLIUS	CALIFORNIA
ASTERACEAE	ERIGERON ALLOCOTUS	MONTANA, WYOMING
ASTERACEAE	ERIGERON ARENARIOIDES	UTAH
ASTERACEAE	ERIGERON ARIZONICUS	ARIZONA
ASTERACEAE	ERIGERON BIGELOVII	TEXAS
ASTERACEAE	ERIGERON BLOOMERI VAR. NUDATUS	CALIFORNIA, OREGON
ASTERACEAE	ERIGERON CRONQUISTII	UTAH
ASTERACEAE	ERIGERON FLETTII	WASHINGTON
ASTERACEAE	ERIGERON FLEXUOSUS	CALIFORNIA
ASTERACEAE	ERIGERON GARRETTII	UTAH
ASTERACEAE	ERIGERON HOWELLII	OREGON
ASTERACEAE	ERIGERON HULTENII	ALASKA
ASTERACEAE	ERIGERON LEIBERGII	WASHINGTON
ASTERACEAE	ERIGERON LEMMONII	ARIZONA
ASTERACEAE	ERIGERON LOBATUS	ARIZONA
ASTERACEAE	ERIGERON MANCUS	UTAH
ASTERACEAE	ERIGERON MULTICEPS	CALIFORNIA
ASTERACEAE	ERIGERON OREGANUS	OREGON
ASTERACEAE	ERIGERON OVINUS	NEVADA
ASTERACEAE	ERIGERON PARISHII	CALIFORNIA
ASTERACEAE	ERIGERON PIPERIANUS	WASHINGTON
ASTERACEAE	ERIGERON PRINGLEI	ARIZONA
ASTERACEAE	ERIGERON PULCHELLUS VAR. TOLSTEADII	MINNESOTA
ASTERACEAE	ERIGERON SUPPLEX	CALIFORNIA
ASTERACEAE	ERIGERON UNCIALIS VAR. CONJUGANS	NEVADA
ASTERACEAE	ERIOPHYLLUM LATILOBUM	CALIFORNIA
ASTERACEAE	ERIOPHYLLUM NUBIGENUM VAR. CONGDONII	CALIFORNIA
ASTERACEAE	EUPATORIUM LEUCOLEPIS VAR. NOVAE-ANGLIAE	MASSACHUSETTS, RHODE ISLAND
ASTERACEAE	EUPATORIUM SHASTENSE	CALIFORNIA
ASTERACEAE	GUTIERREZIA CALIFORNICA	CALIFORNIA
ASTERACEAE	HAPLOPAPPUS ABERRANS	IDAHO

THREATENED SPECIES IN THE CONTINENTAL UNITED STATES

FAMILY	SPECIES	STATE
ASTERACEAE	HAPLOPAPPUS BRICKELLIOIDES	NEVADA, CALIFORNIA
ASTERACEAE	HAPLOPAPPUS HALLII	WASHINGTON, OREGON
ASTERACEAE	HAPLOPAPPUS OPHITIDIS	CALIFORNIA
ASTERACEAE	HARTWRIGHTIA FLORIDANA	FLORIDA, GEORGIA
ASTERACEAE	HELENIUM ARIZONICUM	ARIZONA
ASTERACEAE	HELIANTHELLA CASTANEA	CALIFORNIA
ASTERACEAE	HELIANTHUS CARNOSUS	FLORIDA
ASTERACEAE	HELIANTHUS DEBILIS SSP. VESTITUS	FLORIDA
ASTERACEAE	HELIANTHUS DESERTICOLA	ARIZONA, NEVADA, UTAH
ASTERACEAE	HELIANTHUS PRAECOX SSP. HIRTUS	TEXAS
ASTERACEAE	HELIANTHUS SCHWEINITZII	NORTH CAROLINA, SOUTH CAROLINA
ASTERACEAE	HEMIZONIA ARIDA	CALIFORNIA
ASTERACEAE	HEMIZONIA HALLIANA	CALIFORNIA
ASTERACEAE	HETEROTHECA FLEXUOSA	FLORIDA
ASTERACEAE	HIERACIUM LONGIBERBE	OREGON, WASHINGTON
ASTERACEAE	HULSEA VESTITA SSP. INYOENSIS	CALIFORNIA, NEVADA
ASTERACEAE	HYMENOPAPPUS FILIFOLIUS VAR. IDAHOENSIS	IDAHO
ASTERACEAE	HYMENOPAPPUS FILIFOLIUS VAR. TOMENTOSUS	UTAH
ASTERACEAE	HYMENOXYS QUINQUESQUAMATA	ARIZONA
ASTERACEAE	HYMENOXYS SUBINTEGRA	ARIZONA
ASTERACEAE	LASTHENIA LEPTALEA	CALIFORNIA
ASTERACEAE	LASTHENIA MACRANTHA SSP. PRISCA	OREGON
ASTERACEAE	LASTHENIA MINOR SSP. MARITIMA	CALIFORNIA, OREGON, WASHINGTON
ASTERACEAE	LAYIA JONESII	CALIFORNIA
ASTERACEAE	LAYIA LEUCOPAPPA	CALIFORNIA
ASTERACEAE	LAYIA ZIEGLERI	CALIFORNIA
ASTERACEAE	LIATRIS CYMOSA	TEXAS
ASTERACEAE	LIATRIS HELLERI	NORTH CAROLINA, ALABAMA
ASTERACEAE	LIATRIS TENUIS	TEXAS
ASTERACEAE	LUINA SERPENTINA	OREGON
ASTERACEAE	MACHAERANTHERA AMMOPHILA	NEVADA, CALIFORNIA
ASTERACEAE	MACHAERANTHERA GLABRIUSCULA VAR. CONFERTIFOLIA	UTAH
ASTERACEAE	MACHAERANTHERA KINGII	UTAH
ASTERACEAE	MACHAERANTHERA LAGUNENSIS	CALIFORNIA
ASTERACEAE	MACHAERANTHERA MUCRONATA	ARIZONA
ASTERACEAE	MACHAERANTHERA ORCUTTII	CALIFORNIA
ASTERACEAE	MALACOTHRIX SAXATILIS VAR. ARACHNOIDEA	CALIFORNIA
ASTERACEAE	MARSHALLIA RAMOSA	GEORGIA
ASTERACEAE	MELANTHERA PARVIFOLIA	FLORIDA
ASTERACEAE	MICROSERIS HOWELLII	OREGON
ASTERACEAE	MICROSERIS LACINIATA SSP. DETLINGII	OREGON
ASTERACEAE	PARTHENIUM ALPINUM	WYOMING
ASTERACEAE	PENTACHAETA BELLIDIFLORA	CALIFORNIA

THREATENED SPECIES IN THE CONTINENTAL UNITED STATES

FAMILY	SPECIES	STATE
ASTERACEAE	PENTACHAETA EXILIS SSP. AEOLICA	CALIFORNIA
ASTERACEAE	PERITYLE CERNUA	NEW MEXICO
ASTERACEAE	PERITYLE COCHISENSIS	ARIZONA
ASTERACEAE	PERITYLE INYOENSIS	CALIFORNIA
ASTERACEAE	PERITYLE LEMMONII	NEW MEXICO, ARIZONA
ASTERACEAE	PERITYLE MEGALOCEPHALA VAR. INTRICATA	NEVADA
ASTERACEAE	PERITYLE SAXICOLA	ARIZONA
ASTERACEAE	PERITYLE STAUROPHYLLA	NEW MEXICO
ASTERACEAE	PERITYLE VILLOSA	CALIFORNIA
ASTERACEAE	PERITYLE WARNOCKII	TEXAS
ASTERACEAE	PLUMMERA FLORIBUNDA	ARIZONA, NEW MEXICO
ASTERACEAE	POROPHYLLUM GREGGII	TEXAS
ASTERACEAE	PRENANTHES BOOTTII	NEW YORK, NEW HAMPSHIRE, VERMONT, MAINE
ASTERACEAE	PRENANTHES ROANENSIS	KENTUCKY, NORTH CAROLINA, TENNESSEE
ASTERACEAE	PYRROCOMA LIATRIFORMIS	IDAHO, WASHINGTON
ASTERACEAE	PYRROCOMA UNIFLORUS VAR. GOSSYPINUS	CALIFORNIA
ASTERACEAE	RAILLARDELLA MUIRII	CALIFORNIA
ASTERACEAE	RAILLARDELLA PRINGLEI	CALIFORNIA
ASTERACEAE	RUDBECKIA AURICULATA	ALABAMA
ASTERACEAE	RUDBECKIA HELIOPSIDIS	NORTH CAROLINA, SOUTH CAROLINA, GEORGIA, VIRGINIA, ALABAMA
ASTERACEAE	SENECIO BERNARDINUS	CALIFORNIA
ASTERACEAE	SENECIO DIMORPHOPHYLLUS VAR. INTERMEDIUS	UTAH
ASTERACEAE	SENECIO GANDERI	CALIFORNIA
ASTERACEAE	SENECIO HESPERIUS	OREGON
ASTERACEAE	SENECIO LYNCEUS VAR. LEUCOREUS	NEVADA
ASTERACEAE	SENECIO MILLEFOLIUM	NORTH CAROLINA, SOUTH CAROLINA, GEORGIA
ASTERACEAE	SENECIO NEOWEBSTERI	WASHINGTON
ASTERACEAE	SENECIO QUAERENS	NEW MEXICO
ASTERACEAE	SENECIO WARNOCKII	TEXAS
ASTERACEAE	SOLIDAGO HOUGHTONII	MICHIGAN, NEW YORK
ASTERACEAE	SOLIDAGO MOLLIS VAR. ANGUSTATA	TEXAS
ASTERACEAE	SOLIDAGO PULCHRA	NORTH CAROLINA
ASTERACEAE	SOLIDAGO SPITHAMAEA	NORTH CAROLINA, TENNESSEE
ASTERACEAE	SOLIDAGO VERNA	NORTH CAROLINA, SOUTH CAROLINA
ASTERACEAE	TAGETES LEMMONII	ARIZONA
ASTERACEAE	TANACETUM CAMPHORATUM	CALIFORNIA
ASTERACEAE	TANACETUM SIMPLEX	WYOMING
ASTERACEAE	TARAXACUM CALIFORNICUM	CALIFORNIA
ASTERACEAE	TOWNSENDIA JONESII VAR. TUMULOSA	NEVADA
ASTERACEAE	TOWNSENDIA MENSANA	UTAH
ASTERACEAE	TOWNSENDIA MINIMA	UTAH

THREATENED SPECIES IN THE CONTINENTAL UNITED STATES

FAMILY	SPECIES	STATE
ASTERACEAE	TOWNSENDIA ROTHROCKII	COLORADO
ASTERACEAE	VERBESINA CHAPMANNII	FLORIDA
ASTERACEAE	VERBESINA HETEROPHYLLA	FLORIDA
ASTERACEAE	VERNONIA PULCHELLA	GEORGIA
ASTERACEAE	VIGUIERA PORTERI	GEORGIA, ALABAMA
ASTERACEAE	VIGUIERA SOLICEPS	UTAH
ASTERACEAE	WYETHIA RETICULATA	CALIFORNIA
ASTERACEAE	XANTHOCEPHALUM SAROTHRAE VAR. POMARIENSIS	UTAH
BERBERIDACEAE	BERBERIS HIGGINSAE	CALIFORNIA
BERBERIDACEAE	BERBERIS PINNATA SSP. INSULARIS	CALIFORNIA
BERBERIDACEAE	BERBERIS SWASEYI	TEXAS
BETULACEAE	ALNUS MARITIMA	DELAWARE, MARYLAND, OKLAHOMA, VIRGINIA
BETULACEAE	OSTRYA CHISOSENSIS	TEXAS
BORAGINACEAE	AMSINCKIA VERNICOSA VAR. FURCATA	CALIFORNIA
BORAGINACEAE	CRYPTANTHA BARNEBYI	UTAH
BORAGINACEAE	CRYPTANTHA COMPACTA	UTAH, NEVADA
BORAGINACEAE	CRYPTANTHA CRASSIPES	TEXAS
BORAGINACEAE	CRYPTANTHA CRINITA	CALIFORNIA
BORAGINACEAE	CRYPTANTHA ELATA	UTAH, COLORADO
BORAGINACEAE	CRYPTANTHA GANDERI	CALIFORNIA
BORAGINACEAE	CRYPTANTHA GRAHAMII	UTAH
BORAGINACEAE	CRYPTANTHA HYPSOPHILA	IDAHO
BORAGINACEAE	CRYPTANTHA INTERRUPTA	NEVADA
BORAGINACEAE	CRYPTANTHA JOHNSTONII	UTAH
BORAGINACEAE	CRYPTANTHA JONESIANA	UTAH
BORAGINACEAE	CRYPTANTHA MENSANA	UTAH
BORAGINACEAE	CRYPTANTHA PARADOXA	COLORADO, NEW MEXICO, UTAH
BORAGINACEAE	CRYPTANTHA SEMIGLABRA	ARIZONA, UTAH
BORAGINACEAE	CRYPTANTHA THOMPSONII	WASHINGTON
BORAGINACEAE	HACKELIA BREVICULA	CALIFORNIA
BORAGINACEAE	HACKELIA HISPIDA	IDAHO, OREGON, WASHINGTON
BORAGINACEAE	HACKELIA PATENS VAR. SEMIGLABRA	OREGON
BORAGINACEAE	HELIOTROPIUM POLYPHYLLUM VAR. HORIZONTALE	FLORIDA
BORAGINACEAE	ONOSMODIUM HELLERI	TEXAS
BORAGINACEAE	ONOSMODIUM MOLLE	KENTUCKY, TENNESSEE
BORAGINACEAE	PLAGIOBOTHRYS DISTANTIFLORUS	CALIFORNIA
BORAGINACEAE	PLAGIOBOTHRYS GLABER	CALIFORNIA
BORAGINACEAE	PLAGIOBOTHRYS HIRTUS SSP. CORALLICARPA	OREGON
BORAGINACEAE	PLAGIOBOTHRYS HYSTRICULUS	CALIFORNIA
BORAGINACEAE	PLAGIOBOTHRYS SCRIPTUS	CALIFORNIA
BORAGINACEAE	PLAGIOBOTHRYS STRICTUS	CALIFORNIA
BRASSICACEAE	ARABIS ACULEOLATA	OREGON

THREATENED SPECIES IN THE CONTINENTAL UNITED STATES

FAMILY	SPECIES	STATE
BRASSICACEAE	ARABIS BLEPHAROPHYLLA	CALIFORNIA
BRASSICACEAE	ARABIS BREWERI VAR. AUSTINAE	CALIFORNIA
BRASSICACEAE	ARABIS CONSTANCEI	CALIFORNIA
BRASSICACEAE	ARABIS DEMISSA VAR. LANGUIDA	WYOMING, UTAH
BRASSICACEAE	ARABIS DEMISSA VAR. RUSSEOLA	WYOMING, UTAH
BRASSICACEAE	ARABIS GEORGIANA	GEORGIA, ALABAMA
BRASSICACEAE	ARABIS GRACILIPES	ARIZONA
BRASSICACEAE	ARABIS HOFFMANNII	CALIFORNIA
BRASSICACEAE	ARABIS JOHNSTONII	CALIFORNIA
BRASSICACEAE	ARABIS KOEHLERI VAR. STIPITATA	OREGON
BRASSICACEAE	ARABIS MODESTA	CALIFORNIA, OREGON
BRASSICACEAE	ARABIS OREGANA	CALIFORNIA, OREGON
BRASSICACEAE	ARABIS PARISHII	CALIFORNIA
BRASSICACEAE	ARABIS PYGMAEA	CALIFORNIA
BRASSICACEAE	ARABIS SUFFRUTESCENS VAR. HORIZONTALIS	OREGON
BRASSICACEAE	CARDAMINE LONGII	MAINE, VIRGINIA, MARYLAND
BRASSICACEAE	CARDAMINE PENDULIFLORA	OREGON, CALIFORNIA
BRASSICACEAE	CARDAMINE RUPICOLA	MONTANA
BRASSICACEAE	CAULANTHUS AMPLEXICAULIS VAR. BARBARAE	CALIFORNIA
BRASSICACEAE	CAULANTHUS STENOCARPUS	CALIFORNIA
BRASSICACEAE	CAULOSTRAMINA JAEGERI	CALIFORNIA
BRASSICACEAE	DRABA APICULATA VAR. DAVIESIAE	IDAHO, MONTANA
BRASSICACEAE	DRABA ARGYRAEA	IDAHO
BRASSICACEAE	DRABA ASPRELLA VAR. STELLIGERA	ARIZONA
BRASSICACEAE	DRABA ASTEROPHORA VAR. ASTEROPHORA	NEVADA, CALIFORNIA
BRASSICACEAE	DRABA ASTEROPHORA VAR. MACROCARPA	CALIFORNIA
BRASSICACEAE	DRABA CRASSIFOLIA VAR. NEVADENSIS	NEVADA
BRASSICACEAE	DRABA CRUCIATA VAR. CRUCIATA	CALIFORNIA
BRASSICACEAE	DRABA CRUCIATA VAR. INTEGRIFOLIA	CALIFORNIA
BRASSICACEAE	DRABA HOWELLII VAR. CARNOSULA	CALIFORNIA
BRASSICACEAE	DRABA JAEGERI	NEVADA
BRASSICACEAE	DRABA LEMMONII VAR. CYCLOMORPHA	OREGON
BRASSICACEAE	DRABA LEMMONII VAR. INCRASSATA	CALIFORNIA
BRASSICACEAE	DRABA MAGUIREI VAR. BURKEI	UTAH
BRASSICACEAE	DRABA MAGUIREI VAR. MAGUIREI	UTAH
BRASSICACEAE	DRABA MOGOLLONICA	NEW MEXICO
BRASSICACEAE	DRABA NIVALIS VAR. BREVICULA	WYOMING
BRASSICACEAE	DRABA PECTINIPILA	WYOMING, UTAH
BRASSICACEAE	DRABA QUADRICOSTATA	CALIFORNIA
BRASSICACEAE	DRABA SOBOLIFERA	UTAH
BRASSICACEAE	DRABA SPHAEROCARPA	IDAHO
BRASSICACEAE	DRABA STENOLOBA VAR. RAMOSA	CALIFORNIA, NEVADA
BRASSICACEAE	DRABA SUBALPINA	UTAH

THREATENED SPECIES IN THE CONTINENTAL UNITED STATES

FAMILY	SPECIES	STATE
BRASSICACEAE	DRABA ZIONENSIS	UTAH
BRASSICACEAE	ERYSIMUM AMMOPHILUM	CALIFORNIA
BRASSICACEAE	ERYSIMUM INSULARE	CALIFORNIA
BRASSICACEAE	ERYSIMUM MENZIESII	CALIFORNIA
BRASSICACEAE	HALIMOLOBOS PERPLEXA VAR. LEMHIENSIS	IDAHO
BRASSICACEAE	HALIMOLOBOS PERPLEXA VAR. PERPLEXA	IDAHO
BRASSICACEAE	LEAVENWORTHIA ALABAMICA VAR. ALABAMICA	ALABAMA
BRASSICACEAE	LEAVENWORTHIA EXIGUA VAR. EXIGUA	TENNESSEE, GEORGIA
BRASSICACEAE	LEAVENWORTHIA STYLOSA	TENNESSEE
BRASSICACEAE	LEAVENWORTHIA TORULOSA	KENTUCKY, TENNESSEE, ALABAMA
BRASSICACEAE	LEPIDIUM NANUM	NEVADA
BRASSICACEAE	LESQUERELLA ANGUSTIFOLIA	TEXAS, OKLAHOMA
BRASSICACEAE	LESQUERELLA ARCTICA VAR. SCAMMANAE	ALASKA
BRASSICACEAE	LESQUERELLA GARRETTII	UTAH
BRASSICACEAE	LESQUERELLA GLOBOSA	TENNESSEE, KENTUCKY, INDIANA
BRASSICACEAE	LESQUERELLA GOODDINGII	ARIZONA, NEW MEXICO
BRASSICACEAE	LESQUERELLA HITCHCOCKII	NEVADA
BRASSICACEAE	LESQUERELLA KINGII SSP. BERNARDINA	CALIFORNIA
BRASSICACEAE	LESQUERELLA KINGII SSP. DIVERSIFOLIA	OREGON
BRASSICACEAE	LESQUERELLA LESCURII	TENNESSEE
BRASSICACEAE	LESQUERELLA MCVAUGHIANA	TEXAS
BRASSICACEAE	LESQUERELLA RUBICUNDULA	UTAH
BRASSICACEAE	LESQUERELLA THAMNOPHILA	TEXAS
BRASSICACEAE	PHYSARIA BELLII	COLORADO
BRASSICACEAE	PHYSARIA CONDENSATA	WYOMING
BRASSICACEAE	PHYSARIA DIDYMOCARPA VAR. LYRATA	IDAHO
BRASSICACEAE	PHYSARIA GEYERI VAR. PURPUREA	IDAHO
BRASSICACEAE	RORIPPA CALYCINA	MONTANA, NORTH DAKOTA
BRASSICACEAE	RORIPPA COLORADENSIS	COLORADO
BRASSICACEAE	RORIPPA COLUMBIAE	CALIFORNIA, OREGON, WASHINGTON
BRASSICACEAE	RORIPPA SUBUMBELLATA	CALIFORNIA, NEVADA
BRASSICACEAE	SMELOWSKIA PYRIFORMIS	ALASKA
BRASSICACEAE	STANLEYA PINNATA VAR. GIBBEROSA	WYOMING
BRASSICACEAE	STREPTANTHUS BATRACHOPUS	CALIFORNIA
BRASSICACEAE	STREPTANTHUS BRACHIATUS	CALIFORNIA
BRASSICACEAE	STREPTANTHUS BRACTEATUS	TEXAS
BRASSICACEAE	STREPTANTHUS CARINATUS	TEXAS
BRASSICACEAE	STREPTANTHUS CORDATUS VAR. PIUTENSIS	CALIFORNIA
BRASSICACEAE	STREPTANTHUS CUTLERI	TEXAS
BRASSICACEAE	STREPTANTHUS FENESTRATUS	CALIFORNIA
BRASSICACEAE	STREPTANTHUS GLANDULOSUS VAR. HOFFMANII	CALIFORNIA
BRASSICACEAE	STREPTANTHUS GLANDULOSUS VAR. PULCHELLUS	CALIFORNIA
BRASSICACEAE	STREPTANTHUS GRACILIS	CALIFORNIA

THREATENED SPECIES IN THE CONTINENTAL UNITED STATES

FAMILY	SPECIES	STATE
BRASSICACEAE	STREPTANTHUS HISPIDUS	CALIFORNIA
BRASSICACEAE	STREPTANTHUS MORRISONII SSP. ELATUS	CALIFORNIA
BRASSICACEAE	STREPTANTHUS MORRISONII SSP. MORRISONII	CALIFORNIA
BRASSICACEAE	STREPTANTHUS OLIGANTHUS	CALIFORNIA
BRASSICACEAE	THELYPODIUM EUCOSMUM	OREGON
BRASSICACEAE	THELYPODIUM HOWELLII VAR. SPECTABILIS	OREGON
BRASSICACEAE	THELYPODIUM SAGITTATUM VAR. OVALIFOLIUM	NEVADA, UTAH
BRASSICACEAE	THELYPODIUM STENOPETALUM	CALIFORNIA
BRASSICACEAE	THLASPI ARCTICUM	ALASKA
BRASSICACEAE	THLASPI MONTANUM VAR. SISKIYOUENSE	OREGON
BRASSICACEAE	THYSANOCARPUS CONCHULIFERUS	CALIFORNIA
BRASSICACEAE	TROPIDOCARPUM CAPPARIDEUM	CALIFORNIA
BRASSICACEAE	WAREA SESSILIFOLIA	FLORIDA, ALABAMA
CACTACEAE	CEREUS GREGGII	ARIZONA, NEW MEXICO, TEXAS
CACTACEAE	CORYPHANTHA DASYACANTHA VAR. VARICOLOR	TEXAS
CACTACEAE	CORYPHANTHA DUNCANII	NEW MEXICO, TEXAS
CACTACEAE	CORYPHANTHA HESTERI	TEXAS
CACTACEAE	CORYPHANTHA RECURVATA	ARIZONA
CACTACEAE	CORYPHANTHA SCHEERI VAR. ROBUSTISPINA	ARIZONA
CACTACEAE	CORYPHANTHA SULCATA VAR. NICKELSIAE	TEXAS
CACTACEAE	CORYPHANTHA VIVIPARA VAR. ALVERSONII	ARIZONA, CALIFORNIA
CACTACEAE	CORYPHANTHA VIVIPARA VAR. ROSEA	ARIZONA, NEVADA, CALIFORNIA
CACTACEAE	ECHINOCEREUS ENGELMANNII VAR. MUNZII	CALIFORNIA
CACTACEAE	ECHINOCEREUS LEDINGII	ARIZONA
CACTACEAE	ECHINOCEREUS REICHENBACHII VAR. CHISOSENSIS	TEXAS
CACTACEAE	ECHINOCEREUS REICHENBACHII VAR. FITCHII	TEXAS
CACTACEAE	ECHINOCEREUS VIRIDIFLORUS VAR. CORRELLII	TEXAS
CACTACEAE	EPITHELANTHA BOKEI	TEXAS
CACTACEAE	FEROCACTUS ACANTHODES VAR. EASTWOODIAE	ARIZONA
CACTACEAE	MAMMILLARIA ORESTERA	ARIZONA, NEW MEXICO
CACTACEAE	MAMMILLARIA THORNBERI	ARIZONA
CACTACEAE	NEOLLOYDIA ERECTOCENTRA VAR. ACUNENSIS	ARIZONA
CACTACEAE	NEOLLOYDIA ERECTOCENTRA VAR. ERECTOCENTRA	ARIZONA
CACTACEAE	NEOLLOYDIA WARNOCKII	TEXAS
CACTACEAE	OPUNTIA ARENARIA	TEXAS, NEW MEXICO
CACTACEAE	OPUNTIA BASILARIS VAR. BRACHYCLADA	CALIFORNIA
CACTACEAE	OPUNTIA BASILARIS VAR. LONGIAREOLATA	ARIZONA
CACTACEAE	OPUNTIA IMBRICATA VAR. ARGENTEA	TEXAS
CACTACEAE	OPUNTIA MUNZII	CALIFORNIA
CACTACEAE	OPUNTIA PARRYI VAR. SERPENTINA	CALIFORNIA
CACTACEAE	OPUNTIA PHAEACANTHA VAR. FLAVISPINA	ARIZONA
CACTACEAE	OPUNTIA PHAEACANTHA VAR. SUPERBOSPINA	ARIZONA
CACTACEAE	OPUNTIA SPINOSISSIMA	FLORIDA

THREATENED SPECIES IN THE CONTINENTAL UNITED STATES

FAMILY	SPECIES	STATE
CACTACEAE	OPUNTIA TRIACANTHA	FLORIDA
CACTACEAE	OPUNTIA WHIPPLEI VAR. MULTIGENICULATA	ARIZONA, NEVADA, UTAH
CACTACEAE	PEDIOCACTUS PAPYRACANTHUS	NEW MEXICO, ARIZONA
CACTACEAE	PEDIOCACTUS PARADINEI	ARIZONA
CACTACEAE	PEDIOCACTUS PEEBLESIANUS VAR. FICKEISENIAE	ARIZONA
CACTACEAE	SCLEROCACTUS POLYANCISTRUS	CALIFORNIA, NEVADA
CACTACEAE	SCLEROCACTUS PUBISPINUS	NEVADA, UTAH
CACTACEAE	SCLEROCACTUS SPINOSIOR	UTAH, ARIZONA, COLORADO
CACTACEAE	THELOCACTUS BICOLOR VAR. FLAVIDISPINUS	TEXAS
CAMPANULACEAE	CAMPANULA PIPERI	WASHINGTON
CAMPANULACEAE	CAMPANULA REVERCHONII	TEXAS
CAMPANULACEAE	CAMPANULA ROTUNDIFOLIA VAR. SACAJAWEANA	OREGON
CAMPANULACEAE	CAMPANULA SHETLERI	CALIFORNIA
CAMPANULACEAE	CAMPANULA WILKINSIANA	CALIFORNIA
CAMPANULACEAE	LOBELIA GATTINGERI	TENNESSEE
CAMPANULACEAE	NEMACLADUS TWISSELMANNII	CALIFORNIA
CAPPARIDACEAE	CLEOMELLA MONTROSAE	COLORADO
CAPRIFOLIACEAE	SYMPHORICARPOS GUADALUPENSIS	TEXAS
CARYOPHYLLACEAE	ARENARIA FONTINALIS	TENNESSEE, KENTUCKY
CARYOPHYLLACEAE	ARENARIA FRANKLINII VAR. THOMPSONII	OREGON
CARYOPHYLLACEAE	ARENARIA GODFREYI	NORTH CAROLINA, SOUTH CAROLINA, FLORIDA, ALABAMA
CARYOPHYLLACEAE	ARENARIA HOWELLII	CALIFORNIA, OREGON
CARYOPHYLLACEAE	ARENARIA KINGII VAR. ROSEA	NEVADA
CARYOPHYLLACEAE	ARENARIA ROSEI	CALIFORNIA
CARYOPHYLLACEAE	ARENARIA STENOMERES	NEVADA
CARYOPHYLLACEAE	ARENARIA UNIFLORA	ALABAMA, GEORGIA, NORTH CAROLINA, SOUTH CAROLINA
CARYOPHYLLACEAE	CERASTIUM ALEUTICUM	ALASKA
CARYOPHYLLACEAE	PARONYCHIA ARGYROCOMA VAR. ALBIMONTANA	MAINE, MASSACHUSETTS, NEW HAMPSHIRE
CARYOPHYLLACEAE	PARONYCHIA CHORIZANTHOIDES	TEXAS
CARYOPHYLLACEAE	PARONYCHIA DRUMMONDII SSP. PARVIFLORA	TEXAS
CARYOPHYLLACEAE	PARONYCHIA MONTICOLA	TEXAS
CARYOPHYLLACEAE	PARONYCHIA NUDATA	TEXAS
CARYOPHYLLACEAE	PARONYCHIA VIRGINICA VAR. PARKSII	TEXAS
CARYOPHYLLACEAE	PARONYCHIA WILKINSONII	TEXAS
CARYOPHYLLACEAE	SILENE CLOKEYI	NEVADA
CARYOPHYLLACEAE	SILENE PETERSONII VAR. MINOR	UTAH
CARYOPHYLLACEAE	SILENE SCAPOSA VAR. LOBATA	OREGON, IDAHO, NEVADA
CARYOPHYLLACEAE	SILENE SCAPOSA VAR. SCAPOSA	OREGON
CARYOPHYLLACEAE	SILENE SEELYI	WASHINGTON
CHENOPODIACEAE	ATRIPLEX VALLICOLA	CALIFORNIA
CHENOPODIACEAE	ATRIPLEX WELSHII	UTAH
CISTACEAE	HELIANTHEMUM DUMOSUM	MASSACHUSETTS, RHODE ISLAND, NEW

THREATENED SPECIES IN THE CONTINENTAL UNITED STATES

FAMILY	SPECIES	STATE
		YORK, CONNECTICUT
CISTACEAE	HELIANTHEMUM GREENEI	CALIFORNIA
CISTACEAE	HELIANTHEMUM SUFFRUTESCENS	CALIFORNIA
CISTACEAE	LECHEA CERNUA	FLORIDA
CISTACEAE	LECHEA DIVARICATA	FLORIDA
CISTACEAE	LECHEA LAKELAE	FLORIDA
COCHLOSPERMACEAE	AMOREUXIA WRIGHTII	TEXAS
COMMELINACEAE	COMMELINA GIGAS	FLORIDA
COMMELINACEAE	TRADESCANTIA EDWARDSIANA	TEXAS
COMMELINACEAE	TRADESCANTIA WRIGHTII	TEXAS
CONVOLVULACEAE	BONAMIA GRANDIFLORA	FLORIDA
CONVOLVULACEAE	CALYSTEGIA PEIRSONII	CALIFORNIA
CONVOLVULACEAE	DICHONDRA DONNELLIANA	CALIFORNIA
CONVOLVULACEAE	JACQUEMONTIA CURTISSII	FLORIDA
CONVOLVULACEAE	JACQUEMONTIA RECLINATA	FLORIDA
CRASSULACEAE	DUDLEYA BLOCHMANAE SSP. BREVIFOLIA	CALIFORNIA
CRASSULACEAE	DUDLEYA BLOCHMANAE SSP. INSULARIS	CALIFORNIA
CRASSULACEAE	DUDLEYA DENSIFLORA	CALIFORNIA
CRASSULACEAE	DUDLEYA PARVA	CALIFORNIA
CRASSULACEAE	DUDLEYA VARIEGATA	CALIFORNIA
CRASSULACEAE	DUDLEYA VISCIDA	CALIFORNIA
CRASSULACEAE	GRAPTOPETALUM RUSBYI	ARIZONA, NEW MEXICO
CRASSULACEAE	SEDUM ALBOMARGINATUM	CALIFORNIA
CRASSULACEAE	SEDUM LAXUM SSP. EASTWOODIAE	CALIFORNIA
CRASSULACEAE	SEDUM OBLANCEOLATUM	OREGON
CRASSULACEAE	SEDUM PUSILLUM	SOUTH CAROLINA, NORTH CAROLINA, GEORGIA
CROSSOSOMATACEAE	APACHERIA CHIRICAHUENSIS	ARIZONA
CROSSOSOMATACEAE	CROSSOSOMA PARVIFLORUM	ARIZONA
CUCURBITACEAE	CUCURBITA OKEECHOBEENSIS	FLORIDA
CUCURBITACEAE	CUCURBITA TEXANA	TEXAS
CUPRESSACEAE	CUPRESSUS ARIZONICA VAR. STEPHENSONII	CALIFORNIA
CUPRESSACEAE	CUPRESSUS MACROCARPA	CALIFORNIA
CUSCUTACEAE	CUSCUTA HARPERI	ALABAMA, GEORGIA
CYPERACEAE	CAREX AMPLISQUAMA	GEORGIA
CYPERACEAE	CAREX ARAPAHOENSIS	COLORADO, WYOMING, UTAH
CYPERACEAE	CAREX AUSTROCAROLINIANA	GEORGIA, NORTH CAROLINA, SOUTH CAROLINA, TENNESSEE
CYPERACEAE	CAREX BALTZELLII	FLORIDA, ALABAMA
CYPERACEAE	CAREX CHAPMANII	FLORIDA, VIRGINIA, NORTH CAROLINA, SOUTH CAROLINA
CYPERACEAE	CAREX CURATORUM	ARIZONA, UTAH
CYPERACEAE	CAREX FISSA	OKLAHOMA
CYPERACEAE	CAREX INTERRUPTA	OREGON, WASHINGTON

FAMILY	SPECIES	STATE
CYPERACEAE	CAREX JOSSELYNII	MAINE
CYPERACEAE	CAREX MISERA	NORTH CAROLINA, GEORGIA, TENNESSEE
CYPERACEAE	CAREX OBISPOENSIS	CALIFORNIA
CYPERACEAE	CAREX ONUSTA	TEXAS
CYPERACEAE	CAREX ORONENSIS	MAINE
CYPERACEAE	CAREX PARRYANA SSP. IDAHOA	IDAHO, MONTANA
CYPERACEAE	CAREX PAUCIFRUCTA	CALIFORNIA
CYPERACEAE	CAREX PLECTOCARPA	ALASKA, MONTANA
CYPERACEAE	CAREX PURPURIFERA	NORTH CAROLINA, GEORGIA, TENNESSEE, KENTUCKY, ALABAMA
CYPERACEAE	CAREX ROANENSIS	TENNESSEE
CYPERACEAE	CAREX SOCIALIS	ILLINOIS, MISSOURI
CYPERACEAE	CAREX WHITNEYI	CALIFORNIA
CYPERACEAE	CYMOPHYLLUS FRASERI	WEST VIRGINIA, NORTH CAROLINA, PENNSYLVANIA, KENTUCKY, VIRGINIA, TENNESSEE
CYPERACEAE	CYPERUS GRANITOPHILUS	GEORGIA
CYPERACEAE	CYPERUS ONEROSUS	TEXAS
CYPERACEAE	ELEOCHARIS AUSTROTEXANA	TEXAS
CYPERACEAE	RHYNCHOSPORA CULIXA	GEORGIA, FLORIDA
CYPERACEAE	RHYNCHOSPORA GLOBULARIS VAR. SAXICOLA	GEORGIA
CYPERACEAE	RHYNCHOSPORA PUNCTATA	GEORGIA, FLORIDA
CYPERACEAE	SCIRPUS FLACCIDIFOLIUS	VIRGINIA, NORTH CAROLINA
CYPERACEAE	SCIRPUS LONGII	NEW JERSEY, NEW YORK, MAINE, MASSACHUSETTS, CONNECTICUT
DROSERACEAE	DIONAEA MUSCIPULA	NORTH CAROLINA, SOUTH CAROLINA
ERICACEAE	ARCTOSTAPHYLOS CONFERTIFLORA	CALIFORNIA
ERICACEAE	ARCTOSTAPHYLOS CRUZENSIS	CALIFORNIA
ERICACEAE	ARCTOSTAPHYLOS EDMUNDSII VAR. EDMUNDSII	CALIFORNIA
ERICACEAE	ARCTOSTAPHYLOS HOOKERI SSP. MONTANA	CALIFORNIA
ERICACEAE	ARCTOSTAPHYLOS LUCIANA	CALIFORNIA
ERICACEAE	ARCTOSTAPHYLOS MONTARAENSIS	CALIFORNIA
ERICACEAE	ARCTOSTAPHYLOS MONTEREYENSIS	CALIFORNIA
ERICACEAE	ARCTOSTAPHYLOS MORROENSIS	CALIFORNIA
ERICACEAE	ARCTOSTAPHYLOS OTAYENSIS	CALIFORNIA
ERICACEAE	ARCTOSTAPHYLOS PECHOENSIS	CALIFORNIA
ERICACEAE	ARCTOSTAPHYLOS SILVICOLA	CALIFORNIA
ERICACEAE	ARCTOSTAPHYLOS STANFORDIANA SSP. HISPIDULA	CALIFORNIA, OREGON
ERICACEAE	ARCTOSTAPHYLOS VIRGATA	CALIFORNIA
ERICACEAE	MONOTROPA BRITTONII	FLORIDA
ERICACEAE	PITYOPUS CALIFORNICUS	OREGON, CALIFORNIA
ERICACEAE	RHODODENDRON AUSTRINUM	GEORGIA, ALABAMA, FLORIDA, MISSISSIPPI
ERICACEAE	RHODODENDRON BAKERI	KENTUCKY, VIRGINIA, TENNESSEE, GEORGIA, NORTH CAROLINA, ALABAMA
ERICACEAE	RHODODENDRON PRUNIFOLIUM	ALABAMA, GEORGIA

THREATENED SPECIES IN THE CONTINENTAL UNITED STATES

FAMILY	SPECIES	STATE
ERICACEAE	RHODODENDRON VASEYI	NORTH CAROLINA
ERICACEAE	VACCINIUM COCCINIUM	CALIFORNIA, OREGON
ERICACEAE	VACCINIUM VACILLANS VAR. MISSOURIENSE	MISSOURI
ERIOCAULACEAE	LACHNOCAULON BEYRICHIANUM	NORTH CAROLINA, SOUTH CAROLINA, GEORGIA, FLORIDA
EUPHORBIACEAE	ARGYTHAMNIA BLODGETTII	FLORIDA
EUPHORBIACEAE	CHAMAESYCE CUMULICOLA	FLORIDA
EUPHORBIACEAE	CHAMAESYCE DELTOIDEA SSP. DELTOIDEA	FLORIDA
EUPHORBIACEAE	CHAMAESYCE PORTERIANA VAR. PORTERIANA	FLORIDA
EUPHORBIACEAE	DITAXIS CALIFORNICA	CALIFORNIA
EUPHORBIACEAE	EUPHORBIA DISCOIDALIS	FLORIDA
EUPHORBIACEAE	EUPHORBIA EXSERTA	FLORIDA
EUPHORBIACEAE	EUPHORBIA HOOVERI	CALIFORNIA
EUPHORBIACEAE	EUPHORBIA JEJUNA	TEXAS
EUPHORBIACEAE	EUPHORBIA NEPHRADENIA	UTAH
EUPHORBIACEAE	EUPHORBIA PERENNANS	TEXAS
EUPHORBIACEAE	EUPHORBIA PLATYSPERMA	CALIFORNIA, ARIZONA
EUPHORBIACEAE	EUPHORBIA ROEMERIANA	TEXAS
EUPHORBIACEAE	EUPHORBIA STRICTIOR	TEXAS
EUPHORBIACEAE	EUPHORBIA TELEPHIOIDES	FLORIDA
EUPHORBIACEAE	MANIHOT DAVISIAE	ARIZONA
EUPHORBIACEAE	PHYLLANTHUS LIEBMANIANUS SSP. PLATYLEPIS	FLORIDA
EUPHORBIACEAE	PHYLLANTHUS PENTAPHYLLUS SSP. FLORIDANUS	FLORIDA
EUPHORBIACEAE	STILLINGIA SYLVATICA SSP. TENUIS	FLORIDA
EUPHORBIACEAE	TETRACOCCUS ILICIFOLIUS	CALIFORNIA
EUPHORBIACEAE	TITHYMALUS AUSTRINUS	FLORIDA
EUPHORBIACEAE	TRAGIA NIGRICANS	TEXAS
EUPHORBIACEAE	TRAGIA SAXICOLA	FLORIDA
FABACEAE	AMORPHA BRACHYCARPA	MISSOURI
FABACEAE	AMORPHA ROEMERIANA	TEXAS
FABACEAE	ASTRAGALUS ACCUMBENS	NEW MEXICO
FABACEAE	ASTRAGALUS AEQUALIS	NEVADA
FABACEAE	ASTRAGALUS ALTUS	NEW MEXICO
FABACEAE	ASTRAGALUS ALVORDENSIS	OREGON, NEVADA
FABACEAE	ASTRAGALUS AMPULLARIUS	UTAH, ARIZONA
FABACEAE	ASTRAGALUS APPLEGATII	OREGON
FABACEAE	ASTRAGALUS BARNEBYI	ARIZONA, UTAH
FABACEAE	ASTRAGALUS BRAUNTONII	CALIFORNIA
FABACEAE	ASTRAGALUS CALLITHRIX	NEVADA, UTAH
FABACEAE	ASTRAGALUS CASTANEIFORMIS VAR. CONSOBRINUS	UTAH
FABACEAE	ASTRAGALUS CHLOODES	UTAH
FABACEAE	ASTRAGALUS CIMAE VAR. CIMAE	CALIFORNIA
FABACEAE	ASTRAGALUS CONVALLARIUS VAR. FINITIMUS	NEVADA, UTAH

THREATENED SPECIES IN THE CONTINENTAL UNITED STATES

FAMILY	SPECIES	STATE
FABACEAE	ASTRAGALUS COTTAMII	UTAH, ARIZONA
FABACEAE	ASTRAGALUS COTTONII	WASHINGTON
FABACEAE	ASTRAGALUS DEANEI	CALIFORNIA
FABACEAE	ASTRAGALUS DETRITALIS	UTAH, COLORADO
FABACEAE	ASTRAGALUS DRABELLIFORMIS	WYOMING
FABACEAE	ASTRAGALUS DUCHESNENSIS	UTAH
FABACEAE	ASTRAGALUS ENSIFORMIS	ARIZONA, UTAH
FABACEAE	ASTRAGALUS GEYERI VAR. TRIQUETRUS	ARIZONA, NEVADA
FABACEAE	ASTRAGALUS LANCEARIUS	UTAH, ARIZONA
FABACEAE	ASTRAGALUS LENTIGINOSUS VAR. AMBIGUUS	ARIZONA
FABACEAE	ASTRAGALUS LENTIGINOSUS VAR. LATUS	NEVADA
FABACEAE	ASTRAGALUS LIMNOCHARIS	UTAH
FABACEAE	ASTRAGALUS LOANUS	UTAH
FABACEAE	ASTRAGALUS MALACOIDES	UTAH
FABACEAE	ASTRAGALUS MINTHORNIAE VAR. GRACILIOR	UTAH
FABACEAE	ASTRAGALUS MOHAVENSIS VAR. HEMIGYRUS	CALIFORNIA, NEVADA
FABACEAE	ASTRAGALUS MOLLISSIMUS VAR. MARCIDUS	TEXAS
FABACEAE	ASTRAGALUS MONUMENTALIS	UTAH
FABACEAE	ASTRAGALUS MULFORDAE	WASHINGTON, IDAHO, OREGON
FABACEAE	ASTRAGALUS MUSIMONUM	NEVADA
FABACEAE	ASTRAGALUS OOPHORUS VAR. CLOKEYANUS	NEVADA
FABACEAE	ASTRAGALUS OOPHORUS VAR. LONCHOCALYX	NEVADA, UTAH
FABACEAE	ASTRAGALUS PARDALINUS	UTAH
FABACEAE	ASTRAGALUS PAUPERCULUS	CALIFORNIA
FABACEAE	ASTRAGALUS PAYSONII	IDAHO, WYOMING
FABACEAE	ASTRAGALUS PSEUDIODANTHUS	CALIFORNIA, NEVADA
FABACEAE	ASTRAGALUS PTEROCARPUS	NEVADA
FABACEAE	ASTRAGALUS PUNICEUS VAR. GERTRUDIS	NEW MEXICO
FABACEAE	ASTRAGALUS RAFAELENSIS	UTAH
FABACEAE	ASTRAGALUS SABULOSUS	UTAH
FABACEAE	ASTRAGALUS SAURINUS	UTAH
FABACEAE	ASTRAGALUS SOLITARIUS	OREGON
FABACEAE	ASTRAGALUS STOCKSII	UTAH
FABACEAE	ASTRAGALUS SUBVESTITUS	CALIFORNIA
FABACEAE	ASTRAGALUS TENNESSEENSIS	TENNESSEE, ALABAMA, ILLINOIS
FABACEAE	ASTRAGALUS TITANOPHILUS	ARIZONA
FABACEAE	ASTRAGALUS TOQUIMANUS	NEVADA
FABACEAE	ASTRAGALUS TRASKIAE	CALIFORNIA
FABACEAE	ASTRAGALUS TROGLODYTUS	ARIZONA
FABACEAE	ASTRAGALUS VEXILLIFLEXUS VAR. NUBILUS	IDAHO
FABACEAE	ASTRAGALUS WETHERILLII	UTAH, COLORADO
FABACEAE	ASTRAGALUS WOODRUFFII	UTAH
FABACEAE	BAPTISIA CALYCOSA	FLORIDA

THREATENED SPECIES IN THE CONTINENTAL UNITED STATES

FAMILY	SPECIES	STATE
FABACEAE	BAPTISIA HIRSUTA	FLORIDA
FABACEAE	BAPTISIA MEGACARPA	ALABAMA, FLORIDA
FABACEAE	BAPTISIA SIMPLICIFOLIA	FLORIDA
FABACEAE	CAESALPINIA BRACHYCARPA	TEXAS
FABACEAE	CASSIA RIPLEYANA	TEXAS
FABACEAE	CLADRASTIS LUTEA	ALABAMA, GEORGIA, NORTH CAROLINA, TENNESSEE, ARKANSAS, ILLINOIS, INDIANA, KENTUCKY, MISSOURI, OKLAHOMA
FABACEAE	CLITORIA FRAGRANS	FLORIDA
FABACEAE	COURSETIA AXILLARIS	TEXAS
FABACEAE	DALEA BARTONII	TEXAS
FABACEAE	DALEA KINGII	NEVADA
FABACEAE	DALEA THOMPSONAE	ARIZONA, UTAH
FABACEAE	DESMODIUM LINDHEIMERI	TEXAS
FABACEAE	ERRAZURIZIA ROTUNDATA	ARIZONA
FABACEAE	HEDYSARUM BOREALE VAR. GREMIALE	UTAH
FABACEAE	LATHYRUS HOLOCHLORUS	OREGON
FABACEAE	LOTUS ARGOPHYLLUS VAR. NIVEUS	CALIFORNIA
FABACEAE	LOTUS NUTTALIANUS	CALIFORNIA
FABACEAE	LUPINUS ARIDUS SSP. ASHLANDENSIS	OREGON
FABACEAE	LUPINUS BIDDLEI	OREGON
FABACEAE	LUPINUS CERVINUS	CALIFORNIA
FABACEAE	LUPINUS CITRINUS VAR. CITRINUS	CALIFORNIA
FABACEAE	LUPINUS CITRINUS VAR. DEFLEXUS	CALIFORNIA
FABACEAE	LUPINUS CUTLERI	ARIZONA
FABACEAE	LUPINUS DEDECKERAE	CALIFORNIA
FABACEAE	LUPINUS DURANII	CALIFORNIA
FABACEAE	LUPINUS HOLMGRENANUS	CALIFORNIA, NEVADA
FABACEAE	LUPINUS JONESII	UTAH
FABACEAE	LUPINUS MARIANUS	UTAH
FABACEAE	LUPINUS MUCRONULATUS	OREGON
FABACEAE	LUPINUS NIPOMENSIS	CALIFORNIA
FABACEAE	LUPINUS SERICATUS	CALIFORNIA
FABACEAE	LUPINUS SPECTABILIS	CALIFORNIA
FABACEAE	LUPINUS WESTIANUS	FLORIDA
FABACEAE	OXYTROPIS CAMPESTRIS VAR. CHARTACEA	WISCONSIN
FABACEAE	OXYTROPIS JONESII	UTAH
FABACEAE	OXYTROPIS KOKRINENSIS	ALASKA
FABACEAE	OXYTROPIS OBNAPIFORMIS	UTAH, COLORADO, WYOMING
FABACEAE	PETALOSTEMUM GATTINGERI	TENNESSEE, GEORGIA, ALABAMA
FABACEAE	PETALOSTEMUM SCARIOSUM	NEW MEXICO
FABACEAE	PSORALEA PARIENSIS	UTAH
FABACEAE	PSORALEA SUBACAULIS	TENNESSEE, ALABAMA
FABACEAE	RHYNCHOSIA CINEREA	FLORIDA

THREATENED SPECIES IN THE CONTINENTAL UNITED STATES

FAMILY	SPECIES	STATE
FABACEAE	SOPHORA ARIZONICA	ARIZONA
FABACEAE	SOPHORA GYPSOPHILA VAR. GUADALUPENSIS	TEXAS
FABACEAE	TEPHROSIA MOHRII	FLORIDA
FABACEAE	THERMOPSIS MACROPHYLLA VAR. AGNINA	CALIFORNIA
FABACEAE	TRIFOLIUM BOLANDERI	CALIFORNIA
FABACEAE	TRIFOLIUM DEDECKERAE	CALIFORNIA
FABACEAE	TRIFOLIUM OWYHEENSE	OREGON
FABACEAE	TRIFOLIUM PLUMOSUM VAR. AMPLIFOLIUM	IDAHO
FAGACEAE	QUERCUS GEORGIANA	SOUTH CAROLINA, GEORGIA, ALABAMA
FAGACEAE	QUERCUS OGLETHORPENSIS	SOUTH CAROLINA, GEORGIA
FAGACEAE	QUERCUS PARVULA	CALIFORNIA
FAGACEAE	QUERCUS SHUMARDII VAR. ACERIFOLIA	ARKANSAS
FAGACEAE	QUERCUS TOMENTELLA	CALIFORNIA
FUMARIACEAE	CORYDALIS CASEANA SSP. BRACHYCARPA	UTAH
FUMARIACEAE	CORYDALIS CASEANA SSP. HASTATA	IDAHO
FUMARIACEAE	DICENTRA NEVADENSIS	CALIFORNIA
GENTIANACEAE	FRASERA COLORADENSIS	COLORADO, OKLAHOMA
GENTIANACEAE	FRASERA IDAHOENSIS	IDAHO
GENTIANACEAE	FRASERA UMPQUAENSIS	OREGON, CALIFORNIA
GENTIANACEAE	GENTIANA ALEUTICA	ALASKA
GENTIANACEAE	GENTIANA FREMONTII	CALIFORNIA
GERANIACEAE	GERANIUM MARGINALE	UTAH
HALORAGACEAE	MYRIOPHYLLUM LAXUM	NORTH CAROLINA, SOUTH CAROLINA, GEORGIA, FLORIDA
HAMAMELIDACEAE	FOTHERGILLA GARDENI	GEORGIA, ALABAMA
HYDROPHYLLACEAE	NAMA RETRORSUM	ARIZONA, UTAH
HYDROPHYLLACEAE	NAMA XYLOPODUM	TEXAS
HYDROPHYLLACEAE	PHACELIA AMABILIS	CALIFORNIA
HYDROPHYLLACEAE	PHACELIA CEPHALOTES	UTAH, ARIZONA
HYDROPHYLLACEAE	PHACELIA CONSTANCEI	UTAH, ARIZONA
HYDROPHYLLACEAE	PHACELIA DALESIANA	CALIFORNIA
HYDROPHYLLACEAE	PHACELIA DEMISSA VAR. HETEROTRICHA	UTAH
HYDROPHYLLACEAE	PHACELIA DIVARICATA VAR. INSULARIS	CALIFORNIA
HYDROPHYLLACEAE	PHACELIA DUBIA VAR. GEORGIANA	GEORGIA, ALABAMA
HYDROPHYLLACEAE	PHACELIA GLABERRIMA	NEVADA
HYDROPHYLLACEAE	PHACELIA GREENEI	CALIFORNIA
HYDROPHYLLACEAE	PHACELIA MUSTELINA	CALIFORNIA, NEVADA
HYDROPHYLLACEAE	PHACELIA NOVENMILLENSIS	CALIFORNIA
HYDROPHYLLACEAE	PHACELIA OROGENES	CALIFORNIA
HYDROPHYLLACEAE	PHACELIA PECKII	OREGON
HYDROPHYLLACEAE	PHACELIA PHACELIOIDES	CALIFORNIA
HYDROPHYLLACEAE	PHACELIA RAFAELENSIS	UTAH, ARIZONA
HYDROPHYLLACEAE	PHACELIA SERRATA	ARIZONA

THREATENED SPECIES IN THE CONTINENTAL UNITED STATES

FAMILY	SPECIES	STATE
HYDROPHYLLACEAE	PHACELIA UTAHENSIS	UTAH
HYDROPHYLLACEAE	PHACELIA VERNA	OREGON
HYPERICACEAE	HYPERICUM EDISONIANUM	FLORIDA
HYPERICACEAE	HYPERICUM SPHAEROCARPUM VAR. TURGIDUM	ALABAMA, KENTUCKY, TENNESSEE
ILLICIACEAE	ILLICIUM PARVIFLORUM	FLORIDA
IRIDACEAE	IRIS LACUSTRIS	WISCONSIN, MICHIGAN
IRIDACEAE	IRIS TENAX VAR. GORMANII	OREGON
IRIDACEAE	NEMASTYLIS FLORIDANA	FLORIDA
IRIDACEAE	SPHENOSTIGMA COELESTINA	FLORIDA
ISOETACEAE	ISOETES EATONII	NEW HAMPSHIRE, NEW JERSEY, MASSACHUSETTS, CONNECTICUT
ISOETACEAE	ISOETES FOVEOLATA	NEW HAMPSHIRE
ISOETACEAE	ISOETES MELANOSPORA	GEORGIA, SOUTH CAROLINA
ISOETACEAE	ISOETES VIRGINICA	VIRGINIA, GEORGIA
JUNCACEAE	JUNCUS CAESARIENSIS	NEW JERSEY, MARYLAND, VIRGINIA
JUNCACEAE	JUNCUS GYMNOCARPUS	TENNESSEE, NORTH CAROLINA, FLORIDA, PENNSYLVANIA, SOUTH CAROLINA, MISSISSIPPI, ALABAMA
JUNCACEAE	JUNCUS SLWOOKOORUM	ALASKA
LAMIACEAE	AGASTACHE CUSICKII	OREGON
LAMIACEAE	AGASTACHE PARVIFOLIA	CALIFORNIA
LAMIACEAE	CALAMINTHA ASHEI	FLORIDA
LAMIACEAE	CALAMINTHA DENTATUM	FLORIDA, GEORGIA
LAMIACEAE	CONRADINA GRANDIFLORA	FLORIDA
LAMIACEAE	DICERANDRA ODORATISSIMA	SOUTH CAROLINA, GEORGIA, FLORIDA
LAMIACEAE	HEDEOMA APICULATUM	TEXAS
LAMIACEAE	MONARDELLA BENITENSIS	CALIFORNIA
LAMIACEAE	MONARDELLA CRISPA	CALIFORNIA
LAMIACEAE	MONARDELLA HYPOLEUCA SSP. LANATA	CALIFORNIA
LAMIACEAE	MONARDELLA LINOIDES SSP. OBLONGA	CALIFORNIA
LAMIACEAE	MONARDELLA ROBISONII	CALIFORNIA
LAMIACEAE	PHYSOSTEGIA LEPTOPHYLLUM	FLORIDA
LAMIACEAE	PHYSOSTEGIA VERONICIFORMIS	GEORGIA
LAMIACEAE	PYCNANTHEMUM FLORIDANUM	FLORIDA
LAMIACEAE	PYCNANTHEMUM MONOTRICHUM	VIRGINIA
LAMIACEAE	SALVIA PENSTEMONOIDES	TEXAS
LAMIACEAE	SATUREJA CHANDLERI	CALIFORNIA
LAMIACEAE	SCUTELLARIA FLORIDANA	FLORIDA
LAMIACEAE	SCUTELLARIA MONTANA	GEORGIA, TENNESSEE
LAMIACEAE	SCUTELLARIA OVATA SSP. PSEUDOARGUTA	WEST VIRGINIA
LAMIACEAE	SCUTELLARIA THIERETII	LOUISIANA
LAMIACEAE	STACHYS LYTHROIDES	FLORIDA
LAMIACEAE	SYNANDRA HISPIDULA	VIRGINIA, WEST VIRGINIA, ILLINOIS, TENNESSEE, NORTH CAROLINA, KENTUCKY, ALABAMA

THREATENED SPECIES IN THE CONTINENTAL UNITED STATES

FAMILY	SPECIES	STATE
LAURACEAE	LITSEA AESTIVALIS	NORTH CAROLINA, SOUTH CAROLINA, GEORGIA, FLORIDA, MISSISSIPPI
LAURACEAE	PERSEA BORBONIA VAR. HUMILIS	FLORIDA
LEITNERIACEAE	LEITNERIA FLORIDANA	ARKANSAS, FLORIDA, GEORGIA, MISSOURI, TEXAS
LENNOACEAE	PHOLISMA ARENARIUM	CALIFORNIA
LENTIBULARIACEAE	PINGUICULA IONANTHA	FLORIDA
LILIACEAE	AGAVE CHISOENSIS	TEXAS
LILIACEAE	AGAVE UTAHENSIS VAR. EBORISPINA	NEVADA
LILIACEAE	AGAVE UTAHENSIS VAR. KAIBABENSIS	ARIZONA
LILIACEAE	AGAVE UTAHENSIS VAR. NEVADENSIS	CALIFORNIA
LILIACEAE	ALLIUM GOODDINGII	ARIZONA, NEW MEXICO
LILIACEAE	ALLIUM HOFFMANII	CALIFORNIA
LILIACEAE	ALLIUM PLEIANTHUM	OREGON
LILIACEAE	ALLIUM ROBINSONII	OREGON, WASHINGTON
LILIACEAE	ALLIUM TOLMIEI VAR. PERSIMILE	IDAHO
LILIACEAE	ALLIUM YOSEMITENSE	CALIFORNIA
LILIACEAE	ANTHERICUM CHANDLERI	TEXAS
LILIACEAE	CALOCHORTUS DUNNII	CALIFORNIA
LILIACEAE	CALOCHORTUS LONGEBARBATUS VAR. LONGEBARBATUS	CALIFORNIA, OREGON, WASHINGTON
LILIACEAE	CALOCHORTUS NITIDUS	WASHINGTON, IDAHO, OREGON
LILIACEAE	CALOCHORTUS OBISPOENSIS	CALIFORNIA
LILIACEAE	CALOCHORTUS STRIATUS	CALIFORNIA, NEVADA
LILIACEAE	CAMASSIA CUSICKII	OREGON
LILIACEAE	DICHELOSTEMMA LACUNA-VERNALIS	CALIFORNIA
LILIACEAE	ERYTHRONIUM OREGONUM	OREGON, WASHINGTON
LILIACEAE	ERYTHRONIUM TUOLUMNENSE	CALIFORNIA
LILIACEAE	FRITILLARIA BRANDEGEI	CALIFORNIA
LILIACEAE	FRITILLARIA FALCATA	CALIFORNIA
LILIACEAE	FRITILLARIA GENTNERI	OREGON
LILIACEAE	FRITILLARIA PLURIFLORA	CALIFORNIA
LILIACEAE	FRITILLARIA STRIATA	CALIFORNIA
LILIACEAE	HYMENOCALLIS LATIFOLIA	FLORIDA
LILIACEAE	LILIUM GRAYII	NORTH CAROLINA, TENNESSEE, VIRGINIA, MARYLAND
LILIACEAE	LILIUM VOLLMERI	CALIFORNIA, OREGON
LILIACEAE	LILIUM WASHINGTONIANUM VAR. MINUS	CALIFORNIA
LILIACEAE	LILIUM WIGGINSII	OREGON, CALIFORNIA
LILIACEAE	MUILLA CLEVELANDII	CALIFORNIA
LILIACEAE	MUILLA CORONATA	CALIFORNIA
LILIACEAE	NOLINA ARENICOLA	TEXAS
LILIACEAE	POLIANTHES MACULOSA	TEXAS
LILIACEAE	SCHOENOLIRION BRACTEOSUM	OREGON, CALIFORNIA
LILIACEAE	TRILLIUM PUSILLUM VAR. OZARKANUM	MISSOURI, ARKANSAS

THREATENED SPECIES IN THE CONTINENTAL UNITED STATES

FAMILY	SPECIES	STATE
LILIACEAE	TRILLIUM PUSILLUM VAR. PUSILLUM	ALABAMA, NORTH CAROLINA, SOUTH CAROLINA, TENNESSEE
LILIACEAE	TRILLIUM TEXANUM	TEXAS
LILIACEAE	TRITELEIA CLEMENTINA	CALIFORNIA
LILIACEAE	TRITELEIA LEMMONAE	ARIZONA
LILIACEAE	VERATRUM WOODII	ARKANSAS, FLORIDA, ILLINOIS, INDIANA, IOWA, MISSOURI, NORTH CAROLINA, OKLAHOMA, TEXAS
LILIACEAE	YUCCA TOFTIAE	UTAH
LILIACEAE	ZEPHYRANTHES SIMPSONII	FLORIDA
LILIACEAE	ZEPHYRANTHES TREATIAE	FLORIDA
LIMNANTHACEAE	LIMNANTHES DOUGLASII VAR. SULPHUREA	CALIFORNIA
LIMNANTHACEAE	LIMNANTHES FLOCCOSA SSP. BELLINGERIANA	OREGON, CALIFORNIA
LIMNANTHACEAE	LIMNANTHES GRACILIS VAR. GRACILIS	OREGON
LINACEAE	HESPEROLINON ADENOPHYLLUM	CALIFORNIA
LINACEAE	HESPEROLINON BICARPELLATUM	CALIFORNIA
LINACEAE	HESPEROLINON BREWERI	CALIFORNIA
LINACEAE	HESPEROLINON DRYMARIOIDES	CALIFORNIA
LINACEAE	LINUM SULCATUM VAR. HARPERI	ALABAMA, GEORGIA, FLORIDA
LOASACEAE	MENTZELIA ARGILLOSA	UTAH
LOASACEAE	MENTZELIA HIRSUTISSIMA VAR. STENOPHYLLA	CALIFORNIA
LOASACEAE	MENTZELIA MOLLIS	OREGON
LOASACEAE	PETALONYX THURBERI SSP. GILMANII	CALIFORNIA
LORANTHACEAE	ARCEUTHOBIUM APACHENSE	ARIZONA
LYTHRACEAE	HEIMIA LONGIPES	TEXAS
LYTHRACEAE	LYTHRUM OVALIFOLIUM	TEXAS
MAGNOLIACEAE	MAGNOLIA ASHEI	FLORIDA
MALVACEAE	ABUTILON MARSHII	TEXAS
MALVACEAE	CALLIRHOE PAPAVER VAR. BUSHII	MISSOURI, ARKANSAS, OKLAHOMA
MALVACEAE	KOSTELETZKYA SMILACIFOLIA	FLORIDA, ALABAMA
MALVACEAE	MALACOTHAMNUS FASCICULATUS VAR. NESIOTICUS	CALIFORNIA
MALVACEAE	MALACOTHAMNUS PALMERI VAR. LUCIANUS	CALIFORNIA
MALVACEAE	MALACOTHAMNUS PALMERI VAR. PALMERI	CALIFORNIA
MALVACEAE	MALACOTHAMNUS PARISHII	CALIFORNIA
MALVACEAE	SIDA RUBROMARGINATA	FLORIDA
MALVACEAE	SIDALCEA CUSICKII	OREGON
MALVACEAE	SIDALCEA HICKMANII SSP. ANOMALA	CALIFORNIA
MALVACEAE	SIDALCEA HICKMANII SSP. HICKMANII	CALIFORNIA
MALVACEAE	SIDALCEA HICKMANII SSP. PARISHII	CALIFORNIA
MALVACEAE	SIDALCEA HICKMANII SSP. VIRIDIS	CALIFORNIA
MALVACEAE	SIDALCEA KECKII	CALIFORNIA
MALVACEAE	SIDALCEA OREGANA VAR. CALVA	WASHINGTON
MALVACEAE	SIDALCEA OREGANA SSP. HYDROPHILA	CALIFORNIA
MALVACEAE	SIDALCEA PEDATA	CALIFORNIA

THREATENED SPECIES IN THE CONTINENTAL UNITED STATES

FAMILY	SPECIES	STATE
MALVACEAE	SIDALCEA ROBUSTA	CALIFORNIA
MALVACEAE	SPHAERALCEA CAESPITOSA	UTAH
MALVACEAE	SPHAERALCEA RUSBYI SSP. EREMICOLA	CALIFORNIA
MELASTOMATACEAE	RHEXIA SALICIFOLIA	FLORIDA, ALABAMA, TEXAS
MYRICACEAE	MYRICA HARTWEGII	CALIFORNIA
MYRTACEAE	MYRCIANTHES FRAGRANS VAR. SIMPSONII	FLORIDA
NYCTAGINACEAE	ABRONIA ORBICULATA	NEVADA
NYCTAGINACEAE	ACLEISANTHES CRASSIFOLIA	TEXAS
NYCTAGINACEAE	BOERHAAVIA MATHISIANA	TEXAS
NYCTAGINACEAE	MIRABILIS PUDICA	NEVADA
NYMPHAEACEAE	NUPHAR LUTEUM SSP. ULVACEUM	FLORIDA
OLEACEAE	CHIONANTHUS PYGMAEUS	FLORIDA
OLEACEAE	FRAXINUS ANOMALA VAR. LOWELLII	ARIZONA
OLEACEAE	FRAXINUS CUSPIDATA VAR. MACROPETALA	ARIZONA, NEVADA
ONAGRACEAE	CAMISSONIA BENITENSIS	CALIFORNIA
ONAGRACEAE	CAMISSONIA CONFERTIFLORA	ARIZONA
ONAGRACEAE	CAMISSONIA EXILIS	ARIZONA
ONAGRACEAE	CAMISSONIA GOULDII	ARIZONA, UTAH
ONAGRACEAE	CAMISSONIA SPECUICOLA SSP. HESPERIA	ARIZONA
ONAGRACEAE	CLARKIA AMOENA VAR. PACIFICA	OREGON
ONAGRACEAE	CLARKIA AUSTRALIS	CALIFORNIA
ONAGRACEAE	CLARKIA BILOBA SSP. AUSTRALIS	CALIFORNIA
ONAGRACEAE	CLARKIA ROSTRATA	CALIFORNIA
ONAGRACEAE	EPILOBIUM NEVADENSE	NEVADA, UTAH
ONAGRACEAE	EPILOBIUM NIVIUM	CALIFORNIA
ONAGRACEAE	EPILOBIUM OREGANUM	OREGON
ONAGRACEAE	GAURA DEMAREEI	ARKANSAS
ONAGRACEAE	OENOTHERA ORGANENSIS	NEW MEXICO
OPHIOGLOSSACEAE	BOTRYCHIUM PUMICOLA	OREGON, CALIFORNIA
OPHIOGLOSSACEAE	OPHIOGLOSSUM CALIFORNICUM	CALIFORNIA
OPHIOGLOSSACEAE	OPHIOGLOSSUM PALMATUM	FLORIDA
ORCHIDACEAE	CYPRIPEDIUM ARIETINUM	NEW YORK, MICHIGAN, MASSACHUSETTS, VERMONT, WISCONSIN, MINNESOTA, NEW HAMPSHIRE, MAINE, CONNECTICUT
ORCHIDACEAE	CYPRIPEDIUM CALIFORNICUM	CALIFORNIA, OREGON
ORCHIDACEAE	CYPRIPEDIUM CANDIDUM	NEW YORK, NORTH DAKOTA, NEW JERSEY, PENNSYLVANIA, KENTUCKY, MISSOURI, ILLINOIS, INDIANA, IOWA, MICHIGAN, MINNESOTA, NEBRASKA, OHIO, SOUTH DAKOTA, WISCONSIN
ORCHIDACEAE	ENCYCLIA BOOTHIANA VAR. ERYTHRONIOIDES	FLORIDA
ORCHIDACEAE	HEXALECTRIS GRANDIFLORA	TEXAS
ORCHIDACEAE	HEXALECTRIS NITIDA	TEXAS
ORCHIDACEAE	HEXALECTRIS REVOLUTA	TEXAS
ORCHIDACEAE	LEPANTHOPSIS MELANANTHA	FLORIDA
ORCHIDACEAE	LISTERA AURICULATA	NEW HAMPSHIRE, NEW YORK, MICHIGAN,

THREATENED SPECIES IN THE CONTINENTAL UNITED STATES

FAMILY	SPECIES	STATE
		WISCONSIN, MINNESOTA, MAINE, VERMONT
ORCHIDACEAE	PLATANTHERA FLAVA	FLORIDA, MARYLAND, TEXAS, TENNESSEE, IOWA, MAINE, MASSACHUSETTS, KENTUCKY, ARKANSAS, MISSOURI, ILLINOIS, INDIANA, WISCONSIN, MICHIGAN, MINNESOTA, NEW HAMPSHIRE, NEW JERSEY, NEW YORK, NORTH CAROLINA, OHIO, RHODE ISLAND, SOUTH CAROLINA, VERMONT, VIRGINIA, WEST VIRGINIA, GEORGIA, PENNSYLVANIA, DELAWARE
ORCHIDACEAE	PLATANTHERA INTEGRA	FLORIDA, NORTH CAROLINA, TEXAS, NEW JERSEY, SOUTH CAROLINA, TENNESSEE, ALABAMA, GEORGIA, LOUISIANA, MISSISSIPPI
ORCHIDACEAE	PLATANTHERA LEUCOPHAEA	NEW YORK, OHIO, INDIANA, ILLINOIS, LOUISIANA, KANSAS, MAINE, MICHIGAN, MISSOURI, ARKANSAS, NEBRASKA, SOUTH DAKOTA, NORTH DAKOTA, MINNESOTA, WISCONSIN, IOWA
ORCHIDACEAE	PLATANTHERA PERAMOENA	NEW JERSEY, ILLINOIS, MISSOURI, DELAWARE, MARYLAND, GEORGIA, SOUTH CAROLINA, ALABAMA, MISSISSIPPI, TENNESSEE, WEST VIRGINIA, KENTUCKY, ARKANSAS, INDIANA, OHIO, PENNSYLVANIA, NEW YORK, VIRGINIA, NORTH CAROLINA
ORCHIDACEAE	PLATANTHERA UNALASCENSIS SSP. MARITIMA	CALIFORNIA, OREGON, WASHINGTON
ORCHIDACEAE	SPIRANTHES POLYANTHA	FLORIDA
OROBANCHACEAE	OROBANCHE PARISHII SSP. BRACHYLOBA	CALIFORNIA
PAPAVERACEAE	ARGEMONE ARIZONICA	ARIZONA
PAPAVERACEAE	ARGEMONE MUNITA SSP. ROBUSTA	CALIFORNIA
PAPAVERACEAE	ESCHSCHOLZIA PROCERA	CALIFORNIA
PAPAVERACEAE	ESCHSCHOLZIA RAMOSA	CALIFORNIA
PAPAVERACEAE	PAPAVER WALPOLEI	ALASKA
PEDALIACEAE	PROBOSCIDEA SABULOSA	TEXAS
PINACEAE	PINUS TORREYANA	CALIFORNIA
POACEAE	AGROSTIS ARISTIGLUMIS	CALIFORNIA
POACEAE	AGROSTIS CLIVICOLA VAR. PUNTA-REYESENSIS	CALIFORNIA
POACEAE	AGROSTIS HOWELLII	OREGON
POACEAE	AGROSTIS ROSSIAE	WYOMING
POACEAE	ALOPECURUS AEQUALIS VAR. SONOMENSIS	CALIFORNIA
POACEAE	ARISTIDA SIMPLICIFLORA	FLORIDA, MISSISSIPPI, ALABAMA
POACEAE	BROMUS TEXENSIS	TEXAS
POACEAE	CALAMAGROSTIS CAINII	TENNESSEE
POACEAE	CALAMAGROSTIS CRASSIGLUMIS	ALASKA, CALIFORNIA, WASHINGTON
POACEAE	CALAMAGROSTIS FOLIOSA	CALIFORNIA
POACEAE	CALAMAGROSTIS PORTERI	NEW YORK, VIRGINIA, WEST VIRGINIA, NORTH CAROLINA, PENNSYLVANIA
POACEAE	CALAMAGROSTIS TWEEDYI	IDAHO, WASHINGTON, MONTANA
POACEAE	CALAMOVILFA BREVIPILIS VAR. BREVIPILIS	NORTH CAROLINA, SOUTH CAROLINA, NEW JERSEY, VIRGINIA
POACEAE	CHLORIS TEXENSIS	TEXAS
POACEAE	CTENIUM FLORIDANUM	FLORIDA, GEORGIA
POACEAE	ERAGROSTIS TRACYI	FLORIDA

THREATENED SPECIES IN THE CONTINENTAL UNITED STATES

FAMILY	SPECIES	STATE
POACEAE	ERIOCHLOA MICHAUXII VAR. SIMPSONII	FLORIDA
POACEAE	FESTUCA LIGULATA	TEXAS
POACEAE	GYMNOPOGON FLORIDANUS	FLORIDA
POACEAE	HYSTRIX CALIFORNICA	CALIFORNIA
POACEAE	MANISURIS TUBERCULOSA	ALABAMA, FLORIDA
POACEAE	MUHLENBERGIA TORREYANA	NEW JERSEY, DELAWARE, TENNESSEE, GEORGIA, KENTUCKY
POACEAE	PANICUM ACULEATUM	DISTRICT OF COLUMBIA, RHODE ISLAND, NEW YORK, VIRGINIA, NORTH CAROLINA
POACEAE	PANICUM LITHOPHILUM	SOUTH CAROLINA, GEORGIA
POACEAE	PANICUM NUDICAULE	FLORIDA, ALABAMA, MISSISSIPPI
POACEAE	PANICUM PINETORUM	FLORIDA
POACEAE	POA CURTIFOLIA	WASHINGTON
POACEAE	POA PALUDIGENA	NEW YORK, MICHIGAN, WISCONSIN, PENNSYLVANIA, INDIANA, OHIO
POACEAE	PTILAGROSTIS PORTERI	COLORADO
POACEAE	PUCCINELLIA PARISHII	CALIFORNIA, ARIZONA, NEW MEXICO
POACEAE	SCHIZACHYRIUM NIVEUM	FLORIDA, GEORGIA
POACEAE	SPOROBOLUS NEGLECTUS VAR. OZARKANUS	MISSOURI
POACEAE	SPOROBOLUS TERETIFOLIUS	NORTH CAROLINA, SOUTH CAROLINA, GEORGIA
POACEAE	WILLKOMMIA TEXANA	TEXAS
POLEMONIACEAE	COLLOMIA MAZAMA	OREGON
POLEMONIACEAE	ERIASTRUM BRANDEGEAE	CALIFORNIA
POLEMONIACEAE	GILIA MCVICKERAE	UTAH
POLEMONIACEAE	GILIA NYENSIS	NEVADA
POLEMONIACEAE	GILIA RIPLEYI	CALIFORNIA, NEVADA
POLEMONIACEAE	GILIA TENUIFLORA SSP. HOFFMANNII	CALIFORNIA
POLEMONIACEAE	IPOMOPSIS GLOBULARIS	COLORADO
POLEMONIACEAE	LINANTHUS BELLUS	CALIFORNIA
POLEMONIACEAE	LINANTHUS MACULATUS	CALIFORNIA
POLEMONIACEAE	LINANTHUS ORCUTTII SSP. PACIFICUS	CALIFORNIA
POLEMONIACEAE	NAVARRETIA PROLIFERA SSP. LUTEA	CALIFORNIA
POLEMONIACEAE	PHLOX CARYOPHYLLA	COLORADO, NEW MEXICO
POLEMONIACEAE	PHLOX CLUTEANA	UTAH, ARIZONA
POLEMONIACEAE	PHLOX GLADIFORMIS	UTAH, NEVADA
POLEMONIACEAE	PHLOX GRAHAMII	UTAH
POLEMONIACEAE	PHLOX JONESII	UTAH
POLEMONIACEAE	PHLOX MOLLIS	IDAHO
POLEMONIACEAE	PHLOX OKLAHOMENSIS	KANSAS, OKLAHOMA, TEXAS
POLEMONIACEAE	POLEMONIUM NEVADENSE	NEVADA
POLEMONIACEAE	POLEMONIUM PECTINATUM	WASHINGTON
POLYGALACEAE	POLYGALA BOYKINII VAR. SPARSIFOLIA	FLORIDA
POLYGALACEAE	POLYGALA PILIOPHORA	ARIZONA
POLYGONACEAE	CHORIZANTHE BREWERI	CALIFORNIA

THREATENED SPECIES IN THE CONTINENTAL UNITED STATES

FAMILY	SPECIES	STATE
POLYGONACEAE	CHORIZANTHE HOWELLII	CALIFORNIA
POLYGONACEAE	CHORIZANTHE INSIGNIS	CALIFORNIA
POLYGONACEAE	CHORIZANTHE RECTISPINA	CALIFORNIA
POLYGONACEAE	CHORIZANTHE STATICOIDES SSP. CHRYSACANTHA	CALIFORNIA
POLYGONACEAE	ERIOGONUM AMPULLACEUM	CALIFORNIA
POLYGONACEAE	ERIOGONUM APACHENSE	ARIZONA
POLYGONACEAE	ERIOGONUM BEATLEYAE	CALIFORNIA, NEVADA
POLYGONACEAE	ERIOGONUM BIFURCATUM	CALIFORNIA, NEVADA
POLYGONACEAE	ERIOGONUM BRANDEGEI	COLORADO
POLYGONACEAE	ERIOGONUM CLAVELLATUM	UTAH, COLORADO
POLYGONACEAE	ERIOGONUM CONCINNUM	NEVADA
POLYGONACEAE	ERIOGONUM CONGDONII	CALIFORNIA
POLYGONACEAE	ERIOGONUM CONTIGUUM	CALIFORNIA, NEVADA
POLYGONACEAE	ERIOGONUM CORRELLII	TEXAS
POLYGONACEAE	ERIOGONUM CUSICKII	OREGON
POLYGONACEAE	ERIOGONUM DENSUM	NEW MEXICO
POLYGONACEAE	ERIOGONUM DESERTICOLA	CALIFORNIA
POLYGONACEAE	ERIOGONUM EREMICOLA	CALIFORNIA
POLYGONACEAE	ERIOGONUM EREMICUM	UTAH
POLYGONACEAE	ERIOGONUM GOSSYPINUM	CALIFORNIA
POLYGONACEAE	ERIOGONUM HEERMANNII VAR. FLOCCOSUM	CALIFORNIA, NEVADA
POLYGONACEAE	ERIOGONUM HEERMANNII VAR. SUBRACEMOSUM	ARIZONA
POLYGONACEAE	ERIOGONUM HOFFMANNII VAR. HOFFMANNII	CALIFORNIA
POLYGONACEAE	ERIOGONUM HOFFMANNII VAR. ROBUSTIUS	CALIFORNIA
POLYGONACEAE	ERIOGONUM HOLMGRENII	NEVADA
POLYGONACEAE	ERIOGONUM JAMESII VAR. RUPICOLA	UTAH
POLYGONACEAE	ERIOGONUM KELLOGGII	CALIFORNIA
POLYGONACEAE	ERIOGONUM KENNEDYI SSP. AUSTROMONTANUM	CALIFORNIA
POLYGONACEAE	ERIOGONUM LAGOPUS	MONTANA, WYOMING
POLYGONACEAE	ERIOGONUM LATENS	CALIFORNIA
POLYGONACEAE	ERIOGONUM MICROTHECUM VAR. PANAMINTENSE	CALIFORNIA
POLYGONACEAE	ERIOGONUM NANUM	UTAH
POLYGONACEAE	ERIOGONUM NATUM	UTAH
POLYGONACEAE	ERIOGONUM NORTONII	CALIFORNIA
POLYGONACEAE	ERIOGONUM NOVONUDUM	OREGON
POLYGONACEAE	ERIOGONUM OSTLUNDII	UTAH
POLYGONACEAE	ERIOGONUM OVALIFOLIUM VAR. CAELESTRINUM	NEVADA
POLYGONACEAE	ERIOGONUM PANGUICENSE VAR. ALPESTRE	UTAH
POLYGONACEAE	ERIOGONUM PARVIFOLIUM VAR. PAYNEI	CALIFORNIA
POLYGONACEAE	ERIOGONUM PENDULUM	CALIFORNIA, OREGON
POLYGONACEAE	ERIOGONUM RIPLEYI	ARIZONA
POLYGONACEAE	ERIOGONUM RUBRICAULE	NEVADA
POLYGONACEAE	ERIOGONUM SAURINUM	UTAH, COLORADO

THREATENED SPECIES IN THE CONTINENTAL UNITED STATES

FAMILY	SPECIES	STATE
POLYGONACEAE	ERIOGONUM SCOPULORUM	OREGON
POLYGONACEAE	ERIOGONUM SISKIYOUENSE	CALIFORNIA
POLYGONACEAE	ERIOGONUM TEMBLORENSE	CALIFORNIA
POLYGONACEAE	ERIOGONUM THOMPSONAE VAR. ALBIFLORUM	UTAH
POLYGONACEAE	ERIOGONUM THOMPSONAE VAR. THOMPSONAE	UTAH, ARIZONA
POLYGONACEAE	ERIOGONUM TWISSELMANNII	CALIFORNIA
POLYGONACEAE	ERIOGONUM UMBELLATUM VAR. HYPOLEIUM	WASHINGTON
POLYGONACEAE	ERIOGONUM VESTITUM	CALIFORNIA
POLYGONACEAE	ERIOGONUM VIRIDULUM	UTAH, COLORADO
POLYGONACEAE	GILMANIA LUTEOLA	CALIFORNIA
POLYGONACEAE	OXYTHECA WATSONII	NEVADA
POLYGONACEAE	PERSICARIA PALUDICOLA	FLORIDA
POLYGONACEAE	POLYGONELLA MACROPHYLLA	FLORIDA
POLYGONACEAE	POLYGONUM BIDWELLIAE	CALIFORNIA
POLYGONACEAE	POLYGONUM FUSIFORME	ARIZONA, CALIFORNIA
POLYGONACEAE	POLYGONUM MARINENSE	CALIFORNIA
POLYGONACEAE	POLYGONUM STRIATULUM	TEXAS
POLYPODIACEAE	CHEILANTHES PRINGLEI	ARIZONA, NEW MEXICO
POLYPODIACEAE	CHEILANTHES PYRAMIDALIS VAR. ARIZONICA	ARIZONA
POLYPODIACEAE	NOTHOLAENA LEMMONII	ARIZONA, NEW MEXICO
POLYPODIACEAE	NOTHOLAENA SCHAFFNERI VAR. NEALLEYI	TEXAS
POLYPODIACEAE	POLYSTICHUM KRUCKEBERGII	WASHINGTON, OREGON
PORTULACACEAE	CLAYTONIA FLAVA	IDAHO
PORTULACACEAE	CLAYTONIA LANCEOLATA VAR. CHRYSANTHA	WASHINGTON
PORTULACACEAE	CLAYTONIA LANCEOLATA VAR. PEIRSONII	CALIFORNIA
PORTULACACEAE	CLAYTONIA MEGARHIZA VAR. NIVALIS	WASHINGTON
PORTULACACEAE	LEWISIA COLUMBIANA VAR. WALLOWENSIS	OREGON, IDAHO
PORTULACACEAE	LEWISIA DISEPALA	CALIFORNIA
PORTULACACEAE	LEWISIA PYGMAEA SSP. LONGIPETALA	CALIFORNIA
PORTULACACEAE	LEWISIA SERRATA	CALIFORNIA
PORTULACACEAE	LEWISIA STEBBINSII	CALIFORNIA
PORTULACACEAE	LEWISIA TWEEDYI	WASHINGTON
PORTULACACEAE	TALINUM CALCARICUM	ALABAMA, TENNESSEE
PORTULACACEAE	TALINUM MENGESII	ALABAMA, GEORGIA, TENNESSEE
PORTULACACEAE	TALINUM OKANOGANENSE	WASHINGTON
POTAMOGETONACEAE	POTAMOGETON HILLII	OHIO, VERMONT, MICHIGAN, PENNSYLVANIA, NEW YORK
PRIMULACEAE	DODECATHEON FRENCHII	ARKANSAS, KENTUCKY, ILLINOIS
PRIMULACEAE	DODECATHEON POETICUM	WASHINGTON, OREGON
PRIMULACEAE	DOUGLASIA LAEVIGATA VAR. LAEVIGATA	WASHINGTON, OREGON
PRIMULACEAE	LYSIMACHIA ASPERULAEFOLIA	NORTH CAROLINA, SOUTH CAROLINA
PRIMULACEAE	PRIMULA MAGUIREI	UTAH
PRIMULACEAE	PRIMULA SPECUICOLA	UTAH, ARIZONA

THREATENED SPECIES IN THE CONTINENTAL UNITED STATES

FAMILY	SPECIES	STATE
RAFFLESIACEAE	PILOSTYLES THURBERI	ARIZONA, CALIFORNIA
RANUNCULACEAE	ANEMONE EDWARDSIANA VAR. EDWARDSIANA	TEXAS
RANUNCULACEAE	ANEMONE EDWARDSIANA VAR. PETRAEA	TEXAS
RANUNCULACEAE	AQUILEGIA DESERTORUM	ARIZONA
RANUNCULACEAE	AQUILEGIA JONESII	MONTANA, WYOMING
RANUNCULACEAE	AQUILEGIA LARAMIENSIS	WYOMING
RANUNCULACEAE	CIMICIFUGA ARIZONICA	ARIZONA
RANUNCULACEAE	CIMICIFUGA LACINIATA	OREGON, WASHINGTON
RANUNCULACEAE	CIMICIFUGA RUBIFOLIA	VIRGINIA, TENNESSEE, NORTH CAROLINA
RANUNCULACEAE	CLEMATIS HIRSUTISSIMA VAR. ARIZONICA	ARIZONA
RANUNCULACEAE	CLEMATIS OCCIDENTALIS VAR. DISSECTA	WASHINGTON
RANUNCULACEAE	DELPHINIUM HUTCHINSONAE	CALIFORNIA
RANUNCULACEAE	DELPHINIUM MULTIPLEX	WASHINGTON
RANUNCULACEAE	DELPHINIUM NEWTONIANUM	ARKANSAS
RANUNCULACEAE	DELPHINIUM NUTTALLIANUM VAR. LINEAPETALUM	WASHINGTON
RANUNCULACEAE	DELPHINIUM VARIEGATUM SSP. THORNEI	CALIFORNIA
RANUNCULACEAE	DELPHINIUM VIRIDESCENS	WASHINGTON
RANUNCULACEAE	DELPHINIUM XANTHOLEUCUM	WASHINGTON
RANUNCULACEAE	HYDRASTIS CANADENSIS	ALABAMA, ARKANSAS, CONNECTICUT, DELAWARE, GEORGIA, ILLINOIS, INDIANA, KENTUCKY, MARYLAND, MICHIGAN, MINNESOTA, MISSOURI, NEBRASKA, NEW YORK, NORTH CAROLINA, OHIO, OKLAHOMA, PENNSYLVANIA, TENNESSEE, VERMONT, VIRGINIA, WEST VIRGINIA, WISCONSIN, MISSISSIPPI
RANUNCULACEAE	RANUNCULUS RECONDITUS	OREGON, WASHINGTON
RANUNCULACEAE	RANUNCULUS SUBCORDATUS	NORTH CAROLINA
RHAMNACEAE	CEANOTHUS GLORIOSUS VAR. PORRECTUS	CALIFORNIA
RHAMNACEAE	CEANOTHUS PROSTRATUS VAR. LAXUS	CALIFORNIA
RHAMNACEAE	CEANOTHUS RIGIDUS	CALIFORNIA
RHAMNACEAE	CEANOTHUS RODERICKII	CALIFORNIA
RHAMNACEAE	COLUBRINA CALIFORNICA	ARIZONA, CALIFORNIA
RHAMNACEAE	SAGERETIA MINUTIFLORA	ALABAMA, FLORIDA, GEORGIA, MISSISSIPPI, SOUTH CAROLINA
ROSACEAE	AGRIMONIA INCISA	SOUTH CAROLINA, FLORIDA, MISSISSIPPI
ROSACEAE	HORKELIA TULARENSIS	CALIFORNIA
ROSACEAE	IVESIA ARGYROCOMA	CALIFORNIA
ROSACEAE	IVESIA PICKERINGII	CALIFORNIA
ROSACEAE	PETROPHYTUM HENDERSONII	WASHINGTON
ROSACEAE	POTENTILLA MULTIFOLIOLATA	ARIZONA
ROSACEAE	POTENTILLA PATELLIFERA	CALIFORNIA
ROSACEAE	PRUNUS ALLEGHANIENSIS	PENNSYLVANIA, WEST VIRGINIA, CONNECTICUT
ROSACEAE	PRUNUS HAVARDII	TEXAS
ROSACEAE	PRUNUS MINUTIFLORA	TEXAS
ROSACEAE	PRUNUS MURRAYANA	TEXAS

FAMILY	SPECIES	STATE
ROSACEAE	PRUNUS TEXANA	TEXAS
ROSACEAE	ROSA STELLATA	TEXAS, NEW MEXICO, ARIZONA
ROSACEAE	VAUQUELINIA PAUCIFLORA	ARIZONA
ROSACEAE	WALDSTEINIA LOBATA	GEORGIA, SOUTH CAROLINA
RUBIACEAE	GALIUM BUXIFOLIUM	CALIFORNIA
RUBIACEAE	GALIUM CLEMENTIS	CALIFORNIA
RUBIACEAE	GALIUM CORRELLII	TEXAS
RUBIACEAE	GALIUM HYPOTRICHIUM VAR. TOMENTELLUM	CALIFORNIA
RUBIACEAE	GALIUM SERPENTICUM SSP. WARNERENSE	CALIFORNIA
RUBIACEAE	HEDYOTIS PURPUREA VAR. MONTANA	NORTH CAROLINA, TENNESSEE
RUBIACEAE	PINCKNEYA PUBENS	GEORGIA, FLORIDA, SOUTH CAROLINA
RUTACEAE	CHOISYA ARIZONICA	ARIZONA
RUTACEAE	CHOISYA MOLLIS	ARIZONA
SALICACEAE	SALIX ARIZONICA	ARIZONA
SALICACEAE	SALIX FLUVIATILIS	OREGON, WASHINGTON
SANTALACEAE	BUCKLEYA DISTICHOPHYLLA	VIRGINIA, NORTH CAROLINA, TENNESSEE
SANTALACEAE	NESTRONIA UMBELLULA	VIRGINIA, ALABAMA, GEORGIA, NORTH CAROLINA, SOUTH CAROLINA
SARRACENIACEAE	SARRACENIA ALABAMENSIS SSP. WHERRYI	ALABAMA, MISSISSIPPI
SARRACENIACEAE	SARRACENIA OREOPHILA	ALABAMA
SAXIFRAGACEAE	HEUCHERA ARKANSANA	ARKANSAS
SAXIFRAGACEAE	HEUCHERA BREVISTAMINEA	CALIFORNIA
SAXIFRAGACEAE	HEUCHERA DURANII	CALIFORNIA, NEVADA
SAXIFRAGACEAE	HEUCHERA HISPIDA	VIRGINIA, WEST VIRGINIA
SAXIFRAGACEAE	PHILADELPHUS ERNESTII	TEXAS
SAXIFRAGACEAE	PHILADELPHUS TEXENSIS VAR. TEXENSIS	TEXAS
SAXIFRAGACEAE	RIBES CANTHARIFORME	CALIFORNIA
SAXIFRAGACEAE	SAXIFRAGA ALEUTICA	ALASKA
SAXIFRAGACEAE	SAXIFRAGA CAREYANA	NORTH CAROLINA, TENNESSEE, VIRGINIA
SAXIFRAGACEAE	SAXIFRAGA CAROLINIANA	VIRGINIA, NORTH CAROLINA, TENNESSEE, KENTUCKY
SAXIFRAGACEAE	SAXIFRAGA OCCIDENTALIS VAR. LATIPETIOLATA	OREGON
SAXIFRAGACEAE	SULLIVANTIA HAPEMANII	COLORADO, MONTANA, WYOMING
SAXIFRAGACEAE	SULLIVANTIA OREGANA	WASHINGTON, OREGON
SAXIFRAGACEAE	SULLIVANTIA RENIFOLIA	ILLINOIS, IOWA, MINNESOTA, MISSOURI, WISCONSIN
SAXIFRAGACEAE	SULLIVANTIA SULLIVANTII	INDIANA, KENTUCKY, OHIO
SCHISANDRACEAE	SCHISANDRA GLABRA	NORTH CAROLINA, SOUTH CAROLINA, GEORGIA, FLORIDA, TENNESSEE, ALABAMA, ARKANSAS, LOUISIANA, MISSISSIPPI
SCHIZAEACEAE	SCHIZAEA PUSILLA	NEW JERSEY, NEW YORK
SCROPHULARIACEAE	AGALINIS PURPUREA VAR. CARTERI	FLORIDA
SCROPHULARIACEAE	ANTIRRHINUM SUBCORDATUM	CALIFORNIA
SCROPHULARIACEAE	AUREOLARIA PATULA	KENTUCKY, GEORGIA, TENNESSEE
SCROPHULARIACEAE	CASTILLEJA BREVILOBATA	CALIFORNIA, OREGON

THREATENED SPECIES IN THE CONTINENTAL UNITED STATES

FAMILY	SPECIES	STATE
SCROPHULARIACEAE	CASTILLEJA CHRYSANTHA	OREGON
SCROPHULARIACEAE	CASTILLEJA CINEREA	CALIFORNIA
SCROPHULARIACEAE	CASTILLEJA CRYPTANTHA	WASHINGTON
SCROPHULARIACEAE	CASTILLEJA CULBERTSONII	CALIFORNIA
SCROPHULARIACEAE	CASTILLEJA ELONGATA	TEXAS
SCROPHULARIACEAE	CASTILLEJA EWANII	CALIFORNIA
SCROPHULARIACEAE	CASTILLEJA FRATERNA	OREGON
SCROPHULARIACEAE	CASTILLEJA GLANDULIFERA	OREGON
SCROPHULARIACEAE	CASTILLEJA GLEASONII	CALIFORNIA
SCROPHULARIACEAE	CASTILLEJA KAIBABENSIS	ARIZONA
SCROPHULARIACEAE	CASTILLEJA LASSENENSIS	CALIFORNIA
SCROPHULARIACEAE	CASTILLEJA LATIFOLIA SSP. MENDOCINENSIS	CALIFORNIA
SCROPHULARIACEAE	CASTILLEJA LINOIDES	NEVADA
SCROPHULARIACEAE	CASTILLEJA MOLLIS	CALIFORNIA
SCROPHULARIACEAE	CASTILLEJA NEGLECTA	CALIFORNIA
SCROPHULARIACEAE	CASTILLEJA PARVIFLORA VAR. OLYMPICA	WASHINGTON
SCROPHULARIACEAE	CASTILLEJA PARVULA VAR. PARVULA	UTAH
SCROPHULARIACEAE	CASTILLEJA SCABRIDA	UTAH
SCROPHULARIACEAE	CASTILLEJA STEENENSIS	OREGON
SCROPHULARIACEAE	CASTILLEJA XANTHOTRICHA	OREGON
SCROPHULARIACEAE	CHELONE OBLIQUA VAR. SPECIOSA	ARKANSAS, ILLINOIS, INDIANA, IOWA, MICHIGAN, MINNESOTA, MISSOURI
SCROPHULARIACEAE	COLLINSIA ANTONINA SSP. ANTONINA	CALIFORNIA
SCROPHULARIACEAE	COLLINSIA ANTONINA SSP. PURPUREA	CALIFORNIA
SCROPHULARIACEAE	CORDYLANTHUS EREMICUS SSP. EREMICUS	CALIFORNIA
SCROPHULARIACEAE	CORDYLANTHUS MARITIMUS SSP. PALUSTRIS	CALIFORNIA, OREGON
SCROPHULARIACEAE	CORDYLANTHUS MOLLIS SSP. HISPIDUS	CALIFORNIA
SCROPHULARIACEAE	CORDYLANTHUS TECOPENSIS	CALIFORNIA, NEVADA
SCROPHULARIACEAE	DIPLACUS ARIDIS	CALIFORNIA
SCROPHULARIACEAE	GERARDIA ACUTA	CONNECTICUT, MASSACHUSETTS, NEW YORK, RHODE ISLAND
SCROPHULARIACEAE	MAURANDYA PETROPHILA	CALIFORNIA
SCROPHULARIACEAE	MICRANTHEMUM MICRANTHEMOIDES	NEW YORK, NEW JERSEY, PENNSYLVANIA, DELAWARE, MARYLAND, DISTRICT OF COLUMBIA, VIRGINIA
SCROPHULARIACEAE	MIMULUS EXIGUUS	CALIFORNIA
SCROPHULARIACEAE	MIMULUS GLABRATUS VAR. MICHIGANENSIS	MICHIGAN
SCROPHULARIACEAE	MIMULUS JUNGERMANNIOIDES	WASHINGTON, OREGON
SCROPHULARIACEAE	MIMULUS PICTUS	CALIFORNIA
SCROPHULARIACEAE	MIMULUS PURPUREUS	CALIFORNIA
SCROPHULARIACEAE	MIMULUS RUPICOLA	CALIFORNIA
SCROPHULARIACEAE	ORTHOCARPUS FLORIBUNDUS	CALIFORNIA
SCROPHULARIACEAE	ORTHOCARPUS LASIORHYNCHUS	CALIFORNIA
SCROPHULARIACEAE	PEDICULARIS HOWELLII	CALIFORNIA, OREGON
SCROPHULARIACEAE	PEDICULARIS RAINIERENSIS	WASHINGTON

THREATENED SPECIES IN THE CONTINENTAL UNITED STATES

FAMILY	SPECIES	STATE
SCROPHULARIACEAE	PENSTEMON ABIETINUS	UTAH
SCROPHULARIACEAE	PENSTEMON ACAULIS	UTAH, WYOMING
SCROPHULARIACEAE	PENSTEMON ALAMOSENSIS	NEW MEXICO
SCROPHULARIACEAE	PENSTEMON ARENARIUS	NEVADA
SCROPHULARIACEAE	PENSTEMON ATWOODII	UTAH
SCROPHULARIACEAE	PENSTEMON BICOLOR SSP. BICOLOR	NEVADA
SCROPHULARIACEAE	PENSTEMON BICOLOR SSP. ROSEUS	NEVADA, ARIZONA
SCROPHULARIACEAE	PENSTEMON CAESPITOSUS VAR. SUFFRUTICOSUS	UTAH
SCROPHULARIACEAE	PENSTEMON CALCAREUS	CALIFORNIA
SCROPHULARIACEAE	PENSTEMON CALIFORNICUS	CALIFORNIA
SCROPHULARIACEAE	PENSTEMON CARYI	MONTANA, WYOMING
SCROPHULARIACEAE	PENSTEMON CINICOLA	CALIFORNIA, OREGON
SCROPHULARIACEAE	PENSTEMON COBAEA VAR. PURPUREUS	MISSOURI, ARKANSAS
SCROPHULARIACEAE	PENSTEMON COMPACTUS	UTAH
SCROPHULARIACEAE	PENSTEMON DEGENERI	COLORADO
SCROPHULARIACEAE	PENSTEMON DISSECTUS	GEORGIA
SCROPHULARIACEAE	PENSTEMON ELEGANTULUS	OREGON, IDAHO
SCROPHULARIACEAE	PENSTEMON FILIFORMIS	CALIFORNIA
SCROPHULARIACEAE	PENSTEMON HUMILIS VAR. BREVIFOLIUS	UTAH
SCROPHULARIACEAE	PENSTEMON HUMILIS VAR. OBTUSIFOLIUS	UTAH
SCROPHULARIACEAE	PENSTEMON LEIOPHYLLUS	UTAH
SCROPHULARIACEAE	PENSTEMON MODESTUS	NEVADA
SCROPHULARIACEAE	PENSTEMON NANUS	UTAH
SCROPHULARIACEAE	PENSTEMON PAPILLATUS	CALIFORNIA
SCROPHULARIACEAE	PENSTEMON PARVUS	UTAH
SCROPHULARIACEAE	PENSTEMON PECKII	OREGON
SCROPHULARIACEAE	PENSTEMON PUDICUS	NEVADA
SCROPHULARIACEAE	PENSTEMON STEPHENSII	CALIFORNIA
SCROPHULARIACEAE	PENSTEMON THOMPSONIAE SSP. JAEGERI	NEVADA
SCROPHULARIACEAE	PENSTEMON THURBERI VAR. ANESTIUS	NEVADA
SCROPHULARIACEAE	PENSTEMON TRACYI	CALIFORNIA
SCROPHULARIACEAE	PENSTEMON UINTAHENSIS	UTAH
SCROPHULARIACEAE	PENSTEMON VIRGATUS SSP. PSEUDOPUTUS	ARIZONA
SCROPHULARIACEAE	PENSTEMON WARDII	UTAH
SCROPHULARIACEAE	PENSTEMON WASHINGTONENSIS	WASHINGTON
SCROPHULARIACEAE	SCHWALBEA AMERICANA	CONNECTICUT, DELAWARE, KENTUCKY, LOUISIANA, MARYLAND, MASSACHUSETTS, MISSISSIPPI, NEW JERSEY, NEW YORK, TENNESSEE, SOUTH CAROLINA
SCROPHULARIACEAE	SCROPHULARIA ATRATA	CALIFORNIA
SCROPHULARIACEAE	SYNTHYRIS CANBYI	MONTANA
SCROPHULARIACEAE	SYNTHYRIS HENDERSONII	IDAHO
SCROPHULARIACEAE	SYNTHYRIS MISSURICA SSP. STELLATA	OREGON
SCROPHULARIACEAE	SYNTHYRIS PINNATIFIDA VAR. LANUGINOSA	WASHINGTON

THREATENED SPECIES IN THE CONTINENTAL UNITED STATES

FAMILY	SPECIES	STATE
SCROPHULARIACEAE	SYNTHYRIS PLATYCARPA	IDAHO
SCROPHULARIACEAE	SYNTHYRIS SCHIZANTHA	WASHINGTON, OREGON
SCROPHULARIACEAE	VERONICA COPELANDII	CALIFORNIA
SELAGINELLACEAE	SELAGINELLA UTAHENSIS	UTAH, NEVADA
SOLANACEAE	LYCIUM BERBERIOIDES	TEXAS
SOLANACEAE	LYCIUM TEXANUM	TEXAS
SOLANACEAE	ORYCTES NEVADENSIS	CALIFORNIA, IDAHO, NEVADA
SOLANACEAE	PHYSALIS VISCOSA VAR. ELLIOTII	FLORIDA
SOLANACEAE	SOLANUM TENUILOBATUM	CALIFORNIA
STEMONACEAE	CROOMIA PAUCIFLORA	FLORIDA, ALABAMA, GEORGIA
STYRACACEAE	STYRAX YOUNGAE	TEXAS
VALERIANACEAE	VALERIANA COLUMBIANA	WASHINGTON
VALERIANACEAE	VALERIANA TEXANA	TEXAS
VALERIANACEAE	VALERIANELLA FLORIFERA	TEXAS
VERBENACEAE	VERBENA MARITIMA	FLORIDA
VIOLACEAE	VIOLA ADUNCA VAR. CASCADENSIS	OREGON, WASHINGTON
VIOLACEAE	VIOLA CHARLESTONENSIS	NEVADA, UTAH
VIOLACEAE	VIOLA EGGLESTONII	KENTUCKY, TENNESSEE, ALABAMA, GEORGIA
VIOLACEAE	VIOLA FLETTII	WASHINGTON
VIOLACEAE	VIOLA LANCEOLATA SSP. OCCIDENTALIS	CALIFORNIA, OREGON
VIOLACEAE	VIOLA TOMENTOSA	CALIFORNIA
XYRIDACEAE	XYRIS DRUMMONDII	ALABAMA
XYRIDACEAE	XYRIS ISOETIFOLIA	FLORIDA
XYRIDACEAE	XYRIS LONGISEPALA	ALABAMA, FLORIDA
XYRIDACEAE	XYRIS SCABRIFOLIA	FLORIDA, GEORGIA, MISSISSIPPI, ALABAMA

State Lists of Endangered and Threatened
Species of the Continental United States

These state lists of endangered and threatened plants are the species, subspecies, and varieties of plants on the recommended national list, arranged alphabetically according to the state's name. Since many of the species are found in more than one state, they are listed in all of the states in which they occur. Species presumed to be extinct are listed after the endangered species for each state.

The selection of species on the national lists has been based on different criteria than the separate lists developed by the various states (see Bibliography). As indicated earlier, the national lists are based on the total range and abundance of a species. Rarity and distribution data from various local state lists were considered in developing the national lists, but the state categories are, by definition, limited. A species that is endangered within one state's borders but that is abundant in another state would not be included in the national lists.

The disjunct or isolated occurrence of a species, far removed from areas where it is abundant, is biologically interesting and often of taxonomic and/or ecological importance. It was not possible to include these disjunct distributions in the present lists unless the species is rare within its total distribution. Nearly all species are rare at the periphery of their range, and this is especially true in disjunct populations.

Individual State governments should attach more importance to the protection of *nationally* endangered and threatened species in their state, than to locally endangered and threatened species which are common elsewhere. The latter, of course, do have significance in local lists for the states involved.

These lists of nationally endangered and threatened plant species should be of assistance to the various states in the establishment of state natural areas and parks for protecting and preserving these species as well as for developing needed state laws for plant protection.

STATE LISTS OF ENDANGERED, EXTINCT AND THREATENED SPECIES IN THE CONTINENTAL UNITED STATES

STATE	STATUS	FAMILY	SPECIES
ALABAMA	ENDANGERED	ARISTOLOCHIACEAE	HEXASTYLIS SPECIOSA
ALABAMA	ENDANGERED	ASCLEPIADACEAE	MATELEA ALABAMENSIS
ALABAMA	ENDANGERED	ASTERACEAE	HELIANTHUS SMITHII
ALABAMA	ENDANGERED	ASTERACEAE	JAMESIANTHUS ALABAMENSIS
ALABAMA	ENDANGERED	ASTERACEAE	MARSHALLIA MOHRI
ALABAMA	ENDANGERED	BRASSICACEAE	ARABIS PERSTELLATA VAR. PERSTELLATA
ALABAMA	ENDANGERED	BRASSICACEAE	LEAVENWORTHIA ALABAMICA VAR. BRACHYSTYLA
ALABAMA	ENDANGERED	BRASSICACEAE	LEAVENWORTHIA CRASSA VAR. CRASSA
ALABAMA	ENDANGERED	BRASSICACEAE	LEAVENWORTHIA CRASSA VAR. ELONGATA
ALABAMA	ENDANGERED	BRASSICACEAE	LEAVENWORTHIA EXIGUA VAR. LUTEA
ALABAMA	ENDANGERED	BRASSICACEAE	LESQUERELLA DENSIPILA
ALABAMA	ENDANGERED	BRASSICACEAE	LESQUERELLA LYRATA
ALABAMA	ENDANGERED	CAPRIFOLIACEAE	VIBURNUM BRACTEATUM
ALABAMA	ENDANGERED	CARYOPHYLLACEAE	ARENARIA ALABAMENSIS
ALABAMA	ENDANGERED	CRASSULACEAE	SEDUM NEVII
ALABAMA	ENDANGERED	CYPERACEAE	RHYNCHOSPORA CRINIPES
ALABAMA	ENDANGERED	EUPHORBIACEAE	CROTON ALABAMENSIS
ALABAMA	ENDANGERED	EUPHORBIACEAE	CROTON ELLIOTTII
ALABAMA	ENDANGERED	FABACEAE	PETALOSTEMUM FOLIOSUM
ALABAMA	ENDANGERED	LAMIACEAE	PYCNANTHEMUM CURVIPES
ALABAMA	ENDANGERED	LAURACEAE	LINDERA MELISSIFOLIA
ALABAMA	ENDANGERED	LILIACEAE	HYMENOCALLIS CORONARIA
ALABAMA	ENDANGERED	LILIACEAE	LILIUM IRIDOLLAE
ALABAMA	ENDANGERED	LILIACEAE	SCHOENOLIRION TEXANUM
ALABAMA	ENDANGERED	POACEAE	ANDROPOGON ARCTATUS
ALABAMA	ENDANGERED	POLEMONIACEAE	PHLOX PULCHRA
ALABAMA	ENDANGERED	POLYGONACEAE	ERIOGONUM LONGIFOLIUM VAR. HARPERI
ALABAMA	ENDANGERED	POLYPODIACEAE	LEPTOGRAMMA PILOSA VAR. ALABAMENSIS
ALABAMA	ENDANGERED	PORTULACACEAE	TALINUM APPALACHIANUM
ALABAMA	ENDANGERED	RANUNCULACEAE	CLEMATIS GATTINGERI
ALABAMA	ENDANGERED	RANUNCULACEAE	DELPHINIUM ALABAMICUM
ALABAMA	ENDANGERED	ROSACEAE	NEVIUSIA ALABAMENSIS
ALABAMA	ENDANGERED	SARRACENIACEAE	SARRACENIA ALABAMENSIS SSP. ALABAMENSIS
ALABAMA	THREATENED	APIACEAE	PTILIMNIUM FLUVIATILE
ALABAMA	THREATENED	APIACEAE	PTILIMNIUM NODOSUM
ALABAMA	THREATENED	ARALIACEAE	PANAX QUINQUEFOLIUS
ALABAMA	THREATENED	ARECACEAE	RHAPIDOPHYLLUM HYSTRIX
ALABAMA	THREATENED	ASTERACEAE	BRICKELLIA CORDIFOLIA
ALABAMA	THREATENED	ASTERACEAE	CACALIA DIVERSIFOLIA
ALABAMA	THREATENED	ASTERACEAE	COREOPSIS PULCHRA
ALABAMA	THREATENED	ASTERACEAE	ECHINACEA LAEVIGATA
ALABAMA	THREATENED	ASTERACEAE	LIATRIS HELLERI
ALABAMA	THREATENED	ASTERACEAE	RUDBECKIA AURICULATA

STATE LISTS OF ENDANGERED, EXTINCT AND THREATENED SPECIES IN THE CONTINENTAL UNITED STATES

STATE	STATUS	FAMILY	SPECIES
ALABAMA	THREATENED	ASTERACEAE	RUDBECKIA HELIOPSIDIS
ALABAMA	THREATENED	ASTERACEAE	VIGUIERA PORTERI
ALABAMA	THREATENED	BRASSICACEAE	ARABIS GEORGIANA
ALABAMA	THREATENED	BRASSICACEAE	LEAVENWORTHIA ALABAMICA VAR. ALABAMICA
ALABAMA	THREATENED	BRASSICACEAE	LEAVENWORTHIA TORULOSA
ALABAMA	THREATENED	BRASSICACEAE	WAREA SESSILIFOLIA
ALABAMA	THREATENED	CARYOPHYLLACEAE	ARENARIA GODFREYI
ALABAMA	THREATENED	CARYOPHYLLACEAE	ARENARIA UNIFLORA
ALABAMA	THREATENED	CUSCUTACEAE	CUSCUTA HARPERI
ALABAMA	THREATENED	CYPERACEAE	CAREX BALTZELLII
ALABAMA	THREATENED	CYPERACEAE	CAREX PURPURIFERA
ALABAMA	THREATENED	ERICACEAE	RHODODENDRON AUSTRINUM
ALABAMA	THREATENED	ERICACEAE	RHODODENDRON BAKERI
ALABAMA	THREATENED	ERICACEAE	RHODODENDRON PRUNIFOLIUM
ALABAMA	THREATENED	FABACEAE	ASTRAGALUS TENNESSEENSIS
ALABAMA	THREATENED	FABACEAE	BAPTISIA MEGACARPA
ALABAMA	THREATENED	FABACEAE	CLADRASTIS LUTEA
ALABAMA	THREATENED	FABACEAE	PETALOSTEMUM GATTINGERI
ALABAMA	THREATENED	FABACEAE	PSORALEA SUBACAULIS
ALABAMA	THREATENED	FAGACEAE	QUERCUS GEORGIANA
ALABAMA	THREATENED	HAMAMELIDACEAE	FOTHERGILLA GARDENI
ALABAMA	THREATENED	HYDROPHYLLACEAE	PHACELIA DUBIA VAR. GEORGIANA
ALABAMA	THREATENED	HYPERICACEAE	HYPERICUM SPHAEROCARPUM VAR. TURGIDUM
ALABAMA	THREATENED	JUNCACEAE	JUNCUS GYMNOCARPUS
ALABAMA	THREATENED	LAMIACEAE	SYNANDRA HISPIDULA
ALABAMA	THREATENED	LILIACEAE	TRILLIUM PUSILLUM VAR. PUSILLUM
ALABAMA	THREATENED	LINACEAE	LINUM SULCATUM VAR. HARPERI
ALABAMA	THREATENED	MALVACEAE	KOSTELETZKYA SMILACIFOLIA
ALABAMA	THREATENED	MELASTOMATACEAE	RHEXIA SALICIFOLIA
ALABAMA	THREATENED	ORCHIDACEAE	PLATANTHERA INTEGRA
ALABAMA	THREATENED	ORCHIDACEAE	PLATANTHERA PERAMOENA
ALABAMA	THREATENED	POACEAE	ARISTIDA SIMPLICIFLORA
ALABAMA	THREATENED	POACEAE	MANISURIS TUBERCULOSA
ALABAMA	THREATENED	POACEAE	PANICUM NUDICAULE
ALABAMA	THREATENED	PORTULACACEAE	TALINUM CALCARICUM
ALABAMA	THREATENED	PORTULACACEAE	TALINUM MENGESII
ALABAMA	THREATENED	RANUNCULACEAE	HYDRASTIS CANADENSIS
ALABAMA	THREATENED	RHAMNACEAE	SAGERETIA MINUTIFLORA
ALABAMA	THREATENED	SANTALACEAE	NESTRONIA UMBELLULA
ALABAMA	THREATENED	SARRACENIACEAE	SARRACENIA ALABAMENSIS SSP. WHERRYI
ALABAMA	THREATENED	SARRACENIACEAE	SARRACENIA OREOPHILA
ALABAMA	THREATENED	SCHISANDRACEAE	SCHISANDRA GLABRA
ALABAMA	THREATENED	STEMONACEAE	CROOMIA PAUCIFLORA

STATE LISTS OF ENDANGERED, EXTINCT AND THREATENED SPECIES IN THE CONTINENTAL UNITED STATES

STATE	STATUS	FAMILY	SPECIES
ALABAMA	THREATENED	VIOLACEAE	VIOLA EGGLESTONII
ALABAMA	THREATENED	XYRIDACEAE	XYRIS DRUMMONDII
ALABAMA	THREATENED	XYRIDACEAE	XYRIS LONGISEPALA
ALABAMA	THREATENED	XYRIDACEAE	XYRIS SCABRIFOLIA
ALASKA	ENDANGERED	BORAGINACEAE	CRYPTANTHA SHACKLETTEANA
ALASKA	ENDANGERED	BRASSICACEAE	SMELOWSKIA BOREALIS VAR. VILLOSA
ALASKA	ENDANGERED	CYPERACEAE	CAREX JACOBI-PETERI
ALASKA	ENDANGERED	FABACEAE	OXYTROPIS KOBUKENSIS
ALASKA	ENDANGERED	POLYGONACEAE	ERIOGONUM FLAVUM VAR. AQUILINUM
ALASKA	ENDANGERED	POLYPODIACEAE	POLYSTICHUM ALEUTICUM
ALASKA	THREATENED	ASTERACEAE	ARTEMISIA ALEUTICA
ALASKA	THREATENED	ASTERACEAE	ERIGERON HULTENII
ALASKA	THREATENED	BRASSICACEAE	LESQUERELLA ARCTICA VAR. SCAMMANAE
ALASKA	THREATENED	BRASSICACEAE	SMELOWSKIA PYRIFORMIS
ALASKA	THREATENED	BRASSICACEAE	THLASPI ARCTICUM
ALASKA	THREATENED	CARYOPHYLLACEAE	CERASTIUM ALEUTICUM
ALASKA	THREATENED	CYPERACEAE	CAREX PLECTOCARPA
ALASKA	THREATENED	FABACEAE	OXYTROPIS KOKRINENSIS
ALASKA	THREATENED	GENTIANACEAE	GENTIANA ALEUTICA
ALASKA	THREATENED	JUNCACEAE	JUNCUS SLWOOKOORUM
ALASKA	THREATENED	PAPAVERACEAE	PAPAVER WALPOLEI
ALASKA	THREATENED	POACEAE	CALAMAGROSTIS CRASSIGLUMIS
ALASKA	THREATENED	SAXIFRAGACEAE	SAXIFRAGA ALEUTICA
ARIZONA	ENDANGERED	ANACARDIACEAE	RHUS KEARNEYI
ARIZONA	ENDANGERED	ASTERACEAE	ERIGERON ERIOPHYLLUS
ARIZONA	ENDANGERED	ASTERACEAE	ERIGERON KUSCHEI
ARIZONA	ENDANGERED	ASTERACEAE	GALINSOGA SEMICALVA VAR. PERCALVA
ARIZONA	ENDANGERED	ASTERACEAE	HAPLOPAPPUS SALICINUS
ARIZONA	ENDANGERED	ASTERACEAE	PECTIS RUSBYI
ARIZONA	ENDANGERED	ASTERACEAE	PERITYLE GILENSIS VAR. SALENSIS
ARIZONA	ENDANGERED	ASTERACEAE	SENECIO FRANCISCANUS
ARIZONA	ENDANGERED	ASTERACEAE	STEPHANOMERIA SCHOTTII
ARIZONA	ENDANGERED	BERBERIDACEAE	BERBERIS HARRISONIANA
ARIZONA	ENDANGERED	BORAGINACEAE	CRYPTANTHA ATWOODII
ARIZONA	ENDANGERED	BRASSICACEAE	DRABA ASPRELLA VAR. ASPRELLA
ARIZONA	ENDANGERED	BRASSICACEAE	DRABA ASPRELLA VAR. KAIBABENSIS
ARIZONA	ENDANGERED	BRASSICACEAE	SISYMBRIUM KEARNEYI
ARIZONA	ENDANGERED	BRASSICACEAE	STREPTANTHUS LEMMONII
ARIZONA	ENDANGERED	CACTACEAE	ECHINOCACTUS HORIZONTHALONIUS VAR. NICHOLII
ARIZONA	ENDANGERED	CACTACEAE	ECHINOCEREUS TRIGLOCHIDIATUS VAR. ARIZONICUS
ARIZONA	ENDANGERED	CACTACEAE	OPUNTIA BASILARIS VAR. TRELEASEI
ARIZONA	ENDANGERED	CACTACEAE	PEDIOCACTUS BRADYI
ARIZONA	ENDANGERED	CACTACEAE	PEDIOCACTUS PEEBLESIANUS VAR. PEEBLESIANUS

STATE LISTS OF ENDANGERED, EXTINCT AND THREATENED SPECIES IN THE CONTINENTAL UNITED STATES

STATE	STATUS	FAMILY	SPECIES
ARIZONA	ENDANGERED	CACTACEAE	PEDIOCACTUS SILERI
ARIZONA	ENDANGERED	CAPPARIDACEAE	CLEOME MULTICAULIS
ARIZONA	ENDANGERED	CARYOPHYLLACEAE	SILENE RECTIRAMEA
ARIZONA	ENDANGERED	CHENOPODIACEAE	ATRIPLEX GRIFFITHSII
ARIZONA	ENDANGERED	CONVOLVULACEAE	IPOMOEA EGREGIA
ARIZONA	ENDANGERED	CONVOLVULACEAE	IPOMOEA LEMMONI
ARIZONA	ENDANGERED	CRASSULACEAE	DUDLEYA COLLOMAE
ARIZONA	ENDANGERED	CRASSULACEAE	GRAPTOPETALUM BARTRAMII
ARIZONA	ENDANGERED	CYPERACEAE	CAREX SPECUICOLA
ARIZONA	ENDANGERED	FABACEAE	ASTRAGALUS BEATHII
ARIZONA	ENDANGERED	FABACEAE	ASTRAGALUS CREMNOPHYLAX
ARIZONA	ENDANGERED	FABACEAE	ASTRAGALUS LENTIGINOSUS VAR. MARICOPAE
ARIZONA	ENDANGERED	FABACEAE	ASTRAGALUS XIPHOIDES
ARIZONA	ENDANGERED	FABACEAE	PSORALEA EPIPSILA
ARIZONA	ENDANGERED	HYDROPHYLLACEAE	PHACELIA FILIFORMIS
ARIZONA	ENDANGERED	HYDROPHYLLACEAE	PHACELIA WELSHII
ARIZONA	ENDANGERED	LILIACEAE	AGAVE ARIZONICA
ARIZONA	ENDANGERED	LILIACEAE	AGAVE MCKELVEYANA
ARIZONA	ENDANGERED	LILIACEAE	AGAVE SCHOTTII VAR. TRELEASEI
ARIZONA	ENDANGERED	LILIACEAE	AGAVE TOUMEYANA VAR. BELLA
ARIZONA	ENDANGERED	LOASACEAE	MENTZELIA NITENS VAR. LEPTOCAULIS
ARIZONA	ENDANGERED	MALVACEAE	SPHAERALCEA FENDLERI VAR. ALBESCENS
ARIZONA	ENDANGERED	OLEACEAE	FRAXINUS GOODDINGII
ARIZONA	ENDANGERED	ONAGRACEAE	CAMISSONIA MEGALANTHA
ARIZONA	ENDANGERED	ONAGRACEAE	CAMISSONIA SPECUICOLA SSP. SPECUICOLA
ARIZONA	ENDANGERED	PAPAVERACEAE	ARCTOMECON HUMILIS
ARIZONA	ENDANGERED	POACEAE	SPOROBOLUS PATENS
ARIZONA	ENDANGERED	POLYGONACEAE	ERIOGONUM CAPILLARE
ARIZONA	ENDANGERED	POLYGONACEAE	ERIOGONUM DARROVII
ARIZONA	ENDANGERED	POLYGONACEAE	ERIOGONUM ERICIFOLIUM VAR. ERICIFOLIUM
ARIZONA	ENDANGERED	POLYGONACEAE	ERIOGONUM MORTONIANUM
ARIZONA	ENDANGERED	POLYGONACEAE	ERIOGONUM THOMPSONAE VAR. ATWOODII
ARIZONA	ENDANGERED	POLYGONACEAE	ERIOGONUM ZIONIS VAR. COCCINEUM
ARIZONA	ENDANGERED	POLYGONACEAE	RUMEX ORTHONEURUS
ARIZONA	ENDANGERED	POLYPODIACEAE	ASPLENIUM ANDREWSII
ARIZONA	ENDANGERED	RANUNCULACEAE	RANUNCULUS INAMOENUS VAR. SUBAFFINIS
ARIZONA	ENDANGERED	ROSACEAE	COWANIA SUBINTEGRA
ARIZONA	ENDANGERED	RUBIACEAE	GALIUM COLLOMAE
ARIZONA	ENDANGERED	SCROPHULARIACEAE	CASTILLEJA CRUENTA
ARIZONA	ENDANGERED	SCROPHULARIACEAE	LIMOSELLA PUBIFLORA
ARIZONA	ENDANGERED	SCROPHULARIACEAE	PENSTEMON CLUTEI
ARIZONA	ENDANGERED	SCROPHULARIACEAE	PENSTEMON DISCOLOR
ARIZONA	EXTINCT	APOCYNACEAE	APOCYNUM JONESII

STATE LISTS OF ENDANGERED, EXTINCT AND THREATENED SPECIES IN THE CONTINENTAL UNITED STATES

STATE	STATUS	FAMILY	SPECIES
ARIZONA	EXTINCT	ASTERACEAE	GREENELLA DISCOIDEA
ARIZONA	THREATENED	APOCYNACEAE	AMSONIA PALMERI
ARIZONA	THREATENED	APOCYNACEAE	AMSONIA PEEBLESII
ARIZONA	THREATENED	ASCLEPIADACEAE	ASCLEPIAS CUTLERI
ARIZONA	THREATENED	ASTERACEAE	ASTER LEMMONII
ARIZONA	THREATENED	ASTERACEAE	CIRSIUM RYDBERGII
ARIZONA	THREATENED	ASTERACEAE	ENCELIA FRUTESCENS VAR. RESINOSA
ARIZONA	THREATENED	ASTERACEAE	ERIGERON ARIZONICUS
ARIZONA	THREATENED	ASTERACEAE	ERIGERON LEMMONII
ARIZONA	THREATENED	ASTERACEAE	ERIGERON LOBATUS
ARIZONA	THREATENED	ASTERACEAE	ERIGERON PRINGLEI
ARIZONA	THREATENED	ASTERACEAE	HELENIUM ARIZONICUM
ARIZONA	THREATENED	ASTERACEAE	HELIANTHUS DESERTICOLA
ARIZONA	THREATENED	ASTERACEAE	HYMENOXYS QUINQUESQUAMATA
ARIZONA	THREATENED	ASTERACEAE	HYMENOXYS SUBINTEGRA
ARIZONA	THREATENED	ASTERACEAE	MACHAERANTHERA MUCRONATA
ARIZONA	THREATENED	ASTERACEAE	PERITYLE COCHISENSIS
ARIZONA	THREATENED	ASTERACEAE	PERITYLE LEMMONII
ARIZONA	THREATENED	ASTERACEAE	PERITYLE SAXICOLA
ARIZONA	THREATENED	ASTERACEAE	PLUMMERA FLORIBUNDA
ARIZONA	THREATENED	ASTERACEAE	TAGETES LEMMONII
ARIZONA	THREATENED	BORAGINACEAE	CRYPTANTHA SEMIGLABRA
ARIZONA	THREATENED	BRASSICACEAE	ARABIS GRACILIPES
ARIZONA	THREATENED	BRASSICACEAE	DRABA ASPRELLA VAR. STELLIGERA
ARIZONA	THREATENED	BRASSICACEAE	LESQUERELLA GOODDINGII
ARIZONA	THREATENED	CACTACEAE	CEREUS GREGGII
ARIZONA	THREATENED	CACTACEAE	CORYPHANTHA RECURVATA
ARIZONA	THREATENED	CACTACEAE	CORYPHANTHA SCHEERI VAR. ROBUSTISPINA
ARIZONA	THREATENED	CACTACEAE	CORYPHANTHA VIVIPARA VAR. ALVERSONII
ARIZONA	THREATENED	CACTACEAE	CORYPHANTHA VIVIPARA VAR. ROSEA
ARIZONA	THREATENED	CACTACEAE	ECHINOCEREUS LEDINGII
ARIZONA	THREATENED	CACTACEAE	FEROCACTUS ACANTHODES VAR. EASTWOODIAE
ARIZONA	THREATENED	CACTACEAE	MAMMILLARIA ORESTERA
ARIZONA	THREATENED	CACTACEAE	MAMMILLARIA THORNBERI
ARIZONA	THREATENED	CACTACEAE	NEOLLOYDIA ERECTOCENTRA VAR. ACUNENSIS
ARIZONA	THREATENED	CACTACEAE	NEOLLOYDIA ERECTOCENTRA VAR. ERECTOCENTRA
ARIZONA	THREATENED	CACTACEAE	OPUNTIA BASILARIS VAR. LONGIAREOLATA
ARIZONA	THREATENED	CACTACEAE	OPUNTIA PHAEACANTHA VAR. FLAVISPINA
ARIZONA	THREATENED	CACTACEAE	OPUNTIA PHAEACANTHA VAR. SUPERBOSPINA
ARIZONA	THREATENED	CACTACEAE	OPUNTIA WHIPPLEI VAR. MULTIGENICULATA
ARIZONA	THREATENED	CACTACEAE	PEDIOCACTUS PAPYRACANTHUS
ARIZONA	THREATENED	CACTACEAE	PEDIOCACTUS PARADINEI
ARIZONA	THREATENED	CACTACEAE	PEDIOCACTUS PEEBLESIANUS VAR. FICKEISENIAE

STATE LISTS OF ENDANGERED, EXTINCT AND THREATENED SPECIES IN THE CONTINENTAL UNITED STATES

STATE	STATUS	FAMILY	SPECIES
ARIZONA	THREATENED	CACTACEAE	SCLEROCACTUS SPINOSIOR
ARIZONA	THREATENED	CRASSULACEAE	GRAPTOPETALUM RUSBYI
ARIZONA	THREATENED	CROSSOSOMATACEAE	APACHERIA CHIRICAHUENSIS
ARIZONA	THREATENED	CROSSOSOMATACEAE	CROSSOSOMA PARVIFLORUM
ARIZONA	THREATENED	CYPERACEAE	CAREX CURATORUM
ARIZONA	THREATENED	EUPHORBIACEAE	EUPHORBIA PLATYSPERMA
ARIZONA	THREATENED	EUPHORBIACEAE	MANIHOT DAVISIAE
ARIZONA	THREATENED	FABACEAE	ASTRAGALUS AMPULLARIUS
ARIZONA	THREATENED	FABACEAE	ASTRAGALUS BARNEBYI
ARIZONA	THREATENED	FABACEAE	ASTRAGALUS COTTAMII
ARIZONA	THREATENED	FABACEAE	ASTRAGALUS ENSIFORMIS
ARIZONA	THREATENED	FABACEAE	ASTRAGALUS GEYERI VAR. TRIQUETRUS
ARIZONA	THREATENED	FABACEAE	ASTRAGALUS LANCEARIUS
ARIZONA	THREATENED	FABACEAE	ASTRAGALUS LENTIGINOSUS VAR. AMBIGUUS
ARIZONA	THREATENED	FABACEAE	ASTRAGALUS TITANOPHILUS
ARIZONA	THREATENED	FABACEAE	ASTRAGALUS TROGLODYTUS
ARIZONA	THREATENED	FABACEAE	DALEA THOMPSONAE
ARIZONA	THREATENED	FABACEAE	ERRAZURIZIA ROTUNDATA
ARIZONA	THREATENED	FABACEAE	LUPINUS CUTLERI
ARIZONA	THREATENED	FABACEAE	SOPHORA ARIZONICA
ARIZONA	THREATENED	HYDROPHYLLACEAE	NAMA RETRORSUM
ARIZONA	THREATENED	HYDROPHYLLACEAE	PHACELIA CEPHALOTES
ARIZONA	THREATENED	HYDROPHYLLACEAE	PHACELIA CONSTANCEI
ARIZONA	THREATENED	HYDROPHYLLACEAE	PHACELIA RAFAELENSIS
ARIZONA	THREATENED	HYDROPHYLLACEAE	PHACELIA SERRATA
ARIZONA	THREATENED	LILIACEAE	AGAVE UTAHENSIS VAR. KAIBABENSIS
ARIZONA	THREATENED	LILIACEAE	ALLIUM GOODDINGII
ARIZONA	THREATENED	LILIACEAE	TRITELEIA LEMMONAE
ARIZONA	THREATENED	LORANTHACEAE	ARCEUTHOBIUM APACHENSE
ARIZONA	THREATENED	OLEACEAE	FRAXINUS ANOMALA VAR. LOWELLII
ARIZONA	THREATENED	OLEACEAE	FRAXINUS CUSPIDATA VAR. MACROPETALA
ARIZONA	THREATENED	ONAGRACEAE	CAMISSONIA CONFERTIFLORA
ARIZONA	THREATENED	ONAGRACEAE	CAMISSONIA EXILIS
ARIZONA	THREATENED	ONAGRACEAE	CAMISSONIA GOULDII
ARIZONA	THREATENED	ONAGRACEAE	CAMISSONIA SPECUICOLA SSP. HESPERIA
ARIZONA	THREATENED	PAPAVERACEAE	ARGEMONE ARIZONICA
ARIZONA	THREATENED	POACEAE	PUCCINELLIA PARISHII
ARIZONA	THREATENED	POLEMONIACEAE	PHLOX CLUTEANA
ARIZONA	THREATENED	POLYGALACEAE	POLYGALA PILIOPHORA
ARIZONA	THREATENED	POLYGONACEAE	ERIOGONUM APACHENSE
ARIZONA	THREATENED	POLYGONACEAE	ERIOGONUM HEERMANNII VAR. SUBRACEMOSUM
ARIZONA	THREATENED	POLYGONACEAE	ERIOGONUM RIPLEYI
ARIZONA	THREATENED	POLYGONACEAE	ERIOGONUM THOMPSONAE VAR. THOMPSONAE

STATE LISTS OF ENDANGERED, EXTINCT AND THREATENED SPECIES IN THE CONTINENTAL UNITED STATES

STATE	STATUS	FAMILY	SPECIES
ARIZONA	THREATENED	POLYGONACEAE	POLYGONUM FUSIFORME
ARIZONA	THREATENED	POLYPODIACEAE	CHEILANTHES PRINGLEI
ARIZONA	THREATENED	POLYPODIACEAE	CHEILANTHES PYRAMIDALIS VAR. ARIZONICA
ARIZONA	THREATENED	POLYPODIACEAE	NOTHOLAENA LEMMONII
ARIZONA	THREATENED	PRIMULACEAE	PRIMULA SPECUICOLA
ARIZONA	THREATENED	RAFFLESIACEAE	PILOSTYLES THURBERI
ARIZONA	THREATENED	RANUNCULACEAE	AQUILEGIA DESERTORUM
ARIZONA	THREATENED	RANUNCULACEAE	CIMICIFUGA ARIZONICA
ARIZONA	THREATENED	RANUNCULACEAE	CLEMATIS HIRSUTISSIMA VAR. ARIZONICA
ARIZONA	THREATENED	RHAMNACEAE	COLUBRINA CALIFORNICA
ARIZONA	THREATENED	ROSACEAE	POTENTILLA MULTIFOLIOLATA
ARIZONA	THREATENED	ROSACEAE	ROSA STELLATA
ARIZONA	THREATENED	ROSACEAE	VAUQUELINIA PAUCIFLORA
ARIZONA	THREATENED	RUTACEAE	CHOISYA ARIZONICA
ARIZONA	THREATENED	RUTACEAE	CHOISYA MOLLIS
ARIZONA	THREATENED	SALICACEAE	SALIX ARIZONICA
ARIZONA	THREATENED	SCROPHULARIACEAE	CASTILLEJA KAIBABENSIS
ARIZONA	THREATENED	SCROPHULARIACEAE	PENSTEMON BICOLOR SSP. ROSEUS
ARIZONA	THREATENED	SCROPHULARIACEAE	PENSTEMON VIRGATUS SSP. PSEUDOPUTUS
ARKANSAS	ENDANGERED	BRASSICACEAE	DRABA APRICA
ARKANSAS	ENDANGERED	BRASSICACEAE	STREPTANTHUS SQUAMIFORMIS
ARKANSAS	ENDANGERED	CARYOPHYLLACEAE	GEOCARPON MINIMUM
ARKANSAS	ENDANGERED	ERIOCAULACEAE	ERIOCAULON KORNICKIANUM
ARKANSAS	ENDANGERED	FABACEAE	AMORPHA OUACHITENSIS
ARKANSAS	ENDANGERED	FAGACEAE	CASTANEA PUMILA VAR. OZARKENSIS
ARKANSAS	ENDANGERED	LAURACEAE	LINDERA MELISSIFOLIA
ARKANSAS	ENDANGERED	LILIACEAE	SCHOENOLIRION TEXANUM
ARKANSAS	ENDANGERED	ONAGRACEAE	OENOTHERA SESSILIS
ARKANSAS	ENDANGERED	ROSACEAE	NEVIUSIA ALABAMENSIS
ARKANSAS	THREATENED	ARALIACEAE	PANAX QUINQUEFOLIUS
ARKANSAS	THREATENED	FABACEAE	CLADRASTIS LUTEA
ARKANSAS	THREATENED	FAGACEAE	QUERCUS SHUMARDII VAR. ACERIFOLIA
ARKANSAS	THREATENED	LEITNERIACEAE	LEITNERIA FLORIDANA
ARKANSAS	THREATENED	LILIACEAE	TRILLIUM PUSILLUM VAR. OZARKANUM
ARKANSAS	THREATENED	LILIACEAE	VERATRUM WOODII
ARKANSAS	THREATENED	MALVACEAE	CALLIRHOE PAPAVER VAR. BUSHII
ARKANSAS	THREATENED	ONAGRACEAE	GAURA DEMAREEI
ARKANSAS	THREATENED	ORCHIDACEAE	PLATANTHERA FLAVA
ARKANSAS	THREATENED	ORCHIDACEAE	PLATANTHERA LEUCOPHAEA
ARKANSAS	THREATENED	ORCHIDACEAE	PLATANTHERA PERAMOENA
ARKANSAS	THREATENED	PRIMULACEAE	DODECATHEON FRENCHII
ARKANSAS	THREATENED	RANUNCULACEAE	DELPHINIUM NEWTONIANUM
ARKANSAS	THREATENED	RANUNCULACEAE	HYDRASTIS CANADENSIS

STATE LISTS OF ENDANGERED, EXTINCT AND THREATENED SPECIES IN THE CONTINENTAL UNITED STATES

STATE	STATUS	FAMILY	SPECIES
ARKANSAS	THREATENED	SAXIFRAGACEAE	HEUCHERA ARKANSANA
ARKANSAS	THREATENED	SCHISANDRACEAE	SCHISANDRA GLABRA
ARKANSAS	THREATENED	SCROPHULARIACEAE	CHELONE OBLIQUA VAR. SPECIOSA
ARKANSAS	THREATENED	SCROPHULARIACEAE	PENSTEMON COBAEA VAR. PURPUREUS
CALIFORNIA	ENDANGERED	APIACEAE	ANGELICA CALLII
CALIFORNIA	ENDANGERED	APIACEAE	ERYNGIUM ARISTULATUM VAR. PARISHII
CALIFORNIA	ENDANGERED	APIACEAE	ERYNGIUM RACEMOSUM
CALIFORNIA	ENDANGERED	APIACEAE	LOMATIUM RAVENII
CALIFORNIA	ENDANGERED	APIACEAE	SANICULA MARITIMA
CALIFORNIA	ENDANGERED	APIACEAE	SANICULA SAXATILIS
CALIFORNIA	ENDANGERED	APIACEAE	SANICULA TRACYI
CALIFORNIA	ENDANGERED	ASTERACEAE	ASTER CHILENSIS VAR. LENTUS
CALIFORNIA	ENDANGERED	ASTERACEAE	BLENNOSPERMA BAKERI
CALIFORNIA	ENDANGERED	ASTERACEAE	BLENNOSPERMA NANUM VAR. ROBUSTUM
CALIFORNIA	ENDANGERED	ASTERACEAE	CIRSIUM CRASSICAULE
CALIFORNIA	ENDANGERED	ASTERACEAE	CIRSIUM FONTINALE VAR. FONTINALE
CALIFORNIA	ENDANGERED	ASTERACEAE	CIRSIUM FONTINALE SSP. OBISPOENSE
CALIFORNIA	ENDANGERED	ASTERACEAE	CIRSIUM HYDROPHILUM VAR. HYDROPHILUM
CALIFORNIA	ENDANGERED	ASTERACEAE	CIRSIUM LONCHOLEPIS
CALIFORNIA	ENDANGERED	ASTERACEAE	CIRSIUM RHOTHOPHILUM
CALIFORNIA	ENDANGERED	ASTERACEAE	ENCELIOPSIS COVILLEI
CALIFORNIA	ENDANGERED	ASTERACEAE	ERIGERON CALVUS
CALIFORNIA	ENDANGERED	ASTERACEAE	ERIGERON DELICATUS
CALIFORNIA	ENDANGERED	ASTERACEAE	ERIGERON FOLIOSUS VAR. BLOCHMANAE
CALIFORNIA	ENDANGERED	ASTERACEAE	ERIOPHYLLUM LANATUM VAR. HALLII
CALIFORNIA	ENDANGERED	ASTERACEAE	ERIOPHYLLUM MOHAVENSE
CALIFORNIA	ENDANGERED	ASTERACEAE	GRINDELIA FRAXINO-PRATENSIS
CALIFORNIA	ENDANGERED	ASTERACEAE	HAPLOPAPPUS CANUS
CALIFORNIA	ENDANGERED	ASTERACEAE	HAPLOPAPPUS EASTWOODIAE
CALIFORNIA	ENDANGERED	ASTERACEAE	HELIANTHUS EXILIS
CALIFORNIA	ENDANGERED	ASTERACEAE	HELIANTHUS NIVEUS SSP. TEPHRODES
CALIFORNIA	ENDANGERED	ASTERACEAE	HEMIZONIA CONJUGENS
CALIFORNIA	ENDANGERED	ASTERACEAE	HEMIZONIA FLORIBUNDA
CALIFORNIA	ENDANGERED	ASTERACEAE	HEMIZONIA MINTHORNII
CALIFORNIA	ENDANGERED	ASTERACEAE	HEMIZONIA MOHAVENSIS
CALIFORNIA	ENDANGERED	ASTERACEAE	HOLOCARPHA MACRADENIA
CALIFORNIA	ENDANGERED	ASTERACEAE	LASTHENIA BURKEI
CALIFORNIA	ENDANGERED	ASTERACEAE	LASTHENIA CONJUGENS
CALIFORNIA	ENDANGERED	ASTERACEAE	LAYIA DISCOIDEA
CALIFORNIA	ENDANGERED	ASTERACEAE	MICROSERIS DECIPIENS
CALIFORNIA	ENDANGERED	ASTERACEAE	PALAFOXIA ARIDA VAR. GIGANTEA
CALIFORNIA	ENDANGERED	ASTERACEAE	PENTACHAETA LYONII
CALIFORNIA	ENDANGERED	ASTERACEAE	PSEUDOBAHIA BAHIAEFOLIA

STATE LISTS OF ENDANGERED, EXTINCT AND THREATENED SPECIES IN THE CONTINENTAL UNITED STATES

STATE	STATUS	FAMILY	SPECIES
CALIFORNIA	ENDANGERED	ASTERACEAE	PSEUDOBAHIA PEIRSONII
CALIFORNIA	ENDANGERED	ASTERACEAE	SENECIO LAYNEAE
CALIFORNIA	ENDANGERED	ASTERACEAE	TRACYINA ROSTRATA
CALIFORNIA	ENDANGERED	BERBERIDACEAE	BERBERIS NEVINII
CALIFORNIA	ENDANGERED	BERBERIDACEAE	BERBERIS SONNEI
CALIFORNIA	ENDANGERED	BORAGINACEAE	AMSINCKIA GRANDIFLORA
CALIFORNIA	ENDANGERED	BORAGINACEAE	CRYPTANTHA ROOSIORUM
CALIFORNIA	ENDANGERED	BORAGINACEAE	PLAGIOBOTHRYS DIFFUSUS
CALIFORNIA	ENDANGERED	BRASSICACEAE	ARABIS BREWERI VAR. PECUNIARIA
CALIFORNIA	ENDANGERED	BRASSICACEAE	ARABIS MCDONALDIANA
CALIFORNIA	ENDANGERED	BRASSICACEAE	ERYSIMUM CAPITATUM VAR. ANGUSTATUM
CALIFORNIA	ENDANGERED	BRASSICACEAE	ERYSIMUM FRANCISCANUM VAR. FRANCISCANUM
CALIFORNIA	ENDANGERED	BRASSICACEAE	ERYSIMUM TERETIFOLIUM
CALIFORNIA	ENDANGERED	BRASSICACEAE	SIBARA FILIFOLIA
CALIFORNIA	ENDANGERED	BRASSICACEAE	SMELOWSKIA OVALIS VAR. CONGESTA
CALIFORNIA	ENDANGERED	BRASSICACEAE	STREPTANTHUS ALBIDUS SSP. ALBIDUS
CALIFORNIA	ENDANGERED	BRASSICACEAE	STREPTANTHUS CALLISTUS
CALIFORNIA	ENDANGERED	BRASSICACEAE	STREPTANTHUS FARNSWORTHIANUS
CALIFORNIA	ENDANGERED	BRASSICACEAE	STREPTANTHUS MORRISONII SSP. HIRTIFLORUS
CALIFORNIA	ENDANGERED	BRASSICACEAE	STREPTANTHUS NIGER
CALIFORNIA	ENDANGERED	CACTACEAE	ECHINOCEREUS ENGELMANNII VAR. HOWEI
CALIFORNIA	ENDANGERED	CACTACEAE	FEROCACTUS VIRIDESCENS
CALIFORNIA	ENDANGERED	CACTACEAE	OPUNTIA BASILARIS VAR. TRELEASEI
CALIFORNIA	ENDANGERED	CAMPANULACEAE	CAMPANULA CALIFORNICA
CALIFORNIA	ENDANGERED	CAMPANULACEAE	LEGENERE LIMOSA
CALIFORNIA	ENDANGERED	CARYOPHYLLACEAE	ARENARIA URSINA
CALIFORNIA	ENDANGERED	CARYOPHYLLACEAE	SILENE INVISA
CALIFORNIA	ENDANGERED	CARYOPHYLLACEAE	SILENE MARMORENSIS
CALIFORNIA	ENDANGERED	CELASTRACEAE	FORSELLESIA PUNGENS VAR. GLABRA
CALIFORNIA	ENDANGERED	CHENOPODIACEAE	NITROPHILA MOHAVENSIS
CALIFORNIA	ENDANGERED	CONVOLVULACEAE	CALYSTEGIA STEBBINSII
CALIFORNIA	ENDANGERED	CONVOLVULACEAE	DICHONDRA OCCIDENTALIS
CALIFORNIA	ENDANGERED	CRASSULACEAE	DUDLEYA BETTINAE
CALIFORNIA	ENDANGERED	CRASSULACEAE	DUDLEYA CANDELABRUM
CALIFORNIA	ENDANGERED	CRASSULACEAE	DUDLEYA CYMOSA SSP. MARCESCENS
CALIFORNIA	ENDANGERED	CRASSULACEAE	DUDLEYA MULTICAULIS
CALIFORNIA	ENDANGERED	CRASSULACEAE	DUDLEYA NESIOTICA
CALIFORNIA	ENDANGERED	CRASSULACEAE	DUDLEYA STOLONIFERA
CALIFORNIA	ENDANGERED	CRASSULACEAE	DUDLEYA TRASKIAE
CALIFORNIA	ENDANGERED	CRASSULACEAE	PARVISEDUM LEIOCARPUM
CALIFORNIA	ENDANGERED	CRASSULACEAE	SEDUM RADIATUM SSP. DEPAUPERATUM
CALIFORNIA	ENDANGERED	CUPRESSACEAE	CUPRESSUS GOVENIANA VAR. ABRAMSIANA
CALIFORNIA	ENDANGERED	CUSCUTACEAE	CUSCUTA HOWELLIANA

STATE LISTS OF ENDANGERED, EXTINCT AND THREATENED SPECIES IN THE CONTINENTAL UNITED STATES

STATE	STATUS	FAMILY	SPECIES
CALIFORNIA	ENDANGERED	CYPERACEAE	CAREX ALBIDA
CALIFORNIA	ENDANGERED	CYPERACEAE	CAREX TOMPKINSII
CALIFORNIA	ENDANGERED	CYPERACEAE	RHYNCHOSPORA CALIFORNICA
CALIFORNIA	ENDANGERED	ERICACEAE	ARCTOSTAPHYLOS AURICULATA
CALIFORNIA	ENDANGERED	ERICACEAE	ARCTOSTAPHYLOS BAKERI
CALIFORNIA	ENDANGERED	ERICACEAE	ARCTOSTAPHYLOS DENSIFLORA
CALIFORNIA	ENDANGERED	ERICACEAE	ARCTOSTAPHYLOS EDMUNDSII VAR. PARVIFOLIA
CALIFORNIA	ENDANGERED	ERICACEAE	ARCTOSTAPHYLOS GLANDULOSA SSP. CRASSIFOLIA
CALIFORNIA	ENDANGERED	ERICACEAE	ARCTOSTAPHYLOS GLUTINOSA
CALIFORNIA	ENDANGERED	ERICACEAE	ARCTOSTAPHYLOS HOOKERI SSP. HEARSTIORUM
CALIFORNIA	ENDANGERED	ERICACEAE	ARCTOSTAPHYLOS HOOKERI SSP. RAVENII
CALIFORNIA	ENDANGERED	ERICACEAE	ARCTOSTAPHYLOS IMBRICATA
CALIFORNIA	ENDANGERED	ERICACEAE	ARCTOSTAPHYLOS MYRTIFOLIA
CALIFORNIA	ENDANGERED	ERICACEAE	ARCTOSTAPHYLOS PACIFICA
CALIFORNIA	ENDANGERED	ERICACEAE	ARCTOSTAPHYLOS PALLIDA
CALIFORNIA	ENDANGERED	ERICACEAE	ARCTOSTAPHYLOS PUMILA
CALIFORNIA	ENDANGERED	ERICACEAE	ARCTOSTAPHYLOS REFUGIOENSIS
CALIFORNIA	ENDANGERED	EUPHORBIACEAE	CROTON WIGGINSII
CALIFORNIA	ENDANGERED	FABACEAE	ASTRAGALUS AGNICIDUS
CALIFORNIA	ENDANGERED	FABACEAE	ASTRAGALUS CLARIANUS
CALIFORNIA	ENDANGERED	FABACEAE	ASTRAGALUS FUNEREUS
CALIFORNIA	ENDANGERED	FABACEAE	ASTRAGALUS JAEGERIANUS
CALIFORNIA	ENDANGERED	FABACEAE	ASTRAGALUS JOHANNIS-HOWELLII
CALIFORNIA	ENDANGERED	FABACEAE	ASTRAGALUS LENTIGINOSUS VAR. MICANS
CALIFORNIA	ENDANGERED	FABACEAE	ASTRAGALUS LENTIGINOSUS VAR. SESQUIMETRALIS
CALIFORNIA	ENDANGERED	FABACEAE	ASTRAGALUS MONOENSIS
CALIFORNIA	ENDANGERED	FABACEAE	ASTRAGALUS RAVENII
CALIFORNIA	ENDANGERED	FABACEAE	ASTRAGALUS TENER VAR. TITI
CALIFORNIA	ENDANGERED	FABACEAE	LATHYRUS HITCHCOCKIANUS
CALIFORNIA	ENDANGERED	FABACEAE	LATHYRUS JEPSONII SSP. JEPSONII
CALIFORNIA	ENDANGERED	FABACEAE	LOTUS SCOPARIUS SSP. TRASKIAE
CALIFORNIA	ENDANGERED	FABACEAE	LUPINUS GUADALUPENSIS
CALIFORNIA	ENDANGERED	FABACEAE	LUPINUS LUDOVICIANUS
CALIFORNIA	ENDANGERED	FABACEAE	LUPINUS MILO-BAKERI
CALIFORNIA	ENDANGERED	FABACEAE	LUPINUS TIDESTROMII VAR. LAYNEAE
CALIFORNIA	ENDANGERED	FABACEAE	LUPINUS TIDESTROMII VAR. TIDESTROMII
CALIFORNIA	ENDANGERED	FABACEAE	LUPINUS TRACYI
CALIFORNIA	ENDANGERED	FABACEAE	TRIFOLIUM AMOENUM
CALIFORNIA	ENDANGERED	FABACEAE	TRIFOLIUM LEMMONII
CALIFORNIA	ENDANGERED	FABACEAE	TRIFOLIUM POLYODON
CALIFORNIA	ENDANGERED	FABACEAE	TRIFOLIUM TRICHOCALYX
CALIFORNIA	ENDANGERED	FUMARIACEAE	DICENTRA OCHROLEUCA
CALIFORNIA	ENDANGERED	GENTIANACEAE	CENTAURIUM NAMOPHILUM

STATE LISTS OF ENDANGERED, EXTINCT AND THREATENED SPECIES IN THE CONTINENTAL UNITED STATES

STATE	STATUS	FAMILY	SPECIES
CALIFORNIA	ENDANGERED	HYDROPHYLLACEAE	ERIODICTYON ALTISSIMUM
CALIFORNIA	ENDANGERED	HYDROPHYLLACEAE	ERIODICTYON CAPITATUM
CALIFORNIA	ENDANGERED	HYDROPHYLLACEAE	PHACELIA COOKEI
CALIFORNIA	ENDANGERED	IRIDACEAE	IRIS TENAX SSP. KLAMATHENSIS
CALIFORNIA	ENDANGERED	JUGLANDACEAE	JUGLANS HINDSII
CALIFORNIA	ENDANGERED	LAMIACEAE	ACANTHOMINTHA ILICIFOLIA
CALIFORNIA	ENDANGERED	LAMIACEAE	ACANTHOMINTHA OBOVATA SSP. DUTTONII
CALIFORNIA	ENDANGERED	LAMIACEAE	MONARDELLA LINOIDES SSP. VIMINEA
CALIFORNIA	ENDANGERED	LAMIACEAE	MONARDELLA MACRANTHA VAR. HALLII
CALIFORNIA	ENDANGERED	LAMIACEAE	MONARDELLA PRINGLEI
CALIFORNIA	ENDANGERED	LAMIACEAE	MONARDELLA UNDULATA VAR. FRUTESCENS
CALIFORNIA	ENDANGERED	LAMIACEAE	POGOGYNE ABRAMSII
CALIFORNIA	ENDANGERED	LAMIACEAE	POGOGYNE CLAREANA
CALIFORNIA	ENDANGERED	LAMIACEAE	POGOGYNE DOUGLASII SSP. PARVIFLORA
CALIFORNIA	ENDANGERED	LAMIACEAE	POGOGYNE NUDIUSCULA
CALIFORNIA	ENDANGERED	LAMIACEAE	SALVIA COLUMBARIAE VAR. ZIEGLERI
CALIFORNIA	ENDANGERED	LAMIACEAE	TRICHOSTEMA AUSTROMONTANUM SSP. COMPACTUM
CALIFORNIA	ENDANGERED	LENNOACEAE	AMMOBROMA SONORAE
CALIFORNIA	ENDANGERED	LILIACEAE	ALLIUM HICKMANII
CALIFORNIA	ENDANGERED	LILIACEAE	BLOOMERIA HUMILIS
CALIFORNIA	ENDANGERED	LILIACEAE	BRODIAEA CORONARIA VAR. ROSEA
CALIFORNIA	ENDANGERED	LILIACEAE	BRODIAEA FILIFOLIA
CALIFORNIA	ENDANGERED	LILIACEAE	BRODIAEA ORCUTTII
CALIFORNIA	ENDANGERED	LILIACEAE	BRODIAEA PALLIDA
CALIFORNIA	ENDANGERED	LILIACEAE	CALOCHORTUS CLAVATUS SSP. RECURVIFOLIUS
CALIFORNIA	ENDANGERED	LILIACEAE	CALOCHORTUS COERULEUS VAR. WESTONII
CALIFORNIA	ENDANGERED	LILIACEAE	CALOCHORTUS GREENEI
CALIFORNIA	ENDANGERED	LILIACEAE	CALOCHORTUS PERSISTENS
CALIFORNIA	ENDANGERED	LILIACEAE	CALOCHORTUS TIBURONENSIS
CALIFORNIA	ENDANGERED	LILIACEAE	CHLOROGALUM GRANDIFLORUM
CALIFORNIA	ENDANGERED	LILIACEAE	CHLOROGALUM PURPUREUM VAR. PURPUREUM
CALIFORNIA	ENDANGERED	LILIACEAE	CHLOROGALUM PURPUREUM VAR. REDUCTUM
CALIFORNIA	ENDANGERED	LILIACEAE	ERYTHRONIUM GRANDIFLORUM SSP. PUSATERII
CALIFORNIA	ENDANGERED	LILIACEAE	FRITILLARIA PHAEANTHERA
CALIFORNIA	ENDANGERED	LILIACEAE	FRITILLARIA RODERICKII
CALIFORNIA	ENDANGERED	LILIACEAE	LILIUM OCCIDENTALE
CALIFORNIA	ENDANGERED	LILIACEAE	LILIUM PITKINENSE
CALIFORNIA	ENDANGERED	LILIACEAE	NOLINA INTERRATA
CALIFORNIA	ENDANGERED	LIMNANTHACEAE	LIMNANTHES BAKERI
CALIFORNIA	ENDANGERED	LIMNANTHACEAE	LIMNANTHES GRACILIS VAR. PARISHII
CALIFORNIA	ENDANGERED	LIMNANTHACEAE	LIMNANTHES VINCULANS
CALIFORNIA	ENDANGERED	LINACEAE	HESPEROLINON CONGESTUM
CALIFORNIA	ENDANGERED	LINACEAE	HESPEROLINON DIDYMOCARPUM

STATE LISTS OF ENDANGERED, EXTINCT AND THREATENED SPECIES IN THE CONTINENTAL UNITED STATES

STATE	STATUS	FAMILY	SPECIES
CALIFORNIA	ENDANGERED	LOASACEAE	MENTZELIA LEUCOPHYLLA
CALIFORNIA	ENDANGERED	MALVACEAE	HIBISCUS CALIFORNICUS
CALIFORNIA	ENDANGERED	MALVACEAE	LAVATERA ASSURGENTIFLORA
CALIFORNIA	ENDANGERED	MALVACEAE	MALACOTHAMNUS CLEMENTINUS
CALIFORNIA	ENDANGERED	MALVACEAE	MALACOTHAMNUS PALMERI VAR. INVOLUCRATUS
CALIFORNIA	ENDANGERED	MALVACEAE	SIDALCEA OREGANA SSP. VALIDA
CALIFORNIA	ENDANGERED	NYCTAGINACEAE	ABRONIA ALPINA
CALIFORNIA	ENDANGERED	ONAGRACEAE	CLARKIA BOREALIS SSP. ARIDA
CALIFORNIA	ENDANGERED	ONAGRACEAE	CLARKIA FRANCISCANA
CALIFORNIA	ENDANGERED	ONAGRACEAE	CLARKIA IMBRICATA
CALIFORNIA	ENDANGERED	ONAGRACEAE	CLARKIA LINGULATA
CALIFORNIA	ENDANGERED	ONAGRACEAE	CLARKIA MOSQUINII SSP. MOSQUINII
CALIFORNIA	ENDANGERED	ONAGRACEAE	CLARKIA SPECIOSA SSP. IMMACULATA
CALIFORNIA	ENDANGERED	ONAGRACEAE	OENOTHERA AVITA SSP. EUREKENSIS
CALIFORNIA	ENDANGERED	ONAGRACEAE	OENOTHERA DELTOIDES VAR. HOWELLII
CALIFORNIA	ENDANGERED	PAPAVERACEAE	ARCTOMECON MERRIAMII
CALIFORNIA	ENDANGERED	POACEAE	AGROSTIS BLASDALEI VAR. MARINENSIS
CALIFORNIA	ENDANGERED	POACEAE	AGROSTIS HENDERSONII
CALIFORNIA	ENDANGERED	POACEAE	NEOSTAPFIA COLUSANA
CALIFORNIA	ENDANGERED	POACEAE	ORCUTTIA CALIFORNICA VAR. CALIFORNICA
CALIFORNIA	ENDANGERED	POACEAE	ORCUTTIA CALIFORNICA VAR. INAEQUALIS
CALIFORNIA	ENDANGERED	POACEAE	ORCUTTIA CALIFORNICA VAR. VISCIDA
CALIFORNIA	ENDANGERED	POACEAE	ORCUTTIA GREENEI
CALIFORNIA	ENDANGERED	POACEAE	ORCUTTIA MUCRONATA
CALIFORNIA	ENDANGERED	POACEAE	ORCUTTIA PILOSA
CALIFORNIA	ENDANGERED	POACEAE	ORCUTTIA TENUIS
CALIFORNIA	ENDANGERED	POACEAE	PANICUM THERMALE
CALIFORNIA	ENDANGERED	POACEAE	PLEUROPOGON HOOVERIANUS
CALIFORNIA	ENDANGERED	POACEAE	POA ATROPURPUREA
CALIFORNIA	ENDANGERED	POACEAE	POA FIBRATA
CALIFORNIA	ENDANGERED	POACEAE	POA NAPENSIS
CALIFORNIA	ENDANGERED	POACEAE	STIPA LEMMONII VAR. PUBESCENS
CALIFORNIA	ENDANGERED	POACEAE	SWALLENIA ALEXANDRAE
CALIFORNIA	ENDANGERED	POLEMONIACEAE	ERIASTRUM TRACYI
CALIFORNIA	ENDANGERED	POLEMONIACEAE	NAVARRETIA PAUCIFLORA
CALIFORNIA	ENDANGERED	POLEMONIACEAE	NAVARRETIA PLIEANTHA
CALIFORNIA	ENDANGERED	POLEMONIACEAE	NAVARRETIA SETILOBA
CALIFORNIA	ENDANGERED	POLEMONIACEAE	PHLOX HIRSUTA
CALIFORNIA	ENDANGERED	POLYGONACEAE	CHORIZANTHE LEPTOCERAS
CALIFORNIA	ENDANGERED	POLYGONACEAE	CHORIZANTHE ORCUTTIANA
CALIFORNIA	ENDANGERED	POLYGONACEAE	CHORIZANTHE PARRYI VAR. FERNANDINA
CALIFORNIA	ENDANGERED	POLYGONACEAE	CHORIZANTHE SPINOSA
CALIFORNIA	ENDANGERED	POLYGONACEAE	DEDECKERA EUREKENSIS

STATE LISTS OF ENDANGERED, EXTINCT AND THREATENED SPECIES IN THE CONTINENTAL UNITED STATES

STATE	STATUS	FAMILY	SPECIES
CALIFORNIA	ENDANGERED	POLYGONACEAE	ERIOGONUM ALPINUM
CALIFORNIA	ENDANGERED	POLYGONACEAE	ERIOGONUM APRICUM VAR. APRICUM
CALIFORNIA	ENDANGERED	POLYGONACEAE	ERIOGONUM APRICUM VAR. PROSTRATUM
CALIFORNIA	ENDANGERED	POLYGONACEAE	ERIOGONUM BREEDLOVEI
CALIFORNIA	ENDANGERED	POLYGONACEAE	ERIOGONUM BUTTERWORTHIANUM
CALIFORNIA	ENDANGERED	POLYGONACEAE	ERIOGONUM CANINUM
CALIFORNIA	ENDANGERED	POLYGONACEAE	ERIOGONUM CROCATUM
CALIFORNIA	ENDANGERED	POLYGONACEAE	ERIOGONUM DICLINUM
CALIFORNIA	ENDANGERED	POLYGONACEAE	ERIOGONUM ERICIFOLIUM VAR. THORNEI
CALIFORNIA	ENDANGERED	POLYGONACEAE	ERIOGONUM GIGANTEUM VAR. COMPACTUM
CALIFORNIA	ENDANGERED	POLYGONACEAE	ERIOGONUM GILMANII
CALIFORNIA	ENDANGERED	POLYGONACEAE	ERIOGONUM GRANDE VAR. TIMORUM
CALIFORNIA	ENDANGERED	POLYGONACEAE	ERIOGONUM HIRTELLUM
CALIFORNIA	ENDANGERED	POLYGONACEAE	ERIOGONUM INTRAFRACTUM
CALIFORNIA	ENDANGERED	POLYGONACEAE	ERIOGONUM KENNEDYI VAR. PINICOLA
CALIFORNIA	ENDANGERED	POLYGONACEAE	ERIOGONUM MICROTHECUM VAR. JOHNSTONII
CALIFORNIA	ENDANGERED	POLYGONACEAE	ERIOGONUM NUDUM VAR. MURINUM
CALIFORNIA	ENDANGERED	POLYGONACEAE	ERIOGONUM OVALIFOLIUM VAR. VINEUM
CALIFORNIA	ENDANGERED	POLYGONACEAE	ERIOGONUM PARVIFOLIUM VAR. LUCIDUM
CALIFORNIA	ENDANGERED	POLYGONACEAE	ERIOGONUM TRUNCATUM
CALIFORNIA	ENDANGERED	POLYGONACEAE	ERIOGONUM UMBELLATUM VAR. MINUS
CALIFORNIA	ENDANGERED	POLYGONACEAE	ERIOGONUM UMBELLATUM VAR. TORREYANUM
CALIFORNIA	ENDANGERED	POLYGONACEAE	ERIOGONUM WRIGHTII VAR. OLANCHENSE
CALIFORNIA	ENDANGERED	POLYPODIACEAE	CHEILANTHES FIBRILLOSA
CALIFORNIA	ENDANGERED	PORTULACACEAE	CALYPTRIDIUM PULCHELLUM
CALIFORNIA	ENDANGERED	RANUNCULACEAE	DELPHINIUM BAKERI
CALIFORNIA	ENDANGERED	RANUNCULACEAE	DELPHINIUM KINKIENSE
CALIFORNIA	ENDANGERED	RANUNCULACEAE	DELPHINIUM LUTEUM
CALIFORNIA	ENDANGERED	RHAMNACEAE	CEANOTHUS FERRISAE
CALIFORNIA	ENDANGERED	RHAMNACEAE	CEANOTHUS HEARSTIORUM
CALIFORNIA	ENDANGERED	RHAMNACEAE	CEANOTHUS MARITIMUS
CALIFORNIA	ENDANGERED	RHAMNACEAE	CEANOTHUS MASONII
CALIFORNIA	ENDANGERED	ROSACEAE	CERCOCARPUS TRASKIAE
CALIFORNIA	ENDANGERED	ROSACEAE	HORKELIA WILDERAE
CALIFORNIA	ENDANGERED	ROSACEAE	IVESIA CALLIDA
CALIFORNIA	ENDANGERED	ROSACEAE	POTENTILLA HICKMANII
CALIFORNIA	ENDANGERED	RUBIACEAE	GALIUM ANGUSTIFOLIUM SSP. BORREGOENSE
CALIFORNIA	ENDANGERED	RUBIACEAE	GALIUM CALIFORNICUM SSP. LUCIENSE
CALIFORNIA	ENDANGERED	RUBIACEAE	GALIUM CALIFORNICUM SSP. PRIMUM
CALIFORNIA	ENDANGERED	RUBIACEAE	GALIUM CALIFORNICUM SSP. SIERRAE
CALIFORNIA	ENDANGERED	RUBIACEAE	GALIUM CATALINENSE SSP. ACRISPUM
CALIFORNIA	ENDANGERED	RUBIACEAE	GALIUM GLABRESCENS SSP. MODOCENSE
CALIFORNIA	ENDANGERED	RUBIACEAE	GALIUM GRANDE

STATE LISTS OF ENDANGERED, EXTINCT AND THREATENED SPECIES IN THE CONTINENTAL UNITED STATES

STATE	STATUS	FAMILY	SPECIES
CALIFORNIA	ENDANGERED	RUBIACEAE	GALIUM HARDHAMAE
CALIFORNIA	ENDANGERED	RUBIACEAE	GALIUM HILENDIAE SSP. KINGSTONENSE
CALIFORNIA	ENDANGERED	RUBIACEAE	GALIUM SERPENTICUM SSP. SCOTTICUM
CALIFORNIA	ENDANGERED	SAXIFRAGACEAE	BENSONIELLA OREGANA
CALIFORNIA	ENDANGERED	SCROPHULARIACEAE	CASTILLEJA GRISEA
CALIFORNIA	ENDANGERED	SCROPHULARIACEAE	CASTILLEJA ULIGINOSA
CALIFORNIA	ENDANGERED	SCROPHULARIACEAE	CORDYLANTHUS BRUNNEUS SSP. CAPILLARIS
CALIFORNIA	ENDANGERED	SCROPHULARIACEAE	CORDYLANTHUS EREMICUS SSP. BERNARDINUS
CALIFORNIA	ENDANGERED	SCROPHULARIACEAE	CORDYLANTHUS MARITIMUS SSP. MARITIMUS
CALIFORNIA	ENDANGERED	SCROPHULARIACEAE	CORDYLANTHUS NIDULARIUS
CALIFORNIA	ENDANGERED	SCROPHULARIACEAE	CORDYLANTHUS RIGIDUS SSP. LITTORALIS
CALIFORNIA	ENDANGERED	SCROPHULARIACEAE	CORDYLANTHUS TENUIS SSP. PALLESCENS
CALIFORNIA	ENDANGERED	SCROPHULARIACEAE	GRATIOLA HETEROSEPALA
CALIFORNIA	ENDANGERED	SCROPHULARIACEAE	MIMULUS GUTTATUS SSP. ARENICOLA
CALIFORNIA	ENDANGERED	SCROPHULARIACEAE	MIMULUS PYGMAEUS
CALIFORNIA	ENDANGERED	SCROPHULARIACEAE	MIMULUS WHIPPLEI
CALIFORNIA	ENDANGERED	SCROPHULARIACEAE	ORTHOCARPUS CASTILLEJOIDES VAR. HUMBOLDTIENSIS
CALIFORNIA	ENDANGERED	SCROPHULARIACEAE	ORTHOCARPUS SUCCULENTIS
CALIFORNIA	ENDANGERED	SCROPHULARIACEAE	PEDICULARIS DUDLEYI
CALIFORNIA	ENDANGERED	SCROPHULARIACEAE	PENSTEMON PERSONATUS
CALIFORNIA	ENDANGERED	STERCULIACEAE	FREMONTODENDRON DECUMBENS
CALIFORNIA	EXTINCT	APIACEAE	LILAEOPSIS MASONII
CALIFORNIA	EXTINCT	ASTERACEAE	CALYCADENIA FREMONTII
CALIFORNIA	EXTINCT	ASTERACEAE	ERIOPHYLLUM NUBIGENUM VAR. NUBIGENUM
CALIFORNIA	EXTINCT	ASTERACEAE	HELIANTHUS NUTTALLII VAR. PARISHII
CALIFORNIA	EXTINCT	CAMPANULACEAE	GITHOPSIS FILICAULIS
CALIFORNIA	EXTINCT	CAMPANULACEAE	GITHOPSIS LATIFOLIA
CALIFORNIA	EXTINCT	CHENOPODIACEAE	ATRIPLEX TULARENSIS
CALIFORNIA	EXTINCT	ERICACEAE	ARCTOSTAPHYLOS HOOKERI SSP. FRANCISCANA
CALIFORNIA	EXTINCT	FABACEAE	ASTRAGALUS PYCNOSTACHYUS SSP. LANOSISSIMUS
CALIFORNIA	EXTINCT	FABACEAE	LOTUS ARGOPHYLLUS VAR. ADSURGENS
CALIFORNIA	EXTINCT	HYDROCHARITACEAE	ELODEA BRANDEGEAE
CALIFORNIA	EXTINCT	HYDROPHYLLACEAE	PHACELIA CINEREA
CALIFORNIA	EXTINCT	JUNCACEAE	JUNCUS LEIOSPERMUS
CALIFORNIA	EXTINCT	LAMIACEAE	MONARDELLA LEUCOCEPHALA
CALIFORNIA	EXTINCT	LILIACEAE	CALOCHORTUS MONANTHUS
CALIFORNIA	EXTINCT	MALVACEAE	MALACOTHAMNUS ABBOTTII
CALIFORNIA	EXTINCT	MALVACEAE	MALACOTHAMNUS MENDOCINENSIS
CALIFORNIA	EXTINCT	MALVACEAE	SIDALCEA COVILLEI
CALIFORNIA	EXTINCT	ONAGRACEAE	CLARKIA MOSQUINII SSP. XEROPHILA
CALIFORNIA	EXTINCT	OROBANCHACEAE	OROBANCHE VALIDA
CALIFORNIA	EXTINCT	POACEAE	DISSANTHELIUM CALIFORNICUM
CALIFORNIA	EXTINCT	POLYGONACEAE	CHORIZANTHE VALIDA

STATE LISTS OF ENDANGERED, EXTINCT AND THREATENED SPECIES IN THE CONTINENTAL UNITED STATES

STATE	STATUS	FAMILY	SPECIES
CALIFORNIA	EXTINCT	POLYGONACEAE	POLYGONUM MONTEREYENSE
CALIFORNIA	EXTINCT	ROSACEAE	POTENTILLA MULTIJUGA
CALIFORNIA	EXTINCT	SAXIFRAGACEAE	LITHOPHRAGMA MAXIMUM
CALIFORNIA	EXTINCT	SCROPHULARIACEAE	CASTILLEJA LESCHKEANA
CALIFORNIA	EXTINCT	SCROPHULARIACEAE	CORDYLANTHUS MOLLIS SSP. MOLLIS
CALIFORNIA	EXTINCT	SCROPHULARIACEAE	CORDYLANTHUS PALMATUS
CALIFORNIA	EXTINCT	SCROPHULARIACEAE	MIMULUS BRANDEGEI
CALIFORNIA	EXTINCT	SCROPHULARIACEAE	MIMULUS TRASKIAE
CALIFORNIA	EXTINCT	SCROPHULARIACEAE	ORTHOCARPUS PACHYSTACHYUS
CALIFORNIA	EXTINCT	SOLANACEAE	LYCIUM HASSEI
CALIFORNIA	EXTINCT	SOLANACEAE	LYCIUM VERRUCOSUM
CALIFORNIA	THREATENED	APIACEAE	CICUTA BOLANDERI
CALIFORNIA	THREATENED	APIACEAE	CYMOPTERUS DESERTICOLA
CALIFORNIA	THREATENED	APIACEAE	ERYNGIUM PINNATISECTUM
CALIFORNIA	THREATENED	APIACEAE	LOMATIUM CONGDONII
CALIFORNIA	THREATENED	APIACEAE	LOMATIUM FOENICULACEUM SSP. INYOENSE
CALIFORNIA	THREATENED	APIACEAE	LOMATIUM HOWELLII
CALIFORNIA	THREATENED	APIACEAE	LOMATIUM PECKIANUM
CALIFORNIA	THREATENED	APIACEAE	LOMATIUM RIGIDUM
CALIFORNIA	THREATENED	APIACEAE	PERIDERIDIA BACIGALUPII
CALIFORNIA	THREATENED	APIACEAE	PERIDERIDIA GAIRDNERI SSP. GAIRDNERI
CALIFORNIA	THREATENED	APIACEAE	PERIDERIDIA LEPTOCARPA
CALIFORNIA	THREATENED	APIACEAE	PERIDERIDIA PRINGLEI
CALIFORNIA	THREATENED	APIACEAE	SANICULA HOFFMANNII
CALIFORNIA	THREATENED	APIACEAE	SANICULA PECKIANA
CALIFORNIA	THREATENED	APIACEAE	TAUSCHIA GLAUCA
CALIFORNIA	THREATENED	APIACEAE	TAUSCHIA HOWELLII
CALIFORNIA	THREATENED	ASTERACEAE	AMBROSIA PUMILA
CALIFORNIA	THREATENED	ASTERACEAE	ANTENNARIA SUFFRUTESCENS
CALIFORNIA	THREATENED	ASTERACEAE	ARNICA VENOSA
CALIFORNIA	THREATENED	ASTERACEAE	ARNICA VISCOSA
CALIFORNIA	THREATENED	ASTERACEAE	ASTER BRICKELLIOIDES
CALIFORNIA	THREATENED	ASTERACEAE	ASTER PEIRSONII
CALIFORNIA	THREATENED	ASTERACEAE	BENITOA OCCIDENTALIS
CALIFORNIA	THREATENED	ASTERACEAE	BRICKELLIA KNAPPIANA
CALIFORNIA	THREATENED	ASTERACEAE	CHAENACTIS PARISHII
CALIFORNIA	THREATENED	ASTERACEAE	CIRSIUM CAMPYLON
CALIFORNIA	THREATENED	ASTERACEAE	CIRSIUM HYDROPHILUM VAR. VASEYI
CALIFORNIA	THREATENED	ASTERACEAE	COREOPSIS HAMILTONII
CALIFORNIA	THREATENED	ASTERACEAE	ERIGERON AEQUIFOLIUS
CALIFORNIA	THREATENED	ASTERACEAE	ERIGERON BLOOMERI VAR. NUDATUS
CALIFORNIA	THREATENED	ASTERACEAE	ERIGERON FLEXUOSUS
CALIFORNIA	THREATENED	ASTERACEAE	ERIGERON MULTICEPS

STATE LISTS OF ENDANGERED, EXTINCT AND THREATENED SPECIES IN THE CONTINENTAL UNITED STATES

STATE	STATUS	FAMILY	SPECIES
CALIFORNIA	THREATENED	ASTERACEAE	ERIGERON PARISHII
CALIFORNIA	THREATENED	ASTERACEAE	ERIGERON SUPPLEX
CALIFORNIA	THREATENED	ASTERACEAE	ERIOPHYLLUM LATILOBUM
CALIFORNIA	THREATENED	ASTERACEAE	ERIOPHYLLUM NUBIGENUM VAR. CONGDONII
CALIFORNIA	THREATENED	ASTERACEAE	EUPATORIUM SHASTENSE
CALIFORNIA	THREATENED	ASTERACEAE	GUTIERREZIA CALIFORNICA
CALIFORNIA	THREATENED	ASTERACEAE	HAPLOPAPPUS BRICKELLIOIDES
CALIFORNIA	THREATENED	ASTERACEAE	HAPLOPAPPUS OPHITIDIS
CALIFORNIA	THREATENED	ASTERACEAE	HELIANTHELLA CASTANEA
CALIFORNIA	THREATENED	ASTERACEAE	HEMIZONIA ARIDA
CALIFORNIA	THREATENED	ASTERACEAE	HEMIZONIA HALLIANA
CALIFORNIA	THREATENED	ASTERACEAE	HULSEA VESTITA SSP. INYOENSIS
CALIFORNIA	THREATENED	ASTERACEAE	LASTHENIA LEPTALEA
CALIFORNIA	THREATENED	ASTERACEAE	LASTHENIA MINOR SSP. MARITIMA
CALIFORNIA	THREATENED	ASTERACEAE	LAYIA JONESII
CALIFORNIA	THREATENED	ASTERACEAE	LAYIA LEUCOPAPPA
CALIFORNIA	THREATENED	ASTERACEAE	LAYIA ZIEGLERI
CALIFORNIA	THREATENED	ASTERACEAE	MACHAERANTHERA AMMOPHILA
CALIFORNIA	THREATENED	ASTERACEAE	MACHAERANTHERA LAGUNENSIS
CALIFORNIA	THREATENED	ASTERACEAE	MACHAERANTHERA ORCUTTII
CALIFORNIA	THREATENED	ASTERACEAE	MALACOTHRIX SAXATILIS VAR. ARACHNOIDEA
CALIFORNIA	THREATENED	ASTERACEAE	PENTACHAETA BELLIDIFLORA
CALIFORNIA	THREATENED	ASTERACEAE	PENTACHAETA EXILIS SSP. AEOLICA
CALIFORNIA	THREATENED	ASTERACEAE	PERITYLE INYOENSIS
CALIFORNIA	THREATENED	ASTERACEAE	PERITYLE VILLOSA
CALIFORNIA	THREATENED	ASTERACEAE	PYRROCOMA UNIFLORUS VAR. GOSSYPINUS
CALIFORNIA	THREATENED	ASTERACEAE	RAILLARDELLA MUIRII
CALIFORNIA	THREATENED	ASTERACEAE	RAILLARDELLA PRINGLEI
CALIFORNIA	THREATENED	ASTERACEAE	SENECIO BERNARDINUS
CALIFORNIA	THREATENED	ASTERACEAE	SENECIO GANDERI
CALIFORNIA	THREATENED	ASTERACEAE	TANACETUM CAMPHORATUM
CALIFORNIA	THREATENED	ASTERACEAE	TARAXACUM CALIFORNICUM
CALIFORNIA	THREATENED	ASTERACEAE	WYETHIA RETICULATA
CALIFORNIA	THREATENED	BERBERIDACEAE	BERBERIS HIGGINSAE
CALIFORNIA	THREATENED	BERBERIDACEAE	BERBERIS PINNATA SSP. INSULARIS
CALIFORNIA	THREATENED	BORAGINACEAE	AMSINCKIA VERNICOSA VAR. FURCATA
CALIFORNIA	THREATENED	BORAGINACEAE	CRYPTANTHA CRINITA
CALIFORNIA	THREATENED	BORAGINACEAE	CRYPTANTHA GANDERI
CALIFORNIA	THREATENED	BORAGINACEAE	HACKELIA BREVICULA
CALIFORNIA	THREATENED	BORAGINACEAE	PLAGIOBOTHRYS DISTANTIFLORUS
CALIFORNIA	THREATENED	BORAGINACEAE	PLAGIOBOTHRYS GLABER
CALIFORNIA	THREATENED	BORAGINACEAE	PLAGIOBOTHRYS HYSTRICULUS
CALIFORNIA	THREATENED	BORAGINACEAE	PLAGIOBOTHRYS SCRIPTUS

STATE LISTS OF ENDANGERED, EXTINCT AND THREATENED SPECIES IN THE CONTINENTAL UNITED STATES

STATE	STATUS	FAMILY	SPECIES
CALIFORNIA	THREATENED	BORAGINACEAE	PLAGIOBOTHRYS STRICTUS
CALIFORNIA	THREATENED	BRASSICACEAE	ARABIS BLEPHAROPHYLLA
CALIFORNIA	THREATENED	BRASSICACEAE	ARABIS BREWERI VAR. AUSTINAE
CALIFORNIA	THREATENED	BRASSICACEAE	ARABIS CONSTANCEI
CALIFORNIA	THREATENED	BRASSICACEAE	ARABIS HOFFMANNII
CALIFORNIA	THREATENED	BRASSICACEAE	ARABIS JOHNSTONII
CALIFORNIA	THREATENED	BRASSICACEAE	ARABIS MODESTA
CALIFORNIA	THREATENED	BRASSICACEAE	ARABIS OREGANA
CALIFORNIA	THREATENED	BRASSICACEAE	ARABIS PARISHII
CALIFORNIA	THREATENED	BRASSICACEAE	ARABIS PYGMAEA
CALIFORNIA	THREATENED	BRASSICACEAE	CARDAMINE PENDULIFLORA
CALIFORNIA	THREATENED	BRASSICACEAE	CAULANTHUS AMPLEXICAULIS VAR. BARBARAE
CALIFORNIA	THREATENED	BRASSICACEAE	CAULANTHUS STENOCARPUS
CALIFORNIA	THREATENED	BRASSICACEAE	CAULOSTRAMINA JAEGERI
CALIFORNIA	THREATENED	BRASSICACEAE	DRABA ASTEROPHORA VAR. ASTEROPHORA
CALIFORNIA	THREATENED	BRASSICACEAE	DRABA ASTEROPHORA VAR. MACROCARPA
CALIFORNIA	THREATENED	BRASSICACEAE	DRABA CRUCIATA VAR. CRUCIATA
CALIFORNIA	THREATENED	BRASSICACEAE	DRABA CRUCIATA VAR. INTEGRIFOLIA
CALIFORNIA	THREATENED	BRASSICACEAE	DRABA HOWELLII VAR. CARNOSULA
CALIFORNIA	THREATENED	BRASSICACEAE	DRABA LEMMONII VAR. INCRASSATA
CALIFORNIA	THREATENED	BRASSICACEAE	DRABA QUADRICOSTATA
CALIFORNIA	THREATENED	BRASSICACEAE	DRABA STENOLOBA VAR. RAMOSA
CALIFORNIA	THREATENED	BRASSICACEAE	ERYSIMUM AMMOPHILUM
CALIFORNIA	THREATENED	BRASSICACEAE	ERYSIMUM INSULARE
CALIFORNIA	THREATENED	BRASSICACEAE	ERYSIMUM MENZIESII
CALIFORNIA	THREATENED	BRASSICACEAE	LESQUERELLA KINGII SSP. BERNARDINA
CALIFORNIA	THREATENED	BRASSICACEAE	RORIPPA COLUMBIAE
CALIFORNIA	THREATENED	BRASSICACEAE	RORIPPA SUBUMBELLATA
CALIFORNIA	THREATENED	BRASSICACEAE	STREPTANTHUS BATRACHOPUS
CALIFORNIA	THREATENED	BRASSICACEAE	STREPTANTHUS BRACHIATUS
CALIFORNIA	THREATENED	BRASSICACEAE	STREPTANTHUS CORDATUS VAR. PIUTENSIS
CALIFORNIA	THREATENED	BRASSICACEAE	STREPTANTHUS FENESTRATUS
CALIFORNIA	THREATENED	BRASSICACEAE	STREPTANTHUS GLANDULOSUS VAR. HOFFMANII
CALIFORNIA	THREATENED	BRASSICACEAE	STREPTANTHUS GLANDULOSUS VAR. PULCHELLUS
CALIFORNIA	THREATENED	BRASSICACEAE	STREPTANTHUS GRACILIS
CALIFORNIA	THREATENED	BRASSICACEAE	STREPTANTHUS HISPIDUS
CALIFORNIA	THREATENED	BRASSICACEAE	STREPTANTHUS MORRISONII SSP. ELATUS
CALIFORNIA	THREATENED	BRASSICACEAE	STREPTANTHUS MORRISONII SSP. MORRISONII
CALIFORNIA	THREATENED	BRASSICACEAE	STREPTANTHUS OLIGANTHUS
CALIFORNIA	THREATENED	BRASSICACEAE	THELYPODIUM STENOPETALUM
CALIFORNIA	THREATENED	BRASSICACEAE	THYSANOCARPUS CONCHULIFERUS
CALIFORNIA	THREATENED	BRASSICACEAE	TROPIDOCARPUM CAPPARIDEUM
CALIFORNIA	THREATENED	CACTACEAE	CORYPHANTHA VIVIPARA VAR. ALVERSONII

STATE LISTS OF ENDANGERED, EXTINCT AND THREATENED SPECIES IN THE CONTINENTAL UNITED STATES

STATE	STATUS	FAMILY	SPECIES
CALIFORNIA	THREATENED	CACTACEAE	CORYPHANTHA VIVIPARA VAR. ROSEA
CALIFORNIA	THREATENED	CACTACEAE	ECHINOCEREUS ENGELMANNII VAR. MUNZII
CALIFORNIA	THREATENED	CACTACEAE	OPUNTIA BASILARIS VAR. BRACHYCLADA
CALIFORNIA	THREATENED	CACTACEAE	OPUNTIA MUNZII
CALIFORNIA	THREATENED	CACTACEAE	OPUNTIA PARRYI VAR. SERPENTINA
CALIFORNIA	THREATENED	CACTACEAE	SCLEROCACTUS POLYANCISTRUS
CALIFORNIA	THREATENED	CAMPANULACEAE	CAMPANULA SHETLERI
CALIFORNIA	THREATENED	CAMPANULACEAE	CAMPANULA WILKINSIANA
CALIFORNIA	THREATENED	CAMPANULACEAE	NEMACLADUS TWISSELMANNII
CALIFORNIA	THREATENED	CARYOPHYLLACEAF	ARENARIA HOWELLII
CALIFORNIA	THREATENED	CARYOPHYLLACEAE	ARENARIA ROSEI
CALIFORNIA	THREATENED	CHENOPODIACEAE	ATRIPLEX VALLICOLA
CALIFORNIA	THREATENED	CISTACEAE	HELIANTHEMUM GREENEI
CALIFORNIA	THREATENED	CISTACEAE	HELIANTHEMUM SUFFRUTESCENS
CALIFORNIA	THREATENED	CONVOLVULACEAE	CALYSTEGIA PEIRSONII
CALIFORNIA	THREATENED	CONVOLVULACEAE	DICHONDRA DONNELLIANA
CALIFORNIA	THREATENED	CRASSULACEAE	DUDLEYA BLOCHMANAE SSP. BREVIFOLIA
CALIFORNIA	THREATENED	CRASSULACEAE	DUDLEYA BLOCHMANAE SSP. INSULARIS
CALIFORNIA	THREATENED	CRASSULACEAE	DUDLEYA DENSIFLORA
CALIFORNIA	THREATENED	CRASSULACEAE	DUDLEYA PARVA
CALIFORNIA	THREATENED	CRASSULACEAE	DUDLEYA VARIEGATA
CALIFORNIA	THREATENED	CRASSULACEAE	DUDLEYA VISCIDA
CALIFORNIA	THREATENED	CRASSULACEAE	SEDUM ALBOMARGINATUM
CALIFORNIA	THREATENED	CRASSULACEAE	SEDUM LAXUM SSP. EASTWOODIAE
CALIFORNIA	THREATENED	CUPRESSACEAE	CUPRESSUS ARIZONICA VAR. STEPHENSONII
CALIFORNIA	THREATENED	CUPRESSACEAE	CUPRESSUS MACROCARPA
CALIFORNIA	THREATENED	CYPERACEAE	CAREX OBISPOENSIS
CALIFORNIA	THREATENED	CYPERACEAE	CAREX PAUCIFRUCTA
CALIFORNIA	THREATENED	CYPERACEAE	CAREX WHITNEYI
CALIFORNIA	THREATENED	ERICACEAE	ARCTOSTAPHYLOS CONFERTIFLORA
CALIFORNIA	THREATENED	ERICACEAE	ARCTOSTAPHYLOS CRUZENSIS
CALIFORNIA	THREATENED	ERICACEAE	ARCTOSTAPHYLOS EDMUNDSII VAR. EDMUNDSII
CALIFORNIA	THREATENED	ERICACEAE	ARCTOSTAPHYLOS HOOKERI SSP. MONTANA
CALIFORNIA	THREATENED	ERICACEAE	ARCTOSTAPHYLOS LUCIANA
CALIFORNIA	THREATENED	ERICACEAE	ARCTOSTAPHYLOS MONTARAENSIS
CALIFORNIA	THREATENED	ERICACEAE	ARCTOSTAPHYLOS MONTEREYENSIS
CALIFORNIA	THREATENED	ERICACEAE	ARCTOSTAPHYLOS MORROENSIS
CALIFORNIA	THREATENED	ERICACEAE	ARCTOSTAPHYLOS OTAYENSIS
CALIFORNIA	THREATENED	ERICACEAE	ARCTOSTAPHYLOS PECHOENSIS
CALIFORNIA	THREATENED	ERICACEAE	ARCTOSTAPHYLOS SILVICOLA
CALIFORNIA	THREATENED	ERICACEAE	ARCTOSTAPHYLOS STANFORDIANA SSP. HISPIDULA
CALIFORNIA	THREATENED	ERICACEAE	ARCTOSTAPHYLOS VIRGATA
CALIFORNIA	THREATENED	ERICACEAE	PITYOPUS CALIFORNICUS

STATE LISTS OF ENDANGERED, EXTINCT AND THREATENED SPECIES IN THE CONTINENTAL UNITED STATES

STATE	STATUS	FAMILY	SPECIES
CALIFORNIA	THREATENED	ERICACEAE	VACCINIUM COCCINIUM
CALIFORNIA	THREATENED	EUPHORBIACEAE	DITAXIS CALIFORNICA
CALIFORNIA	THREATENED	EUPHORBIACEAE	EUPHORBIA HOOVERI
CALIFORNIA	THREATENED	EUPHORBIACEAE	EUPHORBIA PLATYSPERMA
CALIFORNIA	THREATENED	EUPHORBIACEAE	TETRACOCCUS ILICIFOLIUS
CALIFORNIA	THREATENED	FABACEAE	ASTRAGALUS BRAUNTONII
CALIFORNIA	THREATENED	FABACEAE	ASTRAGALUS CIMAE VAR. CIMAE
CALIFORNIA	THREATENED	FABACEAE	ASTRAGALUS DEANEI
CALIFORNIA	THREATENED	FABACEAE	ASTRAGALUS MOHAVENSIS VAR. HEMIGYRUS
CALIFORNIA	THREATENED	FABACEAE	ASTRAGALUS PAUPERCULUS
CALIFORNIA	THREATENED	FABACEAE	ASTRAGALUS PSEUDIODANTHUS
CALIFORNIA	THREATENED	FABACEAE	ASTRAGALUS SUBVESTITUS
CALIFORNIA	THREATENED	FABACEAE	ASTRAGALUS TRASKIAE
CALIFORNIA	THREATENED	FABACEAE	LOTUS ARGOPHYLLUS VAR. NIVEUS
CALIFORNIA	THREATENED	FABACEAE	LOTUS NUTTALIANUS
CALIFORNIA	THREATENED	FABACEAE	LUPINUS CERVINUS
CALIFORNIA	THREATENED	FABACEAE	LUPINUS CITRINUS VAR. CITRINUS
CALIFORNIA	THREATENED	FABACEAE	LUPINUS CITRINUS VAR. DEFLEXUS
CALIFORNIA	THREATENED	FABACEAE	LUPINUS DEDECKERAE
CALIFORNIA	THREATENED	FABACEAE	LUPINUS DURANII
CALIFORNIA	THREATENED	FABACEAE	LUPINUS HOLMGRENANUS
CALIFORNIA	THREATENED	FABACEAE	LUPINUS NIPOMENSIS
CALIFORNIA	THREATENED	FABACEAE	LUPINUS SERICATUS
CALIFORNIA	THREATENED	FABACEAE	LUPINUS SPECTABILIS
CALIFORNIA	THREATENED	FABACEAE	THERMOPSIS MACROPHYLLA VAR. AGNINA
CALIFORNIA	THREATENED	FABACEAE	TRIFOLIUM BOLANDERI
CALIFORNIA	THREATENED	FABACEAE	TRIFOLIUM DEDECKERAE
CALIFORNIA	THREATENED	FAGACEAE	QUERCUS PARVULA
CALIFORNIA	THREATENED	FAGACEAE	QUERCUS TOMENTELLA
CALIFORNIA	THREATENED	FUMARIACEAE	DICENTRA NEVADENSIS
CALIFORNIA	THREATENED	GENTIANACEAE	FRASERA UMPQUAENSIS
CALIFORNIA	THREATENED	GENTIANACEAE	GENTIANA FREMONTII
CALIFORNIA	THREATENED	HYDROPHYLLACEAE	PHACELIA AMABILIS
CALIFORNIA	THREATENED	HYDROPHYLLACEAE	PHACELIA DALESIANA
CALIFORNIA	THREATENED	HYDROPHYLLACEAE	PHACELIA DIVARICATA VAR. INSULARIS
CALIFORNIA	THREATENED	HYDROPHYLLACEAE	PHACELIA GREENEI
CALIFORNIA	THREATENED	HYDROPHYLLACEAE	PHACELIA MUSTELINA
CALIFORNIA	THREATENED	HYDROPHYLLACEAE	PHACELIA NOVENMILLENSIS
CALIFORNIA	THREATENED	HYDROPHYLLACEAE	PHACELIA OROGENES
CALIFORNIA	THREATENED	HYDROPHYLLACEAE	PHACELIA PHACELIOIDES
CALIFORNIA	THREATENED	LAMIACEAE	AGASTACHE PARVIFOLIA
CALIFORNIA	THREATENED	LAMIACEAE	MONARDELLA BENITENSIS
CALIFORNIA	THREATENED	LAMIACEAE	MONARDELLA CRISPA

STATE LISTS OF ENDANGERED, EXTINCT AND THREATENED SPECIES IN THE CONTINENTAL UNITED STATES

STATE	STATUS	FAMILY	SPECIES
CALIFORNIA	THREATENED	LAMIACEAE	MONARDELLA HYPOLEUCA SSP. LANATA
CALIFORNIA	THREATENED	LAMIACEAE	MONARDELLA LINOIDES SSP. OBLONGA
CALIFORNIA	THREATENED	LAMIACEAE	MONARDELLA ROBISONII
CALIFORNIA	THREATENED	LAMIACEAE	SATUREJA CHANDLERI
CALIFORNIA	THREATENED	LENNOACEAE	PHOLISMA ARENARIUM
CALIFORNIA	THREATENED	LILIACEAE	AGAVE UTAHENSIS VAR. NEVADENSIS
CALIFORNIA	THREATENED	LILIACEAE	ALLIUM HOFFMANII
CALIFORNIA	THREATENED	LILIACEAE	ALLIUM YOSEMITENSE
CALIFORNIA	THREATENED	LILIACEAE	CALOCHORTUS DUNNII
CALIFORNIA	THREATENED	LILIACEAE	CALOCHORTUS LONGEBARBATUS VAR. LONGEBARBATUS
CALIFORNIA	THREATENED	LILIACEAE	CALOCHORTUS OBISPOENSIS
CALIFORNIA	THREATENED	LILIACEAE	CALOCHORTUS STRIATUS
CALIFORNIA	THREATENED	LILIACEAE	DICHELOSTEMMA LACUNA-VERNALIS
CALIFORNIA	THREATENED	LILIACEAE	ERYTHRONIUM TUOLUMNENSE
CALIFORNIA	THREATENED	LILIACEAE	FRITILLARIA BRANDEGEI
CALIFORNIA	THREATENED	LILIACEAE	FRITILLARIA FALCATA
CALIFORNIA	THREATENED	LILIACEAE	FRITILLARIA PLURIFLORA
CALIFORNIA	THREATENED	LILIACEAE	FRITILLARIA STRIATA
CALIFORNIA	THREATENED	LILIACEAE	LILIUM VOLLMERI
CALIFORNIA	THREATENED	LILIACEAE	LILIUM WASHINGTONIANUM VAR. MINUS
CALIFORNIA	THREATENED	LILIACEAE	LILIUM WIGGINSII
CALIFORNIA	THREATENED	LILIACEAE	MUILLA CLEVELANDII
CALIFORNIA	THREATENED	LILIACEAE	MUILLA CORONATA
CALIFORNIA	THREATENED	LILIACEAE	SCHOENOLIRION BRACTEOSUM
CALIFORNIA	THREATENED	LILIACEAE	TRITELEIA CLEMENTINA
CALIFORNIA	THREATENED	LIMNANTHACEAE	LIMNANTHES DOUGLASII VAR. SULPHUREA
CALIFORNIA	THREATENED	LIMNANTHACEAE	LIMNANTHES FLOCCOSA SSP. BELLINGERIANA
CALIFORNIA	THREATENED	LINACEAE	HESPEROLINON ADENOPHYLLUM
CALIFORNIA	THREATENED	LINACEAE	HESPEROLINON BICARPELLATUM
CALIFORNIA	THREATENED	LINACEAE	HESPEROLINON BREWERI
CALIFORNIA	THREATENED	LINACEAE	HESPEROLINON DRYMARIOIDES
CALIFORNIA	THREATENED	LOASACEAE	MENTZELIA HIRSUTISSIMA VAR. STENOPHYLLA
CALIFORNIA	THREATENED	LOASACEAE	PETALONYX THURBERI SSP. GILMANII
CALIFORNIA	THREATENED	MALVACEAE	MALACOTHAMNUS FASCICULATUS VAR. NESIOTICUS
CALIFORNIA	THREATENED	MALVACEAE	MALACOTHAMNUS PALMERI VAR. LUCIANUS
CALIFORNIA	THREATENED	MALVACEAE	MALACOTHAMNUS PALMERI VAR. PALMERI
CALIFORNIA	THREATENED	MALVACEAE	MALACOTHAMNUS PARISHII
CALIFORNIA	THREATENED	MALVACEAE	SIDALCEA HICKMANII SSP. ANOMALA
CALIFORNIA	THREATENED	MALVACEAE	SIDALCEA HICKMANII SSP. HICKMANII
CALIFORNIA	THREATENED	MALVACEAE	SIDALCEA HICKMANII SSP. PARISHII
CALIFORNIA	THREATENED	MALVACEAE	SIDALCEA HICKMANII SSP. VIRIDIS
CALIFORNIA	THREATENED	MALVACEAE	SIDALCEA KECKII
CALIFORNIA	THREATENED	MALVACEAE	SIDALCEA OREGANA SSP. HYDROPHILA

STATE LISTS OF ENDANGERED, EXTINCT AND THREATENED SPECIES IN THE CONTINENTAL UNITED STATES

STATE	STATUS	FAMILY	SPECIES
CALIFORNIA	THREATENED	MALVACEAE	SIDALCEA PEDATA
CALIFORNIA	THREATENED	MALVACEAE	SIDALCEA ROBUSTA
CALIFORNIA	THREATENED	MALVACEAE	SPHAERALCEA RUSBYI SSP. EREMICOLA
CALIFORNIA	THREATENED	MYRICACEAE	MYRICA HARTWEGII
CALIFORNIA	THREATENED	ONAGRACEAE	CAMISSONIA BENITENSIS
CALIFORNIA	THREATENED	ONAGRACEAE	CLARKIA AUSTRALIS
CALIFORNIA	THREATENED	ONAGRACEAE	CLARKIA BILOBA SSP. AUSTRALIS
CALIFORNIA	THREATENED	ONAGRACEAE	CLARKIA ROSTRATA
CALIFORNIA	THREATENED	ONAGRACEAE	EPILOBIUM NIVIUM
CALIFORNIA	THREATENED	OPHIOGLOSSACEAE	BOTRYCHIUM PUMICOLA
CALIFORNIA	THREATENED	OPHIOGLOSSACEAE	OPHIOGLOSSUM CALIFORNICUM
CALIFORNIA	THREATENED	ORCHIDACEAE	CYPRIPEDIUM CALIFORNICUM
CALIFORNIA	THREATENED	ORCHIDACEAE	PLATANTHERA UNALASCENSIS SSP. MARITIMA
CALIFORNIA	THREATENED	OROBANCHACEAE	OROBANCHE PARISHII SSP. BRACHYLOBA
CALIFORNIA	THREATENED	PAPAVERACEAE	ARGEMONE MUNITA SSP. ROBUSTA
CALIFORNIA	THREATENED	PAPAVERACEAE	ESCHSCHOLZIA PROCERA
CALIFORNIA	THREATENED	PAPAVERACEAE	ESCHSCHOLZIA RAMOSA
CALIFORNIA	THREATENED	PINACEAE	PINUS TORREYANA
CALIFORNIA	THREATENED	POACEAE	AGROSTIS ARISTIGLUMIS
CALIFORNIA	THREATENED	POACEAE	AGROSTIS CLIVICOLA VAR. PUNTA-REYESENSIS
CALIFORNIA	THREATENED	POACEAE	ALOPECURUS AEQUALIS VAR. SONOMENSIS
CALIFORNIA	THREATENED	POACEAE	CALAMAGROSTIS CRASSIGLUMIS
CALIFORNIA	THREATENED	POACEAE	CALAMAGROSTIS FOLIOSA
CALIFORNIA	THREATENED	POACEAE	HYSTRIX CALIFORNICA
CALIFORNIA	THREATENED	POACEAE	PUCCINELLIA PARISHII
CALIFORNIA	THREATENED	POLEMONIACEAE	ERIASTRUM BRANDEGEAE
CALIFORNIA	THREATENED	POLEMONIACEAE	GILIA RIPLEYI
CALIFORNIA	THREATENED	POLEMONIACEAE	GILIA TENUIFLORA SSP. HOFFMANNII
CALIFORNIA	THREATENED	POLEMONIACEAE	LINANTHUS BELLUS
CALIFORNIA	THREATENED	POLEMONIACEAE	LINANTHUS MACULATUS
CALIFORNIA	THREATENED	POLEMONIACEAE	LINANTHUS ORCUTTII SSP. PACIFICUS
CALIFORNIA	THREATENED	POLEMONIACEAE	NAVARRETIA PROLIFERA SSP. LUTEA
CALIFORNIA	THREATENED	POLYGONACEAE	CHORIZANTHE BREWERI
CALIFORNIA	THREATENED	POLYGONACEAE	CHORIZANTHE HOWELLII
CALIFORNIA	THREATENED	POLYGONACEAE	CHORIZANTHE INSIGNIS
CALIFORNIA	THREATENED	POLYGONACEAE	CHORIZANTHE RECTISPINA
CALIFORNIA	THREATENED	POLYGONACEAE	CHORIZANTHE STATICOIDES SSP. CHRYSACANTHA
CALIFORNIA	THREATENED	POLYGONACEAE	ERIOGONUM AMPULLACEUM
CALIFORNIA	THREATENED	POLYGONACEAE	ERIOGONUM BEATLEYAE
CALIFORNIA	THREATENED	POLYGONACEAE	ERIOGONUM BIFURCATUM
CALIFORNIA	THREATENED	POLYGONACEAE	ERIOGONUM CONGDONII
CALIFORNIA	THREATENED	POLYGONACEAE	ERIOGONUM CONTIGUUM
CALIFORNIA	THREATENED	POLYGONACEAE	ERIOGONUM DESERTICOLA

STATE LISTS OF ENDANGERED, EXTINCT AND THREATENED SPECIES IN THE CONTINENTAL UNITED STATES

STATE	STATUS	FAMILY	SPECIES
CALIFORNIA	THREATENED	POLYGONACEAE	ERIOGONUM EREMICOLA
CALIFORNIA	THREATENED	POLYGONACEAE	ERIOGONUM GOSSYPINUM
CALIFORNIA	THREATENED	POLYGONACEAE	ERIOGONUM HEERMANNII VAR. FLOCCOSUM
CALIFORNIA	THREATENED	POLYGONACEAE	ERIOGONUM HOFFMANNII VAR. HOFFMANNII
CALIFORNIA	THREATENED	POLYGONACEAE	ERIOGONUM HOFFMANNII VAR. ROBUSTIUS
CALIFORNIA	THREATENED	POLYGONACEAE	ERIOGONUM KELLOGGII
CALIFORNIA	THREATENED	POLYGONACEAE	ERIOGONUM KENNEDYI SSP. AUSTROMONTANUM
CALIFORNIA	THREATENED	POLYGONACEAE	ERIOGONUM LATENS
CALIFORNIA	THREATENED	POLYGONACEAE	ERIOGONUM MICROTHECUM VAR. PANAMINTENSE
CALIFORNIA	THREATENED	POLYGONACEAE	ERIOGONUM NORTONII
CALIFORNIA	THREATENED	POLYGONACEAE	ERIOGONUM PARVIFOLIUM VAR. PAYNEI
CALIFORNIA	THREATENED	POLYGONACEAE	ERIOGONUM PENDULUM
CALIFORNIA	THREATENED	POLYGONACEAE	ERIOGONUM SISKIYOUENSE
CALIFORNIA	THREATENED	POLYGONACEAE	ERIOGONUM TEMBLORENSE
CALIFORNIA	THREATENED	POLYGONACEAE	ERIOGONUM TWISSELMANNII
CALIFORNIA	THREATENED	POLYGONACEAE	ERIOGONUM VESTITUM
CALIFORNIA	THREATENED	POLYGONACEAE	GILMANIA LUTEOLA
CALIFORNIA	THREATENED	POLYGONACEAE	POLYGONUM BIDWELLIAE
CALIFORNIA	THREATENED	POLYGONACEAE	POLYGONUM FUSIFORME
CALIFORNIA	THREATENED	POLYGONACEAE	POLYGONUM MARINENSE
CALIFORNIA	THREATENED	PORTULACACEAE	CLAYTONIA LANCEOLATA VAR. PEIRSONII
CALIFORNIA	THREATENED	PORTULACACEAE	LEWISIA DISEPALA
CALIFORNIA	THREATENED	PORTULACACEAE	LEWISIA PYGMAEA SSP. LONGIPETALA
CALIFORNIA	THREATENED	PORTULACACEAE	LEWISIA SERRATA
CALIFORNIA	THREATENED	PORTULACACEAE	LEWISIA STEBBINSII
CALIFORNIA	THREATENED	RAFFLESIACEAE	PILOSTYLES THURBERI
CALIFORNIA	THREATENED	RANUNCULACEAE	DELPHINIUM HUTCHINSONAE
CALIFORNIA	THREATENED	RANUNCULACEAE	DELPHINIUM VARIEGATUM SSP. THORNEI
CALIFORNIA	THREATENED	RHAMNACEAE	CEANOTHUS GLORIOSUS VAR. PORRECTUS
CALIFORNIA	THREATENED	RHAMNACEAE	CEANOTHUS PROSTRATUS VAR. LAXUS
CALIFORNIA	THREATENED	RHAMNACEAE	CEANOTHUS RIGIDUS
CALIFORNIA	THREATENED	RHAMNACEAE	CEANOTHUS RODERICKII
CALIFORNIA	THREATENED	RHAMNACEAE	COLUBRINA CALIFORNICA
CALIFORNIA	THREATENED	ROSACEAE	HORKELIA TULARENSIS
CALIFORNIA	THREATENED	ROSACEAE	IVESIA ARGYROCOMA
CALIFORNIA	THREATENED	ROSACEAE	IVESIA PICKERINGII
CALIFORNIA	THREATENED	ROSACEAE	POTENTILLA PATELLIFERA
CALIFORNIA	THREATENED	RUBIACEAE	GALIUM BUXIFOLIUM
CALIFORNIA	THREATENED	RUBIACEAE	GALIUM CLEMENTIS
CALIFORNIA	THREATENED	RUBIACEAE	GALIUM HYPOTRICHIUM VAR. TOMENTELLUM
CALIFORNIA	THREATENED	RUBIACEAE	GALIUM SERPENTICUM SSP. WARNERENSE
CALIFORNIA	THREATENED	SAXIFRAGACEAE	HEUCHERA BREVISTAMINEA
CALIFORNIA	THREATENED	SAXIFRAGACEAE	HEUCHERA DURANII

STATE LISTS OF ENDANGERED, EXTINCT AND THREATENED SPECIES IN THE CONTINENTAL UNITED STATES

STATE	STATUS	FAMILY	SPECIES
CALIFORNIA	THREATENED	SAXIFRAGACEAE	RIBES CANTHARIFORME
CALIFORNIA	THREATENED	SCROPHULARIACEAE	ANTIRRHINUM SUBCORDATUM
CALIFORNIA	THREATENED	SCROPHULARIACEAE	CASTILLEJA BREVILOBATA
CALIFORNIA	THREATENED	SCROPHULARIACEAE	CASTILLEJA CINEREA
CALIFORNIA	THREATENED	SCROPHULARIACEAE	CASTILLEJA CULBERTSONII
CALIFORNIA	THREATENED	SCROPHULARIACEAE	CASTILLEJA EWANII
CALIFORNIA	THREATENED	SCROPHULARIACEAE	CASTILLEJA GLEASONII
CALIFORNIA	THREATENED	SCROPHULARIACEAE	CASTILLEJA LASSENENSIS
CALIFORNIA	THREATENED	SCROPHULARIACEAE	CASTILLEJA LATIFOLIA SSP. MENDOCINENSIS
CALIFORNIA	THREATENED	SCROPHULARIACEAE	CASTILLEJA MOLLIS
CALIFORNIA	THREATENED	SCROPHULARIACEAE	CASTILLEJA NEGLECTA
CALIFORNIA	THREATENED	SCROPHULARIACEAE	COLLINSIA ANTONINA SSP. ANTONINA
CALIFORNIA	THREATENED	SCROPHULARIACEAE	COLLINSIA ANTONINA SSP. PURPUREA
CALIFORNIA	THREATENED	SCROPHULARIACEAE	CORDYLANTHUS EREMICUS SSP. EREMICUS
CALIFORNIA	THREATENED	SCROPHULARIACEAE	CORDYLANTHUS MARITIMUS SSP. PALUSTRIS
CALIFORNIA	THREATENED	SCROPHULARIACEAE	CORDYLANTHUS MOLLIS SSP. HISPIDUS
CALIFORNIA	THREATENED	SCROPHULARIACEAE	CORDYLANTHUS TECOPENSIS
CALIFORNIA	THREATENED	SCROPHULARIACEAE	DIPLACUS ARIDIS
CALIFORNIA	THREATENED	SCROPHULARIACEAE	MAURANDYA PETROPHILA
CALIFORNIA	THREATENED	SCROPHULARIACEAE	MIMULUS EXIGUUS
CALIFORNIA	THREATENED	SCROPHULARIACEAE.	MIMULUS PICTUS
CALIFORNIA	THREATENED	SCROPHULARIACEAE	MIMULUS PURPUREUS
CALIFORNIA	THREATENED	SCROPHULARIACEAE	MIMULUS RUPICOLA
CALIFORNIA	THREATENED	SCROPHULARIACEAE	ORTHOCARPUS FLORIBUNDUS
CALIFORNIA	THREATENED	SCROPHULARIACEAE	ORTHOCARPUS LASIORHYNCHUS
CALIFORNIA	THREATENED	SCROPHULARIACEAE	PEDICULARIS HOWELLII
CALIFORNIA	THREATENED	SCROPHULARIACEAE	PENSTEMON CALCAREUS
CALIFORNIA	THREATENED	SCROPHULARIACEAE	PENSTEMON CALIFORNICUS
CALIFORNIA	THREATENED	SCROPHULARIACEAE	PENSTEMON CINICOLA
CALIFORNIA	THREATENED	SCROPHULARIACEAE	PENSTEMON FILIFORMIS
CALIFORNIA	THREATENED	SCROPHULARIACEAE	PENSTEMON PAPILLATUS
CALIFORNIA	THREATENED	SCROPHULARIACEAE	PENSTEMON STEPHENSII
CALIFORNIA	THREATENED	SCROPHULARIACEAE	PENSTEMON TRACYI
CALIFORNIA	THREATENED	SCROPHULARIACEAE	SCROPHULARIA ATRATA
CALIFORNIA	THREATENED	SCROPHULARIACEAE	VERONICA COPELANDII
CALIFORNIA	THREATENED	SOLANACEAE	ORYCTES NEVADENSIS
CALIFORNIA	THREATENED	SOLANACEAE	SOLANUM TENUILOBATUM
CALIFORNIA	THREATENED	VIOLACEAE	VIOLA LANCEOLATA SSP. OCCIDENTALIS
CALIFORNIA	THREATENED	VIOLACEAE	VIOLA TOMENTOSA
COLORADO	ENDANGERED	ASTERACEAE	ARTEMISIA ARGILOSA
COLORADO	ENDANGERED	ASTERACEAE	HAPLOPAPPUS FREMONTII SSP. MONOCEPHALUS
COLORADO	ENDANGERED	ASTERACEAE	PARTHENIUM LIGULATUM
COLORADO	ENDANGERED	ASTERACEAE	PARTHENIUM TETRANEURIS

STATE LISTS OF ENDANGERED, EXTINCT AND THREATENED SPECIES IN THE CONTINENTAL UNITED STATES

STATE	STATUS	FAMILY	SPECIES
COLORADO	ENDANGERED	ASTERACEAE	SAUSSUREA WEBERI
COLORADO	ENDANGERED	ASTERACEAE	SENECIO HALLII VAR. DISCOIDEA
COLORADO	ENDANGERED	ASTERACEAE	SENECIO PORTERI
COLORADO	ENDANGERED	BORAGINACEAE	CRYPTANTHA WEBERI
COLORADO	ENDANGERED	BRASSICACEAE	ARABIS CRANDALLII
COLORADO	ENDANGERED	BRASSICACEAE	ARABIS OXYLOBULA
COLORADO	ENDANGERED	BRASSICACEAE	BRAYA HUMILIS SSP. VENTOSA
COLORADO	ENDANGERED	BRASSICACEAE	EUTREMA PENLANDII
COLORADO	ENDANGERED	BRASSICACEAE	LESQUERELLA PRUINOSA
COLORADO	ENDANGERED	CACTACEAE	PEDIOCACTUS KNOWLTONII
COLORADO	ENDANGERED	CACTACEAE	SCLEROCACTUS GLAUCUS
COLORADO	ENDANGERED	CACTACEAE	SCLEROCACTUS MESAE-VERDAE
COLORADO	ENDANGERED	CAPPARIDACEAE	CLEOME MULTICAULIS
COLORADO	ENDANGERED	CARYOPHYLLACEAE	STELLARIA IRRIGUA
COLORADO	ENDANGERED	CHENOPODIACEAE	ATRIPLEX PLEIANTHA
COLORADO	ENDANGERED	FABACEAE	ASTRAGALUS DETERIOR
COLORADO	ENDANGERED	FABACEAE	ASTRAGALUS LUTOSUS
COLORADO	ENDANGERED	FABACEAE	ASTRAGALUS MICROCYMBUS
COLORADO	ENDANGERED	FABACEAE	ASTRAGALUS NATURITENSIS
COLORADO	ENDANGERED	FABACEAE	ASTRAGALUS OOCALYCIS
COLORADO	ENDANGERED	FABACEAE	ASTRAGALUS OSTERHOUTII
COLORADO	ENDANGERED	FABACEAE	ASTRAGALUS SCHMOLLAE
COLORADO	ENDANGERED	HYDROPHYLLACEAE	PHACELIA FORMOSULA
COLORADO	ENDANGERED	HYDROPHYLLACEAE	PHACELIA SUBMUTICA
COLORADO	ENDANGERED	ONAGRACEAE	GAURA NEOMEXICANA SSP. COLORADENSIS
COLORADO	ENDANGERED	POACEAE	FESTUCA DASYCLADA
COLORADO	ENDANGERED	POLEMONIACEAE	GILIA PENSTEMONOIDES
COLORADO	ENDANGERED	POLYGONACEAE	ERIOGONUM EPHEDROIDES
COLORADO	ENDANGERED	POLYGONACEAE	ERIOGONUM PELINOPHILUM
COLORADO	ENDANGERED	POLYPODIACEAE	ASPLENIUM ANDREWSII
COLORADO	ENDANGERED	RANUNCULACEAE	AQUILEGIA BARNEBYI
COLORADO	ENDANGERED	RANUNCULACEAE	AQUILEGIA CAERULEA VAR. DAILEYAE
COLORADO	ENDANGERED	RANUNCULACEAE	AQUILEGIA MICRANTHA VAR. MANCOSANA
COLORADO	ENDANGERED	RANUNCULACEAE	AQUILEGIA SAXIMONTANA
COLORADO	ENDANGERED	ROSACEAE	POTENTILLA RUPINCOLA
COLORADO	ENDANGERED	SCROPHULARIACEAE	MIMULUS GEMMIPARUS
COLORADO	ENDANGERED	SCROPHULARIACEAE	PENSTEMON RETRORSUS
COLORADO	EXTINCT	BORAGINACEAE	CRYPTANTHA APERTA
COLORADO	EXTINCT	BRASSICACEAE	ARABIS GUNNISONIANA
COLORADO	EXTINCT	FABACEAE	ASTRAGALUS HUMILLIMUS
COLORADO	EXTINCT	FABACEAE	ASTRAGALUS LINIFOLIUS
COLORADO	THREATENED	APIACEAE	NEOPARRYA LITHOPHILA
COLORADO	THREATENED	ASTERACEAE	CHAMAECHAENACTIS SCAPOSA

STATE LISTS OF ENDANGERED, EXTINCT AND THREATENED SPECIES IN THE CONTINENTAL UNITED STATES

STATE	STATUS	FAMILY	SPECIES
COLORADO	THREATENED	ASTERACEAE	TOWNSENDIA ROTHROCKII
COLORADO	THREATENED	BORAGINACEAE	CRYPTANTHA ELATA
COLORADO	THREATENED	BORAGINACEAE	CRYPTANTHA PARADOXA
COLORADO	THREATENED	BRASSICACEAE	PHYSARIA BELLII
COLORADO	THREATENED	BRASSICACEAE	RORIPPA COLORADENSIS
COLORADO	THREATENED	CACTACEAE	SCLEROCACTUS SPINOSIOR
COLORADO	THREATENED	CAPPARIDACEAE	CLEOMELLA MONTROSAE
COLORADO	THREATENED	CYPERACEAE	CAREX ARAPAHOENSIS
COLORADO	THREATENED	FABACEAE	ASTRAGALUS DETRITALIS
COLORADO	THREATENED	FABACEAE	ASTRAGALUS WETHERILLII
COLORADO	THREATENED	FABACEAE	OXYTROPIS OBNAPIFORMIS
COLORADO	THREATENED	GENTIANACEAE	FRASERA COLORADENSIS
COLORADO	THREATENED	POACEAE	PTILAGROSTIS PORTERI
COLORADO	THREATENED	POLEMONIACEAE	IPOMOPSIS GLOBULARIS
COLORADO	THREATENED	POLEMONIACEAE	PHLOX CARYOPHYLLA
COLORADO	THREATENED	POLYGONACEAE	ERIOGONUM BRANDEGEI
COLORADO	THREATENED	POLYGONACEAE	ERIOGONUM CLAVELLATUM
COLORADO	THREATENED	POLYGONACEAE	ERIOGONUM SAURINUM
COLORADO	THREATENED	POLYGONACEAE	ERIOGONUM VIRIDULUM
COLORADO	THREATENED	SAXIFRAGACEAE	SULLIVANTIA HAPEMANII
COLORADO	THREATENED	SCROPHULARIACEAE	PENSTEMON DEGENERI
CONNECTICUT	ENDANGERED	ORCHIDACEAE	ISOTRIA MEDEOLOIDES
CONNECTICUT	ENDANGERED	RANUNCULACEAE	TROLLIUS LAXUS
CONNECTICUT	ENDANGERED	ROSACEAE	PRUNUS GRAVESII
CONNECTICUT	THREATENED	ARALIACEAE	PANAX QUINQUEFOLIUS
CONNECTICUT	THREATENED	CISTACEAE	HELIANTHEMUM DUMOSUM
CONNECTICUT	THREATENED	CYPERACEAE	SCIRPUS LONGII
CONNECTICUT	THREATENED	ISOETACEAE	ISOETES EATONII
CONNECTICUT	THREATENED	ORCHIDACEAE	CYPRIPEDIUM ARIETINUM
CONNECTICUT	THREATENED	RANUNCULACEAE	HYDRASTIS CANADENSIS
CONNECTICUT	THREATENED	ROSACEAE	PRUNUS ALLEGHANIENSIS
CONNECTICUT	THREATENED	SCROPHULARIACEAE	GERARDIA ACUTA
CONNECTICUT	THREATENED	SCROPHULARIACEAE	SCHWALBEA AMERICANA
DELAWARE	ENDANGERED	CYPERACEAE	RHYNCHOSPORA KNIESKERNII
DELAWARE	ENDANGERED	RANUNCULACEAE	TROLLIUS LAXUS
DELAWARE	THREATENED	APIACEAE	OXYPOLIS CANBYI
DELAWARE	THREATENED	ARALIACEAE	PANAX QUINQUEFOLIUS
DELAWARE	THREATENED	BETULACEAE	ALNUS MARITIMA
DELAWARE	THREATENED	ORCHIDACEAE	PLATANTHERA FLAVA
DELAWARE	THREATENED	ORCHIDACEAE	PLATANTHERA PERAMOENA
DELAWARE	THREATENED	POACEAE	MUHLENBERGIA TORREYANA
DELAWARE	THREATENED	RANUNCULACEAE	HYDRASTIS CANADENSIS
DELAWARE	THREATENED	SCROPHULARIACEAE	MICRANTHEMUM MICRANTHEMOIDES

STATE LISTS OF ENDANGERED, EXTINCT AND THREATENED SPECIES IN THE CONTINENTAL UNITED STATES

STATE	STATUS	FAMILY	SPECIES
DELAWARE	THREATENED	SCROPHULARIACEAE	SCHWALBEA AMERICANA
DISTRICT OF COLUMBIA	THREATENED	POACEAE	PANICUM ACULEATUM
DISTRICT OF COLUMBIA	THREATENED	SCROPHULARIACEAE	MICRANTHEMUM MICRANTHEMOIDES
FLORIDA	ENDANGERED	ACANTHACEAE	JUSTICIA COOLEYI
FLORIDA	ENDANGERED	ACANTHACEAE	JUSTICIA CRASSIFOLIA
FLORIDA	ENDANGERED	ANNONACEAE	ASIMINA TETRAMERA
FLORIDA	ENDANGERED	APIACEAE	OXYPOLIS GREENMANII
FLORIDA	ENDANGERED	APIACEAE	SIUM FLORIDANUM
FLORIDA	ENDANGERED	ARECACEAE	ROYSTONEA ELATA
FLORIDA	ENDANGERED	ASTERACEAE	ASTER PINIFOLIUS
FLORIDA	ENDANGERED	ASTERACEAE	BALDUINA ATROPURPUREA
FLORIDA	ENDANGERED	ASTERACEAE	LIATRIS OHLINGERAE
FLORIDA	ENDANGERED	BRASSICACEAE	WAREA AMPLEXIFOLIA
FLORIDA	ENDANGERED	BRASSICACEAE	WAREA CARTERI
FLORIDA	ENDANGERED	CACTACEAE	CEREUS ERIOPHORUS VAR. FRAGRANS
FLORIDA	ENDANGERED	CACTACEAE	CEREUS GRACILIS VAR. ABORIGINUM
FLORIDA	ENDANGERED	CACTACEAE	CEREUS GRACILIS VAR. SIMPSONII
FLORIDA	ENDANGERED	CACTACEAE	CEREUS ROBINII VAR. ROBINII
FLORIDA	ENDANGERED	CACTACEAE	CEREUS ROBINII VAR. DEERINGII
FLORIDA	ENDANGERED	CARYOPHYLLACEAE	PARONYCHIA CHARTACEA
FLORIDA	ENDANGERED	CARYOPHYLLACEAE	PARONYCHIA RUGELII VAR. INTERIOR
FLORIDA	ENDANGERED	CARYOPHYLLACEAE	SILENE POLYPETALA
FLORIDA	ENDANGERED	CYCADACEAE	ZAMIA INTEGRIFOLIA
FLORIDA	ENDANGERED	ERICACEAE	MONOTROPSIS REYNOLDSIAE
FLORIDA	ENDANGERED	ERICACEAE	RHODODENDRON MINUS VAR. CHAPMANII
FLORIDA	ENDANGERED	EUPHORBIACEAE	CHAMAESYCE DELTOIDEA SSP. SERPYLLUM
FLORIDA	ENDANGERED	EUPHORBIACEAE	CHAMAESYCE GARBERI
FLORIDA	ENDANGERED	EUPHORBIACEAE	CHAMAESYCE PORTERIANA VAR. KEYENSIS
FLORIDA	ENDANGERED	EUPHORBIACEAE	CHAMAESYCE PORTERIANA VAR. SCOPARIA
FLORIDA	ENDANGERED	EUPHORBIACEAE	CROTON ELLIOTTII
FLORIDA	ENDANGERED	EUPHORBIACEAE	CROTON GLANDULOSUS VAR. SIMPSONII
FLORIDA	ENDANGERED	FABACEAE	CASSIA KEYENSIS
FLORIDA	ENDANGERED	FABACEAE	CENTROSEMA ARENICOLA
FLORIDA	ENDANGERED	FABACEAE	GALACTIA PINETORUM
FLORIDA	ENDANGERED	FABACEAE	VICIA OCALENSIS
FLORIDA	ENDANGERED	GENTIANACEAE	GENTIANA PENNELLIANA
FLORIDA	ENDANGERED	HYPERICACEAE	HYPERICUM CUMULICOLA
FLORIDA	ENDANGERED	LAMIACEAE	CONRADINA BREVIFOLIA
FLORIDA	ENDANGERED	LAMIACEAE	CONRADINA GLABRA
FLORIDA	ENDANGERED	LAMIACEAE	DICERANDRA FRUTESCENS
FLORIDA	ENDANGERED	LAMIACEAE	DICERANDRA IMMACULATA
FLORIDA	ENDANGERED	LAMIACEAE	HEDEOMA GRAVEOLENS
FLORIDA	ENDANGERED	LAMIACEAE	MACBRIDEA ALBA

STATE LISTS OF ENDANGERED, EXTINCT AND THREATENED SPECIES IN THE CONTINENTAL UNITED STATES

STATE	STATUS	FAMILY	SPECIES
FLORIDA	ENDANGERED	LAMIACEAE	SALVIA BLODGETTII
FLORIDA	ENDANGERED	LAURACEAE	LINDERA MELISSIFOLIA
FLORIDA	ENDANGERED	LILIACEAE	HARPEROCALLIS FLAVA
FLORIDA	ENDANGERED	LILIACEAE	HYMENOCALLIS CORONARIA
FLORIDA	ENDANGERED	LILIACEAE	LILIUM IRIDOLLAE
FLORIDA	ENDANGERED	LILIACEAE	NOLINA ATOPOCARPA
FLORIDA	ENDANGERED	LILIACEAE	NOLINA BRITTONIANA
FLORIDA	ENDANGERED	LINACEAE	LINUM ARENICOLA
FLORIDA	ENDANGERED	LINACEAE	LINUM CARTERI VAR. CARTERI
FLORIDA	ENDANGERED	LINACEAE	LINUM CARTERI VAR. SMALLII
FLORIDA	ENDANGERED	LINACEAE	LINUM WESTII
FLORIDA	ENDANGERED	LOGANIACEAE	SPIGELIA GENTIANOIDES
FLORIDA	ENDANGERED	LOGANIACEAE	SPIGELIA LOGANIOIDES
FLORIDA	ENDANGERED	LYTHRACEAE	CUPHEA ASPERA
FLORIDA	ENDANGERED	MELASTOMATACEAE	RHEXIA PARVIFLORA
FLORIDA	ENDANGERED	OLEACEAE	FORESTIERA SEGREGATA VAR. PINETORUM
FLORIDA	ENDANGERED	ORCHIDACEAE	SPIRANTHES LANCEOLATA VAR. PALUDICOLA
FLORIDA	ENDANGERED	ORCHIDACEAE	TRIPHORA CRAIGHEADII
FLORIDA	ENDANGERED	ORCHIDACEAE	TRIPHORA LATIFOLIA
FLORIDA	ENDANGERED	PLUMBAGINACEAE	LIMONIUM CAROLINIANUM VAR. ANGUSTATUM
FLORIDA	ENDANGERED	POACEAE	ANDROPOGON ARCTATUS
FLORIDA	ENDANGERED	POACEAE	ARISTIDA FLORIDANA
FLORIDA	ENDANGERED	POACEAE	CALAMOVILFA CURTISSII
FLORIDA	ENDANGERED	POACEAE	SCHIZACHYRIUM RHIZOMATUM
FLORIDA	ENDANGERED	POACEAE	TRIPSACUM FLORIDANUM
FLORIDA	ENDANGERED	POLYGALACEAE	POLYGALA LEWTONII
FLORIDA	ENDANGERED	POLYGONACEAE	POLYGONELLA CILIATA VAR. BASIRAMIA
FLORIDA	ENDANGERED	POLYGONACEAE	POLYGONELLA MYRIOPHYLLA
FLORIDA	ENDANGERED	RANUNCULACEAE	AQUILEGIA CANADENSIS VAR. AUSTRALIS
FLORIDA	ENDANGERED	RANUNCULACEAE	CLEMATIS MICRANTHA
FLORIDA	ENDANGERED	ROSACEAE	PRUNUS GENICULATA
FLORIDA	ENDANGERED	SALICACEAE	SALIX FLORIDANA
FLORIDA	ENDANGERED	SAXIFRAGACEAE	RIBES ECHINELLUM
FLORIDA	ENDANGERED	SCHIZAEACEAE	SCHIZAEA GERMANII
FLORIDA	ENDANGERED	TAXACEAE	TAXUS FLORIDANA
FLORIDA	ENDANGERED	TAXACEAE	TORREYA TAXIFOLIA
FLORIDA	ENDANGERED	VERBENACEAE	VERBENA TAMPENSIS
FLORIDA	EXTINCT	CAMPANULACEAE	CAMPANULA ROBINSIAE
FLORIDA	EXTINCT	CERATOPHYLLACEAE	CERATOPHYLLUM FLORIDANUM
FLORIDA	EXTINCT	SCROPHULARIACEAE	AGALINIS STENOPHYLLA
FLORIDA	EXTINCT	SOLANACEAE	SOLANUM BAHAMENSE VAR. RUGELII
FLORIDA	THREATENED	ACANTHACEAE	ELYTRARIA CAROLINIENSIS VAR. ANGUSTIFOLIA
FLORIDA	THREATENED	ANNONACEAE	ASIMINA PULCHELLA

STATE LISTS OF ENDANGERED, EXTINCT AND THREATENED SPECIES IN THE CONTINENTAL UNITED STATES

STATE	STATUS	FAMILY	SPECIES
FLORIDA	THREATENED	ANNONACEAE	ASIMINA RUGELII
FLORIDA	THREATENED	APIACEAE	ERYNGIUM CUNEIFOLIUM
FLORIDA	THREATENED	APIACEAE	ZIZIA LATIFOLIA
FLORIDA	THREATENED	AQUIFOLIACEAE	ILEX OPACA VAR. ARENICOLA
FLORIDA	THREATENED	ARECACEAE	RHAPIDOPHYLLUM HYSTRIX
FLORIDA	THREATENED	ASCLEPIADACEAE	ASCLEPIAS VIRIDULA
FLORIDA	THREATENED	ASCLEPIADACEAE	MATELEA FLORIDANA
FLORIDA	THREATENED	ASTERACEAE	ASTER BRACHYPHOLIS
FLORIDA	THREATENED	ASTERACEAE	ASTER PLUMOSUS
FLORIDA	THREATENED	ASTERACEAE	ASTER SPINULOSUS
FLORIDA	THREATENED	ASTERACEAE	BRICKELLIA CORDIFOLIA
FLORIDA	THREATENED	ASTERACEAE	BRICKELLIA EUPATORIOIDES VAR. FLORIDANA
FLORIDA	THREATENED	ASTERACEAE	CACALIA DIVERSIFOLIA
FLORIDA	THREATENED	ASTERACEAE	HARTWRIGHTIA FLORIDANA
FLORIDA	THREATENED	ASTERACEAE	HELIANTHUS CARNOSUS
FLORIDA	THREATENED	ASTERACEAE	HELIANTHUS DEBILIS SSP. VESTITUS
FLORIDA	THREATENED	ASTERACEAE	HETEROTHECA FLEXUOSA
FLORIDA	THREATENED	ASTERACEAE	MELANTHERA PARVIFOLIA
FLORIDA	THREATENED	ASTERACEAE	VERBESINA CHAPMANNII
FLORIDA	THREATENED	ASTERACEAE	VERBESINA HETEROPHYLLA
FLORIDA	THREATENED	BORAGINACEAE	HELIOTROPIUM POLYPHYLLUM VAR. HORIZONTALE
FLORIDA	THREATENED	BRASSICACEAE	WAREA SESSILIFOLIA
FLORIDA	THREATENED	CACTACEAE	OPUNTIA SPINOSISSIMA
FLORIDA	THREATENED	CACTACEAE	OPUNTIA TRIACANTHA
FLORIDA	THREATENED	CARYOPHYLLACEAE	ARENARIA GODFREYI
FLORIDA	THREATENED	CISTACEAE	LECHEA CERNUA
FLORIDA	THREATENED	CISTACEAE	LECHEA DIVARICATA
FLORIDA	THREATENED	CISTACEAE	LECHEA LAKELAE
FLORIDA	THREATENED	COMMELINACEAE	COMMELINA GIGAS
FLORIDA	THREATENED	CONVOLVULACEAE	BONAMIA GRANDIFLORA
FLORIDA	THREATENED	CONVOLVULACEAE	JACQUEMONTIA CURTISSII
FLORIDA	THREATENED	CONVOLVULACEAE	JACQUEMONTIA RECLINATA
FLORIDA	THREATENED	CUCURBITACEAE	CUCURBITA OKEECHOBEENSIS
FLORIDA	THREATENED	CYPERACEAE	CAREX BALTZELLII
FLORIDA	THREATENED	CYPERACEAE	CAREX CHAPMANII
FLORIDA	THREATENED	CYPERACEAE	RHYNCHOSPORA CULIXA
FLORIDA	THREATENED	CYPERACEAE	RHYNCHOSPORA PUNCTATA
FLORIDA	THREATENED	ERICACEAE	MONOTROPA BRITTONII
FLORIDA	THREATENED	ERICACEAE	RHODODENDRON AUSTRINUM
FLORIDA	THREATENED	ERIOCAULACEAE	LACHNOCAULON BEYRICHIANUM
FLORIDA	THREATENED	EUPHORBIACEAE	ARGYTHAMNIA BLODGETTII
FLORIDA	THREATENED	EUPHORBIACEAE	CHAMAESYCE CUMULICOLA
FLORIDA	THREATENED	EUPHORBIACEAE	CHAMAESYCE DELTOIDEA SSP. DELTOIDEA

STATE LISTS OF ENDANGERED, EXTINCT AND THREATENED SPECIES IN THE CONTINENTAL UNITED STATES

STATE	STATUS	FAMILY	SPECIES
FLORIDA	THREATENED	EUPHORBIACEAE	CHAMAESYCE PORTERIANA VAR. PORTERIANA
FLORIDA	THREATENED	EUPHORBIACEAE	EUPHORBIA DISCOIDALIS
FLORIDA	THREATENED	EUPHORBIACEAE	EUPHORBIA EXSERTA
FLORIDA	THREATENED	EUPHORBIACEAE	EUPHORBIA TELEPHIOIDES
FLORIDA	THREATENED	EUPHORBIACEAE	PHYLLANTHUS LIEBMANIANUS SSP. PLATYLEPIS
FLORIDA	THREATENED	EUPHORBIACEAE	PHYLLANTHUS PENTAPHYLLUS SSP. FLORIDANUS
FLORIDA	THREATENED	EUPHORBIACEAE	STILLINGIA SYLVATICA SSP. TENUIS
FLORIDA	THREATENED	EUPHORBIACEAE	TITHYMALUS AUSTRINUS
FLORIDA	THREATENED	EUPHORBIACEAE	TRAGIA SAXICOLA
FLORIDA	THREATENED	FABACEAE	BAPTISIA CALYCOSA
FLORIDA	THREATENED	FABACEAE	BAPTISIA HIRSUTA
FLORIDA	THREATENED	FABACEAE	BAPTISIA MEGACARPA
FLORIDA	THREATENED	FABACEAE	BAPTISIA SIMPLICIFOLIA
FLORIDA	THREATENED	FABACEAE	CLITORIA FRAGRANS
FLORIDA	THREATENED	FABACEAE	LUPINUS WESTIANUS
FLORIDA	THREATENED	FABACEAE	RHYNCHOSIA CINEREA
FLORIDA	THREATENED	FABACEAE	TEPHROSIA MOHRII
FLORIDA	THREATENED	HALORAGACEAE	MYRIOPHYLLUM LAXUM
FLORIDA	THREATENED	HYPERICACEAE	HYPERICUM EDISONIANUM
FLORIDA	THREATENED	ILLICIACEAE	ILLICIUM PARVIFLORUM
FLORIDA	THREATENED	IRIDACEAE	NEMASTYLIS FLORIDANA
FLORIDA	THREATENED	IRIDACEAE	SPHENOSTIGMA COELESTINA
FLORIDA	THREATENED	JUNCACEAE	JUNCUS GYMNOCARPUS
FLORIDA	THREATENED	LAMIACEAE	CALAMINTHA ASHEI
FLORIDA	THREATENED	LAMIACEAE	CALAMINTHA DENTATUM
FLORIDA	THREATENED	LAMIACEAE	CONRADINA GRANDIFLORA
FLORIDA	THREATENED	LAMIACEAE	DICERANDRA ODORATISSIMA
FLORIDA	THREATENED	LAMIACEAE	PHYSOSTEGIA LEPTOPHYLLUM
FLORIDA	THREATENED	LAMIACEAE	PYCNANTHEMUM FLORIDANUM
FLORIDA	THREATENED	LAMIACEAE	SCUTELLARIA FLORIDANA
FLORIDA	THREATENED	LAMIACEAE	STACHYS LYTHROIDES
FLORIDA	THREATENED	LAURACEAE	LITSEA AESTIVALIS
FLORIDA	THREATENED	LAURACEAE	PERSEA BORBONIA VAR. HUMILIS
FLORIDA	THREATENED	LEITNERIACEAE	LEITNERIA FLORIDANA
FLORIDA	THREATENED	LENTIBULARIACEAE	PINGUICULA IONANTHA
FLORIDA	THREATENED	LILIACEAE	HYMENOCALLIS LATIFOLIA
FLORIDA	THREATENED	LILIACEAE	VERATRUM WOODII
FLORIDA	THREATENED	LILIACEAE	ZEPHYRANTHES SIMPSONII
FLORIDA	THREATENED	LILIACEAE	ZEPHYRANTHES TREATIAE
FLORIDA	THREATENED	LINACEAE	LINUM SULCATUM VAR. HARPERI
FLORIDA	THREATENED	MAGNOLIACEAE	MAGNOLIA ASHEI
FLORIDA	THREATENED	MALVACEAE	KOSTELETZKYA SMILACIFOLIA
FLORIDA	THREATENED	MALVACEAE	SIDA RUBROMARGINATA

STATE LISTS OF ENDANGERED, EXTINCT AND THREATENED SPECIES IN THE CONTINENTAL UNITED STATES

STATE	STATUS	FAMILY	SPECIES
FLORIDA	THREATENED	MELASTOMATACEAE	RHEXIA SALICIFOLIA
FLORIDA	THREATENED	MYRTACEAE	MYRCIANTHES FRAGRANS VAR. SIMPSONII
FLORIDA	THREATENED	NYMPHAEACEAE	NUPHAR LUTEUM SSP. ULVACEUM
FLORIDA	THREATENED	OLEACEAE	CHIONANTHUS PYGMAEUS
FLORIDA	THREATENED	OPHIOGLOSSACEAE	OPHIOGLOSSUM PALMATUM
FLORIDA	THREATENED	ORCHIDACEAE	ENCYCLIA BOOTHIANA VAR. ERYTHRONIOIDES
FLORIDA	THREATENED	ORCHIDACEAE	LEPANTHOPSIS MELANANTHA
FLORIDA	THREATENED	ORCHIDACEAE	PLATANTHERA FLAVA
FLORIDA	THREATENED	ORCHIDACEAE	PLATANTHERA INTEGRA
FLORIDA	THREATENED	ORCHIDACEAE	SPIRANTHES POLYANTHA
FLORIDA	THREATENED	POACEAE	ARISTIDA SIMPLICIFLORA
FLORIDA	THREATENED	POACEAE	CTENIUM FLORIDANUM
FLORIDA	THREATENED	POACEAE	ERAGROSTIS TRACYI
FLORIDA	THREATENED	POACEAE	ERIOCHLOA MICHAUXII VAR. SIMPSONII
FLORIDA	THREATENED	POACEAE	GYMNOPOGON FLORIDANUS
FLORIDA	THREATENED	POACEAE	MANISURIS TUBERCULOSA
FLORIDA	THREATENED	POACEAE	PANICUM NUDICAULE
FLORIDA	THREATENED	POACEAE	PANICUM PINETORUM
FLORIDA	THREATENED	POACEAE	SCHIZACHYRIUM NIVEUM
FLORIDA	THREATENED	POLYGALACEAE	POLYGALA BOYKINII VAR. SPARSIFOLIA
FLORIDA	THREATENED	POLYGONACEAE	PERSICARIA PALUDICOLA
FLORIDA	THREATENED	POLYGONACEAE	POLYGONELLA MACROPHYLLA
FLORIDA	THREATENED	RHAMNACEAE	SAGERETIA MINUTIFLORA
FLORIDA	THREATENED	ROSACEAE	AGRIMONIA INCISA
FLORIDA	THREATENED	RUBIACEAE	PINCKNEYA PUBENS
FLORIDA	THREATENED	SCHISANDRACEAE	SCHISANDRA GLABRA
FLORIDA	THREATENED	SCROPHULARIACEAE	AGALINIS PURPUREA VAR. CARTERI
FLORIDA	THREATENED	SOLANACEAE	PHYSALIS VISCOSA VAR. ELLIOTII
FLORIDA	THREATENED	STEMONACEAE	CROOMIA PAUCIFLORA
FLORIDA	THREATENED	VERBENACEAE	VERBENA MARITIMA
FLORIDA	THREATENED	XYRIDACEAE	XYRIS ISOETIFOLIA
FLORIDA	THREATENED	XYRIDACEAE	XYRIS LONGISEPALA
FLORIDA	THREATENED	XYRIDACEAE	XYRIS SCABRIFOLIA
GEORGIA	ENDANGERED	ASTERACEAE	BALDUINA ATROPURPUREA
GEORGIA	ENDANGERED	ASTERACEAE	HELIANTHUS SMITHII
GEORGIA	ENDANGERED	ASTERACEAE	MARSHALLIA MOHRI
GEORGIA	ENDANGERED	BRASSICACEAE	DRABA APRICA
GEORGIA	ENDANGERED	CAPRIFOLIACEAE	VIBURNUM BRACTEATUM
GEORGIA	ENDANGERED	CARYOPHYLLACEAE	PARONYCHIA RUGELII VAR. INTERIOR
GEORGIA	ENDANGERED	CARYOPHYLLACEAE	SILENE POLYPETALA
GEORGIA	ENDANGERED	CYCADACEAE	ZAMIA INTEGRIFOLIA
GEORGIA	ENDANGERED	CYPERACEAE	CAREX BILTMOREANA
GEORGIA	ENDANGERED	CYPERACEAE	FIMBRISTYLIS PERPUSILLA

STATE LISTS OF ENDANGERED, EXTINCT AND THREATENED SPECIES IN THE CONTINENTAL UNITED STATES

STATE	STATUS	FAMILY	SPECIES
GEORGIA	ENDANGERED	DIAPENSIACEAE	SHORTIA GALACIFOLIA VAR. GALACIFOLIA
GEORGIA	ENDANGERED	ERICACEAE	ELLIOTTIA RACEMOSA
GEORGIA	ENDANGERED	EUPHORBIACEAE	CROTON ELLIOTTII
GEORGIA	ENDANGERED	FABACEAE	BAPTISIA ARACHNIFERA
GEORGIA	ENDANGERED	LAMIACEAE	PYCNANTHEMUM CURVIPES
GEORGIA	ENDANGERED	LAURACEAE	LINDERA MELISSIFOLIA
GEORGIA	ENDANGERED	LILIACEAE	HYMENOCALLIS CORONARIA
GEORGIA	ENDANGERED	LILIACEAE	TRILLIUM PERSISTENS
GEORGIA	ENDANGERED	MELASTOMATACEAE	RHEXIA PARVIFLORA
GEORGIA	ENDANGERED	POACEAE	PANICUM HIRSTII
GEORGIA	ENDANGERED	PORTULACACEAE	PORTULACA SMALLII
GEORGIA	ENDANGERED	SALICACEAE	SALIX FLORIDANA
GEORGIA	ENDANGERED	SAPOTACEAE	BUMELIA THORNEI
GEORGIA	ENDANGERED	SCROPHULARIACEAE	AMPHIANTHUS PUSILLUS
GEORGIA	ENDANGERED	SCROPHULARIACEAE	LINDERNIA SAXICOLA
GEORGIA	ENDANGERED	TAXACEAE	TORREYA TAXIFOLIA
GEORGIA	EXTINCT	ASTERACEAE	SOLIDAGO PORTERI
GEORGIA	EXTINCT	LAMIACEAE	SCUTELLARIA OCMULGEE
GEORGIA	EXTINCT	SOLANACEAE	SOLANUM CAROLINENSE VAR. HIRSUTUM
GEORGIA	EXTINCT	THEACEAE	FRANKLINIA ALATAMAHA
GEORGIA	THREATENED	ANACARDIACEAE	RHUS MICHAUXII
GEORGIA	THREATENED	APIACEAE	OXYPOLIS CANBYI
GEORGIA	THREATENED	APIACEAE	PTILIMNIUM NODOSUM
GEORGIA	THREATENED	ARALIACEAE	PANAX QUINQUEFOLIUS
GEORGIA	THREATENED	ARECACEAE	RHAPIDOPHYLLUM HYSTRIX
GEORGIA	THREATENED	ASTERACEAE	BRICKELLIA CORDIFOLIA
GEORGIA	THREATENED	ASTERACEAE	CACALIA DIVERSIFOLIA
GEORGIA	THREATENED	ASTERACEAE	COREOPSIS LATIFOLIA
GEORGIA	THREATENED	ASTERACEAE	ECHINACEA LAEVIGATA
GEORGIA	THREATENED	ASTERACEAE	HARTWRIGHTIA FLORIDANA
GEORGIA	THREATENED	ASTERACEAE	MARSHALLIA RAMOSA
GEORGIA	THREATENED	ASTERACEAE	RUDBECKIA HELIOPSIDIS
GEORGIA	THREATENED	ASTERACEAE	SENECIO MILLEFOLIUM
GEORGIA	THREATENED	ASTERACEAE	VERNONIA PULCHELLA
GEORGIA	THREATENED	ASTERACEAE	VIGUIERA PORTERI
GEORGIA	THREATENED	BRASSICACEAE	ARABIS GEORGIANA
GEORGIA	THREATENED	BRASSICACEAE	LEAVENWORTHIA EXIGUA VAR. EXIGUA
GEORGIA	THREATENED	CARYOPHYLLACEAE	ARENARIA UNIFLORA
GEORGIA	THREATENED	CRASSULACEAE	SEDUM PUSILLUM
GEORGIA	THREATENED	CUSCUTACEAE	CUSCUTA HARPERI
GEORGIA	THREATENED	CYPERACEAE	CAREX AMPLISQUAMA
GEORGIA	THREATENED	CYPERACEAE	CAREX AUSTROCAROLINIANA
GEORGIA	THREATENED	CYPERACEAE	CAREX MISERA

STATE LISTS OF ENDANGERED, EXTINCT AND THREATENED SPECIES IN THE CONTINENTAL UNITED STATES

STATE	STATUS	FAMILY	SPECIES
GEORGIA	THREATENED	CYPERACEAE	CAREX PURPURIFERA
GEORGIA	THREATENED	CYPERACEAE	CYPERUS GRANITOPHILUS
GEORGIA	THREATENED	CYPERACEAE	RHYNCHOSPORA CULIXA
GEORGIA	THREATENED	CYPERACEAE	RHYNCHOSPORA GLOBULARIS VAR. SAXICOLA
GEORGIA	THREATENED	CYPERACEAE	RHYNCHOSPORA PUNCTATA
GEORGIA	THREATENED	ERICACEAE	RHODODENDRON AUSTRINUM
GEORGIA	THREATENED	ERICACEAE	RHODODENDRON BAKERI
GEORGIA	THREATENED	ERICACEAE	RHODODENDRON PRUNIFOLIUM
GEORGIA	THREATENED	ERIOCAULACEAE	LACHNOCAULON BEYRICHIANUM
GEORGIA	THREATENED	FABACEAE	CLADRASTIS LUTEA
GEORGIA	THREATENED	FABACEAE	PETALOSTEMUM GATTINGERI
GEORGIA	THREATENED	FAGACEAE	QUERCUS GEORGIANA
GEORGIA	THREATENED	FAGACEAE	QUERCUS OGLETHORPENSIS
GEORGIA	THREATENED	HALORAGACEAE	MYRIOPHYLLUM LAXUM
GEORGIA	THREATENED	HAMAMELIDACEAE	FOTHERGILLA GARDENI
GEORGIA	THREATENED	HYDROPHYLLACEAE	PHACELIA DUBIA VAR. GEORGIANA
GEORGIA	THREATENED	ISOETACEAE	ISOETES MELANOSPORA
GEORGIA	THREATENED	ISOETACEAE	ISOETES VIRGINICA
GEORGIA	THREATENED	LAMIACEAE	CALAMINTHA DENTATUM
GEORGIA	THREATENED	LAMIACEAE	DICERANDRA ODORATISSIMA
GEORGIA	THREATENED	LAMIACEAE	PHYSOSTEGIA VERONICIFORMIS
GEORGIA	THREATENED	LAMIACEAE	SCUTELLARIA MONTANA
GEORGIA	THREATENED	LAURACEAE	LITSEA AESTIVALIS
GEORGIA	THREATENED	LEITNERIACEAE	LEITNERIA FLORIDANA
GEORGIA	THREATENED	LINACEAE	LINUM SULCATUM VAR. HARPERI
GEORGIA	THREATENED	ORCHIDACEAE	PLATANTHERA FLAVA
GEORGIA	THREATENED	ORCHIDACEAE	PLATANTHERA INTEGRA
GEORGIA	THREATENED	ORCHIDACEAE	PLATANTHERA PERAMOENA
GEORGIA	THREATENED	POACEAE	CTENIUM FLORIDANUM
GEORGIA	THREATENED	POACEAE	MUHLENBERGIA TORREYANA
GEORGIA	THREATENED	POACEAE	PANICUM LITHOPHILUM
GEORGIA	THREATENED	POACEAE	SCHIZACHYRIUM NIVEUM
GEORGIA	THREATENED	POACEAE	SPOROBOLUS TERETIFOLIUS
GEORGIA	THREATENED	PORTULACACEAE	TALINUM MENGESII
GEORGIA	THREATENED	RANUNCULACEAE	HYDRASTIS CANADENSIS
GEORGIA	THREATENED	RHAMNACEAE	SAGERETIA MINUTIFLORA
GEORGIA	THREATENED	ROSACEAE	WALDSTEINIA LOBATA
GEORGIA	THREATENED	RUBIACEAE	PINCKNEYA PUBENS
GEORGIA	THREATENED	SANTALACEAE	NESTRONIA UMBELLULA
GEORGIA	THREATENED	SCHISANDRACEAE	SCHISANDRA GLABRA
GEORGIA	THREATENED	SCROPHULARIACEAE	AUREOLARIA PATULA
GEORGIA	THREATENED	SCROPHULARIACEAE	PENSTEMON DISSECTUS
GEORGIA	THREATENED	STEMONACEAE	CROOMIA PAUCIFLORA

STATE LISTS OF ENDANGERED, EXTINCT AND THREATENED SPECIES IN THE CONTINENTAL UNITED STATES

STATE	STATUS	FAMILY	SPECIES
GEORGIA	THREATENED	VIOLACEAE	VIOLA EGGLESTONII
GEORGIA	THREATENED	XYRIDACEAE	XYRIS SCABRIFOLIA
IDAHO	ENDANGERED	ASTERACEAE	ANTENNARIA ARCUATA
IDAHO	ENDANGERED	ASTERACEAE	ERIGERON LATUS
IDAHO	ENDANGERED	ASTERACEAE	GRINDELIA HOWELLII
IDAHO	ENDANGERED	ASTERACEAE	PYRROCOMA RADIATUS
IDAHO	ENDANGERED	BORAGINACEAE	DASYNOTUS DAUBENMIREI
IDAHO	ENDANGERED	BORAGINACEAE	HACKELIA DAVISII
IDAHO	ENDANGERED	BRASSICACEAE	CARDAMINE CONSTANCEI
IDAHO	ENDANGERED	BRASSICACEAE	LEPIDIUM DAVISII
IDAHO	ENDANGERED	BRASSICACEAE	THELYPODIUM REPANDUM
IDAHO	ENDANGERED	CARYOPHYLLACEAE	SILENE SPALDINGII
IDAHO	ENDANGERED	CYPERACEAE	CAREX ABORIGINUM
IDAHO	ENDANGERED	FABACEAE	ASTRAGALUS AMNIS-AMISSI
IDAHO	ENDANGERED	FABACEAE	ASTRAGALUS ATRATUS VAR. INSEPTUS
IDAHO	ENDANGERED	FABACEAE	ASTRAGALUS PURSHII VAR. OPHIOGENES
IDAHO	ENDANGERED	FABACEAE	ASTRAGALUS STERILIS
IDAHO	ENDANGERED	LILIACEAE	ALLIUM AASEAE
IDAHO	ENDANGERED	LILIACEAE	TOFIELDIA GLUTINOSA SSP. ABSONA
IDAHO	ENDANGERED	NYCTAGINACEAE	MIRABILIS MACFARLANEI
IDAHO	ENDANGERED	ONAGRACEAE	OENOTHERA PSAMMOPHILA
IDAHO	ENDANGERED	POLEMONIACEAE	PHLOX IDAHONIS
IDAHO	ENDANGERED	PRIMULACEAE	PRIMULA CUSICKIANA
IDAHO	ENDANGERED	SCROPHULARIACEAE	CASTILLEJA CHRISTII
IDAHO	ENDANGERED	SCROPHULARIACEAE	PENSTEMON LEMHIENSIS
IDAHO	THREATENED	APIACEAE	LOMATIUM FOENICULACEUM SSP. INYOENSE
IDAHO	THREATENED	APIACEAE	LOMATIUM HENDERSONII
IDAHO	THREATENED	APIACEAE	LOMATIUM ROLLINSII
IDAHO	THREATENED	APIACEAE	LOMATIUM SERPENTINUM
IDAHO	THREATENED	ASTERACEAE	ARTEMISIA PAPPOSA
IDAHO	THREATENED	ASTERACEAE	ASTER JESSICAE
IDAHO	THREATENED	ASTERACEAE	CHAENACTIS EVERMANNII
IDAHO	THREATENED	ASTERACEAE	CIRSIUM DAVISII
IDAHO	THREATENED	ASTERACEAE	HAPLOPAPPUS ABERRANS
IDAHO	THREATENED	ASTERACEAE	HYMENOPAPPUS FILIFOLIUS VAR. IDAHOENSIS
IDAHO	THREATENED	ASTERACEAE	PYRROCOMA LIATRIFORMIS
IDAHO	THREATENED	BORAGINACEAE	CRYPTANTHA HYPSOPHILA
IDAHO	THREATENED	BORAGINACEAE	HACKELIA HISPIDA
IDAHO	THREATENED	BRASSICACEAE	DRABA APICULATA VAR. DAVIESIAE
IDAHO	THREATENED	BRASSICACEAE	DRABA ARGYRAEA
IDAHO	THREATENED	BRASSICACEAE	DRABA SPHAEROCARPA
IDAHO	THREATENED	BRASSICACEAE	HALIMOLOBOS PERPLEXA VAR. LEMHIENSIS
IDAHO	THREATENED	BRASSICACEAE	HALIMOLOBOS PERPLEXA VAR. PERPLEXA

STATE LISTS OF ENDANGERED, EXTINCT AND THREATENED SPECIES IN THE CONTINENTAL UNITED STATES

STATE	STATUS	FAMILY	SPECIES
IDAHO	THREATENED	BRASSICACEAE	PHYSARIA DIDYMOCARPA VAR. LYRATA
IDAHO	THREATENED	BRASSICACEAE	PHYSARIA GEYERI VAR. PURPUREA
IDAHO	THREATENED	CARYOPHYLLACEAE	SILENE SCAPOSA VAR. LOBATA
IDAHO	THREATENED	CYPERACEAE	CAREX PARRYANA SSP. IDAHOA
IDAHO	THREATENED	FABACEAE	ASTRAGALUS MULFORDAE
IDAHO	THREATENED	FABACEAE	ASTRAGALUS PAYSONII
IDAHO	THREATENED	FABACEAE	ASTRAGALUS VEXILLIFLEXUS VAR. NUBILUS
IDAHO	THREATENED	FABACEAE	TRIFOLIUM PLUMOSUM VAR. AMPLIFOLIUM
IDAHO	THREATENED	FUMARIACEAE	CORYDALIS CASEANA SSP. HASTATA
IDAHO	THREATENED	GENTIANACEAE	FRASERA IDAHOENSIS
IDAHO	THREATENED	LILIACEAE	ALLIUM TOLMIEI VAR. PERSIMILE
IDAHO	THREATENED	LILIACEAE	CALOCHORTUS NITIDUS
IDAHO	THREATENED	POACEAE	CALAMAGROSTIS TWEEDYI
IDAHO	THREATENED	POLEMONIACEAE	PHLOX MOLLIS
IDAHO	THREATENED	PORTULACACEAE	CLAYTONIA FLAVA
IDAHO	THREATENED	PORTULACACEAE	LEWISIA COLUMBIANA VAR. WALLOWENSIS
IDAHO	THREATENED	SCROPHULARIACEAE	PENSTEMON ELEGANTULUS
IDAHO	THREATENED	SCROPHULARIACEAE	SYNTHYRIS HENDERSONII
IDAHO	THREATENED	SCROPHULARIACEAE	SYNTHYRIS PLATYCARPA
IDAHO	THREATENED	SOLANACEAE	ORYCTES NEVADENSIS
ILLINOIS	ENDANGERED	ASCLEPIADACEAE	ASCLEPIAS MEADII
ILLINOIS	ENDANGERED	CYPERACEAE	CYPERUS GRAYIOIDES
ILLINOIS	ENDANGERED	FABACEAE	APIOS PRICEANA
ILLINOIS	ENDANGERED	FABACEAE	LESPEDEZA LEPTOSTACHYA
ILLINOIS	ENDANGERED	FABACEAE	PETALOSTEMUM FOLIOSUM
ILLINOIS	ENDANGERED	MALVACEAE	ILIAMNA REMOTA
ILLINOIS	ENDANGERED	ORCHIDACEAE	ISOTRIA MEDEOLOIDES
ILLINOIS	EXTINCT	BURMANNIACEAE	THISMIA AMERICANA
ILLINOIS	THREATENED	ARALIACEAE	PANAX QUINQUEFOLIUS
ILLINOIS	THREATENED	ASTERACEAE	BOLTONIA ASTEROIDES VAR. DECURRENS
ILLINOIS	THREATENED	ASTERACEAE	CIRSIUM PITCHERI
ILLINOIS	THREATENED	CYPERACEAE	CAREX SOCIALIS
ILLINOIS	THREATENED	FABACEAE	ASTRAGALUS TENNESSEENSIS
ILLINOIS	THREATENED	FABACEAE	CLADRASTIS LUTEA
ILLINOIS	THREATENED	LAMIACEAE	SYNANDRA HISPIDULA
ILLINOIS	THREATENED	LILIACEAE	VERATRUM WOODII
ILLINOIS	THREATENED	ORCHIDACEAE	CYPRIPEDIUM CANDIDUM
ILLINOIS	THREATENED	ORCHIDACEAE	PLATANTHERA FLAVA
ILLINOIS	THREATENED	ORCHIDACEAE	PLATANTHERA LEUCOPHAEA
ILLINOIS	THREATENED	ORCHIDACEAE	PLATANTHERA PERAMOENA
ILLINOIS	THREATENED	PRIMULACEAE	DODECATHEON FRENCHII
ILLINOIS	THREATENED	RANUNCULACEAE	HYDRASTIS CANADENSIS
ILLINOIS	THREATENED	SAXIFRAGACEAE	SULLIVANTIA RENIFOLIA

STATE LISTS OF ENDANGERED, EXTINCT AND THREATENED SPECIES IN THE CONTINENTAL UNITED STATES

STATE	STATUS	FAMILY	SPECIES
ILLINOIS	THREATENED	SCROPHULARIACEAE	CHELONE OBLIQUA VAR. SPECIOSA
INDIANA	ENDANGERED	ASCLEPIADACEAE	ASCLEPIAS MEADII
INDIANA	ENDANGERED	MALVACEAE	ILIAMNA REMOTA
INDIANA	EXTINCT	FABACEAE	PSORALEA STIPULATA
INDIANA	THREATENED	ARALIACEAE	PANAX QUINQUEFOLIUS
INDIANA	THREATENED	ASTERACEAE	CIRSIUM PITCHERI
INDIANA	THREATENED	BRASSICACEAE	LESQUERELLA GLOBOSA
INDIANA	THREATENED	FABACEAE	CLADRASTIS LUTEA
INDIANA	THREATENED	LILIACEAE	VERATRUM WOODII
INDIANA	THREATENED	ORCHIDACEAE	CYPRIPEDIUM CANDIDUM
INDIANA	THREATENED	ORCHIDACEAE	PLATANTHERA FLAVA
INDIANA	THREATENED	ORCHIDACEAE	PLATANTHERA LEUCOPHAEA
INDIANA	THREATENED	ORCHIDACEAE	PLATANTHERA PERAMOENA
INDIANA	THREATENED	POACEAE	POA PALUDIGENA
INDIANA	THREATENED	RANUNCULACEAE	HYDRASTIS CANADENSIS
INDIANA	THREATENED	SAXIFRAGACEAE	SULLIVANTIA SULLIVANTII
INDIANA	THREATENED	SCROPHULARIACEAE	CHELONE OBLIQUA VAR. SPECIOSA
IOWA	ENDANGERED	ASCLEPIADACEAE	ASCLEPIAS MEADII
IOWA	ENDANGERED	FABACEAE	LESPEDEZA LEPTOSTACHYA
IOWA	ENDANGERED	RANUNCULACEAE	ACONITUM NOVEBORACENSE
IOWA	THREATENED	ARALIACEAE	PANAX QUINQUEFOLIUS
IOWA	THREATENED	LILIACEAE	VERATRUM WOODII
IOWA	THREATENED	ORCHIDACEAE	CYPRIPEDIUM CANDIDUM
IOWA	THREATENED	ORCHIDACEAE	PLATANTHERA FLAVA
IOWA	THREATENED	ORCHIDACEAE	PLATANTHERA LEUCOPHAEA
IOWA	THREATENED	SAXIFRAGACEAE	SULLIVANTIA RENIFOLIA
IOWA	THREATENED	SCROPHULARIACEAE	CHELONE OBLIQUA VAR. SPECIOSA
KANSAS	ENDANGERED	ASCLEPIADACEAE	ASCLEPIAS MEADII
KANSAS	THREATENED	ORCHIDACEAE	PLATANTHERA LEUCOPHAEA
KANSAS	THREATENED	POLEMONIACEAE	PHLOX OKLAHOMENSIS
KENTUCKY	ENDANGERED	ASTERACEAE	EUPATORIUM RESINOSUM VAR. KENTUCKIENSE
KENTUCKY	ENDANGERED	ASTERACEAE	SOLIDAGO ALBOPILOSA
KENTUCKY	ENDANGERED	ASTERACEAE	SOLIDAGO SHORTII
KENTUCKY	ENDANGERED	BRASSICACEAE	ARABIS PERSTELLATA VAR. PERSTELLATA
KENTUCKY	ENDANGERED	BRASSICACEAE	LEAVENWORTHIA EXIGUA VAR. LACINIATA
KENTUCKY	ENDANGERED	FABACEAE	APIOS PRICEANA
KENTUCKY	ENDANGERED	LAMIACEAE	CONRADINA VERTICILLATA
KENTUCKY	THREATENED	ARALIACEAE	PANAX QUINQUEFOLIUS
KENTUCKY	THREATENED	ASTERACEAE	PRENANTHES ROANENSIS
KENTUCKY	THREATENED	BORAGINACEAE	ONOSMODIUM MOLLE
KENTUCKY	THREATENED	BRASSICACEAE	LEAVENWORTHIA TORULOSA
KENTUCKY	THREATENED	BRASSICACEAE	LESQUERELLA GLOBOSA
KENTUCKY	THREATENED	CARYOPHYLLACEAE	ARENARIA FONTINALIS

STATE LISTS OF ENDANGERED, EXTINCT AND THREATENED SPECIES IN THE CONTINENTAL UNITED STATES

STATE	STATUS	FAMILY	SPECIES
KENTUCKY	THREATENED	CYPERACEAE	CAREX PURPURIFERA
KENTUCKY	THREATENED	CYPERACEAE	CYMOPHYLLUS FRASERI
KENTUCKY	THREATENED	ERICACEAE	RHODODENDRON BAKERI
KENTUCKY	THREATENED	FABACEAE	CLADRASTIS LUTEA
KENTUCKY	THREATENED	HYPERICACEAE	HYPERICUM SPHAEROCARPUM VAR. TURGIDUM
KENTUCKY	THREATENED	LAMIACEAE	SYNANDRA HISPIDULA
KENTUCKY	THREATENED	ORCHIDACEAE	CYPRIPEDIUM CANDIDUM
KENTUCKY	THREATENED	ORCHIDACEAE	PLATANTHERA FLAVA
KENTUCKY	THREATENED	ORCHIDACEAE	PLATANTHERA PERAMOENA
KENTUCKY	THREATENED	POACEAE	MUHLENBERGIA TORREYANA
KENTUCKY	THREATENED	PRIMULACEAE	DODECATHEON FRENCHII
KENTUCKY	THREATENED	RANUNCULACEAE	HYDRASTIS CANADENSIS
KENTUCKY	THREATENED	SAXIFRAGACEAE	SAXIFRAGA CAROLINIANA
KENTUCKY	THREATENED	SAXIFRAGACEAE	SULLIVANTIA SULLIVANTII
KENTUCKY	THREATENED	SCROPHULARIACEAE	AUREOLARIA PATULA
KENTUCKY	THREATENED	SCROPHULARIACEAE	SCHWALBEA AMERICANA
KENTUCKY	THREATENED	VIOLACEAE	VIOLA EGGLESTONII
LOUISIANA	ENDANGERED	ASTERACEAE	COREOPSIS INTERMEDIA
LOUISIANA	EXTINCT	ISOETACEAE	ISOETES LOUISIANENSIS
LOUISIANA	EXTINCT	SCROPHULARIACEAE	AGALINIS CADDOENSIS
LOUISIANA	EXTINCT	SCROPHULARIACEAE	CASTILLEJA LUDOVICIANA
LOUISIANA	THREATENED	APIACEAE	LILAEOPSIS CAROLINENSIS
LOUISIANA	THREATENED	APOCYNACEAE	AMSONIA GLABERRIMA
LOUISIANA	THREATENED	ARALIACEAE	PANAX QUINQUEFOLIUS
LOUISIANA	THREATENED	LAMIACEAE	SCUTELLARIA THIERETII
LOUISIANA	THREATENED	ORCHIDACEAE	PLATANTHERA INTEGRA
LOUISIANA	THREATENED	ORCHIDACEAE	PLATANTHERA LEUCOPHAEA
LOUISIANA	THREATENED	SCHISANDRACEAE	SCHISANDRA GLABRA
LOUISIANA	THREATENED	SCROPHULARIACEAE	SCHWALBEA AMERICANA
MAINE	ENDANGERED	CYPERACEAE	CAREX ELACHYCARPA
MAINE	ENDANGERED	ORCHIDACEAE	ISOTRIA MEDEOLOIDES
MAINE	ENDANGERED	POACEAE	CALAMAGROSTIS INEXPANSA VAR. NOVAE-ANGLIAE
MAINE	ENDANGERED	RANUNCULACEAE	TROLLIUS LAXUS
MAINE	ENDANGERED	SCROPHULARIACEAE	MIMULUS RINGENS VAR. COLPOPHILUS
MAINE	ENDANGERED	SCROPHULARIACEAE	PEDICULARIS FURBISHIAE
MAINE	THREATENED	ARALIACEAE	PANAX QUINQUEFOLIUS
MAINE	THREATENED	ASTERACEAE	PRENANTHES BOOTTII
MAINE	THREATENED	BRASSICACEAE	CARDAMINE LONGII
MAINE	THREATENED	CARYOPHYLLACEAE	PARONYCHIA ARGYROCOMA VAR. ALBIMONTANA
MAINE	THREATENED	CYPERACEAE	CAREX JOSSELYNII
MAINE	THREATENED	CYPERACEAE	CAREX ORONENSIS
MAINE	THREATENED	CYPERACEAE	SCIRPUS LONGII
MAINE	THREATENED	ORCHIDACEAE	CYPRIPEDIUM ARIETINUM

STATE LISTS OF ENDANGERED, EXTINCT AND THREATENED SPECIES IN THE CONTINENTAL UNITED STATES

STATE	STATUS	FAMILY	SPECIES
MAINE	THREATENED	ORCHIDACEAE	LISTERA AURICULATA
MAINE	THREATENED	ORCHIDACEAE	PLATANTHERA FLAVA
MAINE	THREATENED	ORCHIDACEAE	PLATANTHERA LEUCOPHAEA
MARYLAND	ENDANGERED	LILIACEAE	TRILLIUM PUSILLUM VAR. VIRGINIANUM
MARYLAND	THREATENED	APIACEAE	PTILIMNIUM FLUVIATILE
MARYLAND	THREATENED	ARALIACEAE	PANAX QUINQUEFOLIUS
MARYLAND	THREATENED	BETULACEAE	ALNUS MARITIMA
MARYLAND	THREATENED	BRASSICACEAE	CARDAMINE LONGII
MARYLAND	THREATENED	JUNCACEAE	JUNCUS CAESARIENSIS
MARYLAND	THREATENED	LILIACEAE	LILIUM GRAYII
MARYLAND	THREATENED	ORCHIDACEAE	PLATANTHERA FLAVA
MARYLAND	THREATENED	ORCHIDACEAE	PLATANTHERA PERAMOENA
MARYLAND	THREATENED	RANUNCULACEAE	HYDRASTIS CANADENSIS
MARYLAND	THREATENED	SCROPHULARIACEAE	MICRANTHEMUM MICRANTHEMOIDES
MARYLAND	THREATENED	SCROPHULARIACEAE	SCHWALBEA AMERICANA
MASSACHUSETTS	ENDANGERED	ORCHIDACEAE	ISOTRIA MEDEOLOIDES
MASSACHUSETTS	EXTINCT	JUNCACEAE	JUNCUS PERVETUS
MASSACHUSETTS	THREATENED	ARALIACEAE	PANAX QUINQUEFOLIUS
MASSACHUSETTS	THREATENED	ASTERACEAE	EUPATORIUM LEUCOLEPIS VAR. NOVAE-ANGLIAE
MASSACHUSETTS	THREATENED	CARYOPHYLLACEAE	PARONYCHIA ARGYROCOMA VAR. ALBIMONTANA
MASSACHUSETTS	THREATENED	CISTACEAE	HELIANTHEMUM DUMOSUM
MASSACHUSETTS	THREATENED	CYPERACEAE	SCIRPUS LONGII
MASSACHUSETTS	THREATENED	ISOETACEAE	ISOETES EATONII
MASSACHUSETTS	THREATENED	ORCHIDACEAE	CYPRIPEDIUM ARIETINUM
MASSACHUSETTS	THREATENED	ORCHIDACEAE	PLATANTHERA FLAVA
MASSACHUSETTS	THREATENED	SCROPHULARIACEAE	GERARDIA ACUTA
MASSACHUSETTS	THREATENED	SCROPHULARIACEAE	SCHWALBEA AMERICANA
MICHIGAN	ENDANGERED	ORCHIDACEAE	ISOTRIA MEDEOLOIDES
MICHIGAN	ENDANGERED	POLYPODIACEAE	PHYLLITIS SCOLOPENDRIUM VAR. AMERICANUM
MICHIGAN	THREATENED	ARALIACEAE	PANAX QUINQUEFOLIUS
MICHIGAN	THREATENED	ASTERACEAE	CIRSIUM PITCHERI
MICHIGAN	THREATENED	ASTERACEAE	SOLIDAGO HOUGHTONII
MICHIGAN	THREATENED	IRIDACEAE	IRIS LACUSTRIS
MICHIGAN	THREATENED	ORCHIDACEAE	CYPRIPEDIUM ARIETINUM
MICHIGAN	THREATENED	ORCHIDACEAE	CYPRIPEDIUM CANDIDUM
MICHIGAN	THREATENED	ORCHIDACEAE	LISTERA AURICULATA
MICHIGAN	THREATENED	ORCHIDACEAE	PLATANTHERA FLAVA
MICHIGAN	THREATENED	ORCHIDACEAE	PLATANTHERA LEUCOPHAEA
MICHIGAN	THREATENED	POACEAE	POA PALUDIGENA
MICHIGAN	THREATENED	POTAMOGETONACEAE	POTAMOGETON HILLII
MICHIGAN	THREATENED	RANUNCULACEAE	HYDRASTIS CANADENSIS
MICHIGAN	THREATENED	SCROPHULARIACEAE	CHELONE OBLIQUA VAR. SPECIOSA
MICHIGAN	THREATENED	SCROPHULARIACEAE	MIMULUS GLABRATUS VAR. MICHIGANENSIS

STATE LISTS OF ENDANGERED, EXTINCT AND THREATENED SPECIES IN THE CONTINENTAL UNITED STATES

STATE	STATUS	FAMILY	SPECIES
MINNESOTA	ENDANGERED	FABACEAE	LESPEDEZA LEPTOSTACHYA
MINNESOTA	ENDANGERED	LILIACEAE	ERYTHRONIUM PROPULLANS
MINNESOTA	ENDANGERED	POLEMONIACEAE	POLEMONIUM OCCIDENTALE VAR. LACUSTRE
MINNESOTA	THREATENED	ARALIACEAE	PANAX QUINQUEFOLIUS
MINNESOTA	THREATENED	ASTERACEAE	ERIGERON PULCHELLUS VAR. TOLSTEADII
MINNESOTA	THREATENED	ORCHIDACEAE	CYPRIPEDIUM ARIETINUM
MINNESOTA	THREATENED	ORCHIDACEAE	CYPRIPEDIUM CANDIDUM
MINNESOTA	THREATENED	ORCHIDACEAE	LISTERA AURICULATA
MINNESOTA	THREATENED	ORCHIDACEAE	PLATANTHERA FLAVA
MINNESOTA	THREATENED	ORCHIDACEAE	PLATANTHERA LEUCOPHAEA
MINNESOTA	THREATENED	RANUNCULACEAE	HYDRASTIS CANADENSIS
MINNESOTA	THREATENED	SAXIFRAGACEAE	SULLIVANTIA RENIFOLIA
MINNESOTA	THREATENED	SCROPHULARIACEAE	CHELONE OBLIQUA VAR. SPECIOSA
MISSISSIPPI	ENDANGERED	FABACEAE	APIOS PRICEANA
MISSISSIPPI	ENDANGERED	LAURACEAE	LINDERA MELISSIFOLIA
MISSISSIPPI	THREATENED	ARALIACEAE	PANAX QUINQUEFOLIUS
MISSISSIPPI	THREATENED	ARECACEAE	RHAPIDOPHYLLUM HYSTRIX
MISSISSIPPI	THREATENED	ASTERACEAE	ASTER VERUTIFOLIUS
MISSISSIPPI	THREATENED	ERICACEAE	RHODODENDRON AUSTRINUM
MISSISSIPPI	THREATENED	JUNCACEAE	JUNCUS GYMNOCARPUS
MISSISSIPPI	THREATENED	LAURACEAE	LITSEA AESTIVALIS
MISSISSIPPI	THREATENED	ORCHIDACEAE	PLATANTHERA INTEGRA
MISSISSIPPI	THREATENED	ORCHIDACEAE	PLATANTHERA PERAMOENA
MISSISSIPPI	THREATENED	POACEAE	ARISTIDA SIMPLICIFLORA
MISSISSIPPI	THREATENED	POACEAE	PANICUM NUDICAULE
MISSISSIPPI	THREATENED	RANUNCULACEAE	HYDRASTIS CANADENSIS
MISSISSIPPI	THREATENED	RHAMNACEAE	SAGERETIA MINUTIFLORA
MISSISSIPPI	THREATENED	ROSACEAE	AGRIMONIA INCISA
MISSISSIPPI	THREATENED	SARRACENIACEAE	SARRACENIA ALABAMENSIS SSP. WHERRYI
MISSISSIPPI	THREATENED	SCHISANDRACEAE	SCHISANDRA GLABRA
MISSISSIPPI	THREATENED	SCROPHULARIACEAE	SCHWALBEA AMERICANA
MISSISSIPPI	THREATENED	XYRIDACEAE	XYRIS SCABRIFOLIA
MISSOURI	ENDANGERED	ASCLEPIADACEAE	ASCLEPIAS MEADII
MISSOURI	ENDANGERED	BRASSICACEAE	DRABA APRICA
MISSOURI	ENDANGERED	BRASSICACEAE	LESQUERELLA FILIFORMIS
MISSOURI	ENDANGERED	CARYOPHYLLACEAE	GEOCARPON MINIMUM
MISSOURI	ENDANGERED	FAGACEAE	CASTANEA PUMILA VAR. OZARKENSIS
MISSOURI	ENDANGERED	LAURACEAE	LINDERA MELISSIFOLIA
MISSOURI	ENDANGERED	POACEAE	CALAMAGROSTIS INSPERATA
MISSOURI	ENDANGERED	ROSACEAE	NEVIUSIA ALABAMENSIS
MISSOURI	ENDANGERED	SAXIFRAGACEAE	HEUCHERA MISSOURIENSIS
MISSOURI	THREATENED	ARALIACEAE	PANAX QUINQUEFOLIUS
MISSOURI	THREATENED	ASTERACEAE	BOLTONIA ASTEROIDES VAR. DECURRENS

STATE LISTS OF ENDANGERED, EXTINCT AND THREATENED SPECIES IN THE CONTINENTAL UNITED STATES

STATE	STATUS	FAMILY	SPECIES
MISSOURI	THREATENED	CYPERACEAE	CAREX SOCIALIS
MISSOURI	THREATENED	ERICACEAE	VACCINIUM VACILLANS VAR. MISSOURIENSE
MISSOURI	THREATENED	FABACEAE	AMORPHA BRACHYCARPA
MISSOURI	THREATENED	FABACEAE	CLADRASTIS LUTEA
MISSOURI	THREATENED	LEITNERIACEAE	LEITNERIA FLORIDANA
MISSOURI	THREATENED	LILIACEAE	TRILLIUM PUSILLUM VAR. OZARKANUM
MISSOURI	THREATENED	LILIACEAE	VERATRUM WOODII
MISSOURI	THREATENED	MALVACEAE	CALLIRHOE PAPAVER VAR. BUSHII
MISSOURI	THREATENED	ORCHIDACEAE	CYPRIPEDIUM CANDIDUM
MISSOURI	THREATENED	ORCHIDACEAE	PLATANTHERA FLAVA
MISSOURI	THREATENED	ORCHIDACEAE	PLATANTHERA LEUCOPHAEA
MISSOURI	THREATENED	ORCHIDACEAE	PLATANTHERA PERAMOENA
MISSOURI	THREATENED	POACEAE	SPOROBOLUS NEGLECTUS VAR. OZARKANUS
MISSOURI	THREATENED	RANUNCULACEAE	HYDRASTIS CANADENSIS
MISSOURI	THREATENED	SAXIFRAGACEAE	SULLIVANTIA RENIFOLIA
MISSOURI	THREATENED	SCROPHULARIACEAE	CHELONE OBLIQUA VAR. SPECIOSA
MISSOURI	THREATENED	SCROPHULARIACEAE	PENSTEMON COBAEA VAR. PURPUREUS
MONTANA	ENDANGERED	ASTERACEAE	GRINDELIA HOWELLII
MONTANA	ENDANGERED	ASTERACEAE	SAUSSUREA WEBERI
MONTANA	ENDANGERED	CARYOPHYLLACEAE	SILENE SPALDINGII
MONTANA	ENDANGERED	POACEAE	TRISETUM ORTHOCHAETUM
MONTANA	ENDANGERED	POLEMONIACEAE	PHLOX MISSOULENSIS
MONTANA	ENDANGERED	SCROPHULARIACEAE	PENSTEMON LEMHIENSIS
MONTANA	THREATENED	ASTERACEAE	ERIGERON ALLOCOTUS
MONTANA	THREATENED	BRASSICACEAE	CARDAMINE RUPICOLA
MONTANA	THREATENED	BRASSICACEAE	DRABA APICULATA VAR. DAVIESIAE
MONTANA	THREATENED	BRASSICACEAE	RORIPPA CALYCINA
MONTANA	THREATENED	CYPERACEAE	CAREX PARRYANA SSP. IDAHOA
MONTANA	THREATENED	CYPERACEAE	CAREX PLECTOCARPA
MONTANA	THREATENED	POACEAE	CALAMAGROSTIS TWEEDYI
MONTANA	THREATENED	POLYGONACEAE	ERIOGONUM LAGOPUS
MONTANA	THREATENED	RANUNCULACEAE	AQUILEGIA JONESII
MONTANA	THREATENED	SAXIFRAGACEAE	SULLIVANTIA HAPEMANII
MONTANA	THREATENED	SCROPHULARIACEAE	PENSTEMON CARYI
MONTANA	THREATENED	SCROPHULARIACEAE	SYNTHYRIS CANBYI
NEBRASKA	THREATENED	ARALIACEAE	PANAX QUINQUEFOLIUS
NEBRASKA	THREATENED	ORCHIDACEAE	CYPRIPEDIUM CANDIDUM
NEBRASKA	THREATENED	ORCHIDACEAE	PLATANTHERA LEUCOPHAEA
NEBRASKA	THREATENED	RANUNCULACEAE	HYDRASTIS CANADENSIS
NEVADA	ENDANGERED	APIACEAE	CYMOPTERUS NIVALIS
NEVADA	ENDANGERED	ASCLEPIADACEAE	ASCLEPIAS EASTWOODIANA
NEVADA	ENDANGERED	ASTERACEAE	CIRSIUM CLOKEYI
NEVADA	ENDANGERED	ASTERACEAE	GRINDELIA FRAXINO-PRATENSIS

STATE LISTS OF ENDANGERED, EXTINCT AND THREATENED SPECIES IN THE CONTINENTAL UNITED STATES

STATE	STATUS	FAMILY	SPECIES
NEVADA	ENDANGERED	ASTERACEAE	MACHAERANTHERA LEUCANTHEMIFOLIA
NEVADA	ENDANGERED	ASTERACEAE	TANACETUM COMPACTUM
NEVADA	ENDANGERED	BRASSICACEAE	DRABA ARIDA
NEVADA	ENDANGERED	BRASSICACEAE	DRABA PAUCIFRUCTA
NEVADA	ENDANGERED	EUPHORBIACEAE	CROTON WIGGINSII
NEVADA	ENDANGERED	EUPHORBIACEAE	DITAXIS DIVERSIFLORA
NEVADA	ENDANGERED	FABACEAE	ASTRAGALUS BEATLEYAE
NEVADA	ENDANGERED	FABACEAE	ASTRAGALUS FUNEREUS
NEVADA	ENDANGERED	FABACEAE	ASTRAGALUS LENTIGINOSUS VAR. SESQUIMETRALIS
NEVADA	ENDANGERED	FABACEAE	ASTRAGALUS NYENSIS
NEVADA	ENDANGERED	FABACEAE	ASTRAGALUS PHOENIX
NEVADA	ENDANGERED	FABACEAE	ASTRAGALUS PORRECTUS
NEVADA	ENDANGERED	FABACEAE	ASTRAGALUS ROBBINSII VAR. OCCIDENTALIS
NEVADA	ENDANGERED	FABACEAE	ASTRAGALUS SERENOI VAR. SORDESCENS
NEVADA	ENDANGERED	FABACEAE	ASTRAGALUS UNCIALIS
NEVADA	ENDANGERED	FABACEAE	LATHYRUS HITCHCOCKIANUS
NEVADA	ENDANGERED	FABACEAE	TRIFOLIUM ANDERSONII SSP. BEATLEYAE
NEVADA	ENDANGERED	FABACEAE	TRIFOLIUM LEMMONII
NEVADA	ENDANGERED	GENTIANACEAE	CENTAURIUM NAMOPHILUM
NEVADA	ENDANGERED	GENTIANACEAE	FRASERA GYPSICOLA
NEVADA	ENDANGERED	GENTIANACEAE	FRASERA PAHUTENSIS
NEVADA	ENDANGERED	GERANIACEAE	GERANIUM TOQUIMENSE
NEVADA	ENDANGERED	HYDROPHYLLACEAE	PHACELIA BEATLEYAE
NEVADA	ENDANGERED	LOASACEAE	MENTZELIA LEUCOPHYLLA
NEVADA	ENDANGERED	ONAGRACEAE	CAMISSONIA MEGALANTHA
NEVADA	ENDANGERED	ONAGRACEAE	CAMISSONIA NEVADENSIS
NEVADA	ENDANGERED	PAPAVERACEAE	ARCTOMECON MERRIAMII
NEVADA	ENDANGERED	POLYGONACEAE	ERIOGONUM ANEMOPHILUM
NEVADA	ENDANGERED	POLYGONACEAE	ERIOGONUM ARGOPHYLLUM
NEVADA	ENDANGERED	POLYGONACEAE	ERIOGONUM DARROVII
NEVADA	ENDANGERED	POLYGONACEAE	ERIOGONUM LEMMONII
NEVADA	ENDANGERED	POLYGONACEAE	ERIOGONUM VISCIDULUM
NEVADA	ENDANGERED	PORTULACACEAE	LEWISIA MAGUIREI
NEVADA	ENDANGERED	PRIMULACEAE	PRIMULA CAPILLARIS
NEVADA	ENDANGERED	PRIMULACEAE	PRIMULA NEVADENSIS
NEVADA	ENDANGERED	ROSACEAE	IVESIA CRYPTOCAULIS
NEVADA	ENDANGERED	ROSACEAE	IVESIA EREMICA
NEVADA	ENDANGERED	RUBIACEAE	GALIUM HILENDIAE SSP. KINGSTONENSE
NEVADA	ENDANGERED	SCROPHULARIACEAE	CASTILLEJA SALSUGINOSA
NEVADA	ENDANGERED	SCROPHULARIACEAE	PENSTEMON DECURVUS
NEVADA	ENDANGERED	SCROPHULARIACEAE	PENSTEMON KECKII
NEVADA	ENDANGERED	SCROPHULARIACEAE	PENSTEMON NYEENSIS
NEVADA	ENDANGERED	SCROPHULARIACEAE	PENSTEMON PAHUTENSIS

STATE LISTS OF ENDANGERED, EXTINCT AND THREATENED SPECIES IN THE CONTINENTAL UNITED STATES

STATE	STATUS	FAMILY	SPECIES
NEVADA	ENDANGERED	SCROPHULARIACEAE	PENSTEMON RUBICUNDUS
NEVADA	ENDANGERED	SCROPHULARIACEAE	SYNTHYRIS RANUNCULINA
NEVADA	EXTINCT	BORAGINACEAE	CRYPTANTHA INSOLITA
NEVADA	EXTINCT	BORAGINACEAE	MERTENSIA TOIYABENSIS
NEVADA	EXTINCT	BRASSICACEAE	SMELOWSKIA HOLMGRENII
NEVADA	EXTINCT	HYDROCHARITACEAE	ELODEA NEVADENSIS
NEVADA	THREATENED	APIACEAE	ANGELICA SCABRIDA
NEVADA	THREATENED	APIACEAE	ANGELICA WHEELERI
NEVADA	THREATENED	APIACEAE	CYMOPTERUS CORRUGATUS
NEVADA	THREATENED	APIACEAE	LOMATIUM FOENICULACEUM SSP. INYOENSE
NEVADA	THREATENED	ASTERACEAE	ANTENNARIA SOLICEPS
NEVADA	THREATENED	ASTERACEAE	ENCELIOPSIS NUDICAULIS VAR. CORRUGATA
NEVADA	THREATENED	ASTERACEAE	ERIGERON OVINUS
NEVADA	THREATENED	ASTERACEAE	ERIGERON UNCIALIS VAR. CONJUGANS
NEVADA	THREATENED	ASTERACEAE	HAPLOPAPPUS BRICKELLIOIDES
NEVADA	THREATENED	ASTERACEAE	HELIANTHUS DESERTICOLA
NEVADA	THREATENED	ASTERACEAE	HULSEA VESTITA SSP. INYOENSIS
NEVADA	THREATENED	ASTERACEAE	MACHAERANTHERA AMMOPHILA
NEVADA	THREATENED	ASTERACEAE	PERITYLE MEGALOCEPHALA VAR. INTRICATA
NEVADA	THREATENED	ASTERACEAE	SENECIO LYNCEUS VAR. LEUCOREUS
NEVADA	THREATENED	ASTERACEAE	TOWNSENDIA JONESII VAR. TUMULOSA
NEVADA	THREATENED	BORAGINACEAE	CRYPTANTHA COMPACTA
NEVADA	THREATENED	BORAGINACEAE	CRYPTANTHA INTERRUPTA
NEVADA	THREATENED	BRASSICACEAE	DRABA ASTEROPHORA VAR. ASTEROPHORA
NEVADA	THREATENED	BRASSICACEAE	DRABA CRASSIFOLIA VAR. NEVADENSIS
NEVADA	THREATENED	BRASSICACEAE	DRABA JAEGERI
NEVADA	THREATENED	BRASSICACEAE	DRABA STENOLOBA VAR. RAMOSA
NEVADA	THREATENED	BRASSICACEAE	LEPIDIUM NANUM
NEVADA	THREATENED	BRASSICACEAE	LESQUERELLA HITCHCOCKII
NEVADA	THREATENED	BRASSICACEAE	RORIPPA SUBUMBELLATA
NEVADA	THREATENED	BRASSICACEAE	THELYPODIUM SAGITTATUM VAR. OVALIFOLIUM
NEVADA	THREATENED	CACTACEAE	CORYPHANTHA VIVIPARA VAR. ROSEA
NEVADA	THREATENED	CACTACEAE	OPUNTIA WHIPPLEI VAR. MULTIGENICULATA
NEVADA	THREATENED	CACTACEAE	SCLEROCACTUS POLYANCISTRUS
NEVADA	THREATENED	CACTACEAE	SCLEROCACTUS PUBISPINUS
NEVADA	THREATENED	CARYOPHYLLACEAE	ARENARIA KINGII VAR. ROSEA
NEVADA	THREATENED	CARYOPHYLLACEAE	ARENARIA STENOMERES
NEVADA	THREATENED	CARYOPHYLLACEAE	SILENE CLOKEYI
NEVADA	THREATENED	CARYOPHYLLACEAE	SILENE SCAPOSA VAR. LOBATA
NEVADA	THREATENED	FABACEAE	ASTRAGALUS AEQUALIS
NEVADA	THREATENED	FABACEAE	ASTRAGALUS ALVORDENSIS
NEVADA	THREATENED	FABACEAE	ASTRAGALUS CALLITHRIX
NEVADA	THREATENED	FABACEAE	ASTRAGALUS CONVALLARIUS VAR. FINITIMUS

STATE LISTS OF ENDANGERED, EXTINCT AND THREATENED SPECIES IN THE CONTINENTAL UNITED STATES

STATE	STATUS	FAMILY	SPECIES
NEVADA	THREATENED	FABACEAE	ASTRAGALUS GEYERI VAR. TRIQUETRUS
NEVADA	THREATENED	FABACEAE	ASTRAGALUS LENTIGINOSUS VAR. LATUS
NEVADA	THREATENED	FABACEAE	ASTRAGALUS MOHAVENSIS VAR. HEMIGYRUS
NEVADA	THREATENED	FABACEAE	ASTRAGALUS MUSIMONUM
NEVADA	THREATENED	FABACEAE	ASTRAGALUS OOPHORUS VAR. CLOKEYANUS
NEVADA	THREATENED	FABACEAE	ASTRAGALUS OOPHORUS VAR. LONCHOCALYX
NEVADA	THREATENED	FABACEAE	ASTRAGALUS PSEUDIODANTHUS
NEVADA	THREATENED	FABACEAE	ASTRAGALUS PTEROCARPUS
NEVADA	THREATENED	FABACEAE	ASTRAGALUS TOQUIMANUS
NEVADA	THREATENED	FABACEAE	DALEA KINGII
NEVADA	THREATENED	FABACEAE	LUPINUS HOLMGRENANUS
NEVADA	THREATENED	HYDROPHYLLACEAE	PHACELIA GLABERRIMA
NEVADA	THREATENED	HYDROPHYLLACEAE	PHACELIA MUSTELINA
NEVADA	THREATENED	LILIACEAE	AGAVE UTAHENSIS VAR. EBORISPINA
NEVADA	THREATENED	LILIACEAE	CALOCHORTUS STRIATUS
NEVADA	THREATENED	NYCTAGINACEAE	ABRONIA ORBICULATA
NEVADA	THREATENED	NYCTAGINACEAE	MIRABILIS PUDICA
NEVADA	THREATENED	OLEACEAE	FRAXINUS CUSPIDATA VAR. MACROPETALA
NEVADA	THREATENED	ONAGRACEAE	EPILOBIUM NEVADENSE
NEVADA	THREATENED	POLEMONIACEAE	GILIA NYENSIS
NEVADA	THREATENED	POLEMONIACEAE	GILIA RIPLEYI
NEVADA	THREATENED	POLEMONIACEAE	PHLOX GLADIFORMIS
NEVADA	THREATENED	POLEMONIACEAE	POLEMONIUM NEVADENSE
NEVADA	THREATENED	POLYGONACEAE	ERIOGONUM BEATLEYAE
NEVADA	THREATENED	POLYGONACEAE	ERIOGONUM BIFURCATUM
NEVADA	THREATENED	POLYGONACEAE	ERIOGONUM CONCINNUM
NEVADA	THREATENED	POLYGONACEAE	ERIOGONUM CONTIGUUM
NEVADA	THREATENED	POLYGONACEAE	ERIOGONUM HEERMANNII VAR. FLOCCOSUM
NEVADA	THREATENED	POLYGONACEAE	ERIOGONUM HOLMGRENII
NEVADA	THREATENED	POLYGONACEAE	ERIOGONUM OVALIFOLIUM VAR. CAELESTRINUM
NEVADA	THREATENED	POLYGONACEAE	ERIOGONUM RUBRICAULE
NEVADA	THREATENED	POLYGONACEAE	OXYTHECA WATSONII
NEVADA	THREATENED	SAXIFRAGACEAE	HEUCHERA DURANII
NEVADA	THREATENED	SCROPHULARIACEAE	CASTILLEJA LINOIDES
NEVADA	THREATENED	SCROPHULARIACEAE	CORDYLANTHUS TECOPENSIS
NEVADA	THREATENED	SCROPHULARIACEAE	PENSTEMON ARENARIUS
NEVADA	THREATENED	SCROPHULARIACEAE	PENSTEMON BICOLOR SSP. BICOLOR
NEVADA	THREATENED	SCROPHULARIACEAE	PENSTEMON BICOLOR SSP. ROSEUS
NEVADA	THREATENED	SCROPHULARIACEAE	PENSTEMON MODESTUS
NEVADA	THREATENED	SCROPHULARIACEAE	PENSTEMON PUDICUS
NEVADA	THREATENED	SCROPHULARIACEAE	PENSTEMON THOMPSONIAE SSP. JAEGERI
NEVADA	THREATENED	SCROPHULARIACEAE	PENSTEMON THURBERI VAR. ANESTIUS
NEVADA	THREATENED	SELAGINELLACEAE	SELAGINELLA UTAHENSIS

STATE LISTS OF ENDANGERED, EXTINCT AND THREATENED SPECIES IN THE CONTINENTAL UNITED STATES

STATE	STATUS	FAMILY	SPECIES
NEVADA	THREATENED	SOLANACEAE	ORYCTES NEVADENSIS
NEVADA	THREATENED	VIOLACEAE	VIOLA CHARLESTONENSIS
NEW HAMPSHIRE	ENDANGERED	FABACEAE	ASTRAGALUS ROBBINSII VAR. JESUPI
NEW HAMPSHIRE	ENDANGERED	ORCHIDACEAE	ISOTRIA MEDEOLOIDES
NEW HAMPSHIRE	ENDANGERED	POACEAE	CALAMAGROSTIS INEXPANSA VAR. NOVAE-ANGLIAE
NEW HAMPSHIRE	ENDANGERED	RANUNCULACEAE	TROLLIUS LAXUS
NEW HAMPSHIRE	ENDANGERED	ROSACEAE	GEUM PECKII
NEW HAMPSHIRE	ENDANGERED	ROSACEAE	POTENTILLA ROBBINSIANA
NEW HAMPSHIRE	THREATENED	ARALIACEAE	PANAX QUINQUEFOLIUS
NEW HAMPSHIRE	THREATENED	ASTERACEAE	PRENANTHES BOOTTII
NEW HAMPSHIRE	THREATENED	CARYOPHYLLACEAE	PARONYCHIA ARGYROCOMA VAR. ALBIMONTANA
NEW HAMPSHIRE	THREATENED	ISOETACEAE	ISOETES EATONII
NEW HAMPSHIRE	THREATENED	ISOETACEAE	ISOETES FOVEOLATA
NEW HAMPSHIRE	THREATENED	ORCHIDACEAE	CYPRIPEDIUM ARIETINUM
NEW HAMPSHIRE	THREATENED	ORCHIDACEAE	LISTERA AURICULATA
NEW HAMPSHIRE	THREATENED	ORCHIDACEAE	PLATANTHERA FLAVA
NEW JERSEY	ENDANGERED	CYPERACEAE	RHYNCHOSPORA KNIESKERNII
NEW JERSEY	ENDANGERED	ORCHIDACEAE	ISOTRIA MEDEOLOIDES
NEW JERSEY	ENDANGERED	POACEAE	PANICUM HIRSTII
NEW JERSEY	ENDANGERED	RANUNCULACEAE	TROLLIUS LAXUS
NEW JERSEY	THREATENED	ARALIACEAE	PANAX QUINQUEFOLIUS
NEW JERSEY	THREATENED	CYPERACEAE	SCIRPUS LONGII
NEW JERSEY	THREATENED	ISOETACEAE	ISOETES EATONII
NEW JERSEY	THREATENED	JUNCACEAE	JUNCUS CAESARIENSIS
NEW JERSEY	THREATENED	ORCHIDACEAE	CYPRIPEDIUM CANDIDUM
NEW JERSEY	THREATENED	ORCHIDACEAE	PLATANTHERA FLAVA
NEW JERSEY	THREATENED	ORCHIDACEAE	PLATANTHERA INTEGRA
NEW JERSEY	THREATENED	ORCHIDACEAE	PLATANTHERA PERAMOENA
NEW JERSEY	THREATENED	POACEAE	CALAMOVILFA BREVIPILIS VAR. BREVIPILIS
NEW JERSEY	THREATENED	POACEAE	MUHLENBERGIA TORREYANA
NEW JERSEY	THREATENED	SCHIZAEACEAE	SCHIZAEA PUSILLA
NEW JERSEY	THREATENED	SCROPHULARIACEAE	MICRANTHEMUM MICRANTHEMOIDES
NEW JERSEY	THREATENED	SCROPHULARIACEAE	SCHWALBEA AMERICANA
NEW MEXICO	ENDANGERED	ASTERACEAE	ASTER BLEPHAROPHYLLUS
NEW MEXICO	ENDANGERED	ASTERACEAE	ERIGERON RHIZOMATUS
NEW MEXICO	ENDANGERED	ASTERACEAE	HAPLOPAPPUS SPINULOSUS SSP. LAEVIS
NEW MEXICO	ENDANGERED	ASTERACEAE	HELIANTHUS PARADOXUS
NEW MEXICO	ENDANGERED	BRASSICACEAE	LESQUERELLA AUREA
NEW MEXICO	ENDANGERED	BRASSICACEAE	LESQUERELLA LATA
NEW MEXICO	ENDANGERED	BRASSICACEAE	LESQUERELLA VALIDA
NEW MEXICO	ENDANGERED	CACTACEAE	CORYPHANTHA SNEEDII VAR. LEEI
NEW MEXICO	ENDANGERED	CACTACEAE	CORYPHANTHA SNEEDII VAR. SNEEDII
NEW MEXICO	ENDANGERED	CACTACEAE	ECHINOCEREUS FENDLERI VAR. KUENZLERI

STATE LISTS OF ENDANGERED, EXTINCT AND THREATENED SPECIES IN THE CONTINENTAL UNITED STATES

STATE	STATUS	FAMILY	SPECIES
NEW MEXICO	ENDANGERED	CACTACEAE	ECHINOCEREUS LLOYDII
NEW MEXICO	ENDANGERED	CACTACEAE	PEDIOCACTUS KNOWLTONII
NEW MEXICO	ENDANGERED	CACTACEAE	SCLEROCACTUS MESAE-VERDAE
NEW MEXICO	ENDANGERED	CAPPARIDACEAE	CLEOME MULTICAULIS
NEW MEXICO	ENDANGERED	CARYOPHYLLACEAE	SILENE PLANKII
NEW MEXICO	ENDANGERED	CHENOPODIACEAE	ATRIPLEX GRIFFITHSII
NEW MEXICO	ENDANGERED	FABACEAE	ASTRAGALUS CASTETTERI
NEW MEXICO	ENDANGERED	FABACEAE	ASTRAGALUS OOCALYCIS
NEW MEXICO	ENDANGERED	FABACEAE	ASTRAGALUS SILICEUS
NEW MEXICO	ENDANGERED	NYCTAGINACEAE	ABRONIA BIGELOVII
NEW MEXICO	ENDANGERED	PAPAVERACEAE	ARGEMONE PLEIACANTHA SSP. PINNATISECTA
NEW MEXICO	ENDANGERED	POLYGALACEAE	POLYGALA RIMULICOLA
NEW MEXICO	ENDANGERED	POLYGONACEAE	ERIOGONUM GYPSOPHILUM
NEW MEXICO	ENDANGERED	RANUNCULACEAE	AQUILEGIA CHAPLINEI
NEW MEXICO	ENDANGERED	ROSACEAE	POTENTILLA SIERRA-BLANCAE
NEW MEXICO	ENDANGERED	SCROPHULARIACEAE	SCROPHULARIA MACRANTHA
NEW MEXICO	EXTINCT	ASTERACEAE	HELIANTHUS PRAETERMISSUS
NEW MEXICO	THREATENED	APIACEAE	ALETES FILIFOLIUS
NEW MEXICO	THREATENED	ASTERACEAE	CHAETOPAPPA HERSHEYI
NEW MEXICO	THREATENED	ASTERACEAE	CIRSIUM VINACEUM
NEW MEXICO	THREATENED	ASTERACEAE	PERITYLE CERNUA
NEW MEXICO	THREATENED	ASTERACEAE	PERITYLE LEMMONII
NEW MEXICO	THREATENED	ASTERACEAE	PERITYLE STAUROPHYLLA
NEW MEXICO	THREATENED	ASTERACEAE	PLUMMERA FLORIBUNDA
NEW MEXICO	THREATENED	ASTERACEAE	SENECIO QUAERENS
NEW MEXICO	THREATENED	BORAGINACEAE	CRYPTANTHA PARADOXA
NEW MEXICO	THREATENED	BRASSICACEAE	DRABA MOGOLLONICA
NEW MEXICO	THREATENED	BRASSICACEAE	LESQUERELLA GOODDINGII
NEW MEXICO	THREATENED	CACTACEAE	CEREUS GREGGII
NEW MEXICO	THREATENED	CACTACEAE	CORYPHANTHA DUNCANII
NEW MEXICO	THREATENED	CACTACEAE	MAMMILLARIA ORESTERA
NEW MEXICO	THREATENED	CACTACEAE	OPUNTIA ARENARIA
NEW MEXICO	THREATENED	CACTACEAE	PEDIOCACTUS PAPYRACANTHUS
NEW MEXICO	THREATENED	CRASSULACEAE	GRAPTOPETALUM RUSBYI
NEW MEXICO	THREATENED	FABACEAE	ASTRAGALUS ACCUMBENS
NEW MEXICO	THREATENED	FABACEAE	ASTRAGALUS ALTUS
NEW MEXICO	THREATENED	FABACEAE	ASTRAGALUS PUNICEUS VAR. GERTRUDIS
NEW MEXICO	THREATENED	FABACEAE	PETALOSTEMUM SCARIOSUM
NEW MEXICO	THREATENED	LILIACEAE	ALLIUM GOODDINGII
NEW MEXICO	THREATENED	ONAGRACEAE	OENOTHERA ORGANENSIS
NEW MEXICO	THREATENED	POACEAE	PUCCINELLIA PARISHII
NEW MEXICO	THREATENED	POLEMONIACEAE	PHLOX CARYOPHYLLA
NEW MEXICO	THREATENED	POLYGONACEAE	ERIOGONUM DENSUM

STATE LISTS OF ENDANGERED, EXTINCT AND THREATENED SPECIES IN THE CONTINENTAL UNITED STATES

STATE	STATUS	FAMILY	SPECIES
NEW MEXICO	THREATENED	POLYPODIACEAE	CHEILANTHES PRINGLEI
NEW MEXICO	THREATENED	POLYPODIACEAE	NOTHOLAENA LEMMONII
NEW MEXICO	THREATENED	ROSACEAE	ROSA STELLATA
NEW MEXICO	THREATENED	SCROPHULARIACEAE	PENSTEMON ALAMOSENSIS
NEW YORK	ENDANGERED	CYPERACEAE	SCIRPUS ANCISTROCHAETUS
NEW YORK	ENDANGERED	ORCHIDACEAE	ISOTRIA MEDEOLOIDES
NEW YORK	ENDANGERED	POACEAE	CALAMAGROSTIS PERPLEXA
NEW YORK	ENDANGERED	POLYPODIACEAE	PHYLLITIS SCOLOPENDRIUM VAR. AMERICANUM
NEW YORK	ENDANGERED	RANUNCULACEAE	ACONITUM NOVEBORACENSE
NEW YORK	ENDANGERED	RANUNCULACEAE	TROLLIUS LAXUS
NEW YORK	THREATENED	ARALIACEAE	PANAX QUINQUEFOLIUS
NEW YORK	THREATENED	ASTERACEAE	PRENANTHES BOOTTII
NEW YORK	THREATENED	ASTERACEAE	SOLIDAGO HOUGHTONII
NEW YORK	THREATENED	CISTACEAE	HELIANTHEMUM DUMOSUM
NEW YORK	THREATENED	CYPERACEAE	SCIRPUS LONGII
NEW YORK	THREATENED	ORCHIDACEAE	CYPRIPEDIUM ARIETINUM
NEW YORK	THREATENED	ORCHIDACEAE	CYPRIPEDIUM CANDIDUM
NEW YORK	THREATENED	ORCHIDACEAE	LISTERA AURICULATA
NEW YORK	THREATENED	ORCHIDACEAE	PLATANTHERA FLAVA
NEW YORK	THREATENED	ORCHIDACEAE	PLATANTHERA LEUCOPHAEA
NEW YORK	THREATENED	ORCHIDACEAE	PLATANTHERA PERAMOENA
NEW YORK	THREATENED	POACEAE	CALAMAGROSTIS PORTERI
NEW YORK	THREATENED	POACEAE	PANICUM ACULEATUM
NEW YORK	THREATENED	POACEAE	POA PALUDIGENA
NEW YORK	THREATENED	POTAMOGETONACEAE	POTAMOGETON HILLII
NEW YORK	THREATENED	RANUNCULACEAE	HYDRASTIS CANADENSIS
NEW YORK	THREATENED	SCHIZAEACEAE	SCHIZAEA PUSILLA
NEW YORK	THREATENED	SCROPHULARIACEAE	GERARDIA ACUTA
NEW YORK	THREATENED	SCROPHULARIACEAE	MICRANTHEMUM MICRANTHEMOIDES
NEW YORK	THREATENED	SCROPHULARIACEAE	SCHWALBEA AMERICANA
NORTH CAROLINA	ENDANGERED	ALISMATACEAE	SAGITTARIA FASCICULATA
NORTH CAROLINA	ENDANGERED	ARISTOLOCHIACEAE	HEXASTYLIS NANIFLORA
NORTH CAROLINA	ENDANGERED	BRASSICACEAE	CARDAMINE MICRANTHERA
NORTH CAROLINA	ENDANGERED	CISTACEAE	HUDSONIA ERICOIDES SSP. MONTANA
NORTH CAROLINA	ENDANGERED	CYPERACEAE	CAREX BILTMOREANA
NORTH CAROLINA	ENDANGERED	DIAPENSIACEAE	PYXIDANTHERA BARBULATA VAR. BREVIFOLIA
NORTH CAROLINA	ENDANGERED	DIAPENSIACEAE	SHORTIA GALACIFOLIA VAR. BREVISTYLA
NORTH CAROLINA	ENDANGERED	DIAPENSIACEAE	SHORTIA GALACIFOLIA VAR. GALACIFOLIA
NORTH CAROLINA	ENDANGERED	ERICACEAE	KALMIA CUNEATA
NORTH CAROLINA	ENDANGERED	LAURACEAE	LINDERA MELISSIFOLIA
NORTH CAROLINA	ENDANGERED	LILIACEAE	TRILLIUM PUSILLUM VAR. VIRGINIANUM
NORTH CAROLINA	ENDANGERED	ORCHIDACEAE	ISOTRIA MEDEOLOIDES
NORTH CAROLINA	ENDANGERED	POACEAE	GLYCERIA NUBIGENA

STATE LISTS OF ENDANGERED, EXTINCT AND THREATENED SPECIES IN THE CONTINENTAL UNITED STATES

STATE	STATUS	FAMILY	SPECIES
NORTH CAROLINA	ENDANGERED	POACEAE	PANICUM MUNDUM
NORTH CAROLINA	ENDANGERED	PORTULACACEAE	PORTULACA SMALLII
NORTH CAROLINA	ENDANGERED	RANUNCULACEAE	THALICTRUM COOLEYI
NORTH CAROLINA	ENDANGERED	ROSACEAE	GEUM GENICULATUM
NORTH CAROLINA	ENDANGERED	ROSACEAE	GEUM RADIATUM
NORTH CAROLINA	ENDANGERED	SARRACENIACEAE	SARRACENIA JONESII
NORTH CAROLINA	ENDANGERED	SCROPHULARIACEAE	LINDERNIA SAXICOLA
NORTH CAROLINA	EXTINCT	ASTERACEAE	SOLIDAGO PORTERI
NORTH CAROLINA	EXTINCT	FABACEAE	PSORALEA MACROPHYLLA
NORTH CAROLINA	THREATENED	ANACARDIACEAE	RHUS MICHAUXII
NORTH CAROLINA	THREATENED	APIACEAE	LILAEOPSIS CAROLINENSIS
NORTH CAROLINA	THREATENED	APIACEAE	PTILIMNIUM FLUVIATILE
NORTH CAROLINA	THREATENED	ARALIACEAE	PANAX QUINQUEFOLIUS
NORTH CAROLINA	THREATENED	ARISTOLOCHIACEAE	HEXASTYLIS CONTRACTA
NORTH CAROLINA	THREATENED	ARISTOLOCHIACEAE	HEXASTYLIS LEWISII
NORTH CAROLINA	THREATENED	ASTERACEAE	CACALIA RUGELIA
NORTH CAROLINA	THREATENED	ASTERACEAE	COREOPSIS LATIFOLIA
NORTH CAROLINA	THREATENED	ASTERACEAE	ECHINACEA LAEVIGATA
NORTH CAROLINA	THREATENED	ASTERACEAE	HELIANTHUS SCHWEINITZII
NORTH CAROLINA	THREATENED	ASTERACEAE	LIATRIS HELLERI
NORTH CAROLINA	THREATENED	ASTERACEAE	PRENANTHES ROANENSIS
NORTH CAROLINA	THREATENED	ASTERACEAE	RUDBECKIA HELIOPSIDIS
NORTH CAROLINA	THREATENED	ASTERACEAE	SENECIO MILLEFOLIUM
NORTH CAROLINA	THREATENED	ASTERACEAE	SOLIDAGO PULCHRA
NORTH CAROLINA	THREATENED	ASTERACEAE	SOLIDAGO SPITHAMAEA
NORTH CAROLINA	THREATENED	ASTERACEAE	SOLIDAGO VERNA
NORTH CAROLINA	THREATENED	CARYOPHYLLACEAE	ARENARIA GODFREYI
NORTH CAROLINA	THREATENED	CARYOPHYLLACEAE	ARENARIA UNIFLORA
NORTH CAROLINA	THREATENED	CRASSULACEAE	SEDUM PUSILLUM
NORTH CAROLINA	THREATENED	CYPERACEAE	CAREX AUSTROCAROLINIANA
NORTH CAROLINA	THREATENED	CYPERACEAE	CAREX CHAPMANII
NORTH CAROLINA	THREATENED	CYPERACEAE	CAREX MISERA
NORTH CAROLINA	THREATENED	CYPERACEAE	CAREX PURPURIFERA
NORTH CAROLINA	THREATENED	CYPERACEAE	CYMOPHYLLUS FRASERI
NORTH CAROLINA	THREATENED	CYPERACEAE	SCIRPUS FLACCIDIFOLIUS
NORTH CAROLINA	THREATENED	DROSERACEAE	DIONAEA MUSCIPULA
NORTH CAROLINA	THREATENED	ERICACEAE	RHODODENDRON BAKERI
NORTH CAROLINA	THREATENED	ERICACEAE	RHODODENDRON VASEYI
NORTH CAROLINA	THREATENED	ERIOCAULACEAE	LACHNOCAULON BEYRICHIANUM
NORTH CAROLINA	THREATENED	FABACEAE	CLADRASTIS LUTEA
NORTH CAROLINA	THREATENED	HALORAGACEAE	MYRIOPHYLLUM LAXUM
NORTH CAROLINA	THREATENED	JUNCACEAE	JUNCUS GYMNOCARPUS
NORTH CAROLINA	THREATENED	LAMIACEAE	SYNANDRA HISPIDULA

STATE LISTS OF ENDANGERED, EXTINCT AND THREATENED SPECIES IN THE CONTINENTAL UNITED STATES

STATE	STATUS	FAMILY	SPECIES
NORTH CAROLINA	THREATENED	LAURACEAE	LITSEA AESTIVALIS
NORTH CAROLINA	THREATENED	LILIACEAE	LILIUM GRAYII
NORTH CAROLINA	THREATENED	LILIACEAE	TRILLIUM PUSILLUM VAR. PUSILLUM
NORTH CAROLINA	THREATENED	LILIACEAE	VERATRUM WOODII
NORTH CAROLINA	THREATENED	ORCHIDACEAE	PLATANTHERA FLAVA
NORTH CAROLINA	THREATENED	ORCHIDACEAE	PLATANTHERA INTEGRA
NORTH CAROLINA	THREATENED	ORCHIDACEAE	PLATANTHERA PERAMOENA
NORTH CAROLINA	THREATENED	POACEAE	CALAMAGROSTIS PORTERI
NORTH CAROLINA	THREATENED	POACEAE	CALAMOVILFA BREVIPILIS VAR. BREVIPILIS
NORTH CAROLINA	THREATENED	POACEAE	PANICUM ACULEATUM
NORTH CAROLINA	THREATENED	POACEAE	SPOROBOLUS TERETIFOLIUS
NORTH CAROLINA	THREATENED	PRIMULACEAE	LYSIMACHIA ASPERULAEFOLIA
NORTH CAROLINA	THREATENED	RANUNCULACEAE	CIMICIFUGA RUBIFOLIA
NORTH CAROLINA	THREATENED	RANUNCULACEAE	HYDRASTIS CANADENSIS
NORTH CAROLINA	THREATENED	RANUNCULACEAE	RANUNCULUS SUBCORDATUS
NORTH CAROLINA	THREATENED	RUBIACEAE	HEDYOTIS PURPUREA VAR. MONTANA
NORTH CAROLINA	THREATENED	SANTALACEAE	BUCKLEYA DISTICHOPHYLLA
NORTH CAROLINA	THREATENED	SANTALACEAE	NESTRONIA UMBELLULA
NORTH CAROLINA	THREATENED	SAXIFRAGACEAE	SAXIFRAGA CAREYANA
NORTH CAROLINA	THREATENED	SAXIFRAGACEAE	SAXIFRAGA CAROLINIANA
NORTH CAROLINA	THREATENED	SCHISANDRACEAE	SCHISANDRA GLABRA
NORTH DAKOTA	THREATENED	BRASSICACEAE	RORIPPA CALYCINA
NORTH DAKOTA	THREATENED	ORCHIDACEAE	CYPRIPEDIUM CANDIDUM
NORTH DAKOTA	THREATENED	ORCHIDACEAE	PLATANTHERA LEUCOPHAEA
OHIO	ENDANGERED	POACEAE	CALAMAGROSTIS INSPERATA
OHIO	ENDANGERED	POLYGONACEAE	POLYGONUM PENSYLVANICUM VAR. EGLANDULOSUM
OHIO	ENDANGERED	RANUNCULACEAE	TROLLIUS LAXUS
OHIO	THREATENED	ARALIACEAE	PANAX QUINQUEFOLIUS
OHIO	THREATENED	ORCHIDACEAE	CYPRIPEDIUM CANDIDUM
OHIO	THREATENED	ORCHIDACEAE	PLATANTHERA FLAVA
OHIO	THREATENED	ORCHIDACEAE	PLATANTHERA LEUCOPHAEA
OHIO	THREATENED	ORCHIDACEAE	PLATANTHERA PERAMOENA
OHIO	THREATENED	POACEAE	POA PALUDIGENA
OHIO	THREATENED	POTAMOGETONACEAE	POTAMOGETON HILLII
OHIO	THREATENED	RANUNCULACEAE	HYDRASTIS CANADENSIS
OHIO	THREATENED	SAXIFRAGACEAE	SULLIVANTIA SULLIVANTII
OKLAHOMA	ENDANGERED	BRASSICACEAE	DRABA APRICA
OKLAHOMA	ENDANGERED	BRASSICACEAE	LEAVENWORTHIA AUREA
OKLAHOMA	ENDANGERED	BRASSICACEAE	STREPTANTHUS SQUAMIFORMIS
OKLAHOMA	ENDANGERED	CUSCUTACEAE	CUSCUTA ATTENUATA
OKLAHOMA	ENDANGERED	CYPERACEAE	CAREX LATEBRACTEATA
OKLAHOMA	ENDANGERED	ERIOCAULACEAE	ERIOCAULON KORNICKIANUM
OKLAHOMA	ENDANGERED	FABACEAE	AMORPHA OUACHITENSIS

STATE LISTS OF ENDANGERED, EXTINCT AND THREATENED SPECIES IN THE CONTINENTAL UNITED STATES

STATE	STATUS	FAMILY	SPECIES
OKLAHOMA	ENDANGERED	FAGACEAE	CASTANEA PUMILA VAR. OZARKENSIS
OKLAHOMA	ENDANGERED	LILIACEAE	HYPOXIS LONGII
OKLAHOMA	ENDANGERED	POACEAE	CALAMOVILFA ARCUATA
OKLAHOMA	ENDANGERED	POLEMONIACEAE	PHLOX LONGIPILOSA
OKLAHOMA	EXTINCT	FABACEAE	VICIA REVERCHONII
OKLAHOMA	THREATENED	ARALIACEAE	PANAX QUINQUEFOLIUS
OKLAHOMA	THREATENED	BETULACEAE	ALNUS MARITIMA
OKLAHOMA	THREATENED	BRASSICACEAE	LESQUERELLA ANGUSTIFOLIA
OKLAHOMA	THREATENED	CYPERACEAE	CAREX FISSA
OKLAHOMA	THREATENED	FABACEAE	CLADRASTIS LUTEA
OKLAHOMA	THREATENED	GENTIANACEAE	FRASERA COLORADENSIS
OKLAHOMA	THREATENED	LILIACEAE	VERATRUM WOODII
OKLAHOMA	THREATENED	MALVACEAE	CALLIRHOE PAPAVER VAR. BUSHII
OKLAHOMA	THREATENED	POLEMONIACEAE	PHLOX OKLAHOMENSIS
OKLAHOMA	THREATENED	RANUNCULACEAE	HYDRASTIS CANADENSIS
OREGON	ENDANGERED	APIACEAE	LOMATIUM BRADSHAWII
OREGON	ENDANGERED	APIACEAE	LOMATIUM GREENMANII
OREGON	ENDANGERED	APIACEAE	LOMATIUM LAEVIGATUM
OREGON	ENDANGERED	APIACEAE	LOMATIUM SUKSDORFII
OREGON	ENDANGERED	APIACEAE	SANICULA TRACYI
OREGON	ENDANGERED	ASTERACEAE	ASTER GORMANII
OREGON	ENDANGFRED	ASTERACEAE	ASTER VIALIS
OREGON	ENDANGERED	ASTERACEAE	ERIGERON DELICATUS
OREGON	ENDANGERED	ASTERACEAE	PYRROCOMA RADIATUS
OREGON	ENDANGERED	ASTERACEAE	SENECIO PORTERI
OREGON	ENDANGERED	ASTERACEAE	STEPHANOMERIA MALHEURENSIS
OREGON	ENDANGERED	BORAGINACEAE	HACKELIA CRONQUISTII
OREGON	ENDANGFRED	BORAGINACEAE	HACKELIA OPHIOBIA
OREGON	ENDANGFRED	BORAGINACEAE	PLAGIOBOTHRYS HIRTUS SSP. HIRTUS
OREGON	ENDANGERED	BORAGINACEAE	PLAGIOBOTHRYS LAMPROCARPUS
OREGON	ENDANGERED	BRASSICACEAE	ARABIS KOEHLERI VAR. KOEHLERI
OREGON	ENDANGERED	BRASSICACEAE	CARDAMINE PATTERSONII
OREGON	ENDANGERED	CARYOPHYLLACEAE	SILENE DOUGLASII VAR. ORARIA
OREGON	ENDANGERED	CARYOPHYLLACEAE	SILENE SPALDINGII
OREGON	ENDANGERED	CRASSULACEAE	SEDUM MORANII
OREGON	ENDANGERED	CRASSULACEAE	SEDUM RADIATUM SSP. DEPAUPERATUM
OREGON	ENDANGERED	ERICACEAE	KALMIOPSIS LEACHIANA
OREGON	ENDANGERED	FABACEAE	ASTRAGALUS KENTROPHYTA VAR. DOUGLASII
OREGON	ENDANGERED	FABACEAE	ASTRAGALUS PURSHII VAR. OPHIOGENES
OREGON	ENDANGERED	FABACEAE	ASTRAGALUS ROBBINSII VAR. ALPINIFORMIS
OREGON	ENDANGERED	FABACEAE	ASTRAGALUS STERILIS
OREGON	ENDANGERED	FABACEAE	LUPINUS BURKEI SSP. CAFRULEOMONTANUS
OREGON	ENDANGERED	FUMARIACEAE	CORYDALIS AQUAE-GELIDAE

STATE LISTS OF ENDANGERED, EXTINCT AND THREATENED SPECIES IN THE CONTINENTAL UNITED STATES

STATE	STATUS	FAMILY	SPECIES
OREGON	ENDANGERED	GENTIANACEAE	GENTIANA BISETAEA
OREGON	ENDANGERED	HYDROPHYLLACEAE	PHACELIA CAPITATA
OREGON	ENDANGERED	IRIDACEAE	IRIS TENUIS
OREGON	ENDANGERED	LILIACEAE	CALOCHORTUS GREENEI
OREGON	ENDANGERED	LILIACEAE	CALOCHORTUS INDECORUS
OREGON	ENDANGERED	LILIACEAE	CALOCHORTUS LONGEBARBATUS VAR. PECKII
OREGON	ENDANGERED	LILIACEAE	LILIUM OCCIDENTALE
OREGON	ENDANGERED	LIMNANTHACEAE	LIMNANTHES FLOCCOSA SSP. GRANDIFLORA
OREGON	ENDANGERED	LIMNANTHACEAE	LIMNANTHES FLOCCOSA SSP. PUMILA
OREGON	ENDANGERED	LOASACEAE	MENTZELIA PACKARDIAE
OREGON	ENDANGERED	MALVACEAE	SIDALCEA CAMPESTRIS
OREGON	ENDANGERED	MALVACEAE	SIDALCEA NELSONIANA
OREGON	ENDANGERED	NYCTAGINACEAE	MIRABILIS MACFARLANEI
OREGON	ENDANGERED	POACEAE	AGROSTIS HENDERSONII
OREGON	ENDANGERED	POLEMONIACEAE	COLLOMIA MACROCALYX
OREGON	ENDANGERED	POLYGONACEAE	ERIOGONUM CHRYSOPS
OREGON	ENDANGERED	POLYGONACEAE	ERIOGONUM DICLINUM
OREGON	ENDANGERED	PRIMULACEAE	PRIMULA CUSICKIANA
OREGON	ENDANGERED	RANUNCULACEAE	ANEMONE OREGANA VAR. FELIX
OREGON	ENDANGERED	RANUNCULACEAE	DELPHINIUM LEUCOPHAEUM
OREGON	ENDANGERED	RANUNCULACEAE	DELPHINIUM PAVONACEUM
OREGON	ENDANGERED	ROSACEAE	FILIPENDULA OCCIDENTALIS
OREGON	ENDANGERED	SAXIFRAGACEAE	BENSONIELLA OREGANA
OREGON	ENDANGERED	SCROPHULARIACEAE	CASTILLEJA CHLOROTICA
OREGON	ENDANGERED	SCROPHULARIACEAE	CASTILLEJA OWNBEYANA
OREGON	ENDANGERED	SCROPHULARIACEAE	CORDYLANTHUS MARITIMUS SSP. MARITIMUS
OREGON	ENDANGERED	SCROPHULARIACEAE	PENSTEMON BARRETTIAE
OREGON	ENDANGERED	SCROPHULARIACEAE	PENSTEMON GLAUCINUS
OREGON	ENDANGERED	SCROPHULARIACEAE	PENSTEMON SPATULATUS
OREGON	EXTINCT	LILIACEAE	FRITILLARIA ADAMANTINA
OREGON	EXTINCT	POACEAE	PLEUROPOGON OREGONUS
OREGON	EXTINCT	SCROPHULARIACEAE	SYNTHYRIS MISSURICA SSP. HIRSUTA
OREGON	THREATENED	APIACEAE	CYMOPTERUS CORRUGATUS
OREGON	THREATENED	APIACEAE	ERYNGIUM PETIOLATUM
OREGON	THREATENED	APIACEAE	LOMATIUM HENDERSONII
OREGON	THREATENED	APIACEAE	LOMATIUM HOWELLII
OREGON	THREATENED	APIACEAE	LOMATIUM MINUS
OREGON	THREATENED	APIACEAE	LOMATIUM OREGANUM
OREGON	THREATENED	APIACEAE	LOMATIUM PECKIANUM
OREGON	THREATENED	APIACEAE	LOMATIUM ROLLINSII
OREGON	THREATENED	APIACEAE	LOMATIUM SERPENTINUM
OREGON	THREATENED	APIACEAE	PERIDERIDIA ERYTHRORHIZA
OREGON	THREATENED	APIACEAE	RHYSOPTERUS PLURIJUGUS

STATE LISTS OF ENDANGERED, EXTINCT AND THREATENED SPECIES IN THE CONTINENTAL UNITED STATES

STATE	STATUS	FAMILY	SPECIES
OREGON	THREATENED	APIACEAE	SANICULA PECKIANA
OREGON	THREATENED	APIACEAE	TAUSCHIA GLAUCA
OREGON	THREATENED	APIACEAE	TAUSCHIA HOWELLII
OREGON	THREATENED	ASTERACEAE	ANTENNARIA SUFFRUTESCENS
OREGON	THREATENED	ASTERACEAE	ARNICA AMPLEXICAULIS VAR. PIPERI
OREGON	THREATENED	ASTERACEAE	ARNICA VISCOSA
OREGON	THREATENED	ASTERACEAE	ASTER BRICKELLIOIDES
OREGON	THREATENED	ASTERACEAE	ASTER CHILENSIS SSP. HALLII
OREGON	THREATENED	ASTERACEAE	ASTER CURTUS
OREGON	THREATENED	ASTERACEAE	CHAENACTIS NEVII
OREGON	THREATENED	ASTERACEAE	CIRSIUM CILIOLATUM
OREGON	THREATENED	ASTERACEAE	ERIGERON BLOOMERI VAR. NUDATUS
OREGON	THREATENED	ASTERACEAE	ERIGERON HOWELLII
OREGON	THREATENED	ASTERACEAE	ERIGERON OREGANUS
OREGON	THREATENED	ASTERACEAE	HAPLOPAPPUS HALLII
OREGON	THREATENED	ASTERACEAE	HIERACIUM LONGIBERBE
OREGON	THREATENED	ASTERACEAE	LASTHENIA MACRANTHA SSP. PRISCA
OREGON	THREATENED	ASTERACEAE	LASTHENIA MINOR SSP. MARITIMA
OREGON	THREATENED	ASTERACEAE	LUINA SERPENTINA
OREGON	THREATENED	ASTERACEAE	MICROSERIS HOWELLII
OREGON	THREATENED	ASTERACEAE	MICROSERIS LACINIATA SSP. DETLINGII
OREGON	THREATENED	ASTERACEAE	SENECIO HESPERIUS
OREGON	THREATENED	BORAGINACEAE	HACKELIA HISPIDA
OREGON	THREATENED	BORAGINACEAE	HACKELIA PATENS VAR. SEMIGLABRA
OREGON	THREATENED	BORAGINACEAE	PLAGIOBOTHRYS HIRTUS SSP. CORALLICARPA
OREGON	THREATENED	BRASSICACEAE	ARABIS ACULEOLATA
OREGON	THREATENED	BRASSICACEAE	ARABIS KOEHLERI VAR. STIPITATA
OREGON	THREATENED	BRASSICACEAE	ARABIS MODESTA
OREGON	THREATENED	BRASSICACEAE	ARABIS OREGANA
OREGON	THREATENED	BRASSICACEAE	ARABIS SUFFRUTESCENS VAR. HORIZONTALIS
OREGON	THREATENED	BRASSICACEAE	CARDAMINE PENDULIFLORA
OREGON	THREATENED	BRASSICACEAE	DRABA LEMMONII VAR. CYCLOMORPHA
OREGON	THREATENED	BRASSICACEAE	LESQUERELLA KINGII SSP. DIVERSIFOLIA
OREGON	THREATENED	BRASSICACEAE	RORIPPA COLUMBIAE
OREGON	THREATENED	BRASSICACEAE	THELYPODIUM EUCOSMUM
OREGON	THREATENED	BRASSICACEAE	THELYPODIUM HOWELLII VAR. SPECTABILIS
OREGON	THREATENED	BRASSICACEAE	THLASPI MONTANUM VAR. SISKIYOUENSE
OREGON	THREATENED	CAMPANULACEAE	CAMPANULA ROTUNDIFOLIA VAR. SACAJAWEANA
OREGON	THREATENED	CARYOPHYLLACEAE	ARENARIA FRANKLINII VAR. THOMPSONII
OREGON	THREATENED	CARYOPHYLLACEAE	ARENARIA HOWELLII
OREGON	THREATENED	CARYOPHYLLACEAE	SILENE SCAPOSA VAR. LOBATA
OREGON	THREATENED	CARYOPHYLLACEAE	SILENE SCAPOSA VAR. SCAPOSA
OREGON	THREATENED	CRASSULACEAE	SEDUM OBLANCEOLATUM

STATE LISTS OF ENDANGERED, EXTINCT AND THREATENED SPECIES IN THE CONTINENTAL UNITED STATES

STATE	STATUS	FAMILY	SPECIES
OREGON	THREATENED	CYPERACEAE	CAREX INTERRUPTA
OREGON	THREATENED	ERICACEAE	ARCTOSTAPHYLOS STANFORDIANA SSP. HISPIDULA
OREGON	THREATENED	ERICACEAE	PITYOPUS CALIFORNICUS
OREGON	THREATENED	ERICACEAE	VACCINIUM COCCINIUM
OREGON	THREATENED	FABACEAE	ASTRAGALUS ALVORDENSIS
OREGON	THREATENED	FABACEAE	ASTRAGALUS APPLEGATII
OREGON	THREATENED	FABACEAE	ASTRAGALUS MULFORDAE
OREGON	THREATENED	FABACEAE	ASTRAGALUS SOLITARIUS
OREGON	THREATENED	FABACEAE	LATHYRUS HOLOCHLORUS
OREGON	THREATENED	FABACEAE	LUPINUS ARIDUS SSP. ASHLANDENSIS
OREGON	THREATENED	FABACEAE	LUPINUS BIDDLEI
OREGON	THREATENED	FABACEAE	LUPINUS MUCRONULATUS
OREGON	THREATENED	FABACEAE	TRIFOLIUM OWYHEENSE
OREGON	THREATENED	GENTIANACEAE	FRASERA UMPQUAENSIS
OREGON	THREATENED	HYDROPHYLLACEAF	PHACELIA PECKII
OREGON	THREATENED	HYDROPHYLLACEAE	PHACELIA VERNA
OREGON	THREATENED	IRIDACEAE	IRIS TENAX VAR. GORMANII
OREGON	THREATENED	LAMIACEAE	AGASTACHE CUSICKII
OREGON	THREATENED	LILIACEAE	ALLIUM PLEIANTHUM
OREGON	THREATENED	LILIACEAE	ALLIUM ROBINSONII
OREGON	THREATENED	LILIACEAE	CALOCHORTUS LONGEBARBATUS VAR. LONGEBARBATUS
OREGON	THREATENED	LILIACEAE	CALOCHORTUS NITIDUS
OREGON	THREATENED	LILIACEAE	CAMASSIA CUSICKII
OREGON	THREATENED	LILIACEAE	ERYTHRONIUM OREGONUM
OREGON	THREATENED	LILIACEAE	FRITILLARIA GENTNERI
OREGON	THREATENED	LILIACEAE	LILIUM VOLLMERI
OREGON	THREATENED	LILIACEAE	LILIUM WIGGINSII
OREGON	THREATENED	LILIACEAE	SCHOENOLIRION BRACTEOSUM
OREGON	THREATENED	LIMNANTHACEAE	LIMNANTHES FLOCCOSA SSP. BELLINGERIANA
OREGON	THREATENED	LIMNANTHACEAE	LIMNANTHES GRACILIS VAR. GRACILIS
OREGON	THREATENED	LOASACEAE	MENTZELIA MOLLIS
OREGON	THREATENED	MALVACEAE	SIDALCEA CUSICKII
OREGON	THREATENED	ONAGRACEAE	CLARKIA AMOENA VAR. PACIFICA
OREGON	THREATENED	ONAGRACEAE	EPILOBIUM OREGANUM
OREGON	THREATENED	OPHIOGLOSSACEAF	BOTRYCHIUM PUMICOLA
OREGON	THREATENED	ORCHIDACEAE	CYPRIPEDIUM CALIFORNICUM
OREGON	THREATENED	ORCHIDACEAE	PLATANTHERA UNALASCENSIS SSP. MARITIMA
OREGON	THREATENED	POACEAE	AGROSTIS HOWELLII
OREGON	THREATENED	POLEMONIACEAE	COLLOMIA MAZAMA
OREGON	THREATENED	POLYGONACEAE	ERIOGONUM CUSICKII
OREGON	THREATENED	POLYGONACEAE	ERIOGONUM NOVONUDUM
OREGON	THREATENED	POLYGONACEAE	ERIOGONUM PENDULUM
OREGON	THREATENED	POLYGONACEAE	ERIOGONUM SCOPULORUM

STATE LISTS OF ENDANGERED, EXTINCT AND THREATENED SPECIES IN THE CONTINENTAL UNITED STATES

STATE	STATUS	FAMILY	SPECIES
OREGON	THREATENED	POLYPODIACEAE	POLYSTICHUM KRUCKEBERGII
OREGON	THREATENED	PORTULACACEAE	LEWISIA COLUMBIANA VAR. WALLOWENSIS
OREGON	THREATENED	PRIMULACEAE	DODECATHEON POETICUM
OREGON	THREATENED	PRIMULACEAE	DOUGLASIA LAEVIGATA VAR. LAEVIGATA
OREGON	THREATENED	RANUNCULACEAE	CIMICIFUGA LACINIATA
OREGON	THREATENED	RANUNCULACEAE	RANUNCULUS RECONDITUS
OREGON	THREATENED	SALICACEAE	SALIX FLUVIATILIS
OREGON	THREATENED	SAXIFRAGACEAE	SAXIFRAGA OCCIDENTALIS VAR. LATIPETIOLATA
OREGON	THREATENED	SAXIFRAGACEAE	SULLIVANTIA OREGANA
OREGON	THREATENED	SCROPHULARIACEAE	CASTILLEJA BREVILOBATA
OREGON	THREATENED	SCROPHULARIACEAE	CASTILLEJA CHRYSANTHA
OREGON	THREATENED	SCROPHULARIACEAE	CASTILLEJA FRATERNA
OREGON	THREATENED	SCROPHULARIACEAE	CASTILLEJA GLANDULIFERA
OREGON	THREATENED	SCROPHULARIACEAE	CASTILLEJA STEENENSIS
OREGON	THREATENED	SCROPHULARIACEAE	CASTILLEJA XANTHOTRICHA
OREGON	THREATENED	SCROPHULARIACEAE	CORDYLANTHUS MARITIMUS SSP. PALUSTRIS
OREGON	THREATENED	SCROPHULARIACEAE	MIMULUS JUNGERMANNIOIDES
OREGON	THREATENED	SCROPHULARIACEAE	PEDICULARIS HOWELLII
OREGON	THREATENED	SCROPHULARIACEAE	PENSTEMON CINICOLA
OREGON	THREATENED	SCROPHULARIACEAE	PENSTEMON ELEGANTULUS
OREGON	THREATENED	SCROPHULARIACEAE	PENSTEMON PECKII
OREGON	THREATENED	SCROPHULARIACEAE	SYNTHYRIS MISSURICA SSP. STELLATA
OREGON	THREATENED	SCROPHULARIACEAE	SYNTHYRIS SCHIZANTHA
OREGON	THREATENED	VIOLACEAE	VIOLA ADUNCA VAR. CASCADENSIS
OREGON	THREATENED	VIOLACEAE	VIOLA LANCEOLATA SSP. OCCIDENTALIS
PENNSYLVANIA	ENDANGERED	CARYOPHYLLACEAE	CERASTIUM ARVENSE VAR. VILLOSISSIMUM
PENNSYLVANIA	ENDANGERED	CYPERACEAE	SCIRPUS ANCISTROCHAETUS
PENNSYLVANIA	ENDANGERED	ORCHIDACEAE	ISOTRIA MEDEOLOIDES
PENNSYLVANIA	ENDANGERED	RANUNCULACEAE	TROLLIUS LAXUS
PENNSYLVANIA	EXTINCT	HYDROCHARITACEAE	ELODEA SCHWEINITZII
PENNSYLVANIA	THREATENED	ARALIACEAE	PANAX QUINQUEFOLIUS
PENNSYLVANIA	THREATENED	ASTERACEAE	ECHINACEA LAEVIGATA
PENNSYLVANIA	THREATENED	CYPERACEAE	CYMOPHYLLUS FRASERI
PENNSYLVANIA	THREATENED	JUNCACEAE	JUNCUS GYMNOCARPUS
PENNSYLVANIA	THREATENED	ORCHIDACEAE	CYPRIPEDIUM CANDIDUM
PENNSYLVANIA	THREATENED	ORCHIDACEAE	PLATANTHERA FLAVA
PENNSYLVANIA	THREATENED	ORCHIDACEAE	PLATANTHERA PERAMOENA
PENNSYLVANIA	THREATENED	POACEAE	CALAMAGROSTIS PORTERI
PENNSYLVANIA	THREATENED	POACEAE	POA PALUDIGENA
PENNSYLVANIA	THREATENED	POTAMOGETONACEAE	POTAMOGETON HILLII
PENNSYLVANIA	THREATENED	RANUNCULACEAE	HYDRASTIS CANADENSIS
PENNSYLVANIA	THREATENED	ROSACEAE	PRUNUS ALLEGHANIENSIS
PENNSYLVANIA	THREATENED	SCROPHULARIACEAE	MICRANTHEMUM MICRANTHEMOIDES

STATE LISTS OF ENDANGERED, EXTINCT AND THREATENED SPECIES IN THE CONTINENTAL UNITED STATES

STATE	STATUS	FAMILY	SPECIES
RHODE ISLAND	ENDANGERED	ORCHIDACEAE	ISOTRIA MEDEOLOIDES
RHODE ISLAND	THREATENED	ASTERACEAE	EUPATORIUM LEUCOLEPIS VAR. NOVAE-ANGLIAE
RHODE ISLAND	THREATENED	CISTACEAE	HELIANTHEMUM DUMOSUM
RHODE ISLAND	THREATENED	ORCHIDACEAE	PLATANTHERA FLAVA
RHODE ISLAND	THREATENED	POACEAE	PANICUM ACULEATUM
RHODE ISLAND	THREATENED	SCROPHULARIACEAE	GERARDIA ACUTA
SOUTH CAROLINA	ENDANGERED	ALISMATACEAE	SAGITTARIA FASCICULATA
SOUTH CAROLINA	ENDANGERED	ARISTOLOCHIACEAE	HEXASTYLIS NANIFLORA
SOUTH CAROLINA	ENDANGERED	BRASSICACEAE	DRABA APRICA
SOUTH CAROLINA	ENDANGERED	DIAPENSIACEAE	PYXIDANTHERA BARBULATA VAR. BREVIFOLIA
SOUTH CAROLINA	ENDANGERED	DIAPENSIACEAE	SHORTIA GALACIFOLIA VAR. BREVISTYLA
SOUTH CAROLINA	ENDANGERED	DIAPENSIACEAE	SHORTIA GALACIFOLIA VAR. GALACIFOLIA
SOUTH CAROLINA	ENDANGERED	ERICACEAE	KALMIA CUNEATA
SOUTH CAROLINA	ENDANGERED	LAURACEAE	LINDERA MELISSIFOLIA
SOUTH CAROLINA	ENDANGERED	LILIACEAE	HYMENOCALLIS CORONARIA
SOUTH CAROLINA	ENDANGERED	LILIACEAE	TRILLIUM PERSISTENS
SOUTH CAROLINA	ENDANGERED	SARRACENIACEAE	SARRACENIA JONESII
SOUTH CAROLINA	ENDANGERED	SAXIFRAGACEAE	RIBES ECHINELLUM
SOUTH CAROLINA	ENDANGERED	SCROPHULARIACEAE	AMPHIANTHUS PUSILLUS
SOUTH CAROLINA	THREATENED	ANACARDIACEAE	RHUS MICHAUXII
SOUTH CAROLINA	THREATENED	APIACEAE	LILAEOPSIS CAROLINENSIS
SOUTH CAROLINA	THREATENED	APIACEAE	PTILIMNIUM FLUVIATILE
SOUTH CAROLINA	THREATENED	APIACEAE	PTILIMNIUM NODOSUM
SOUTH CAROLINA	THREATENED	ARALIACEAE	PANAX QUINQUEFOLIUS
SOUTH CAROLINA	THREATENED	ASTERACEAE	COREOPSIS LATIFOLIA
SOUTH CAROLINA	THREATENED	ASTERACEAE	ECHINACEA LAEVIGATA
SOUTH CAROLINA	THREATENED	ASTERACEAE	HELIANTHUS SCHWEINITZII
SOUTH CAROLINA	THREATENED	ASTERACEAE	RUDBECKIA HELIOPSIDIS
SOUTH CAROLINA	THREATENED	ASTERACEAE	SENECIO MILLEFOLIUM
SOUTH CAROLINA	THREATENED	ASTERACEAE	SOLIDAGO VERNA
SOUTH CAROLINA	THREATENED	CARYOPHYLLACEAE	ARENARIA GODFREYI
SOUTH CAROLINA	THREATENED	CARYOPHYLLACEAE	ARENARIA UNIFLORA
SOUTH CAROLINA	THREATENED	CRASSULACEAE	SEDUM PUSILLUM
SOUTH CAROLINA	THREATENED	CYPERACEAE	CAREX AUSTROCAROLINIANA
SOUTH CAROLINA	THREATENED	CYPERACEAE	CAREX CHAPMANII
SOUTH CAROLINA	THREATENED	DROSERACEAE	DIONAEA MUSCIPULA
SOUTH CAROLINA	THREATENED	ERIOCAULACEAE	LACHNOCAULON BEYRICHIANUM
SOUTH CAROLINA	THREATENED	FAGACEAE	QUERCUS GEORGIANA
SOUTH CAROLINA	THREATENED	FAGACEAE	QUERCUS OGLETHORPENSIS
SOUTH CAROLINA	THREATENED	HALORAGACEAE	MYRIOPHYLLUM LAXUM
SOUTH CAROLINA	THREATENED	ISOETACEAE	ISOETES MELANOSPORA
SOUTH CAROLINA	THREATENED	JUNCACEAE	JUNCUS GYMNOCARPUS
SOUTH CAROLINA	THREATENED	LAMIACEAE	DICERANDRA ODORATISSIMA

STATE LISTS OF ENDANGERED, EXTINCT AND THREATENED SPECIES IN THE CONTINENTAL UNITED STATES

STATE	STATUS	FAMILY	SPECIES
SOUTH CAROLINA	THREATENED	LAURACEAE	LITSEA AESTIVALIS
SOUTH CAROLINA	THREATENED	LILIACEAE	TRILLIUM PUSILLUM VAR. PUSILLUM
SOUTH CAROLINA	THREATENED	ORCHIDACEAE	PLATANTHERA FLAVA
SOUTH CAROLINA	THREATENED	ORCHIDACEAE	PLATANTHERA INTEGRA
SOUTH CAROLINA	THREATENED	ORCHIDACEAE	PLATANTHERA PERAMOENA
SOUTH CAROLINA	THREATENED	POACEAE	CALAMOVILFA BREVIPILIS VAR. BREVIPILIS
SOUTH CAROLINA	THREATENED	POACEAE	PANICUM LITHOPHILUM
SOUTH CAROLINA	THREATENED	POACEAE	SPOROBOLUS TERETIFOLIUS
SOUTH CAROLINA	THREATENED	PRIMULACEAE	LYSIMACHIA ASPERULAEFOLIA
SOUTH CAROLINA	THREATENED	RHAMNACEAE	SAGERETIA MINUTIFLORA
SOUTH CAROLINA	THREATENED	ROSACEAE	AGRIMONIA INCISA
SOUTH CAROLINA	THREATENED	ROSACEAE	WALDSTEINIA LOBATA
SOUTH CAROLINA	THREATENED	RUBIACEAE	PINCKNEYA PUBENS
SOUTH CAROLINA	THREATENED	SANTALACEAE	NESTRONIA UMBELLULA
SOUTH CAROLINA	THREATENED	SCHISANDRACEAE	SCHISANDRA GLABRA
SOUTH CAROLINA	THREATENED	SCROPHULARIACEAE	SCHWALBEA AMERICANA
SOUTH DAKOTA	THREATENED	ORCHIDACEAE	CYPRIPEDIUM CANDIDUM
SOUTH DAKOTA	THREATENED	ORCHIDACEAE	PLATANTHERA LEUCOPHAEA
TENNESSEE	ENDANGERED	ASTERACEAE	ECHINACEA TENNESSEENSIS
TENNESSEE	ENDANGERED	ASTERACEAE	HETEROTHECA RUTHII
TENNESSEE	ENDANGERED	ASTERACEAE	SILPHIUM BRACHIATUM
TENNESSEE	ENDANGERED	ASTERACEAE	SILPHIUM INTEGRIFOLIUM VAR. GATTINGERI
TENNESSEE	ENDANGERED	BRASSICACEAE	ARABIS PERSTELLATA VAR. AMPLA
TENNESSEE	ENDANGERED	BRASSICACEAE	DENTARIA INCISA
TENNESSEE	ENDANGERED	BRASSICACEAE	LEAVENWORTHIA EXIGUA VAR. LUTEA
TENNESSEE	ENDANGERED	BRASSICACEAE	LESQUERELLA DENSIPILA
TENNESSEE	ENDANGERED	BRASSICACEAE	LESQUERELLA PERFORATA
TENNESSEE	ENDANGERED	BRASSICACEAE	LESQUERELLA STONENSIS
TENNESSEE	ENDANGERED	CRASSULACEAE	SEDUM NEVII
TENNESSEE	ENDANGERED	EUPHORBIACEAE	CROTON ALABAMENSIS
TENNESSEE	ENDANGERED	FABACEAE	APIOS PRICEANA
TENNESSEE	ENDANGERED	FABACEAE	PETALOSTEMUM FOLIOSUM
TENNESSEE	ENDANGERED	LAMIACEAE	CONRADINA VERTICILLATA
TENNESSEE	ENDANGERED	LAMIACEAE	PYCNANTHEMUM CURVIPES
TENNESSEE	ENDANGERED	POACEAE	CALAMOVILFA ARCUATA
TENNESSEE	ENDANGERED	POACEAE	GLYCERIA NUBIGENA
TENNESSEE	ENDANGERED	POLYGONACEAE	ERIOGONUM LONGIFOLIUM VAR. HARPERI
TENNESSEE	ENDANGERED	POLYPODIACEAE	PHYLLITIS SCOLOPENDRIUM VAR. AMERICANUM
TENNESSEE	ENDANGERED	RANUNCULACEAE	CLEMATIS GATTINGERI
TENNESSEE	ENDANGERED	ROSACEAE	GEUM GENICULATUM
TENNESSEE	ENDANGERED	ROSACEAE	GEUM RADIATUM
TENNESSEE	EXTINCT	HYDROCHARITACEAE	ELODEA LINEARIS
TENNESSEE	THREATENED	ARALIACEAE	PANAX QUINQUEFOLIUS

STATE LISTS OF ENDANGERED, EXTINCT AND THREATENED SPECIES IN THE CONTINENTAL UNITED STATES

STATE	STATUS	FAMILY	SPECIES
TENNESSEE	THREATENED	ARISTOLOCHIACEAE	HEXASTYLIS CONTRACTA
TENNESSEE	THREATENED	ASTERACEAE	CACALIA RUGELIA
TENNESSEE	THREATENED	ASTERACEAE	PRENANTHES ROANENSIS
TENNESSEE	THREATENED	ASTERACEAE	SOLIDAGO SPITHAMAEA
TENNESSEE	THREATENED	BORAGINACEAE	ONOSMODIUM MOLLE
TENNESSEE	THREATENED	BRASSICACEAE	LEAVENWORTHIA EXIGUA VAR. EXIGUA
TENNESSEE	THREATENED	BRASSICACEAE	LEAVENWORTHIA STYLOSA
TENNESSEE	THREATENED	BRASSICACEAE	LEAVENWORTHIA TORULOSA
TENNESSEE	THREATENED	BRASSICACEAE	LESQUERELLA GLOBOSA
TENNESSEE	THREATENED	BRASSICACEAE	LESQUERELLA LESCURII
TENNESSEE	THREATENED	CAMPANULACEAE	LOBELIA GATTINGERI
TENNESSEE	THREATENED	CARYOPHYLLACEAE	ARENARIA FONTINALIS
TENNESSEE	THREATENED	CYPERACEAE	CAREX AUSTROCAROLINIANA
TENNESSEE	THREATENED	CYPERACEAE	CAREX MISERA
TENNESSEE	THREATENED	CYPERACEAE	CAREX PURPURIFERA
TENNESSEE	THREATENED	CYPERACEAE	CAREX ROANENSIS
TENNESSEE	THREATENED	CYPERACEAE	CYMOPHYLLUS FRASERI
TENNESSEE	THREATENED	ERICACEAE	RHODODENDRON BAKERI
TENNESSEE	THREATENED	FABACEAE	ASTRAGALUS TENNESSEENSIS
TENNESSEE	THREATENED	FABACEAE	CLADRASTIS LUTEA
TENNESSEE	THREATENED	FABACEAE	PETALOSTEMUM GATTINGERI
TENNESSEE	THREATENED	FABACEAE	PSORALEA SUBACAULIS
TENNESSEE	THREATENED	HYPERICACEAE	HYPERICUM SPHAEROCARPUM VAR. TURGIDUM
TENNESSEE	THREATENED	JUNCACEAE	JUNCUS GYMNOCARPUS
TENNESSEE	THREATENED	LAMIACEAE	SCUTELLARIA MONTANA
TENNESSEE	THREATENED	LAMIACEAE	SYNANDRA HISPIDULA
TENNESSEE	THREATENED	LILIACEAE	LILIUM GRAYII
TENNESSEE	THREATENED	LILIACEAE	TRILLIUM PUSILLUM VAR. PUSILLUM
TENNESSEE	THREATENED	ORCHIDACEAE	PLATANTHERA FLAVA
TENNESSEE	THREATENED	ORCHIDACEAE	PLATANTHERA INTEGRA
TENNESSEE	THREATENED	ORCHIDACEAE	PLATANTHERA PERAMOENA
TENNESSEE	THREATENED	POACEAE	CALAMAGROSTIS CAINII
TENNESSEE	THREATENED	POACEAE	MUHLENBERGIA TORREYANA
TENNESSEE	THREATENED	PORTULACACEAE	TALINUM CALCARICUM
TENNESSEE	THREATENED	PORTULACACEAE	TALINUM MENGESII
TENNESSEE	THREATENED	RANUNCULACEAE	CIMICIFUGA RUBIFOLIA
TENNESSEE	THREATENED	RANUNCULACEAE	HYDRASTIS CANADENSIS
TENNESSEE	THREATENED	RUBIACEAE	HEDYOTIS PURPUREA VAR. MONTANA
TENNESSEE	THREATENED	SANTALACEAE	BUCKLEYA DISTICHOPHYLLA
TENNESSEE	THREATENED	SAXIFRAGACEAE	SAXIFRAGA CAREYANA
TENNESSEE	THREATENED	SAXIFRAGACEAE	SAXIFRAGA CAROLINIANA
TENNESSEE	THREATENED	SCHISANDRACEAE	SCHISANDRA GLABRA
TENNESSEE	THREATENED	SCROPHULARIACEAE	AUREOLARIA PATULA

STATE LISTS OF ENDANGERED, EXTINCT AND THREATENED SPECIES IN THE CONTINENTAL UNITED STATES

STATE	STATUS	FAMILY	SPECIES
TENNESSEE	THREATENED	SCROPHULARIACEAE	SCHWALBEA AMERICANA
TENNESSEE	THREATENED	VIOLACEAE	VIOLA EGGLESTONII
TEXAS	ENDANGERED	ASCLEPIADACEAE	MATELEA EDWARDSENSIS
TEXAS	ENDANGERED	ASCLEPIADACEAE	MATELEA TEXENSIS
TEXAS	ENDANGERED	ASTERACEAE	AMBROSIA CHEIRANTHIFOLIA
TEXAS	ENDANGERED	ASTERACEAE	BRICKELLIA VIEJENSIS
TEXAS	ENDANGERED	ASTERACEAE	COREOPSIS INTERMEDIA
TEXAS	ENDANGERED	ASTERACEAE	DYSSODIA TEPHROLEUCA
TEXAS	ENDANGERED	ASTERACEAE	ERIGERON GEISERI VAR. CALCICOLA
TEXAS	ENDANGERED	ASTERACEAE	GRINDELIA OOLEPIS
TEXAS	ENDANGERED	ASTERACEAE	HELIANTHUS PARADOXUS
TEXAS	ENDANGERED	ASTERACEAE	MACHAERANTHERA AUREA
TEXAS	ENDANGERED	ASTERACEAE	PERITYLE BISETOSA VAR. BISETOSA
TEXAS	ENDANGERED	ASTERACEAE	PERITYLE BISETOSA VAR. SCALARIS
TEXAS	ENDANGERED	ASTERACEAE	PERITYLE CINEREA
TEXAS	ENDANGERED	ASTERACEAE	PERITYLE LINDHEIMERI VAR. HALIMIFOLIA
TEXAS	ENDANGERED	ASTERACEAE	PERITYLE VITREOMONTANA
TEXAS	ENDANGERED	ASTERACEAE	VIGUIERA LUDENS
TEXAS	ENDANGERED	BRASSICACEAE	LEAVENWORTHIA AUREA
TEXAS	ENDANGERED	BRASSICACEAE	LESQUERELLA VALIDA
TEXAS	ENDANGERED	BRASSICACEAE	SELENIA JONESII
TEXAS	ENDANGERED	BRASSICACEAE	STREPTANTHUS SPARSIFLORUS
TEXAS	ENDANGERED	BRASSICACEAE	THELYPODIUM TEXANUM
TEXAS	ENDANGERED	CACTACEAE	ANCISTROCACTUS TOBUSCHII
TEXAS	ENDANGERED	CACTACEAE	CORYPHANTHA MINIMA
TEXAS	ENDANGERED	CACTACEAE	CORYPHANTHA RAMILLOSA
TEXAS	ENDANGERED	CACTACEAE	CORYPHANTHA SNEEDII VAR. SNEEDII
TEXAS	ENDANGERED	CACTACEAE	CORYPHANTHA STROBILIFORMIS VAR. DURISPINA
TEXAS	ENDANGERED	CACTACEAE	ECHINOCEREUS CHLORANTHUS VAR. NEOCAPILLUS
TEXAS	ENDANGERED	CACTACEAE	ECHINOCEREUS LLOYDII
TEXAS	ENDANGERED	CACTACEAE	ECHINOCEREUS REICHENBACHII VAR. ALBERTII
TEXAS	ENDANGERED	CACTACEAE	ECHINOCEREUS RUSSANTHUS
TEXAS	ENDANGERED	CACTACEAE	ECHINOCEREUS VIRIDIFLORUS VAR. DAVISII
TEXAS	ENDANGERED	CACTACEAE	NEOLLOYDIA GAUTII
TEXAS	ENDANGERED	CACTACEAE	NEOLLOYDIA MARIPOSENSIS
TEXAS	ENDANGERED	CAPPARIDACEAE	CLEOME MULTICAULIS
TEXAS	ENDANGERED	CARYOPHYLLACEAE	CERASTIUM CLAWSONII
TEXAS	ENDANGERED	CARYOPHYLLACEAE	PARONYCHIA CONGESTA
TEXAS	ENDANGERED	CARYOPHYLLACEAE	PARONYCHIA MACCARTII
TEXAS	ENDANGERED	CARYOPHYLLACEAE	SILENE PLANKII
TEXAS	ENDANGERED	CHENOPODIACEAE	ATRIPLEX KLEBERGORUM
TEXAS	ENDANGERED	CHENOPODIACEAE	SUAEDA DURIPES
TEXAS	ENDANGERED	CISTACEAE	LECHEA MENSALIS

STATE LISTS OF ENDANGERED, EXTINCT AND THREATENED SPECIES IN THE CONTINENTAL UNITED STATES

STATE	STATUS	FAMILY	SPECIES
TEXAS	ENDANGERED	CRASSULACEAE	LENOPHYLLUM TEXANUM
TEXAS	ENDANGERED	CYPERACEAE	ELEOCHARIS CYLINDRICA
TEXAS	ENDANGERED	ERIOCAULACEAE	ERIOCAULON KORNICKIANUM
TEXAS	ENDANGERED	EUPHORBIACEAE	ANDRACHNE ARIDA
TEXAS	ENDANGERED	EUPHORBIACEAE	ARGYTHAMNIA APHOROIDES
TEXAS	ENDANGERED	EUPHORBIACEAE	ARGYTHAMNIA ARGYRAEA
TEXAS	ENDANGERED	EUPHORBIACEAE	EUPHORBIA FENDLERI VAR. TRILIGULATA
TEXAS	ENDANGERED	EUPHORBIACEAE	EUPHORBIA GOLONDRINA
TEXAS	ENDANGERED	EUPHORBIACEAE	MANIHOT WALKERAE
TEXAS	ENDANGERED	EUPHORBIACEAE	PHYLLANTHUS ERICOIDES
TEXAS	ENDANGERED	FABACEAE	ACACIA EMORYANA
TEXAS	ENDANGERED	FABACEAE	BRONGNIARTIA MINUTIFOLIA
TEXAS	ENDANGERED	FABACEAE	CAESALPINIA DRUMMONDII
TEXAS	ENDANGERED	FABACEAE	CALLIANDRA BIFLORA
TEXAS	ENDANGERED	FABACEAE	GENISTIDIUM DUMOSUM
TEXAS	ENDANGERED	FABACEAE	HOFFMANNSEGGIA TENELLA
TEXAS	ENDANGERED	FABACEAE	PETALOSTEMUM REVERCHONII
TEXAS	ENDANGERED	FABACEAE	PETALOSTEMUM SABINALE
TEXAS	ENDANGERED	FAGACEAE	QUERCUS GRACILIFORMIS
TEXAS	ENDANGERED	FAGACEAE	QUERCUS HINCKLEYI
TEXAS	ENDANGERED	FAGACEAE	QUERCUS TARDIFOLIA
TEXAS	ENDANGERED	FRANKENIACEAE	FRANKENIA JOHNSTONII
TEXAS	ENDANGERED	GENTIANACEAE	BARTONIA TEXANA
TEXAS	ENDANGERED	HYDROPHYLLACEAF	PHACELIA PALLIDA
TEXAS	ENDANGERED	ISOETACEAE	ISOETES LITHOPHYLLA
TEXAS	ENDANGERED	LAMIACEAE	BRAZORIA PULCHERRIMA
TEXAS	ENDANGERED	LAMIACEAE	PHYSOSTEGIA CORRELLII
TEXAS	ENDANGERED	LILIACEAE	POLIANTHES RUNYONII
TEXAS	ENDANGERED	LILIACEAE	SCHOENOLIRION TEXANUM
TEXAS	ENDANGERED	MALVACEAE	CALLIRHOE SCABRIUSCULA
TEXAS	ENDANGERED	MALVACEAE	GAYA VIOLACEA
TEXAS	ENDANGERED	MALVACEAE	HIBISCUS DASYCALYX
TEXAS	ENDANGERED	POACEAE	MUHLENBERGIA VILLOSA
TEXAS	ENDANGERED	POACEAE	POA INVOLUTA
TEXAS	ENDANGERED	POACEAE	ZIZANIA TEXANA
TEXAS	ENDANGERED	POLEMONIACEAE	PHLOX NIVALIS SSP. TEXFNSIS
TEXAS	ENDANGERED	POLEMONIACEAE	POLEMONIUM PAUCIFLORUM SSP. HINCKLEYI
TEXAS	ENDANGERED	POLYGALACEAE	POLYGALA MARAVILLASENSIS
TEXAS	ENDANGERED	POLYGALACEAE	POLYGALA RIMULICOLA
TEXAS	ENDANGERED	POLYGONACEAE	ERIOGONUM NEALLEYI
TEXAS	ENDANGERED	POLYGONACEAE	ERIOGONUM SUFFRUTICOSUM
TEXAS	ENDANGERED	POLYGONACEAE	POLYGONELLA PARKSII
TEXAS	ENDANGERED	POLYGONACEAE	POLYGONUM TEXENSE

STATE LISTS OF ENDANGERED, EXTINCT AND THREATENED SPECIES IN THE CONTINENTAL UNITED STATES

STATE	STATUS	FAMILY	SPECIES
TEXAS	ENDANGERED	POTAMOGETONACEAE	POTAMOGETON CLYSTOCARPUS
TEXAS	ENDANGERED	RANUNCULACEAE	AQUILEGIA CHAPLINEI
TEXAS	ENDANGERED	RANUNCULACEAE	AQUILEGIA HINCKLEYANA
TEXAS	ENDANGERED	RANUNCULACEAE	RANUNCULUS FASCICULARIS VAR. CUNEIFORMIS
TEXAS	ENDANGERED	RHAMNACEAE	COLUBRINA STRICTA
TEXAS	ENDANGERED	RHAMNACEAE	CONDALIA HOOKERI VAR. EDWARDSIANA
TEXAS	ENDANGERED	RUTACEAE	ZANTHOXYLUM PARVUM
TEXAS	ENDANGERED	SALICACEAE	POPULUS HINCKLEYANA
TEXAS	ENDANGERED	SCROPHULARIACEAE	CASTILLEJA CILIATA
TEXAS	ENDANGERED	STYRACACEAE	STYRAX PLATANIFOLIA VAR. STELLATA
TEXAS	ENDANGERED	STYRACACEAE	STYRAX TEXANA
TEXAS	ENDANGERED	URTICACEAE	URTICA CHAMAEDRYOIDES VAR. RUNYONII
TEXAS	ENDANGERED	VALERIANACEAE	VALERIANELLA TEXANA
TEXAS	EXTINCT	AIZOACEAE	SESUVIUM TRIANTHEMOIDES
TEXAS	EXTINCT	ASCLEPIADACEAE	MATELEA RADIATA
TEXAS	EXTINCT	ASTERACEAE	HYMENOXYS TEXANA
TEXAS	EXTINCT	ASTERACEAE	PERITYLE ROTUNDATA
TEXAS	EXTINCT	BRASSICACEAE	THELYPODIUM TENUE
TEXAS	EXTINCT	BROMELIACEAE	HECHTIA TEXENSIS
TEXAS	EXTINCT	CACTACEAE	CORYPHANTHA SCHEERI VAR. UNCINATA
TEXAS	EXTINCT	CACTACEAE	ECHINOCEREUS BLANCKII VAR. ANGUSTICEPS
TEXAS	EXTINCT	CACTACEAE	OPUNTIA STRIGIL VAR. FLEXOSPINA
TEXAS	EXTINCT	CARYOPHYLLACEAE	ARENARIA LIVERMORENSIS
TEXAS	EXTINCT	FABACEAE	VICIA REVERCHONII
TEXAS	EXTINCT	LAMIACEAE	HEDEOMA PILOSUM
TEXAS	EXTINCT	ORCHIDACEAE	SPIRANTHES PARKSII
TEXAS	EXTINCT	SCROPHULARIACEAE	SEYMERIA HAVARDII
TEXAS	EXTINCT	STERCULIACEAE	NEPHROPETALUM PRINGLEI
TEXAS	THREATENED	ACANTHACEAE	DYSCHORISTE CRENULATA
TEXAS	THREATENED	ACANTHACEAE	JUSTICIA RUNYONII
TEXAS	THREATENED	ACANTHACEAE	JUSTICIA WARNOCKII
TEXAS	THREATENED	ACANTHACEAE	JUSTICIA WRIGHTII
TEXAS	THREATENED	ACANTHACEAE	STENANDRIUM FASCICULARIS
TEXAS	THREATENED	ACERACEAE	ACER GRANDIDENTATUM VAR. SINUOSUM
TEXAS	THREATENED	APIACEAE	ALETES FILIFOLIUS
TEXAS	THREATENED	APIACEAE	EURYTAENIA HINCKLEYI
TEXAS	THREATENED	APOCYNACEAE	AMSONIA GLABERRIMA
TEXAS	THREATENED	APOCYNACEAE	AMSONIA REPENS
TEXAS	THREATENED	APOCYNACEAE	AMSONIA THARPII
TEXAS	THREATENED	ASCLEPIADACEAE	MATELEA BREVICORONATA
TEXAS	THREATENED	ASCLEPIADACEAE	MATELEA PARVIFLORA
TEXAS	THREATENED	ASTERACEAE	ASTER SCABRICAULIS
TEXAS	THREATENED	ASTERACEAE	ASTRANTHIUM ROBUSTUM

STATE LISTS OF ENDANGERED, EXTINCT AND THREATENED SPECIES IN THE CONTINENTAL UNITED STATES

STATE	STATUS	FAMILY	SPECIES
TEXAS	THREATENED	ASTERACEAE	BAHIA BIGELOVII
TEXAS	THREATENED	ASTERACEAE	BRICKELLIA BRACHYPHYLLA VAR. HINCKLEYI
TEXAS	THREATENED	ASTERACEAE	BRICKELLIA BRACHYPHYLLA VAR. TERLINGUENSIS
TEXAS	THREATENED	ASTERACEAE	BRICKELLIA DENTATA
TEXAS	THREATENED	ASTERACEAE	BRICKELLIA SHINERI
TEXAS	THREATENED	ASTERACEAE	CHAETOPAPPA HERSHEYI
TEXAS	THREATENED	ASTERACEAE	CIRSIUM TURNERI
TEXAS	THREATENED	ASTERACEAE	ERIGERON BIGELOVII
TEXAS	THREATENED	ASTERACEAE	HELIANTHUS PRAECOX SSP. HIRTUS
TEXAS	THREATENED	ASTERACEAE	LIATRIS CYMOSA
TEXAS	THREATENED	ASTERACEAE	LIATRIS TENUIS
TEXAS	THREATENED	ASTERACEAE	PERITYLE WARNOCKII
TEXAS	THREATENED	ASTERACEAE	POROPHYLLUM GREGGII
TEXAS	THREATENED	ASTERACEAE	SENECIO WARNOCKII
TEXAS	THREATENED	ASTERACEAE	SOLIDAGO MOLLIS VAR. ANGUSTATA
TEXAS	THREATENED	BERBERIDACEAE	BERBERIS SWASEYI
TEXAS	THREATENED	BETULACEAE	OSTRYA CHISOSENSIS
TEXAS	THREATENED	BORAGINACEAE	CRYPTANTHA CRASSIPES
TEXAS	THREATENED	BORAGINACEAE	ONOSMODIUM HELLERI
TEXAS	THREATENED	BRASSICACEAE	LESQUERELLA ANGUSTIFOLIA
TEXAS	THREATENED	BRASSICACEAE	LESQUERELLA MCVAUGHIANA
TEXAS	THREATENED	BRASSICACEAE	LESQUERELLA THAMNOPHILA
TEXAS	THREATENED	BRASSICACEAE	STREPTANTHUS BRACTEATUS
TEXAS	THREATENED	BRASSICACEAE	STREPTANTHUS CARINATUS
TEXAS	THREATENED	BRASSICACEAE	STREPTANTHUS CUTLERI
TEXAS	THREATENED	CACTACEAE	CEREUS GREGGII
TEXAS	THREATENED	CACTACEAE	CORYPHANTHA DASYACANTHA VAR. VARICOLOR
TEXAS	THREATENED	CACTACEAE	CORYPHANTHA DUNCANII
TEXAS	THREATENED	CACTACEAE	CORYPHANTHA HESTERI
TEXAS	THREATENED	CACTACEAE	CORYPHANTHA SULCATA VAR. NICKELSIAE
TEXAS	THREATENED	CACTACEAE	ECHINOCEREUS REICHENBACHII VAR. CHISOSENSIS
TEXAS	THREATENED	CACTACEAE	ECHINOCEREUS REICHENBACHII VAR. FITCHII
TEXAS	THREATENED	CACTACEAE	ECHINOCEREUS VIRIDIFLORUS VAR. CORRELLII
TEXAS	THREATENED	CACTACEAE	EPITHELANTHA BOKEI
TEXAS	THREATENED	CACTACEAE	NEOLLOYDIA WARNOCKII
TEXAS	THREATENED	CACTACEAE	OPUNTIA ARENARIA
TEXAS	THREATENED	CACTACEAE	OPUNTIA IMBRICATA VAR. ARGENTEA
TEXAS	THREATENED	CACTACEAE	THELOCACTUS BICOLOR VAR. FLAVIDISPINUS
TEXAS	THREATENED	CAMPANULACEAE	CAMPANULA REVERCHONII
TEXAS	THREATENED	CAPRIFOLIACEAE	SYMPHORICARPOS GUADALUPENSIS
TEXAS	THREATENED	CARYOPHYLLACEAE	PARONYCHIA CHORIZANTHOIDES
TEXAS	THREATENED	CARYOPHYLLACEAF	PARONYCHIA DRUMMONDII SSP. PARVIFLORA
TEXAS	THREATENED	CARYOPHYLLACEAE	PARONYCHIA MONTICOLA

STATE LISTS OF ENDANGERED, EXTINCT AND THREATENED SPECIES IN THE CONTINENTAL UNITED STATES

STATE	STATUS	FAMILY	SPECIES
TEXAS	THREATENED	CARYOPHYLLACEAE	PARONYCHIA NUDATA
TEXAS	THREATENED	CARYOPHYLLACEAE	PARONYCHIA VIRGINICA VAR. PARKSII
TEXAS	THREATENED	CARYOPHYLLACEAE	PARONYCHIA WILKINSONII
TEXAS	THREATENED	COCHLOSPERMACEAE	AMOREUXIA WRIGHTII
TEXAS	THREATENED	COMMELINACEAE	TRADESCANTIA EDWARDSIANA
TEXAS	THREATENED	COMMELINACEAE	TRADESCANTIA WRIGHTII
TEXAS	THREATENED	CUCURBITACEAE	CUCURBITA TEXANA
TEXAS	THREATENED	CYPERACEAE	CAREX ONUSTA
TEXAS	THREATENED	CYPERACEAE	CYPERUS ONEROSUS
TEXAS	THREATENED	CYPERACEAE	ELEOCHARIS AUSTROTEXANA
TEXAS	THREATENED	EUPHORBIACEAE	EUPHORBIA JEJUNA
TEXAS	THREATENED	EUPHORBIACEAE	EUPHORBIA PERENNANS
TEXAS	THREATENED	EUPHORBIACEAE	EUPHORBIA ROEMERIANA
TEXAS	THREATENED	EUPHORBIACEAE	EUPHORBIA STRICTIOR
TEXAS	THREATENED	EUPHORBIACEAE	TRAGIA NIGRICANS
TEXAS	THREATENED	FABACEAE	AMORPHA ROEMERIANA
TEXAS	THREATENED	FABACEAE	ASTRAGALUS MOLLISSIMUS VAR. MARCIDUS
TEXAS	THREATENED	FABACEAE	CAESALPINIA BRACHYCARPA
TEXAS	THREATENED	FABACEAE	CASSIA RIPLEYANA
TEXAS	THREATENED	FABACEAE	COURSETIA AXILLARIS
TEXAS	THREATENED	FABACEAE	DALEA BARTONII
TEXAS	THREATENED	FABACEAE	DESMODIUM LINDHEIMERI
TEXAS	THREATENED	FABACEAE	SOPHORA GYPSOPHILA VAR. GUADALUPENSIS
TEXAS	THREATENED	HYDROPHYLLACEAE	NAMA XYLOPODUM
TEXAS	THREATENED	LAMIACEAE	HEDEOMA APICULATUM
TEXAS	THREATENED	LAMIACEAE	SALVIA PENSTEMONOIDES
TEXAS	THREATENED	LEITNERIACEAE	LEITNERIA FLORIDANA
TEXAS	THREATENED	LILIACEAE	AGAVE CHISOENSIS
TEXAS	THREATENED	LILIACEAE	ANTHERICUM CHANDLERI
TEXAS	THREATENED	LILIACEAE	NOLINA ARENICOLA
TEXAS	THREATENED	LILIACEAE	POLIANTHES MACULOSA
TEXAS	THREATENED	LILIACEAE	TRILLIUM TEXANUM
TEXAS	THREATENED	LILIACEAE	VERATRUM WOODII
TEXAS	THREATENED	LYTHRACEAE	HEIMIA LONGIPES
TEXAS	THREATENED	LYTHRACEAE	LYTHRUM OVALIFOLIUM
TEXAS	THREATENED	MALVACEAE	ABUTILON MARSHII
TEXAS	THREATENED	MELASTOMATACEAE	RHEXIA SALICIFOLIA
TEXAS	THREATENED	NYCTAGINACEAE	ACLEISANTHES CRASSIFOLIA
TEXAS	THREATENED	NYCTAGINACEAE	BOERHAAVIA MATHISIANA
TEXAS	THREATENED	ORCHIDACEAE	HEXALECTRIS GRANDIFLORA
TEXAS	THREATENED	ORCHIDACEAE	HEXALECTRIS NITIDA
TEXAS	THREATENED	ORCHIDACEAE	HEXALECTRIS REVOLUTA
TEXAS	THREATENED	ORCHIDACEAE	PLATANTHERA FLAVA

STATE LISTS OF ENDANGERED, EXTINCT AND THREATENED SPECIES IN THE CONTINENTAL UNITED STATES

STATE	STATUS	FAMILY	SPECIES
TEXAS	THREATENED	ORCHIDACEAE	PLATANTHERA INTEGRA
TEXAS	THREATENED	PEDALIACEAE	PROBOSCIDEA SABULOSA
TEXAS	THREATENED	POACEAE	BROMUS TEXENSIS
TEXAS	THREATENED	POACEAE	CHLORIS TEXENSIS
TEXAS	THREATENED	POACEAE	FESTUCA LIGULATA
TEXAS	THREATENED	POACEAE	WILLKOMMIA TEXANA
TEXAS	THREATENED	POLEMONIACEAE	PHLOX OKLAHOMENSIS
TEXAS	THREATENED	POLYGONACEAE	ERIOGONUM CORRELLII
TEXAS	THREATENED	POLYGONACEAE	POLYGONUM STRIATULUM
TEXAS	THREATENED	POLYPODIACEAE	NOTHOLAENA SCHAFFNERI VAR. NEALLEYI
TEXAS	THREATENED	RANUNCULACEAE	ANEMONE EDWARDSIANA VAR. EDWARDSIANA
TEXAS	THREATENED	RANUNCULACEAE	ANEMONE EDWARDSIANA VAR. PETRAEA
TEXAS	THREATENED	ROSACEAE	PRUNUS HAVARDII
TEXAS	THREATENED	ROSACEAE	PRUNUS MINUTIFLORA
TEXAS	THREATENED	ROSACEAE	PRUNUS MURRAYANA
TEXAS	THREATENED	ROSACEAE	PRUNUS TEXANA
TEXAS	THREATENED	ROSACEAE	ROSA STELLATA
TEXAS	THREATENED	RUBIACEAE	GALIUM CORRELLII
TEXAS	THREATENED	SAXIFRAGACEAE	PHILADELPHUS ERNESTII
TEXAS	THREATENED	SAXIFRAGACEAE	PHILADELPHUS TEXENSIS VAR. TEXENSIS
TEXAS	THREATENED	SCROPHULARIACEAE	CASTILLEJA ELONGATA
TEXAS	THREATENED	SOLANACEAE	LYCIUM BERBERIOIDES
TEXAS	THREATENED	SOLANACEAE	LYCIUM TEXANUM
TEXAS	THREATENED	STYRACACEAE	STYRAX YOUNGAE
TEXAS	THREATENED	VALERIANACEAE	VALERIANA TEXANA
TEXAS	THREATENED	VALERIANACEAE	VALERIANELLA FLORIFERA
UTAH	ENDANGERED	APIACEAE	CYMOPTERUS MINIMUS
UTAH	ENDANGERED	APOCYNACEAE	CYCLADENIA HUMILIS VAR. JONESII
UTAH	ENDANGERED	ASTERACEAE	ERIGERON FLAGELLARIS VAR. TRILOBATUS
UTAH	ENDANGERED	ASTERACEAE	ERIGERON KACHINENSIS
UTAH	ENDANGERED	ASTERACEAE	ERIGERON MAGUIREI
UTAH	ENDANGERED	ASTERACEAE	ERIGERON RELIGIOSUS
UTAH	ENDANGERED	ASTERACEAE	ERIGERON SIONIS
UTAH	ENDANGERED	ASTERACEAE	GAILLARDIA FLAVA
UTAH	ENDANGERED	ASTERACEAE	HETEROTHECA JONESII
UTAH	ENDANGERED	ASTERACEAE	LYGODESMIA GRANDIFLORA VAR. STRICTA
UTAH	ENDANGERED	ASTERACEAE	PARTHENIUM LIGULATUM
UTAH	ENDANGERED	ASTERACEAE	SPHAEROMERIA RUTHIAE
UTAH	ENDANGERED	ASTERACEAE	TOWNSENDIA APRICA
UTAH	ENDANGERED	BORAGINACEAE	CRYPTANTHA OCHROLEUCA
UTAH	ENDANGERED	BRASSICACEAE	GLAUCOCARPUM SUFFRUTESCENS
UTAH	ENDANGERED	BRASSICACEAE	LEPIDIUM BARNEBYANUM
UTAH	ENDANGERED	BRASSICACEAE	LESQUERELLA TUMULOSA

STATE LISTS OF ENDANGERED, EXTINCT AND THREATENED SPECIES IN THE CONTINENTAL UNITED STATES

STATE	STATUS	FAMILY	SPECIES
UTAH	ENDANGERED	CACTACEAE	ECHINOCEREUS ENGELMANNII VAR. PURPUREUS
UTAH	ENDANGERED	CACTACEAE	PEDIOCACTUS SILERI
UTAH	ENDANGERED	CACTACEAE	SCLEROCACTUS GLAUCUS
UTAH	ENDANGERED	CACTACEAE	SCLEROCACTUS WRIGHTIAE
UTAH	ENDANGERED	FABACEAE	ASTRAGALUS CRONQUISTII
UTAH	ENDANGERED	FABACEAE	ASTRAGALUS HAMILTONII
UTAH	ENDANGERED	FABACEAE	ASTRAGALUS HARRISONII
UTAH	ENDANGERED	FABACEAE	ASTRAGALUS ISELYI
UTAH	ENDANGERED	FABACEAE	ASTRAGALUS LUTOSUS
UTAH	ENDANGERED	FABACEAE	ASTRAGALUS PERIANUS
UTAH	ENDANGERED	FABACEAE	PSORALEA EPIPSILA
UTAH	ENDANGERED	HYDROPHYLLACEAE	PHACELIA ARGILLACEA
UTAH	ENDANGERED	HYDROPHYLLACEAE	PHACELIA INDECORA
UTAH	ENDANGERED	HYDROPHYLLACEAE	PHACELIA MAMMILLARENSIS
UTAH	ENDANGERED	LILIACEAE	ALLIUM PASSEYI
UTAH	ENDANGERED	LILIACEAE	ZIGADENUS VAGINATUS
UTAH	ENDANGERED	NAJADACEAE	NAJAS CAESPITOSA
UTAH	ENDANGERED	NYCTAGINACEAE	HERMIDIUM ALIPES VAR. PALLIDIUM
UTAH	ENDANGERED	ONAGRACEAE	CAMISSONIA MEGALANTHA
UTAH	ENDANGERED	PAPAVERACEAE	ARCTOMECON HUMILIS
UTAH	ENDANGERED	POACEAE	FESTUCA DASYCLADA
UTAH	ENDANGERED	POLEMONIACEAE	GILIA CAESPITOSA
UTAH	ENDANGERED	POLYGONACEAE	ERIOGONUM AMMOPHILUM
UTAH	ENDANGERED	POLYGONACEAE	ERIOGONUM ARETIOIDES
UTAH	ENDANGERED	POLYGONACEAE	ERIOGONUM CORYMBOSUM VAR. DAVIDSEI
UTAH	ENDANGERED	POLYGONACEAE	ERIOGONUM CORYMBOSUM VAR. REVEALIANUM
UTAH	ENDANGERED	POLYGONACEAE	ERIOGONUM CRONQUISTII
UTAH	ENDANGERED	POLYGONACEAE	ERIOGONUM EPHEDROIDES
UTAH	ENDANGERED	POLYGONACEAE	ERIOGONUM HUMIVAGANS
UTAH	ENDANGERED	POLYGONACEAE	ERIOGONUM HYLOPHILUM
UTAH	ENDANGERED	POLYGONACEAE	ERIOGONUM INTERMONTANUM
UTAH	ENDANGERED	POLYGONACEAE	ERIOGONUM LOGANUM
UTAH	ENDANGERED	POLYGONACEAE	ERIOGONUM SMITHII
UTAH	ENDANGERED	POLYGONACEAE	ERIOGONUM ZIONIS VAR. ZIONIS
UTAH	ENDANGERED	POLYPODIACEAE	ASPLENIUM ANDREWSII
UTAH	ENDANGERED	SCROPHULARIACEAE	CASTILLEJA AQUARIENSIS
UTAH	ENDANGERED	SCROPHULARIACEAE	CASTILLEJA REVEALII
UTAH	ENDANGERED	SCROPHULARIACEAE	PENSTEMON CONCINNUS
UTAH	ENDANGERED	SCROPHULARIACEAE	PENSTEMON GRAHAMII
UTAH	EXTINCT	CUSCUTACEAE	CUSCUTA WARNERI
UTAH	EXTINCT	FABACEAE	ASTRAGALUS DESERETICUS
UTAH	EXTINCT	FABACEAE	ASTRAGALUS LENTIGINOSUS VAR. URSINUS
UTAH	EXTINCT	RANUNCULACEAE	RANUNCULUS ACRIFORMIS VAR. AESTIVALIS

STATE LISTS OF ENDANGERED, EXTINCT AND THREATENED SPECIES IN THE CONTINENTAL UNITED STATES

STATE	STATUS	FAMILY	SPECIES
UTAH	EXTINCT	SCROPHULARIACEAE	PENSTEMON GARRETTII
UTAH	THREATENED	APIACEAE	ANGELICA WHEELERI
UTAH	THREATENED	APIACEAE	CYMOPTERUS COULTERI
UTAH	THREATENED	APIACEAE	CYMOPTERUS DUCHESNENSIS
UTAH	THREATENED	APIACEAE	CYMOPTERUS HIGGINSII
UTAH	THREATENED	APIACEAE	CYMOPTERUS ROSEI
UTAH	THREATENED	APIACEAE	LIGUSTICUM PORTERI VAR. BREVILOBUM
UTAH	THREATENED	APIACEAE	LOMATIUM LATILOBUM
UTAH	THREATENED	APIACEAE	LOMATIUM MINIMUM
UTAH	THREATENED	APIACEAE	MUSINEON LINEARE
UTAH	THREATENED	ASCLEPIADACEAE	ASCLEPIAS CUTLERI
UTAH	THREATENED	ASCLEPIADACEAE	ASCLEPIAS RUTHIAE
UTAH	THREATENED	ASTERACEAE	CHAMAECHAENACTIS SCAPOSA
UTAH	THREATENED	ASTERACEAE	CIRSIUM RYDBERGII
UTAH	THREATENED	ASTERACEAE	ERIGERON ABAJOENSIS
UTAH	THREATENED	ASTERACEAE	ERIGERON ARENARIOIDES
UTAH	THREATENED	ASTERACEAE	ERIGERON CRONQUISTII
UTAH	THREATENED	ASTERACEAE	ERIGERON GARRETTII
UTAH	THREATENED	ASTERACEAE	ERIGERON MANCUS
UTAH	THREATENED	ASTERACEAE	HELIANTHUS DESERTICOLA
UTAH	THREATENED	ASTERACEAE	HYMENOPAPPUS FILIFOLIUS VAR. TOMENTOSUS
UTAH	THREATENED	ASTERACEAE	MACHAERANTHERA GLABRIUSCULA VAR. CONFERTIFOLIA
UTAH	THREATENED	ASTERACEAE	MACHAERANTHERA KINGII
UTAH	THREATENED	ASTERACEAE	SENECIO DIMORPHOPHYLLUS VAR. INTERMEDIUS
UTAH	THREATENED	ASTERACEAE	TOWNSENDIA MENSANA
UTAH	THREATENED	ASTERACEAE	TOWNSENDIA MINIMA
UTAH	THREATENED	ASTERACEAE	VIGUIERA SOLICEPS
UTAH	THREATENED	ASTERACEAE	XANTHOCEPHALUM SAROTHRAE VAR. POMARIENSIS
UTAH	THREATENED	BORAGINACEAE	CRYPTANTHA BARNEBYI
UTAH	THREATENED	BORAGINACEAE	CRYPTANTHA COMPACTA
UTAH	THREATENED	BORAGINACEAE	CRYPTANTHA ELATA
UTAH	THREATENED	BORAGINACEAE	CRYPTANTHA GRAHAMII
UTAH	THREATENED	BORAGINACEAE	CRYPTANTHA JOHNSTONII
UTAH	THREATENED	BORAGINACEAE	CRYPTANTHA JONESIANA
UTAH	THREATENED	BORAGINACEAE	CRYPTANTHA MENSANA
UTAH	THREATENED	BORAGINACEAE	CRYPTANTHA PARADOXA
UTAH	THREATENED	BORAGINACEAE	CRYPTANTHA SEMIGLABRA
UTAH	THREATENED	BRASSICACEAE	ARABIS DEMISSA VAR. LANGUIDA
UTAH	THREATENED	BRASSICACEAE	ARABIS DEMISSA VAR. RUSSEOLA
UTAH	THREATENED	BRASSICACEAE	DRABA MAGUIREI VAR. BURKEI
UTAH	THREATENED	BRASSICACEAE	DRABA MAGUIREI VAR. MAGUIREI
UTAH	THREATENED	BRASSICACEAE	DRABA PECTINIPILA
UTAH	THREATENED	BRASSICACEAE	DRABA SOBOLIFERA

STATE LISTS OF ENDANGERED, EXTINCT AND THREATENED SPECIES IN THE CONTINENTAL UNITED STATES

STATE	STATUS	FAMILY	SPECIES
UTAH	THREATENED	BRASSICACEAE	DRABA SUBALPINA
UTAH	THREATENED	BRASSICACEAE	DRABA ZIONENSIS
UTAH	THREATENED	BRASSICACEAE	LESQUERELLA GARRETTII
UTAH	THREATENED	BRASSICACEAE	LESQUERELLA RUBICUNDULA
UTAH	THREATENED	BRASSICACEAE	THELYPODIUM SAGITTATUM VAR. OVALIFOLIUM
UTAH	THREATENED	CACTACEAE	OPUNTIA WHIPPLEI VAR. MULTIGENICULATA
UTAH	THREATENED	CACTACEAE	SCLEROCACTUS PUBISPINUS
UTAH	THREATENED	CACTACEAE	SCLEROCACTUS SPINOSIOR
UTAH	THREATENED	CARYOPHYLLACEAE	SILENE PETERSONII VAR. MINOR
UTAH	THREATENED	CHENOPODIACEAE	ATRIPLEX WELSHII
UTAH	THREATENED	CYPERACEAE	CAREX ARAPAHOENSIS
UTAH	THREATENED	CYPERACEAE	CAREX CURATORUM
UTAH	THREATENED	EUPHORBIACEAE	EUPHORBIA NEPHRADENIA
UTAH	THREATENED	FABACEAE	ASTRAGALUS AMPULLARIUS
UTAH	THREATENED	FABACEAE	ASTRAGALUS BARNEBYI
UTAH	THREATENED	FABACEAE	ASTRAGALUS CALLITHRIX
UTAH	THREATENED	FABACEAE	ASTRAGALUS CASTANEIFORMIS VAR. CONSOBRINUS
UTAH	THREATENED	FABACEAE	ASTRAGALUS CHLOODES
UTAH	THREATENED	FABACEAE	ASTRAGALUS CONVALLARIUS VAR. FINITIMUS
UTAH	THREATENED	FABACEAE	ASTRAGALUS COTTAMII
UTAH	THREATENED	FABACEAE	ASTRAGALUS DETRITALIS
UTAH	THREATENED	FABACEAE	ASTRAGALUS DUCHESNENSIS
UTAH	THREATENED	FABACEAE	ASTRAGALUS ENSIFORMIS
UTAH	THREATENED	FABACEAE	ASTRAGALUS LANCEARIUS
UTAH	THREATENED	FABACEAE	ASTRAGALUS LIMNOCHARIS
UTAH	THREATENED	FABACEAE	ASTRAGALUS LOANUS
UTAH	THREATENED	FABACEAE	ASTRAGALUS MALACOIDES
UTAH	THREATENED	FABACEAE	ASTRAGALUS MINTHORNIAE VAR. GRACILIOR
UTAH	THREATENED	FABACEAE	ASTRAGALUS MONUMENTALIS
UTAH	THREATENED	FABACEAE	ASTRAGALUS OOPHORUS VAR. LONCHOCALYX
UTAH	THREATENED	FABACEAE	ASTRAGALUS PARDALINUS
UTAH	THREATENED	FABACEAE	ASTRAGALUS RAFAELENSIS
UTAH	THREATENED	FABACEAE	ASTRAGALUS SABULOSUS
UTAH	THREATENED	FABACEAE	ASTRAGALUS SAURINUS
UTAH	THREATENED	FABACEAE	ASTRAGALUS STOCKSII
UTAH	THREATENED	FABACEAE	ASTRAGALUS WETHERILLII
UTAH	THREATENED	FABACEAE	ASTRAGALUS WOODRUFFII
UTAH	THREATENED	FABACEAE	DALEA THOMPSONAE
UTAH	THREATENED	FABACEAE	HEDYSARUM BOREALE VAR. GREMIALE
UTAH	THREATENED	FABACEAE	LUPINUS JONESII
UTAH	THREATENED	FABACEAE	LUPINUS MARIANUS
UTAH	THREATENED	FABACEAE	OXYTROPIS JONESII
UTAH	THREATENED	FABACEAE	OXYTROPIS OBNAPIFORMIS

STATE LISTS OF ENDANGERED, EXTINCT AND THREATENED SPECIES IN THE CONTINENTAL UNITED STATES

STATE	STATUS	FAMILY	SPECIES
UTAH	THREATENED	FABACEAE	PSORALEA PARIENSIS
UTAH	THREATENED	FUMARIACEAE	CORYDALIS CASEANA SSP. BRACHYCARPA
UTAH	THREATENED	GERANIACEAE	GERANIUM MARGINALE
UTAH	THREATENED	HYDROPHYLLACEAE	NAMA RETRORSUM
UTAH	THREATENED	HYDROPHYLLACEAE	PHACELIA CEPHALOTES
UTAH	THREATENED	HYDROPHYLLACEAE	PHACELIA CONSTANCEI
UTAH	THREATENED	HYDROPHYLLACEAE	PHACELIA DEMISSA VAR. HETEROTRICHA
UTAH	THREATENED	HYDROPHYLLACEAE	PHACELIA RAFAELENSIS
UTAH	THREATENED	HYDROPHYLLACEAE	PHACELIA UTAHENSIS
UTAH	THREATENED	LILIACEAE	YUCCA TOFTIAE
UTAH	THREATENED	LOASACEAE	MENTZELIA ARGILLOSA
UTAH	THREATENED	MALVACEAE	SPHAERALCEA CAESPITOSA
UTAH	THREATENED	ONAGRACEAE	CAMISSONIA GOULDII
UTAH	THREATENED	ONAGRACEAE	EPILOBIUM NEVADENSE
UTAH	THREATENED	POLEMONIACEAE	GILIA MCVICKERAE
UTAH	THREATENED	POLEMONIACEAE	PHLOX CLUTEANA
UTAH	THREATENED	POLEMONIACEAE	PHLOX GLADIFORMIS
UTAH	THREATENED	POLEMONIACEAE	PHLOX GRAHAMII
UTAH	THREATENED	POLEMONIACEAE	PHLOX JONESII
UTAH	THREATENED	POLYGONACEAE	ERIOGONUM CLAVELLATUM
UTAH	THREATENED	POLYGONACEAE	ERIOGONUM EREMICUM
UTAH	THREATENED	POLYGONACEAE	ERIOGONUM JAMESII VAR. RUPICOLA
UTAH	THREATENED	POLYGONACEAE	ERIOGONUM NANUM
UTAH	THREATENED	POLYGONACEAE	ERIOGONUM NATUM
UTAH	THREATENED	POLYGONACEAE	ERIOGONUM OSTLUNDII
UTAH	THREATENED	POLYGONACEAE	ERIOGONUM PANGUICENSE VAR. ALPESTRE
UTAH	THREATENED	POLYGONACEAE	ERIOGONUM SAURINUM
UTAH	THREATENED	POLYGONACEAE	ERIOGONUM THOMPSONAE VAR. ALBIFLORUM
UTAH	THREATENED	POLYGONACEAE	ERIOGONUM THOMPSONAE VAR. THOMPSONAE
UTAH	THREATENED	POLYGONACEAE	ERIOGONUM VIRIDULUM
UTAH	THREATENED	PRIMULACEAE	PRIMULA MAGUIREI
UTAH	THREATENED	PRIMULACEAE	PRIMULA SPECUICOLA
UTAH	THREATENED	SCROPHULARIACEAE	CASTILLEJA PARVULA VAR. PARVULA
UTAH	THREATENED	SCROPHULARIACEAE	CASTILLEJA SCABRIDA
UTAH	THREATENED	SCROPHULARIACEAE	PENSTEMON ABIETINUS
UTAH	THREATENED	SCROPHULARIACEAE	PENSTEMON ACAULIS
UTAH	THREATENED	SCROPHULARIACEAE	PENSTEMON ATWOODII
UTAH	THREATENED	SCROPHULARIACEAE	PENSTEMON CAESPITOSUS VAR. SUFFRUTICOSUS
UTAH	THREATENED	SCROPHULARIACEAE	PENSTEMON COMPACTUS
UTAH	THREATENED	SCROPHULARIACEAE	PENSTEMON HUMILIS VAR. BREVIFOLIUS
UTAH	THREATENED	SCROPHULARIACEAE	PENSTEMON HUMILIS VAR. OBTUSIFOLIUS
UTAH	THREATENED	SCROPHULARIACEAE	PENSTEMON LEIOPHYLLUS
UTAH	THREATENED	SCROPHULARIACEAE	PENSTEMON NANUS

STATE LISTS OF ENDANGERED, EXTINCT AND THREATENED SPECIES IN THE CONTINENTAL UNITED STATES

STATE	STATUS	FAMILY	SPECIES
UTAH	THREATENED	SCROPHULARIACEAE	PENSTEMON PARVUS
UTAH	THREATENED	SCROPHULARIACEAE	PENSTEMON UINTAHENSIS
UTAH	THREATENED	SCROPHULARIACEAE	PENSTEMON WARDII
UTAH	THREATENED	SELAGINELLACEAE	SELAGINELLA UTAHENSIS
UTAH	THREATENED	VIOLACEAE	VIOLA CHARLESTONENSIS
VERMONT	ENDANGERED	CYPERACEAE	SCIRPUS ANCISTROCHAETUS
VERMONT	ENDANGERED	FABACEAE	ASTRAGALUS ROBBINSII VAR. JESUPI
VERMONT	ENDANGERED	ORCHIDACEAE	ISOTRIA MEDEOLOIDES
VERMONT	ENDANGERED	POACEAE	CALAMAGROSTIS INEXPANSA VAR. NOVAE-ANGLIAE
VERMONT	EXTINCT	FABACEAE	ASTRAGALUS ROBBINSII VAR. ROBBINSII
VERMONT	THREATENED	ARALIACEAE	PANAX QUINQUEFOLIUS
VERMONT	THREATENED	ASTERACEAE	PRENANTHES BOOTTII
VERMONT	THREATENED	ORCHIDACEAE	CYPRIPEDIUM ARIETINUM
VERMONT	THREATENED	ORCHIDACEAE	LISTERA AURICULATA
VERMONT	THREATENED	ORCHIDACEAE	PLATANTHERA FLAVA
VERMONT	THREATENED	POTAMOGETONACEAE	POTAMOGETON HILLII
VERMONT	THREATENED	RANUNCULACEAE	HYDRASTIS CANADENSIS
VIRGINIA	ENDANGERED	ARISTOLOCHIACEAE	HEXASTYLIS NANIFLORA
VIRGINIA	ENDANGERED	BETULACEAE	BETULA UBER
VIRGINIA	ENDANGERED	CISTACEAE	LECHEA MARITIMA VAR. VIRGINICA
VIRGINIA	ENDANGERED	CYPERACEAE	CAREX BILTMOREANA
VIRGINIA	ENDANGERED	CYPERACEAE	SCIRPUS ANCISTROCHAETUS
VIRGINIA	ENDANGERED	LILIACEAE	HYPOXIS LONGII
VIRGINIA	ENDANGERED	LILIACEAE	TRILLIUM PUSILLUM VAR. VIRGINIANUM
VIRGINIA	ENDANGERED	MALVACEAE	ILIAMNA REMOTA
VIRGINIA	ENDANGERED	ORCHIDACEAE	ISOTRIA MEDEOLOIDES
VIRGINIA	ENDANGERED	POACEAE	PANICUM MUNDUM
VIRGINIA	ENDANGERED	RANUNCULACEAE	CLEMATIS ADDISONII
VIRGINIA	ENDANGERED	RANUNCULACEAE	CLEMATIS VITICAULIS
VIRGINIA	ENDANGERED	SCROPHULARIACEAE	BACOPA STRAGULA
VIRGINIA	EXTINCT	SCROPHULARIACEAE	BACOPA SIMULANS
VIRGINIA	THREATENED	ACANTHACEAE	JUSTICIA MORTUIFLUMINIS
VIRGINIA	THREATENED	APIACEAE	LILAEOPSIS CAROLINENSIS
VIRGINIA	THREATENED	ARALIACEAE	PANAX QUINQUEFOLIUS
VIRGINIA	THREATENED	ARISTOLOCHIACEAE	HEXASTYLIS LEWISII
VIRGINIA	THREATENED	ASTERACEAE	ECHINACEA LAEVIGATA
VIRGINIA	THREATENED	ASTERACEAE	RUDBECKIA HELIOPSIDIS
VIRGINIA	THREATENED	BETULACEAE	ALNUS MARITIMA
VIRGINIA	THREATENED	BRASSICACEAE	CARDAMINE LONGII
VIRGINIA	THREATENED	CYPERACEAE	CAREX CHAPMANII
VIRGINIA	THREATENED	CYPERACEAE	CYMOPHYLLUS FRASERI
VIRGINIA	THREATENED	CYPERACEAE	SCIRPUS FLACCIDIFOLIUS
VIRGINIA	THREATENED	ERICACEAE	RHODODENDRON BAKERI

STATE LISTS OF ENDANGERED, EXTINCT AND THREATENED SPECIES IN THE CONTINENTAL UNITED STATES

STATE	STATUS	FAMILY	SPECIES
VIRGINIA	THREATENED	ISOETACEAE	ISOETES VIRGINICA
VIRGINIA	THREATENED	JUNCACEAE	JUNCUS CAESARIENSIS
VIRGINIA	THREATENED	LAMIACEAE	PYCNANTHEMUM MONOTRICHUM
VIRGINIA	THREATENED	LAMIACEAE	SYNANDRA HISPIDULA
VIRGINIA	THREATENED	LILIACEAE	LILIUM GRAYII
VIRGINIA	THREATENED	ORCHIDACEAE	PLATANTHERA FLAVA
VIRGINIA	THREATENED	ORCHIDACEAE	PLATANTHERA PERAMOENA
VIRGINIA	THREATENED	POACEAE	CALAMAGROSTIS PORTERI
VIRGINIA	THREATENED	POACEAE	CALAMOVILFA BREVIPILIS VAR. BREVIPILIS
VIRGINIA	THREATENED	POACEAE	PANICUM ACULEATUM
VIRGINIA	THREATENED	RANUNCULACEAE	CIMICIFUGA RUBIFOLIA
VIRGINIA	THREATENED	RANUNCULACEAE	HYDRASTIS CANADENSIS
VIRGINIA	THREATENED	SANTALACEAE	BUCKLEYA DISTICHOPHYLLA
VIRGINIA	THREATENED	SANTALACEAE	NESTRONIA UMBELLULA
VIRGINIA	THREATENED	SAXIFRAGACEAE	HEUCHERA HISPIDA
VIRGINIA	THREATENED	SAXIFRAGACEAE	SAXIFRAGA CAREYANA
VIRGINIA	THREATENED	SAXIFRAGACEAE	SAXIFRAGA CAROLINIANA
VIRGINIA	THREATENED	SCROPHULARIACEAE	MICRANTHEMUM MICRANTHEMOIDES
WASHINGTON	ENDANGERED	APIACEAE	LOMATIUM LAEVIGATUM
WASHINGTON	ENDANGERED	APIACEAE	LOMATIUM SUKSDORFII
WASHINGTON	ENDANGERED	APIACEAE	LOMATIUM TUBEROSUM
WASHINGTON	ENDANGERED	APIACEAE	TAUSCHIA HOOVERI
WASHINGTON	ENDANGERED	ASTERACEAE	ERIGERON BASALTICUS
WASHINGTON	ENDANGERED	BORAGINACEAE	HACKELIA VENUSTA
WASHINGTON	ENDANGERED	CARYOPHYLLACEAE	SILENE SPALDINGII
WASHINGTON	ENDANGERED	FABACEAE	ASTRAGALUS KENTROPHYTA VAR. DOUGLASII
WASHINGTON	ENDANGERED	FABACEAE	ASTRAGALUS MISELLUS VAR. PAUPER
WASHINGTON	ENDANGERED	FABACEAE	ASTRAGALUS SINUATUS
WASHINGTON	ENDANGERED	FABACEAE	TRIFOLIUM THOMPSONII
WASHINGTON	ENDANGERED	LILIACEAE	ALLIUM DICTUON
WASHINGTON	ENDANGERED	POACEAE	POA PACHYPHOLIS
WASHINGTON	ENDANGERED	RANUNCULACEAE	ANEMONE OREGANA VAR. FELIX
WASHINGTON	ENDANGERED	ROSACEAE	PETROPHYTUM CINERASCENS
WASHINGTON	ENDANGERED	SAXIFRAGACEAE	PARNASSIA KOTZEBUEI VAR. PUMILA
WASHINGTON	ENDANGERED	SCROPHULARIACEAE	PENSTEMON BARRETTIAE
WASHINGTON	EXTINCT	FABACEAE	ASTRAGALUS COLUMBIANUS
WASHINGTON	EXTINCT	HYDROPHYLLACEAE	PHACELIA LENTA
WASHINGTON	THREATENED	APIACEAE	ERYNGIUM PETIOLATUM
WASHINGTON	THREATENED	APIACEAE	LOMATIUM CUSPIDATUM
WASHINGTON	THREATENED	APIACEAE	LOMATIUM SERPENTINUM
WASHINGTON	THREATENED	APIACEAE	LOMATIUM THOMPSONII
WASHINGTON	THREATENED	APIACEAE	TAUSCHIA STRICKLANDII
WASHINGTON	THREATENED	ASTERACEAE	ARNICA AMPLEXICAULIS VAR. PIPERI

STATE LISTS OF ENDANGERED, EXTINCT AND THREATENED SPECIES IN THE CONTINENTAL UNITED STATES

STATE	STATUS	FAMILY	SPECIES
WASHINGTON	THREATENED	ASTERACEAE	ASTER CHILENSIS SSP. HALLII
WASHINGTON	THREATENED	ASTERACEAE	ASTER CURTUS
WASHINGTON	THREATENED	ASTERACEAE	ASTER GLAUCESCENS
WASHINGTON	THREATENED	ASTERACEAE	ASTER JESSICAE
WASHINGTON	THREATENED	ASTERACEAE	BALSAMORHIZA ROSEA
WASHINGTON	THREATENED	ASTERACEAE	CHAENACTIS RAMOSA
WASHINGTON	THREATENED	ASTERACEAE	CHAENACTIS THOMPSONII
WASHINGTON	THREATENED	ASTERACEAE	ERIGERON FLETTII
WASHINGTON	THREATENED	ASTERACEAE	ERIGERON LEIBERGII
WASHINGTON	THREATENED	ASTERACEAE	ERIGERON PIPERIANUS
WASHINGTON	THREATENED	ASTERACEAE	HAPLOPAPPUS HALLII
WASHINGTON	THREATENED	ASTERACEAE	HIERACIUM LONGIBERBE
WASHINGTON	THREATENED	ASTERACEAE	LASTHENIA MINOR SSP. MARITIMA
WASHINGTON	THREATENED	ASTERACEAE	PYRROCOMA LIATRIFORMIS
WASHINGTON	THREATENED	ASTERACEAE	SENECIO NEOWEBSTERI
WASHINGTON	THREATENED	BORAGINACEAE	CRYPTANTHA THOMPSONII
WASHINGTON	THREATENED	BORAGINACEAE	HACKELIA HISPIDA
WASHINGTON	THREATENED	BRASSICACEAE	RORIPPA COLUMBIAE
WASHINGTON	THREATENED	CAMPANULACEAE	CAMPANULA PIPERI
WASHINGTON	THREATENED	CARYOPHYLLACEAE	SILENE SEELYI
WASHINGTON	THREATENED	CYPERACEAE	CAREX INTERRUPTA
WASHINGTON	THREATENED	FABACEAE	ASTRAGALUS COTTONII
WASHINGTON	THREATENED	FABACEAE	ASTRAGALUS MULFORDAE
WASHINGTON	THREATENED	LILIACEAE	ALLIUM ROBINSONII
WASHINGTON	THREATENED	LILIACEAE	CALOCHORTUS LONGEBARBATUS VAR. LONGEBARBATUS
WASHINGTON	THREATENED	LILIACEAE	CALOCHORTUS NITIDUS
WASHINGTON	THREATENED	LILIACEAE	ERYTHRONIUM OREGONUM
WASHINGTON	THREATENED	MALVACEAE	SIDALCEA OREGANA VAR. CALVA
WASHINGTON	THREATENED	ORCHIDACEAE	PLATANTHERA UNALASCENSIS SSP. MARITIMA
WASHINGTON	THREATENED	POACEAE	CALAMAGROSTIS CRASSIGLUMIS
WASHINGTON	THREATENED	POACEAE	CALAMAGROSTIS TWEEDYI
WASHINGTON	THREATENED	POACEAE	POA CURTIFOLIA
WASHINGTON	THREATENED	POLEMONIACEAE	POLEMONIUM PECTINATUM
WASHINGTON	THREATENED	POLYGONACEAE	ERIOGONUM UMBELLATUM VAR. HYPOLEIUM
WASHINGTON	THREATENED	POLYPODIACEAE	POLYSTICHUM KRUCKEBERGII
WASHINGTON	THREATENED	PORTULACACEAE	CLAYTONIA LANCEOLATA VAR. CHRYSANTHA
WASHINGTON	THREATENED	PORTULACACEAE	CLAYTONIA MEGARHIZA VAR. NIVALIS
WASHINGTON	THREATENED	PORTULACACEAE	LEWISIA TWEEDYI
WASHINGTON	THREATENED	PORTULACACEAE	TALINUM OKANOGANENSE
WASHINGTON	THREATENED	PRIMULACEAE	DODECATHEON POETICUM
WASHINGTON	THREATENED	PRIMULACEAE	DOUGLASIA LAEVIGATA VAR. LAEVIGATA
WASHINGTON	THREATENED	RANUNCULACEAE	CIMICIFUGA LACINIATA
WASHINGTON	THREATENED	RANUNCULACEAE	CLEMATIS OCCIDENTALIS VAR. DISSECTA

STATE LISTS OF ENDANGERED, EXTINCT AND THREATENED SPECIES IN THE CONTINENTAL UNITED STATES

STATE	STATUS	FAMILY	SPECIES
WASHINGTON	THREATENED	RANUNCULACEAE	DELPHINIUM MULTIPLEX
WASHINGTON	THREATENED	RANUNCULACEAE	DELPHINIUM NUTTALLIANUM VAR. LINEAPETALUM
WASHINGTON	THREATENED	RANUNCULACEAE	DELPHINIUM VIRIDESCENS
WASHINGTON	THREATENED	RANUNCULACEAE	DELPHINIUM XANTHOLEUCUM
WASHINGTON	THREATENED	RANUNCULACEAE	RANUNCULUS RECONDITUS
WASHINGTON	THREATENED	ROSACEAE	PETROPHYTUM HENDERSONII
WASHINGTON	THREATENED	SALICACEAE	SALIX FLUVIATILIS
WASHINGTON	THREATENED	SAXIFRAGACEAE	SULLIVANTIA OREGANA
WASHINGTON	THREATENED	SCROPHULARIACEAE	CASTILLEJA CRYPTANTHA
WASHINGTON	THREATENED	SCROPHULARIACEAE	CASTILLEJA PARVIFLORA VAR. OLYMPICA
WASHINGTON	THREATENED	SCROPHULARIACEAE	MIMULUS JUNGERMANNIOIDES
WASHINGTON	THREATENED	SCROPHULARIACEAE	PEDICULARIS RAINIERENSIS
WASHINGTON	THREATENED	SCROPHULARIACEAE	PENSTEMON WASHINGTONENSIS
WASHINGTON	THREATENED	SCROPHULARIACEAE	SYNTHYRIS PINNATIFIDA VAR. LANUGINOSA
WASHINGTON	THREATENED	SCROPHULARIACEAE	SYNTHYRIS SCHIZANTHA
WASHINGTON	THREATENED	VALERIANACEAE	VALERIANA COLUMBIANA
WASHINGTON	THREATENED	VIOLACEAE	VIOLA ADUNCA VAR. CASCADENSIS
WASHINGTON	THREATENED	VIOLACEAE	VIOLA FLETTII
WEST VIRGINIA	THREATENED	APIACEAE	PTILIMNIUM FLUVIATILE
WEST VIRGINIA	THREATENED	ARALIACEAE	PANAX QUINQUEFOLIUS
WEST VIRGINIA	THREATENED	CYPERACEAE	CYMOPHYLLUS FRASERI
WEST VIRGINIA	THREATENED	LAMIACEAE	SCUTELLARIA OVATA SSP. PSEUDOARGUTA
WEST VIRGINIA	THREATENED	LAMIACEAE	SYNANDRA HISPIDULA
WEST VIRGINIA	THREATENED	ORCHIDACEAE	PLATANTHERA FLAVA
WEST VIRGINIA	THREATENED	ORCHIDACEAE	PLATANTHERA PERAMOENA
WEST VIRGINIA	THREATENED	POACEAE	CALAMAGROSTIS PORTERI
WEST VIRGINIA	THREATENED	RANUNCULACEAE	HYDRASTIS CANADENSIS
WEST VIRGINIA	THREATENED	ROSACEAE	PRUNUS ALLEGHANIENSIS
WEST VIRGINIA	THREATENED	SAXIFRAGACEAE	HEUCHERA HISPIDA
WISCONSIN	ENDANGERED	ASTERACEAE	GNAPHALIUM OBTUSIFOLIUM VAR. SAXICOLA
WISCONSIN	ENDANGERED	FABACEAE	LESPEDEZA LEPTOSTACHYA
WISCONSIN	ENDANGERED	RANUNCULACEAE	ACONITUM NOVEBORACENSE
WISCONSIN	THREATENED	ARALIACEAE	PANAX QUINQUEFOLIUS
WISCONSIN	THREATENED	ASTERACEAE	CIRSIUM PITCHERI
WISCONSIN	THREATENED	FABACEAE	OXYTROPIS CAMPESTRIS VAR. CHARTACEA
WISCONSIN	THREATENED	IRIDACEAE	IRIS LACUSTRIS
WISCONSIN	THREATENED	ORCHIDACEAE	CYPRIPEDIUM ARIETINUM
WISCONSIN	THREATENED	ORCHIDACEAE	CYPRIPEDIUM CANDIDUM
WISCONSIN	THREATENED	ORCHIDACEAE	LISTERA AURICULATA
WISCONSIN	THREATENED	ORCHIDACEAE	PLATANTHERA FLAVA
WISCONSIN	THREATENED	ORCHIDACEAE	PLATANTHERA LEUCOPHAEA
WISCONSIN	THREATENED	POACEAE	POA PALUDIGENA
WISCONSIN	THREATENED	RANUNCULACEAE	HYDRASTIS CANADENSIS

STATE LISTS OF ENDANGERED, EXTINCT AND THREATENED SPECIES IN THE CONTINENTAL UNITED STATES

STATE	STATUS	FAMILY	SPECIES
WISCONSIN	THREATENED	SAXIFRAGACEAE	SULLIVANTIA RENIFOLIA
WYOMING	ENDANGERED	ASTERACEAE	ANTENNARIA ARCUATA
WYOMING	ENDANGERED	BRASSICACEAE	LESQUERELLA FREMONTII
WYOMING	ENDANGERED	FABACEAE	ASTRAGALUS PROIMANTHUS
WYOMING	EXTINCT	BRASSICACEAE	ARABIS FRUCTICOSA
WYOMING	EXTINCT	BRASSICACEAE	LESQUERELLA MACROCARPA
WYOMING	THREATENED	ASTERACEAE	ARTEMISIA PORTERI
WYOMING	THREATENED	ASTERACEAE	ERIGERON ALLOCOTUS
WYOMING	THREATENED	ASTERACEAE	PARTHENIUM ALPINUM
WYOMING	THREATENED	ASTERACEAE	TANACETUM SIMPLEX
WYOMING	THREATENED	BRASSICACEAE	ARABIS DEMISSA VAR. LANGUIDA
WYOMING	THREATENED	BRASSICACEAE	ARABIS DEMISSA VAR. RUSSEOLA
WYOMING	THREATENED	BRASSICACEAE	DRABA NIVALIS VAR. BREVICULA
WYOMING	THREATENED	BRASSICACEAE	DRABA PECTINIPILA
WYOMING	THREATENED	BRASSICACEAE	PHYSARIA CONDENSATA
WYOMING	THREATENED	BRASSICACEAE	STANLEYA PINNATA VAR. GIBBEROSA
WYOMING	THREATENED	CYPERACEAE	CAREX ARAPAHOENSIS
WYOMING	THREATENED	FABACEAE	ASTRAGALUS DRABELLIFORMIS
WYOMING	THREATENED	FABACEAE	ASTRAGALUS PAYSONII
WYOMING	THREATENED	FABACEAE	OXYTROPIS OBNAPIFORMIS
WYOMING	THREATENED	POACEAE	AGROSTIS ROSSIAE
WYOMING	THREATENED	POLYGONACEAE	ERIOGONUM LAGOPUS
WYOMING	THREATENED	RANUNCULACEAE	AQUILEGIA JONESII
WYOMING	THREATENED	RANUNCULACEAE	AQUILEGIA LARAMIENSIS
WYOMING	THREATENED	SAXIFRAGACEAE	SULLIVANTIA HAPEMANII
WYOMING	THREATENED	SCROPHULARIACEAE	PENSTEMON ACAULIS
WYOMING	THREATENED	SCROPHULARIACEAE	PENSTEMON CARYI

Endangered, Threatened, and Recently Extinct Species of Hawaii

Island floras are particularly vulnerable to threats from man. The Hawaiian Islands flora, greatly altered by human mistreatment, is one of the most fragile in the world. This depredation has resulted in the phenomenon of 50 percent of the entire native flora being placed in the categories of "recently extinct," "endangered," or "seriously threatened" species.

The total indigenous flora of the Hawaiian Islands includes over 2200 species, subspecies, and varieties. Of these, 1113 taxa are listed. The number of endangered plants is 646 or 29.4 percent of the flora; the number of threatened plants is 197 or 8.9 percent. It is certain that, as more data on the flora becomes available, the number of plants having endangered or threatened status will become much larger.

The number of extinct species is very high: 270 taxa, or 12.3 percent, are believed to be extinct. It may be possible, at a future date, to find individuals of a species that has been listed as extinct or to find that the species, while extinct in the wild, is cultivated somewhere in the world.

During preparation of the lists, information from the extensive study by F.R. Fosberg and D. Herbst, "Rare and Endangered Species of Hawaiian Vascular Plants," _Allertonia_ 1(1): 1-72 (March 1975) was utilized, as well as other recent botanical literature and data from Hawaiian botanists.

The ranges given for the various taxa were compiled primarily by reference to Harold St. John's _List_ and _Summary_ _of_ _the_ _Flowering_ _Plants_ _in_ _the_ _Hawaiian_ _Islands_ (1973), and additionally from the data cards of the Endangered Flora Project, correspondence with Hawaiian botanists, and

taxonomic literature. Islands followed by a question-mark in the printout
are cases where it is appropriate to note that the occurrence of a species
on that island has been suggested by a source other than St. John's list.
It is intended that the stated range is the original, maximum-known range
of a plant; in actuality it may not represent the present-day range,
because of extermination within various islands.

In the status column, END indicates endangered; THR, threatened; and
EXT, extinct.

The names of islands on which each species has been found are ab-
breviated in the list as follows:

MAU	Maui
HAW	Hawaii
OAH	Oahu
KAU	Kauai
MOL	Molokai
LAN	Lanai
NII	Niihau
KAH	Kahoolawe
KUR	Kure
LAY	Laysan
LEEWARD	Lehua, Kaula, and islands west of Kaula
LIS	Lisianski
MID	Midway
NEC	Necker
NIH	Nihoa
PRL-HRM	Pearl and Hermes Atoll

ENDANGERED, THREATENED, AND RECENTLY EXTINCT SPECIES OF HAWAII

FAMILY	SPECIES	STATUS	MAU	HAW	OAH	KAU	MOL	LAN	NIH
AMARANTHACEAE	ACHYRANTHES MUTICA	END				KAU			
AMARANTHACEAE	ACHYRANTHES NELSONII	EXT		HAW					
AMARANTHACEAE	ACHYRANTHES SPLENDENS VAR. REFLEXA	END					MOL		
AMARANTHACEAE	ACHYRANTHES SPLENDENS VAR. ROTUNDATA	END			OAH				
AMARANTHACEAE	ACHYRANTHES SPLENDENS VAR. SPLENDENS	END	MAU					LAN	
AMARANTHACEAE	AERVA SERICEA	EXT							
AMARANTHACEAE	AMARANTHUS BROWNII	END							NIH
AMARANTHACEAE	CHARPENTIERA DENSIFLORA	END				KAU			
AMARANTHACEAE	NOTOTRICHIUM HUMILE VAR. HUMILE	END			OAH				
AMARANTHACEAE	NOTOTRICHIUM HUMILE VAR. PARVIFOLIUM	END			OAH				
AMARANTHACEAE	NOTOTRICHIUM HUMILE VAR. SUBRHOMBOIDEUM	END			OAH				
AMARANTHACEAE	NOTOTRICHIUM SANDWICENSE VAR. DECIPIENS	END				KAU			
AMARANTHACEAE	NOTOTRICHIUM SANDWICENSE VAR. DUBIUM	EXT	MAU						
AMARANTHACEAE	NOTOTRICHIUM SANDWICENSE VAR. FORBESII	EXT					MOL		
AMARANTHACEAE	NOTOTRICHIUM SANDWICENSE VAR. HELLERI	END				KAU			
AMARANTHACEAE	NOTOTRICHIUM SANDWICENSE VAR. KOLEKOLENSE	END					MOL		
AMARANTHACEAE	NOTOTRICHIUM SANDWICENSE VAR. LANAIENSE	EXT	MAU					LAN	
AMARANTHACEAE	NOTOTRICHIUM SANDWICENSE VAR. LANCEOLATUM	END		HAW					
AMARANTHACEAE	NOTOTRICHIUM SANDWICENSE VAR. LATIFOLIUM	EXT						LAN	
AMARANTHACEAE	NOTOTRICHIUM SANDWICENSE VAR. LEPTOPODUM	END	MAU						
AMARANTHACEAE	NOTOTRICHIUM SANDWICENSE VAR. LONGISPICATUM	END	MAU				MOL		
AMARANTHACEAE	NOTOTRICHIUM SANDWICENSE VAR. MACROPHYLLUM	END		HAW					
AMARANTHACEAE	NOTOTRICHIUM SANDWICENSE VAR. MAUIENSE	END	MAU						
AMARANTHACEAE	NOTOTRICHIUM SANDWICENSE VAR. OLOKELEANUM	END				KAU			
AMARANTHACEAE	NOTOTRICHIUM SANDWICENSE VAR. PULCHELLOIDES	EXT					MOL		
AMARANTHACEAE	NOTOTRICHIUM SANDWICENSE VAR. PULCHELLUM	EXT					MOL		
AMARANTHACEAE	NOTOTRICHIUM SANDWICENSE VAR. SUBCORDATUM	EXT	MAU	HAW			MOL		
AMARANTHACEAE	NOTOTRICHIUM SANDWICENSE VAR. SYRINGIFOLIUM	EXT			OAH?				
AMARANTHACEAE	NOTOTRICHIUM VIRIDE VAR. OBLONGIFOLIUM	END				KAU			
AMARANTHACEAE	NOTOTRICHIUM VIRIDE VAR. SUBTRUNCATUM	END			OAH				
AMARANTHACEAE	NOTOTRICHIUM VIRIDE VAR. VIRIDE	END				KAU			
APIACEAE	PEUCEDANUM KAUAIENSE	END				KAU			
APIACEAE	PEUCEDANUM SANDWICENSE VAR. SANDWICENSE	END					MOL		
APIACEAE	SANICULA PURPUREA	END	MAU		OAH				
APOCYNACEAE	OCHROSIA COMPTA	END							
APOCYNACEAE	PTERALYXIA CAUMIANA	EXT			OAH				
APOCYNACEAE	PTERALYXIA KAUAIENSIS	END				KAU			
APOCYNACEAE	RAUVOLFIA HELLERI	END				KAU			
APOCYNACEAE	RAUVOLFIA MAUIENSIS	END	MAU						
APOCYNACEAE	RAUVOLFIA MOLOKAIENSIS VAR. PARVIFOLIA	EXT	MAU				MOL		
APOCYNACEAE	RAUVOLFIA REMOTIFLORA	END		HAW					
APOCYNACEAE	RAUVOLFIA SANDWICENSIS VAR. SANDWICENSIS	THR	MAU		OAH	KAU			
APOCYNACEAE	RAUVOLFIA SANDWICENSIS VAR. SUBACUMINATA	THR				KAU			

ENDANGERED, THREATENED, AND RECENTLY EXTINCT SPECIES OF HAWAII

FAMILY	SPECIES	STATUS	RANGE				
ARALIACEAE	CHEIRODENDRON HELLERI VAR. HELLERI	END			KAU		
ARALIACEAE	CHEIRODENDRON HELLERI VAR. MICROCARPUM	END			KAU		
ARALIACEAE	CHEIRODENDRON HELLERI VAR. SODALIUM	END			KAU		
ARALIACEAE	CHEIRODENDRON TRIGYNUM VAR. ROCKII	END				LAN	
ARALIACEAE	CHEIRODENDRON TRIGYNUM VAR. SUBCORDATUM	END	HAW				
ARALIACEAE	MUNROIDENDRON RACEMOSUM VAR. MACDANIELSII	EXT			KAU		
ARALIACEAE	MUNROIDENDRON RACEMOSUM VAR. RACEMOSUM	END			KAU		
ARALIACEAE	REYNOLDSIA DEGENERI	END				MOL	
ARALIACEAE	REYNOLDSIA HILLEBRANDII	END	HAW				
ARALIACEAE	REYNOLDSIA HUEHUENSIS VAR. BREVIPES	END	HAW				
ARALIACEAE	REYNOLDSIA HUEHUENSIS VAR. HUEHUENSIS	END	HAW				
ARALIACEAE	REYNOLDSIA HUEHUENSIS VAR. INTERMEDIA	END	HAW				
ARALIACEAE	REYNOLDSIA MAUIENSIS VAR. MACROCARPA	END	MAU				
ARALIACEAE	REYNOLDSIA MAUIENSIS VAR. MAUIENSIS	END	MAU				
ARALIACEAE	REYNOLDSIA SANDWICENSIS VAR. INTERCEDENS	END		OAH			
ARALIACEAE	REYNOLDSIA SANDWICENSIS VAR. MOLOKAIENSIS	EXT			MOL		
ARALIACEAE	REYNOLDSIA VENUSTA VAR. LANAIENSIS	END				LAN	
ARALIACEAE	REYNOLDSIA VENUSTA VAR. VENUSTA	END		OAH			
ARALIACEAE	TETRAPLASANDRA BISATTENUATA	END			KAU		
ARALIACEAE	TETRAPLASANDRA GYMNOCARPA VAR. PUPUKEENSIS	END		OAH			
ARALIACEAE	TETRAPLASANDRA HAWAIIENSIS VAR. HAWAIIENSIS	THR	HAW				
ARALIACEAE	TETRAPLASANDRA HAWAIIENSIS VAR. MICROCARPA	END				MOL	
ARALIACEAE	TETRAPLASANDRA KAALAE VAR. MULTIPLEX	END		OAH			
ARALIACEAE	TETRAPLASANDRA KAHANANA	END		OAH			
ARALIACEAE	TETRAPLASANDRA KAVAIENSIS VAR. DIPYRENA	END	MAU?HAW?			LAN	
ARALIACEAE	TETRAPLASANDRA KAVAIENSIS VAR. GRANDIS	END	HAW				
ARALIACEAE	TETRAPLASANDRA KAVAIENSIS VAR. INTERCEDENS	END	MAU				
ARALIACEAE	TETRAPLASANDRA KAVAIENSIS VAR. KOLOANA	THR			KAU		
ARALIACEAE	TETRAPLASANDRA KAVAIENSIS VAR. NAHIKUENSIS	END	MAU				
ARALIACEAE	TETRAPLASANDRA KAVAIENSIS VAR. OCCIDUA	END	MAU				
ARALIACEAE	TETRAPLASANDRA KOHALAE	END	HAW				
ARALIACEAE	TETRAPLASANDRA LANAIENSIS	EXT				LAN	
ARALIACEAE	TETRAPLASANDRA LIHUENSIS VAR. GRACILIPES	THR			KAU		
ARALIACEAE	TETRAPLASANDRA LYDGATEI VAR. BRACHYPODA	EXT		OAH			
ARALIACEAE	TETRAPLASANDRA LYDGATEI VAR. CORIACEA	EXT		OAH			
ARALIACEAE	TETRAPLASANDRA LYDGATEI VAR. FORBESII	EXT		OAH			
ARALIACEAE	TETRAPLASANDRA LYDGATEI VAR. LEPTORHACHIS	END		OAH			
ARALIACEAE	TETRAPLASANDRA LYDGATEI VAR. LYDGATEI	EXT		OAH			
ARALIACEAE	TETRAPLASANDRA MEIANDRA VAR. BISOBTUSA	THR	HAW				
ARALIACEAE	TETRAPLASANDRA MEIANDRA VAR. BRYANII	END		OAH			
ARALIACEAE	TETRAPLASANDRA MEIANDRA VAR. DEGENERI	END			KAU		
ARALIACEAE	TETRAPLASANDRA MEIANDRA VAR. HILLEBRANDII	END				LAN	
ARALIACEAE	TETRAPLASANDRA MEIANDRA VAR. HILOENSIS	THR	HAW				

ENDANGERED, THREATENED, AND RECENTLY EXTINCT SPECIES OF HAWAII

FAMILY	SPECIES	STATUS	MAU	HAW	OAH	KAU	MOL	LAN	NII	LAY NIH
ARALIACEAE	TETRAPLASANDRA MEIANDRA VAR. LEPTOMERA	END	MAU							
ARALIACEAE	TETRAPLASANDRA MEIANDRA VAR. MAKALEHANA	END			OAH					
ARALIACEAE	TETRAPLASANDRA MEIANDRA VAR. RHYNCHOCARPOIDES	THR		HAW						
ARALIACEAE	TETRAPLASANDRA MEIANDRA VAR. SIMULANS	THR		HAW						
ARALIACEAE	TETRAPLASANDRA MUNROI	EXT						LAN		
ARALIACEAE	TETRAPLASANDRA OAHUENSIS VAR. ERADIATA	EXT			OAH					
ARALIACEAE	TETRAPLASANDRA OAHUENSIS VAR. FAURIEI	THR			OAH					
ARALIACEAE	TETRAPLASANDRA OAHUENSIS VAR. HAILIENSIS	THR			OAH					
ARALIACEAE	TETRAPLASANDRA OAHUENSIS VAR. LONGIPES	THR			OAH					
ARALIACEAE	TETRAPLASANDRA OAHUENSIS VAR. PSEUDORHACHIS	THR			OAH					
ARALIACEAE	TETRAPLASANDRA PUPUKEENSIS VAR. NITIDA	THR			OAH					
ARALIACEAE	TETRAPLASANDRA PUPUKEENSIS VAR. PUPUKEENSIS	END			OAH					
ARALIACEAE	TETRAPLASANDRA PUPUKEENSIS VAR. VENOSA	THR			OAH					
ARALIACEAE	TETRAPLASANDRA TURBANS	THR			OAH					
ARALIACEAE	TETRAPLASANDRA WAIALEALAE VAR. URCEOLATA	THR				KAU				
ARALIACEAE	TETRAPLASANDRA WAIANENSIS VAR. PALEHUANA	END			OAH					
ARALIACEAE	TETRAPLASANDRA WAIANENSIS VAR. WAIANENSIS	THR			OAH					
ARALIACEAE	TETRAPLASANDRA WAIMEAE VAR. ANGUSTIOR	THR				KAU				
ARECACEAE	PRITCHARDIA AYLMER-ROBINSONII	END							NII	
ARECACEAE	PRITCHARDIA ELLIPTICA	END						LAN		
ARECACEAE	PRITCHARDIA ERIOPHORA	END				KAU				
ARECACEAE	PRITCHARDIA GAUDICHAUDII	END					MOL			
ARECACEAE	PRITCHARDIA HILLEBRANDII	END					MOL			
ARECACEAE	PRITCHARDIA KAALAE VAR. KAALAE	END			OAH					
ARECACEAE	PRITCHARDIA KAALAE VAR. MINIMA	END			OAH					
ARECACEAE	PRITCHARDIA KAHANAE	END			OAH					
ARECACEAE	PRITCHARDIA LANAIENSIS	END						LAN		
ARECACEAE	PRITCHARDIA MUNROII	END					MOL			
ARECACEAE	PRITCHARDIA REMOTA	THR								LAY NIH
ASTERACEAE	ARGYROXIPHIUM KAUENSE	END		HAW						
ASTERACEAE	ARGYROXIPHIUM MACROCEPHALUM	END	MAU	HAW						
ASTERACEAE	ARGYROXIPHIUM VIRESCENS VAR. VIRESCENS	EXT	MAU							
ASTERACEAE	ARTEMISIA SP. (FROM LANAI)	END						LAN		
ASTERACEAE	ASTER SANDWICENSIS	EXT	MAU		OAH	KAU	MOL			
ASTERACEAE	BIDENS ASPLENIOIDES	EXT							NII	
ASTERACEAE	BIDENS CAMPYLOTHECA	THR		HAW	OAH			LAN		
ASTERACEAE	BIDENS CERVICATA	END				KAU				
ASTERACEAE	BIDENS COARTATA	END			OAH					
ASTERACEAE	BIDENS CONJUNCTATA	END	MAU		OAH	KAU				
ASTERACEAE	BIDENS CUNEATA	END			OAH					
ASTERACEAE	BIDENS DEGENERI VAR. APIOIDES	END	MAU		OAH					
ASTERACEAE	BIDENS DEGENERI VAR. DEGENERI	END	MAU		OAH		MOL			
ASTERACEAE	BIDENS DISTANS	EXT						LAN		

ENDANGERED, THREATENED, AND RECENTLY EXTINCT SPECIES OF HAWAII

FAMILY	SPECIES	STATUS	RANGE
ASTERACEAE	BIDENS FORBESII	END	KAU
ASTERACEAE	BIDENS GRACILOIDES	END	OAH
ASTERACEAE	BIDENS HAWAIIENSIS	EXT	HAW
ASTERACEAE	BIDENS MACROCARPA VAR. OVATIFOLIA	EXT	OAH
ASTERACEAE	BIDENS MAGNIDISCA	END	OAH
ASTERACEAE	BIDENS MAUIENSIS VAR. CUNEATOIDES	END	MAU
ASTERACEAE	BIDENS MAUIENSIS VAR. FORBESIANA	END	LAN
ASTERACEAE	BIDENS MAUIENSIS VAR. LANAIENSIS	END	MAU LAN
ASTERACEAE	BIDENS MAUIENSIS VAR. MEDIA	END	LAN
ASTERACEAE	BIDENS MENZIESII VAR. LEPTODONTA	END	MAU HAW
ASTERACEAE	BIDENS MICRANTHA VAR. CADUCA	EXT	MOL
ASTERACEAE	BIDENS MICRANTHA VAR. KAALANA	END	OAH
ASTERACEAE	BIDENS MICRANTHA VAR. LACINIATA	THR	MAU HAW OAH?
ASTERACEAE	BIDENS NAPALIENSIS	END	KAU
ASTERACEAE	BIDENS NEMATOCERA	EXT	MAU? MOL
ASTERACEAE	BIDENS OBTUSILOBA	END	OAH
ASTERACEAE	BIDENS POPULIFOLIA	END	OAH
ASTERACEAE	BIDENS PULCHELLA	EXT	OAH
ASTERACEAE	BIDENS SALICOIDES	END	MOL
ASTERACEAE	BIDENS SANDVICENSIS VAR. SETOSA	END	HAW? KAU
ASTERACEAE	BIDENS SKOTTSBERGII VAR. CONGLUTINATA	END	HAW
ASTERACEAE	BIDENS SKOTTSBERGII VAR. SKOTTSBERGII	END	HAW
ASTERACEAE	BIDENS STOKESII	EXT	NII
ASTERACEAE	BIDENS VALIDA	EXT	KAU
ASTERACEAE	BIDENS WAIMEANA	EXT	KAU
ASTERACEAE	BIDENS WIEBKEI	END	MOL
ASTERACEAE	DUBAUTIA ARBOREA	END	HAW
ASTERACEAE	DUBAUTIA HILLEBRANDII	END	HAW
ASTERACEAE	DUBAUTIA KNUDSENII VAR. DEGENERI	THR	MOL
ASTERACEAE	DUBAUTIA KNUDSENII VAR. KNUDSENII	END	KAU
ASTERACEAE	DUBAUTIA LAEVIGATA VAR. PARVIFOLIA	THR	KAU
ASTERACEAE	DUBAUTIA LATIFOLIA VAR. LATIFOLIA	END	KAU
ASTERACEAE	DUBAUTIA LAXA VAR. BLAKEI	THR	MAU
ASTERACEAE	DUBAUTIA LAXA VAR. WAIANENSIS	END	OAH
ASTERACEAE	DUBAUTIA LONCHOPHYLLA	END	MAU
ASTERACEAE	DUBAUTIA MAGNIFOLIA	EXT	KAU
ASTERACEAE	DUBAUTIA MICROCEPHALA VAR. FORBESII	THR	KAU
ASTERACEAE	DUBAUTIA MICROCEPHALA VAR. MICROCEPHALA	END	KAU
ASTERACEAE	DUBAUTIA MOLOKAIENSIS	END	HAW MOL LAN
ASTERACEAE	DUBAUTIA MONTANA VAR. LONGIFOLIA	END	MAU
ASTERACEAE	DUBAUTIA MONTANA VAR. ROBUSTIOR	END	MAU
ASTERACEAE	DUBAUTIA PLANTAGINEA VAR. ACRIDENTATA	END	MOL
ASTERACEAE	DUBAUTIA PLANTAGINEA VAR. PLANTAGINEA	END	OAH LAN

ENDANGERED, THREATENED, AND RECENTLY EXTINCT SPECIES OF HAWAII

FAMILY	SPECIES	STATUS	RANGE
ASTERACEAE	DUBAUTIA PLATYPHYLLA VAR. LEPTOPHYLLA	END	MAU
ASTERACEAE	DUBAUTIA RETICULATA	END	MAU
ASTERACEAE	DUBAUTIA ROCKII	EXT	MAU
ASTERACEAE	DUBAUTIA SHERFFIANA	END	OAH
ASTERACEAE	DUBAUTIA STRUTHIOLOIDES	EXT	HAW
ASTERACEAE	DUBAUTIA TERNIFOLIA	END	MAU
ASTERACEAE	DUBAUTIA THYRSIFLORA VAR. CERNUA	THR	MAU
ASTERACEAE	DUBAUTIA THYRSIFLORA VAR. THYRSIFLORA	END	MAU
ASTERACEAE	DUBAUTIA WAIALEALAE VAR. MEGAPHYLLA	THR	KAU
ASTERACEAE	GNAPHALIUM SANDWICENSIUM VAR. FLAGELLARE	END	MOL
ASTERACEAE	GNAPHALIUM SANDWICENSIUM VAR. MOLOKAIENSE	END	MOL
ASTERACEAE	HESPEROMANNIA ARBORESCENS SSP. ARBORESCENS	EXT	LAN
ASTERACEAE	HESPEROMANNIA ARBORESCENS SSP. BUSHIANA	END	OAH
ASTERACEAE	HESPEROMANNIA ARBORESCENS SSP. SWEZEYI	END	OAH
ASTERACEAE	HESPEROMANNIA ARBUSCULA SSP. ARBUSCULA	EXT	MAU
ASTERACEAE	HESPEROMANNIA ARBUSCULA SSP. OAHUENSIS	END	OAH
ASTERACEAE	HESPEROMANNIA LYDGATEI	END	KAU
ASTERACEAE	LIPOCHAETA ALATA VAR. ALATA	END	KAU
ASTERACEAE	LIPOCHAETA BRYANII	EXT	KAH
ASTERACEAE	LIPOCHAETA DEGENERI	END	MOL
ASTERACEAE	LIPOCHAETA DELTOIDEA	THR	KAU
ASTERACEAE	LIPOCHAETA DUBIA	THR	OAH
ASTERACEAE	LIPOCHAETA EXIGUA	END	KAU
ASTERACEAE	LIPOCHAETA FAURIEI	END	KAU
ASTERACEAE	LIPOCHAETA FLEXUOSA	EXT	MOL
ASTERACEAE	LIPOCHAETA FORBESII VAR. FORBESII	END	MAU
ASTERACEAE	LIPOCHAETA HETEROPHYLLA VAR. HETEROPHYLLA	THR	MAU MOL LAN
ASTERACEAE	LIPOCHAETA HETEROPHYLLA VAR. MALVACEA	THR	MOL
ASTERACEAE	LIPOCHAETA HETEROPHYLLA VAR. MOLOKAIENSIS	THR	MOL
ASTERACEAE	LIPOCHAETA INTEGRIFOLIA VAR. ARGENTEA	THR	MAU MOL?
ASTERACEAE	LIPOCHAETA INTEGRIFOLIA VAR. GRACILIS	EXT	
ASTERACEAE	LIPOCHAETA INTEGRIFOLIA VAR. MAJOR	END	OAH
ASTERACEAE	LIPOCHAETA INTEGRIFOLIA VAR. MEGACEPHALA	END	OAH
ASTERACEAE	LIPOCHAETA INTERMEDIA	THR	HAW
ASTERACEAE	LIPOCHAETA KAHOOLAWENSIS	EXT	KAH
ASTERACEAE	LIPOCHAETA LAVARUM VAR. CONFERTA	THR	LAN
ASTERACEAE	LIPOCHAETA LAVARUM VAR. HILLEBRANDIANA	THR	MAU HAW
ASTERACEAE	LIPOCHAETA LAVARUM VAR. LONGIFOLIA	END	LAN
ASTERACEAE	LIPOCHAETA LAVARUM VAR. MANELEANA	THR	LAN
ASTERACEAE	LIPOCHAETA LAVARUM VAR. OVATA	THR	MAU
ASTERACEAE	LIPOCHAETA LAVARUM VAR. SALICIFOLIA	EXT	MAU
ASTERACEAE	LIPOCHAETA LAVARUM VAR. SKOTTSBERGII	EXT	MAU
ASTERACEAE	LIPOCHAETA LAVARUM VAR. STEARNSII	THR	LAN

ENDANGERED, THREATENED, AND RECENTLY EXTINCT SPECIES OF HAWAII

FAMILY	SPECIES	STATUS	RANGE
ASTERACEAE	LIPOCHAETA LOBATA VAR. ALBESCENS	EXT	OAH
ASTERACEAE	LIPOCHAETA LOBATA VAR. APREVALLIANA	EXT	OAH
ASTERACEAE	LIPOCHAETA LOBATA VAR. GROSSEDENTATA	THR	OAH
ASTERACEAE	LIPOCHAETA LOBATA VAR. HASTULATA	END	OAH NII?
ASTERACEAE	LIPOCHAETA LOBATA VAR. HASTULATOIDES	THR	MAU
ASTERACEAE	LIPOCHAETA LOBATA VAR. LEPTOPHYLLA	END	OAH
ASTERACEAE	LIPOCHAETA LOBATA VAR. LOBATA	END	OAH
ASTERACEAE	LIPOCHAETA LOBATA VAR. MAKENENSIS	THR	MAU
ASTERACEAE	LIPOCHAETA LOBATA VAR. MAUNALOENSIS	EXT	MOL NII
ASTERACEAE	LIPOCHAETA MICRANTHA	THR	KAU
ASTERACEAE	LIPOCHAETA MINUSCULA	THR	OAH
ASTERACEAE	LIPOCHAETA PERDITA	EXT	HAW
ASTERACEAE	LIPOCHAETA PROFUSA VAR. PROFUSA	EXT	HAW? KAU? NII?
ASTERACEAE	LIPOCHAETA PROFUSA VAR. ROBUSTIOR	END	KAU
ASTERACEAE	LIPOCHAETA REMYI	END	OAH
ASTERACEAE	LIPOCHAETA ROCKII VAR. DISSECTA	EXT	MAU
ASTERACEAE	LIPOCHAETA ROCKII VAR. ROCKII	END	MOL
ASTERACEAE	LIPOCHAETA ROCKII VAR. SUBOVATA	END	MOL
ASTERACEAE	LIPOCHAETA SCABRA	EXT	HAW
ASTERACEAE	LIPOCHAETA SUBCORDATA VAR. MEMBRANACEA	EXT	MAU
ASTERACEAE	LIPOCHAETA SUBCORDATA VAR. POPULIFOLIA	END	LAN
ASTERACEAE	LIPOCHAETA SUCCULENTA VAR. ANGUSTATA	END	KAU
ASTERACEAE	LIPOCHAETA SUCCULENTA VAR. SUCCULENTA	EXT	KAU NII
ASTERACEAE	LIPOCHAETA SUCCULENTA VAR. TRIFIDA	END	MOL
ASTERACEAE	LIPOCHAETA TENUIFOLIA	THR	OAH
ASTERACEAE	LIPOCHAETA TENUIS VAR. SELLINGII	THR	OAH
ASTERACEAE	LIPOCHAETA TENUIS VAR. TENUIS	THR	OAH
ASTERACEAE	LIPOCHAETA TRILOBATA	EXT	HAW
ASTERACEAE	LIPOCHAETA VENOSA	END	HAW
ASTERACEAE	LIPOCHAETA WAIMEANENSIS	THR	KAU
ASTERACEAE	REMYA KAUAIENSIS VAR. KAUAIENSIS	EXT	KAU
ASTERACEAE	REMYA KAUAIENSIS VAR. MAGNIFOLIA	EXT	KAU
ASTERACEAE	REMYA MAUIENSIS	END	MAU
ASTERACEAE	SENECIO SANDVICENSIS	EXT	OAH
ASTERACEAE	TETRAMOLOPIUM ARBUSCULUM	EXT	MAU
ASTERACEAE	TETRAMOLOPIUM ARENARIUM VAR. ARENARIUM	EXT	MAU?HAW
ASTERACEAE	TETRAMOLOPIUM ARENARIUM VAR. CONFERTUM	EXT	HAW
ASTERACEAE	TETRAMOLOPIUM ARENARIUM VAR. DENTATUM	EXT	MAU
ASTERACEAE	TETRAMOLOPIUM CAPILLARE	EXT	MAU
ASTERACEAE	TETRAMOLOPIUM CONSANGUINEUM VAR. CONSANGUINEUM	EXT	HAW KAU
ASTERACEAE	TETRAMOLOPIUM CONSANGUINEUM VAR. LEPTOPHYLLUM	EXT	HAW
ASTERACEAE	TETRAMOLOPIUM CONYZOIDES VAR. CONYZOIDES	EXT	MAU HAW MOL?LAN
ASTERACEAE	TETRAMOLOPIUM CONYZOIDES VAR. DENTATUM	EXT	LAN

ENDANGERED, THREATENED, AND RECENTLY EXTINCT SPECIES OF HAWAII

FAMILY	SPECIES	STATUS	RANGE
ASTERACEAE	TETRAMOLOPIUM FILIFORME	END	OAH
ASTERACEAE	TETRAMOLOPIUM HUMILE VAR. SUBLAEVE	EXT	HAW
ASTERACEAE	TETRAMOLOPIUM LEPIDOTUM VAR. LEPIDOTUM	END	OAH LAN
ASTERACEAE	TETRAMOLOPIUM LEPIDOTUM VAR. LUXURIANS	END	OAH LAN
ASTERACEAE	TETRAMOLOPIUM POLYPHYLLUM	END	OAH
ASTERACEAE	TETRAMOLOPIUM REMYI	EXT	MAU LAN?
ASTERACEAE	TETRAMOLOPIUM ROCKII	END	MOL
ASTERACEAE	TETRAMOLOPIUM TENERRIMUM	EXT	OAH
ASTERACEAE	WILKESIA HOBDYI	END	KAU
BRASSICACEAE	CARDAMINE KONAENSIS	END	MAU HAW
BRASSICACEAE	LEPIDIUM ARBUSCULA	EXT	OAH
BRASSICACEAE	LEPIDIUM BIDENTATUM VAR. O-WAIHIENSE	END	MAU HAW?OAH KAU MOL LAN
BRASSICACEAE	LEPIDIUM BIDENTATUM VAR. REMYI	EXT	HAW
BRASSICACEAE	LEPIDIUM SERRA	END	KAU
CAMPANULACEAE	BRIGHAMIA CITRINA VAR. CITRINA	END	KAU
CAMPANULACEAE	BRIGHAMIA CITRINA VAR. NAPALIENSIS	END	KAU
CAMPANULACEAE	BRIGHAMIA INSIGNIS	END	KAU?MOL?LAN?NII
CAMPANULACEAE	BRIGHAMIA REMYI	EXT	MAU
CAMPANULACEAE	BRIGHAMIA ROCKII	END	MOL
CAMPANULACEAE	CLERMONTIA DREPANOMORPHA	END	HAW
CAMPANULACEAE	CLERMONTIA HALEAKALENSIS	EXT	MAU
CAMPANULACEAE	CLERMONTIA HAWAIIENSIS VAR. HAWAIIENSIS	END	HAW MOL
CAMPANULACEAE	CLERMONTIA KONAENSIS	EXT	HAW
CAMPANULACEAE	CLERMONTIA LINDSEYANA	END	HAW
CAMPANULACEAE	CLERMONTIA LOYANA	END	HAW
CAMPANULACEAE	CLERMONTIA MUNROI	END	MAU LAN
CAMPANULACEAE	CLERMONTIA PELEANA	END	HAW
CAMPANULACEAE	CLERMONTIA PYRULARIA	EXT	HAW
CAMPANULACEAE	CYANEA ANGUSTIFOLIA VAR. LANAIENSIS	THR	MAU LAN
CAMPANULACEAE	CYANEA ANGUSTIFOLIA VAR. RACEMOSA	THR	LAN
CAMPANULACEAE	CYANEA ARBOREA	EXT	MAU
CAMPANULACEAE	CYANEA ASPLENIIFOLIA	EXT	MAU
CAMPANULACEAE	CYANEA BALDWINII	END	LAN
CAMPANULACEAE	CYANEA BRYANII	END	HAW
CAMPANULACEAE	CYANEA CARLSONII	EXT	HAW
CAMPANULACEAE	CYANEA CHOCKII	END	KAU
CAMPANULACEAE	CYANEA COMATA	EXT	MAU
CAMPANULACEAE	CYANEA GIBSONII	EXT	LAN
CAMPANULACEAE	CYANEA GIFFARDII	EXT	HAW
CAMPANULACEAE	CYANEA GRIMESIANA VAR. GRIMESIANA	END	OAH MOL
CAMPANULACEAE	CYANEA GRIMESIANA VAR. HIRSUTIFOLIA	END	OAH
CAMPANULACEAE	CYANEA GRIMESIANA VAR. LYDGATEI	EXT	MAU
CAMPANULACEAE	CYANEA GRIMESIANA VAR. MAUIENSIS	THR	MAU

ENDANGERED, THREATENED, AND RECENTLY EXTINCT SPECIES OF HAWAII

FAMILY	SPECIES	STATUS	RANGE			
CAMPANULACEAE	CYANEA GRIMESIANA VAR. MUNROI	END			LAN	
CAMPANULACEAE	CYANEA KUNTHIANA	EXT				
CAMPANULACEAE	CYANEA LEPTOSTEGIA VAR. LEPTOSTEGIA	END		KAU		
CAMPANULACEAE	CYANEA LEPTOSTEGIA VAR. VELUTINA	THR		KAU		
CAMPANULACEAE	CYANEA LINEARIFOLIA	EXT		KAU		
CAMPANULACEAE	CYANEA MARKSII	END	HAW			
CAMPANULACEAE	CYANEA MCELDOWNEYI	END	MAU			
CAMPANULACEAE	CYANEA NELSONII	EXT	HAW			
CAMPANULACEAE	CYANEA PYCNOCARPA	EXT	HAW			
CAMPANULACEAE	CYANEA REGINA	EXT		OAH		
CAMPANULACEAE	CYANEA SCABRA VAR. VARIABILIS	END	MAU			
CAMPANULACEAE	CYANEA SHIPMANII	END	HAW			
CAMPANULACEAE	CYANEA SOLANACEA	END	MAU?		MOL	
CAMPANULACEAE	CYANEA SOLENOCALYX VAR. SCLENOCALYX	END			MOL	
CAMPANULACEAE	CYANEA SUBMURICATA	THR	HAW			
CAMPANULACEAE	CYANEA SUPERBA VAR. SUPERBA	END		OAH		
CAMPANULACEAE	CYANEA SUPERBA VAR. VELUTINA	THR		OAH		
CAMPANULACEAE	CYANEA TRITOMANTHA VAR. LYDGATEI	END	HAW			
CAMPANULACEAE	CYANEA TRITOMANTHA VAR. TRITOMANTHA	END	HAW			
CAMPANULACEAE	DELISSEA FALLAX	EXT	HAW			
CAMPANULACEAE	DELISSEA LACINIATA VAR. LACINIATA	EXT		OAH		
CAMPANULACEAE	DELISSEA LACINIATA VAR. PARVIFOLIA	EXT		OAH		
CAMPANULACEAE	DELISSEA NIIHAUENSIS	EXT				NII
CAMPANULACEAE	DELISSEA PARVIFLORA	EXT	HA?OAH			
CAMPANULACEAE	DELISSEA RHYTIDOSPERMA	END		KAU		NII?
CAMPANULACEAE	DELISSEA RIVULARIS	THR		KAU		
CAMPANULACEAE	DELISSEA SINUATA VAR. LANAIENSIS	END			LAN	
CAMPANULACEAE	DELISSEA SINUATA VAR. SINUATA	END		OAH		
CAMPANULACEAE	DELISSEA SUBCORDATA VAR. OBTUSIFOLIA	END		OAH		
CAMPANULACEAE	DELISSEA SUBCORDATA VAR. SUBCORDATA	EXT	MAU?	OAH KAU?		
CAMPANULACEAE	DELISSEA UNDULATA VAR. ARGUTIDENTATA	END	HAW			
CAMPANULACEAE	DELISSEA UNDULATA VAR. UNDULATA	EXT	MAU			
CAMPANULACEAE	LOBELIA DUNBARIAE	END			MOL	
CAMPANULACEAE	LOBELIA GAUDICHAUDII VAR. KOOLAUENSIS	END		OAH		
CAMPANULACEAE	LOBELIA HILLEBRANDII VAR. MONOSTACHYA	THR		OAH		
CAMPANULACEAE	LOBELIA HYPOLEUCA VAR. ROCKII	END		OAH		
CAMPANULACEAE	LOBELIA NIIHAUENSIS VAR. FORBESII	END		KAU		
CAMPANULACEAE	LOBELIA NIIHAUENSIS VAR. MERIDIANA	END		OAH		
CAMPANULACEAE	LOBELIA NIIHAUENSIS VAR. NIIHAUENSIS	END				NII
CAMPANULACEAE	LOBELIA OAHUENSIS	END		OAH		
CAMPANULACEAE	LOBELIA REMYI	EXT		OAH		
CAMPANULACEAE	LOBELIA TORTUOSA	END		KAU		
CAMPANULACEAE	ROLLANDIA ANGUSTIFOLIA VAR. OCHREATA	THR		OAH		

ENDANGERED, THREATENED, AND RECENTLY EXTINCT SPECIES OF HAWAII

FAMILY	SPECIES	STATUS	RANGE
CAMPANULACEAE	ROLLANDIA CALYCINA VAR. CALYCINA	THR	OAH
CAMPANULACEAE	ROLLANDIA CALYCINA VAR. KAALAE	THR	OAH
CAMPANULACEAE	ROLLANDIA CRISPA VAR. CRISPA	END	OAH
CAMPANULACEAE	ROLLANDIA DEGENERIANA	THR	OAH
CAMPANULACEAE	ROLLANDIA HUMBOLDTIANA	END	OAH
CAMPANULACEAE	ROLLANDIA LANCEOLATA VAR. LANCEOLATA	THR	OAH
CAMPANULACEAE	ROLLANDIA LANCEOLATA VAR. VIRIDIFLORA	THR	OAH
CAMPANULACEAE	ROLLANDIA PARVIFOLIA	EXT	KAU
CAMPANULACEAE	ROLLANDIA PINNATIFIDA	END	OAH
CAMPANULACEAE	ROLLANDIA PURPURELLIFOLIA	END	OAH
CAMPANULACEAE	ROLLANDIA SESSILIFOLIA	END	OAH
CAMPANULACEAE	ROLLANDIA ST.-JOHNII	END	OAH
CAMPANULACEAE	ROLLANDIA WAIANAEENSIS	THR	OAH
CAMPANULACEAE	TREMATOLOBELIA WIMMERI	EXT	HAW
CAPPARIDACEAE	CAPPARIS SANDWICHIANA VAR. SANDWICHIANA	END	OAH
CAPPARIDACEAE	CLEOME SANDWICENSIS	EXT	OAH
CARYOPHYLLACEAE	ALSINODENDRON OBOVATUM	END	OAH
CARYOPHYLLACEAE	ALSINODENDRON TRINERVE	END	OAH
CARYOPHYLLACEAE	SCHIEDEA ADAMANTIS	END	OAH
CARYOPHYLLACEAE	SCHIEDEA AMPLEXICAULIS	EXT	KAU? NII?
CARYOPHYLLACEAE	SCHIEDEA GLOBOSA VAR. GLOBOSA	END	OAH
CARYOPHYLLACEAE	SCHIEDEA GLOBOSA VAR. GRAMINIFOLIA	END	OAH
CARYOPHYLLACEAE	SCHIEDEA HAWAIIENSIS	EXT	HAW
CARYOPHYLLACEAE	SCHIEDEA HOOKERI VAR. HOOKERI	EXT	HAW?OAH?
CARYOPHYLLACEAE	SCHIEDEA KAALAE VAR. ACUTIFOLIA	THR	OAH
CARYOPHYLLACEAE	SCHIEDEA KAALAE VAR. KAALAE	END	OAH
CARYOPHYLLACEAE	SCHIEDEA KEALIAE	END	OAH
CARYOPHYLLACEAE	SCHIEDEA LIGUSTRINA VAR. NEMATOPODA	THR	OAH
CARYOPHYLLACEAE	SCHIEDEA MANNII	THR	OAH
CARYOPHYLLACEAE	SCHIEDEA MEMBRANACEA	THR	KAU
CARYOPHYLLACEAE	SCHIEDEA MENZIESII VAR. MENZIESII	THR	LAN
CARYOPHYLLACEAE	SCHIEDEA MENZIESII VAR. SPERGULACEA	END	LAN
CARYOPHYLLACEAE	SCHIEDEA PUBESCENS VAR. LANAIENSIS	END	LAN
CARYOPHYLLACEAE	SCHIEDEA SALICARIA	END	MAU
CARYOPHYLLACEAE	SCHIEDEA VERTICILLATA	END	NIH
CARYOPHYLLACEAE	SILENE ALEXANDRI	END	MOL
CARYOPHYLLACEAE	SILENE CRYPTOPETALA	THR	MAU
CARYOPHYLLACEAE	SILENE DEGENERI	THR	MAU
CARYOPHYLLACEAE	SILENE HAWAIIENSIS VAR. HAWAIIENSIS	THR	HAW
CARYOPHYLLACEAE	SILENE HAWAIIENSIS VAR. KAUPOANA	THR	MAU
CARYOPHYLLACEAE	SILENE LANCEOLATA VAR. ANGUSTIFOLIA	THR	
CARYOPHYLLACEAE	SILENE LANCEOLATA VAR. FORBESII	END	MOL
CARYOPHYLLACEAE	SILENE LANCEOLATA VAR. HILLEBRANDII	THR	HAW

ENDANGERED, THREATENED, AND RECENTLY EXTINCT SPECIES OF HAWAII

FAMILY	SPECIES	STATUS	MAU	HAW	OAH	KAU	MOL	LAN	NII	OUTER
CARYOPHYLLACEAE	SILENE LANCEOLATA VAR. LANCEOLATA	THR	MAU	HAW		KAU		LAN		
CHENOPODIACEAE	CHENOPODIUM OAHUENSE VAR. DISCOSPERMUM	THR	MAU							
CHENOPODIACEAE	CHENOPODIUM PEKELOI	THR					MOL			
CONVOLVULACEAE	BONAMIA MENZIESII	END	MAU		OAH	KAU	MOL	LAN		
CONVOLVULACEAE	IPOMOEA CAIRICA VAR. LINEARILOBA	THR					MOL			
CUCURBITACEAE	SICYOS ATOLLENSIS	THR								KUR LAY
CUCURBITACEAE	SICYOS CAUMII	THR								PRL-HRM
CUCURBITACEAE	SICYOS LAMOUREUXII	THR								KUR LIS
CUCURBITACEAE	SICYOS LAYSANENSIS	THR			OAH					LAY
CUCURBITACEAE	SICYOS MAXIMOWICZII	THR			OAH					
CUCURBITACEAE	SICYOS NIHOAENSIS	END								NIH
CUCURBITACEAE	SICYOS NIIHAUENSIS	THR							NII	
CUCURBITACEAE	SICYOS SEMITONSUS	THR								LAY
CYPERACEAE	GAHNIA LANAIENSIS	END						LAN		
EPACRIDACEAE	STYPHELIA TAMEIAMEIAE VAR. HEXAMERA	END			OAH					
EUPHORBIACEAE	CLAOXYLON SANDWICENSE VAR. SANDWICENSE	END	MAU							
EUPHORBIACEAE	DRYPETES PHYLLANTHOIDES	END	MAU	HAW	OAH	KAU	MOL			
EUPHORBIACEAE	EUPHORBIA ARNOTTIANA VAR. ARNOTTIANA	END			OAH					
EUPHORBIACEAE	EUPHORBIA ARNOTTIANA VAR. INTEGRIFOLIA	END	MAU						LAN?	
EUPHORBIACEAE	EUPHORBIA ATROCOCCA VAR. ATROCOCCA	END				KAU				
EUPHORBIACEAE	EUPHORBIA ATROCOCCA VAR. KILAUEANA	END				KAU				
EUPHORBIACEAE	EUPHORBIA ATROCOCCA VAR. KOKEEANA	END				KAU				
EUPHORBIACEAE	EUPHORBIA CELASTROIDES VAR. HALAWANA	END			OAH		MOL			
EUPHORBIACEAE	EUPHORBIA CELASTROIDES VAR. HAUPUANA	END				KAU				
EUPHORBIACEAE	EUPHORBIA CELASTROIDES VAR. HUMBERTII	EXT			OAH?KAU					
EUPHORBIACEAE	EUPHORBIA CELASTROIDES VAR. KAENANA	END			OAH					
EUPHORBIACEAE	EUPHORBIA CELASTROIDES VAR. KEALIANA	END				KAU				
EUPHORBIACEAE	EUPHORBIA CELASTROIDES VAR. KOHALANA	END		HAW						
EUPHORBIACEAE	EUPHORBIA CELASTROIDES VAR. MOOMOMIANA	END					MOL			
EUPHORBIACEAE	EUPHORBIA CELASTROIDES VAR. NELSONII	EXT		HAW						
EUPHORBIACEAE	EUPHORBIA CELASTROIDES VAR. NEMATOPODA	END				KAU				
EUPHORBIACEAE	EUPHORBIA CELASTROIDES VAR. NIUENSIS	EXT			OAH					
EUPHORBIACEAE	EUPHORBIA CELASTROIDES VAR. SAXICOLA	END		HAW						
EUPHORBIACEAE	EUPHORBIA CELASTROIDES VAR. STOKESII	END				KAU			NII	
EUPHORBIACEAE	EUPHORBIA CELASTROIDES VAR. WAIKOLUENSIS	END					MOL			
EUPHORBIACEAE	EUPHORBIA DEGENERI VAR. MOLOKAIENSIS	END					MOL			
EUPHORBIACEAE	EUPHORBIA DEPPEANA	EXT			OAH					
EUPHORBIACEAE	EUPHORBIA HAELEELEANA	END				KAU				
EUPHORBIACEAE	EUPHORBIA HALEMANUI	END				KAU				
EUPHORBIACEAE	EUPHORBIA HILLEBRANDII VAR. PALIKEANA	END			OAH					
EUPHORBIACEAE	EUPHORBIA HILLEBRANDII VAR. WAIMANOANA	END			OAH					
EUPHORBIACEAE	EUPHORBIA MULTIFORMIS VAR. HALEAKALANA	EXT	MAU							
EUPHORBIACEAE	EUPHORBIA MULTIFORMIS VAR. KAALANA	EXT			OAH					

ENDANGERED, THREATENED, AND RECENTLY EXTINCT SPECIES OF HAWAII

FAMILY	SPECIES	STATUS	RANGE
EUPHORBIACEAE	EUPHORBIA MULTIFORMIS VAR. KAPULEIENSIS	END	MOL
EUPHORBIACEAE	EUPHORBIA MULTIFORMIS VAR. MULTIFORMIS	EXT	OAH
EUPHORBIACEAE	EUPHORBIA MULTIFORMIS VAR. PERDITA	EXT	OAH
EUPHORBIACEAE	EUPHORBIA MULTIFORMIS VAR. SPARSIFLORA	END	KAU
EUPHORBIACEAE	EUPHORBIA MULTIFORMIS VAR. TOMENTELLA	EXT	OAH
EUPHORBIACEAE	EUPHORBIA OLOWALUANA VAR. OLOWALUANA	END	MAU
EUPHORBIACEAE	EUPHORBIA REMYI VAR. HANALEIENSIS	EXT	KAU
EUPHORBIACEAE	EUPHORBIA REMYI VAR. KAHILIANA	END	KAU
EUPHORBIACEAE	EUPHORBIA REMYI VAR. KAUAIENSIS	END	KAU
EUPHORBIACEAE	EUPHORBIA REMYI VAR. LEPTOPODA	END	KAU
EUPHORBIACEAE	EUPHORBIA REMYI VAR. LYDGATEI	END	KAU
EUPHORBIACEAE	EUPHORBIA REMYI VAR. MOLESTA	END	KAU
EUPHORBIACEAE	EUPHORBIA REMYI VAR. OLOKELENSIS	END	KAU
EUPHORBIACEAE	EUPHORBIA REMYI VAR. PTEROPODA	END	KAU
EUPHORBIACEAE	EUPHORBIA REMYI VAR. REMYI	EXT	OAH KAU
EUPHORBIACEAE	EUPHORBIA REMYI VAR. WAHIAWANA	END	KAU
EUPHORBIACEAE	EUPHORBIA REMYI VAR. WAIMEANA	END	KAU
EUPHORBIACEAE	EUPHORBIA REMYI VAR. WILKESII	EXT	KAU
EUPHORBIACEAE	EUPHORBIA SKOTTSBERGII VAR. AUDENS	END	MOL
EUPHORBIACEAE	EUPHORBIA SKOTTSBERGII VAR. KALAELOANA	EXT	OAH
EUPHORBIACEAE	EUPHORBIA SKOTTSBERGII VAR. SKOTTSBERGII	EXT	OAH
EUPHORBIACEAE	EUPHORBIA SKOTTSBERGII VAR. VACCINIOIDES	END	MOL
EUPHORBIACEAE	PHYLLANTHUS SANDWICENSIS VAR. DEGENERI	END	MOL LAN
FABACEAE	ACACIA KOAIA	END	MAU HAW KAU?MOL LAN?
FABACEAE	CANAVALIA CENTRALIS	END	KAU
FABACEAE	CANAVALIA FORBESII	END	MAU
FABACEAE	CANAVALIA HALEAKALAENSIS	END	MAU
FABACEAE	CANAVALIA IAOENSIS	END	MAU
FABACEAE	CANAVALIA KAUAIENSIS	END	KAU
FABACEAE	CANAVALIA KAUENSIS	END	HAW
FABACEAE	CANAVALIA LANAIENSIS	END	LAN
FABACEAE	CANAVALIA MAKAHAENSIS	END	KAU
FABACEAE	CANAVALIA MOLOKAIENSIS	END	MOL
FABACEAE	CANAVALIA MUNROI	END	LAN
FABACEAE	CANAVALIA NAPALIENSIS	END	KAU
FABACEAE	CANAVALIA NUALOLOENSIS	END	KAU
FABACEAE	CANAVALIA PENINSULARIS	END	MOL
FABACEAE	CANAVALIA PUBESCENS	END	NII
FABACEAE	CANAVALIA ROCKII	END	LAN
FABACEAE	CANAVALIA SANGUINEA	END	MAU
FABACEAE	CANAVALIA STENOPHYLLA	END	MOL
FABACEAE	MEZONEURON KAVAIENSE	END	MAU HAW OAH KAU
FABACEAE	SESBANIA TOMENTOSA VAR. MOLOKAIENSIS	END	HAW? MOL LAN?

ENDANGERED, THREATENED, AND RECENTLY EXTINCT SPECIES OF HAWAII

FAMILY	SPECIES	STATUS	RANGE
FABACEAE	SESBANIA TOMENTOSA VAR. TOMENTOSA	END	HAW OAH KAU MOL LAN NEC NIH
FABACEAE	SOPHORA CHRYSOPHYLLA VAR. CIRCULARIS	END	HAW
FABACEAE	SOPHORA CHRYSOPHYLLA VAR. ELLIPTICA	END	MAU
FABACEAE	SOPHORA CHRYSOPHYLLA VAR. GLABRATA	EXT	HAW
FABACEAE	SOPHORA CHRYSOPHYLLA VAR. GRISEA	END	HAW OAH KAU
FABACEAE	SOPHORA CHRYSOPHYLLA VAR. KANAIOENSIS	END	MAU
FABACEAE	SOPHORA CHRYSOPHYLLA VAR. KAUENSIS	END	HAW
FABACEAE	SOPHORA CHRYSOPHYLLA VAR. LANAIENSIS	EXT	LAN
FABACEAE	SOPHORA CHRYSOPHYLLA VAR. MAKUAENSIS	END	OAH
FABACEAE	SOPHORA CHRYSOPHYLLA VAR. UNIFOLIATA	EXT	HAW
FABACEAE	VICIA MENZIESII	END	HAW
FABACEAE	VIGNA O-WAHUENSIS	END	OAH KAU MOL
FABACEAE	VIGNA SANDWICENSIS VAR. HETEROPHYLLA	END	MAU HAW
FABACEAE	VIGNA SANDWICENSIS VAR. SANDWICENSIS	END	MAU HAW KAU? LAN
FLACOURTIACEAE	XYLOSMA CRENATUM	END	KAU
FLAGELLARIACEAE	JOINVILLEA ASCENDENS SSP. ASCENDENS	END	OAH KAU
GERANIACEAE	GERANIUM ARBOREUM	END	MAU
GERANIACEAE	GERANIUM CUNEATUM VAR. HOLOLEUCUM	END	HAW
GERANIACEAE	GERANIUM MULTIFLORUM VAR. MULTIFLORUM	EXT	HAW
GERANIACEAE	GERANIUM MULTIFLORUM VAR. OVATIFOLIUM	END	MAU
GERANIACEAE	GERANIUM MULTIFLORUM VAR. SUPERBUM	END	MAU
GESNERIACEAE	CYRTANDRA ALATA	END	OAH
GESNERIACEAE	CYRTANDRA ALNEA	END	OAH
GESNERIACEAE	CYRTANDRA AMBIGUA	END	OAH
GESNERIACEAE	CYRTANDRA AXILLIFLORA	END	OAH
GESNERIACEAE	CYRTANDRA BASIPARTITA	END	OAH
GESNERIACEAE	CYRTANDRA BEGONIIFOLIA	EXT	MAU
GESNERIACEAE	CYRTANDRA BISERRATA	THR	MOL
GESNERIACEAE	CYRTANDRA BREVICORNUTA	END	OAH
GESNERIACEAE	CYRTANDRA BRYANII	END	OAH
GESNERIACEAE	CYRTANDRA CAMPANIFORMIS	END	OAH
GESNERIACEAE	CYRTANDRA CARINATA	EXT	OAH
GESNERIACEAE	CYRTANDRA CAUDATISEPALA	END	OAH
GESNERIACEAE	CYRTANDRA CHARTACEA	END	OAH
GESNERIACEAE	CYRTANDRA CHRISTOPHERSENII	THR	OAH
GESNERIACEAE	CYRTANDRA COLLARIFERA	EXT	OAH
GESNERIACEAE	CYRTANDRA CONRADTII	END	MOL
GESNERIACEAE	CYRTANDRA CORDIFOLIA VAR. BREVIPILITA	EXT	OAH
GESNERIACEAE	CYRTANDRA CORDIFOLIA VAR. GYNOGLABRA	THR	MAU
GESNERIACEAE	CYRTANDRA CRASSIOR	END	OAH
GESNERIACEAE	CYRTANDRA CRENATA	END	OAH
GESNERIACEAE	CYRTANDRA CUPULIFORMIS	END	OAH
GESNERIACEAE	CYRTANDRA DENTATA	END	OAH

ENDANGERED, THREATENED, AND RECENTLY EXTINCT SPECIES OF HAWAII

FAMILY	SPECIES	STATUS	RANGE
GESNERIACEAE	CYRTANDRA ELLIPTICIFOLIA	END	OAH
GESNERIACEAE	CYRTANDRA ELLIPTISEPALA	END	OAH
GESNERIACEAE	CYRTANDRA FERRICOLORATA	END	OAH
GESNERIACEAE	CYRTANDRA FERRUGINOSA	EXT	OAH
GESNERIACEAE	CYRTANDRA FILIPES	THR	MAU
GESNERIACEAE	CYRTANDRA FORBESII	END	OAH
GESNERIACEAE	CYRTANDRA FOSBERGII	END	OAH
GESNERIACEAE	CYRTANDRA FREDERICKII	END	OAH
GESNERIACEAE	CYRTANDRA FUSIFORMIS	END	OAH
GESNERIACEAE	CYRTANDRA GARBERI	EXT	OAH
GESNERIACEAE	CYRTANDRA GEORGIANA	THR	LAN
GESNERIACEAE	CYRTANDRA GIFFARDII	THR	HAW
GESNERIACEAE	CYRTANDRA GLAUCA	END	KAU
GESNERIACEAE	CYRTANDRA GRACILIS	EXT	OAH
GESNERIACEAE	CYRTANDRA GRAYANA VAR. LANAIENSIS	THR	LAN
GESNERIACEAE	CYRTANDRA GROSSECRENATA	EXT	OAH?
GESNERIACEAE	CYRTANDRA HALAWENSIS	THR	MOL
GESNERIACEAE	CYRTANDRA HAWAIENSIS	END	HAW
GESNERIACEAE	CYRTANDRA HIRSUTULA	END	OAH
GESNERIACEAE	CYRTANDRA HOBDYI	END	KAU
GESNERIACEAE	CYRTANDRA HONOLULENSIS	EXT	OAH
GESNERIACEAE	CYRTANDRA HOSAKAE	THR	OAH
GESNERIACEAE	CYRTANDRA INFRAPALLIDA	END	OAH
GESNERIACEAE	CYRTANDRA INTONSA	END	OAH
GESNERIACEAE	CYRTANDRA INTRAPILOSA	END	OAH
GESNERIACEAE	CYRTANDRA INTRAVILLOSA	THR	OAH
GESNERIACEAE	CYRTANDRA KAALAE	END	OAH
GESNERIACEAE	CYRTANDRA KAHANAENSIS	END	OAH
GESNERIACEAE	CYRTANDRA KAHUKUENSIS	END	OAH
GESNERIACEAE	CYRTANDRA KALUANUIENSIS	END	OAH
GESNERIACEAE	CYRTANDRA KANEOHEENSIS	END	OAH
GESNERIACEAE	CYRTANDRA KAUAIENSIS	EXT	KAU
GESNERIACEAE	CYRTANDRA KAULANTHA	THR	OAH
GESNERIACEAE	CYRTANDRA KOOLAUENSIS	END	OAH
GESNERIACEAE	CYRTANDRA LAEVIS	END	OAH
GESNERIACEAE	CYRTANDRA LAXIFLORA	EXT	OAH
GESNERIACEAE	CYRTANDRA LESSONIANA VAR. ANGUSTIFOLIA	EXT	OAH
GESNERIACEAE	CYRTANDRA LESSONIANA VAR. INTRAPUBENS	END	OAH
GESNERIACEAE	CYRTANDRA LIMOSIFLORA	THR	MOL
GESNERIACEAE	CYRTANDRA LINEARIS	EXT	OAH
GESNERIACEAE	CYRTANDRA LONGICALYX	END	OAH
GESNERIACEAE	CYRTANDRA LONGIFOLIA VAR. LONGIFOLIA	END	KAU
GESNERIACEAE	CYRTANDRA LONGIFOLIA VAR. PARALLELA	END	

ENDANGERED, THREATENED, AND RECENTLY EXTINCT SPECIES OF HAWAII

FAMILY	SPECIES	STATUS	RANGE			
GESNERIACEAE	CYRTANDRA LONGILOBA	END		OAH		
GESNERIACEAE	CYRTANDRA LYSIOSEPALA VAR. GRAYI	THR	MAU			
GESNERIACEAE	CYRTANDRA LYSIOSEPALA VAR. HALEAKALENSIS	THR	MAU			
GESNERIACEAE	CYRTANDRA LYSIOSEPALA VAR. LYSIOSEPALA	END	MAU HAW			
GESNERIACEAE	CYRTANDRA MACRANTHA	END				
GESNERIACEAE	CYRTANDRA MALACOPHYLLA VAR. MALACOPHYLLA	THR	MAU?		KAU?	
GESNERIACEAE	CYRTANDRA MANNII	EXT		OAH		
GESNERIACEAE	CYRTANDRA MEGASTIGMATA	END		OAH		
GESNERIACEAE	CYRTANDRA MENZIESII	END	HAW			
GESNERIACEAE	CYRTANDRA MUNROI	THR				LAN
GESNERIACEAE	CYRTANDRA NIUENSIS	END		OAH		
GESNERIACEAE	CYRTANDRA NUBINCOLENS	END		OAH		
GESNERIACEAE	CYRTANDRA OENOBARBA VAR. HERBACEA	THR			KAU	
GESNERIACEAE	CYRTANDRA OENOBARBA VAR. OENOBARBA	END			KAU	
GESNERIACEAE	CYRTANDRA OENOBARBA VAR. PETIOLARIS	THR			KAU	
GESNERIACEAE	CYRTANDRA OLIVACEA	END		OAH		
GESNERIACEAE	CYRTANDRA PALOLOENSIS	END		OAH		
GESNERIACEAE	CYRTANDRA PALUDOSA VAR. HAUPUENSIS	THR			KAU	
GESNERIACEAE	CYRTANDRA PARTITA	EXT		OAH		
GESNERIACEAE	CYRTANDRA PEARSALLII	END		OAH		
GESNERIACEAE	CYRTANDRA PERSTAMINODICA	END		OAH		
GESNERIACEAE	CYRTANDRA PICKERINGII VAR. PICKERINGII	EXT		OAH		
GESNERIACEAE	CYRTANDRA PICKERINGII VAR. WAIHEAE	END	MAU			
GESNERIACEAE	CYRTANDRA PILIGYNA	EXT		OAH		
GESNERIACEAE	CYRTANDRA PLATYPHYLLA VAR. HILOENSIS	THR	HAW			
GESNERIACEAE	CYRTANDRA PLURIFOLIA	END		OAH		
GESNERIACEAE	CYRTANDRA POLYANTHA	END		OAH		
GESNERIACEAE	CYRTANDRA PRUINOSA	END		OAH		
GESNERIACEAE	CYRTANDRA PUBENS	END		OAH		
GESNERIACEAE	CYRTANDRA RAMOSISSIMA	THR	HAW			
GESNERIACEAE	CYRTANDRA ROCKII	END		OAH		
GESNERIACEAE	CYRTANDRA SANDWICENSIS	END		OAH		
GESNERIACEAE	CYRTANDRA SCABRELLA	EXT		OAH		
GESNERIACEAE	CYRTANDRA SKOTTSBERGII	EXT		OAH		
GESNERIACEAE	CYRTANDRA SUBCORDATA	END		OAH		
GESNERIACEAE	CYRTANDRA SUBINTEGRA	EXT		OAH		
GESNERIACEAE	CYRTANDRA SUBRECTA	END		OAH		
GESNERIACEAE	CYRTANDRA SUBUMBELLATA VAR. INTONSA	END		OAH		
GESNERIACEAE	CYRTANDRA TERNATA	END		OAH		
GESNERIACEAE	CYRTANDRA TRIFLORA	EXT	MAU	OAH		
GESNERIACEAE	CYRTANDRA TURBINIFORMIS	END		OAH		
GESNERIACEAE	CYRTANDRA VANIOTA	END		OAH		
GESNERIACEAE	CYRTANDRA VILLICALYX VAR. PUBENTIGYNA	END		OAH		

ENDANGERED, THREATENED, AND RECENTLY EXTINCT SPECIES OF HAWAII

FAMILY	SPECIES	STATUS	MAU	HAW	OAH	KAU	MOL	LAN	NII
GESNERIACEAE	CYRTANDRA VILLOSA	END			OAH				
GESNERIACEAE	CYRTANDRA VILLOSIFLORA	END			OAH				
GESNERIACEAE	CYRTANDRA WAIANUENSIS	EXT			OAH				
GESNERIACEAE	CYRTANDRA WAIOLANI VAR. CAPITATA	EXT			OAH				
GESNERIACEAE	CYRTANDRA WAIOLANI VAR. WAIOLANI	END			OAH				
GESNERIACEAE	CYRTANDRA WAIOMAOENSIS	END			OAH				
GOODENIACEAE	SCAEVOLA CORIACEA	END	MAU	HAW	OAH	KAU	MOL	LAN	NII
GOODENIACEAE	SCAEVOLA GAUDICHAUDII	END	MAU	HAW	OAH	KAU	MOL	LAN	
GOODENIACEAE	SCAEVOLA KILAUEAE VAR. KILAUEAE	END		HAW					
GOODENIACEAE	SCAEVOLA SKOTTSBERGII	THR			OAH				
HALORAGACEAE	GUNNERA KAALENSIS	END			OAH				
HALORAGACEAE	GUNNERA MAKAHAENSIS	END			OAH				
HYMENOPHYLLACEAE	TRICHOMANES DRAYTONIANUM	END	MAU		OAH	KAU	MOL		
JUNCACEAE	LUZULA HAWAIIENSIS VAR. OAHUENSIS	END			OAH				
LAMIACEAE	HAPLOSTACHYS BRYANII VAR. BRYANII	EXT					MOL		
LAMIACEAE	HAPLOSTACHYS BRYANII VAR. MICRODONTA	EXT					MOL		
LAMIACEAE	HAPLOSTACHYS BRYANII VAR. ROBUSTA	EXT					MOL		
LAMIACEAE	HAPLOSTACHYS HAPLOSTACHYA VAR. ANGUSTIFOLIA	EXT		HAW					
LAMIACEAE	HAPLOSTACHYS HAPLOSTACHYA VAR. HAPLOSTACHYA	EXT	MAU						
LAMIACEAE	HAPLOSTACHYS HAPLOSTACHYA VAR. LEPTOSTACHYA	EXT				KAU			
LAMIACEAE	HAPLOSTACHYS LINEARIFOLIA VAR. LINEARIFOLIA	EXT	MAU				MOL		
LAMIACEAE	HAPLOSTACHYS LINEARIFOLIA VAR. ROSMARINIFOLIA	EXT					MOL		
LAMIACEAE	HAPLOSTACHYS MUNROI	EXT						LAN	
LAMIACEAE	HAPLOSTACHYS TRUNCATA	EXT	MAU						
LAMIACEAE	PHYLLOSTEGIA BREVIDENS VAR. AMBIGUA	EXT	MAU						
LAMIACEAE	PHYLLOSTEGIA BREVIDENS VAR. DEGENERI	THR	MAU						
LAMIACEAE	PHYLLOSTEGIA BREVIDENS VAR. HETERODOXA	THR		HAW					
LAMIACEAE	PHYLLOSTEGIA BREVIDENS VAR. HIRSUTULA	EXT	MAU						
LAMIACEAE	PHYLLOSTEGIA BREVIDENS VAR. LONGIPES	EXT	MAU						
LAMIACEAE	PHYLLOSTEGIA BREVIDENS VAR. PUBESCENS	EXT	MAU						
LAMIACEAE	PHYLLOSTEGIA FLORIBUNDA	EXT		HAW					
LAMIACEAE	PHYLLOSTEGIA FORBESII	END		HAW					
LAMIACEAE	PHYLLOSTEGIA GLABRA VAR. LANAIENSIS	EXT						LAN	
LAMIACEAE	PHYLLOSTEGIA HELLERI VAR. IMMINUTA	END						LAN	
LAMIACEAE	PHYLLOSTEGIA HILLEBRANDII	THR	MAU						
LAMIACEAE	PHYLLOSTEGIA HIRSUTA VAR. HIRSUTA	THR	MAU?		OAH				
LAMIACEAE	PHYLLOSTEGIA HIRSUTA VAR. LAXIOR	THR			OAH				
LAMIACEAE	PHYLLOSTEGIA KNUDSENII	END				KAU			
LAMIACEAE	PHYLLOSTEGIA LEDYARDII	EXT		HAW					
LAMIACEAE	PHYLLOSTEGIA LONGIMONTIS	EXT		HAW					
LAMIACEAE	PHYLLOSTEGIA MOLLIS VAR. FAGERLINDII	END				KAU			
LAMIACEAE	PHYLLOSTEGIA MOLLIS VAR. HOCHREUTINERI	END				KAU			
LAMIACEAE	PHYLLOSTEGIA MOLLIS VAR. LYDGATEI	EXT			OAH				

ENDANGERED, THREATENED, AND RECENTLY EXTINCT SPECIES OF HAWAII

FAMILY	SPECIES	STATUS	RANGE			
LAMIACEAE	PHYLLOSTEGIA MOLLIS VAR. MICRANTHA	END				LAN
LAMIACEAE	PHYLLOSTEGIA PARVIFLORA VAR. CANESCENS	EXT	MAU	KAU		
LAMIACEAE	PHYLLOSTEGIA PARVIFLORA VAR. GLABRIUSCULA	EXT	HAW			
LAMIACEAE	PHYLLOSTEGIA PARVIFLORA VAR. HONOLULENSIS	EXT	OAH			
LAMIACEAE	PHYLLOSTEGIA VARIABILIS	END				LAY MID
LAMIACEAE	PHYLLOSTEGIA YAMAGUCHII	THR	OAH			
LAMIACEAE	STENOGYNE AFFINIS VAR. AFFINIS	END	HAW			
LAMIACEAE	STENOGYNE AFFINIS VAR. DEGENERI	END	MAU			
LAMIACEAE	STENOGYNE ANGUSTIFOLIA VAR. ANGUSTIFOLIA	EXT	HAW			
LAMIACEAE	STENOGYNE ANGUSTIFOLIA VAR. HILLEBRANDII	THR		MOL		
LAMIACEAE	STENOGYNE ANGUSTIFOLIA VAR. MAUIENSIS	EXT	MAU			
LAMIACEAE	STENOGYNE ANGUSTIFOLIA VAR. MEEBOLDII	END	HAW			
LAMIACEAE	STENOGYNE ANGUSTIFOLIA VAR. SPATHULATA	EXT	HAW			
LAMIACEAE	STENOGYNE BIFLORA	EXT	HAW			
LAMIACEAE	STENOGYNE CALAMINTHOIDES VAR. OXYODONTA	THR	HAW			
LAMIACEAE	STENOGYNE CINEREA	EXT	MAU			
LAMIACEAE	STENOGYNE CRENATA VAR. CRENATA	END	MAU			
LAMIACEAE	STENOGYNE CRENATA VAR. MURICATA	END	MAU			
LAMIACEAE	STENOGYNE DIFFUSA VAR. DIFFUSA	EXT	HAW			
LAMIACEAE	STENOGYNE DIFFUSA VAR. GLABRA	END	HAW			
LAMIACEAE	STENOGYNE GLABRATA	EXT	MAU			
LAMIACEAE	STENOGYNE HALIAKALAE	END	MAU			
LAMIACEAE	STENOGYNE HIRSUTULA	EXT	HAW			
LAMIACEAE	STENOGYNE KANEHOANA	END	OAH			
LAMIACEAE	STENOGYNE MACRANTHA VAR. GRAYI	EXT	HAW			
LAMIACEAE	STENOGYNE MACRANTHA VAR. LATIFOLIA	EXT	HAW			
LAMIACEAE	STENOGYNE MACRANTHA VAR. MACRANTHA	END	HAW			
LAMIACEAE	STENOGYNE MICROPHYLLA	EXT	HAW			
LAMIACEAE	STENOGYNE OXYGONA	THR	HAW			
LAMIACEAE	STENOGYNE PURPUREA VAR. FORBESII	THR	KAU			
LAMIACEAE	STENOGYNE ROTUNDIFOLIA VAR. OBLONGA	END	MAU			
LAMIACEAE	STENOGYNE RUGOSA VAR. SUBULATA	THR	HAW			
LAMIACEAE	STENOGYNE SALICIFOLIA	THR	HAW			
LAMIACEAE	STENOGYNE SCANDENS	EXT	HAW			
LAMIACEAE	STENOGYNE SCROPHULARIOIDES VAR. NELSONII	EXT	HAW			
LAMIACEAE	STENOGYNE SCROPHULARIOIDES VAR. REMYI	EXT	HAW			
LAMIACEAE	STENOGYNE SCROPHULARIOIDES VAR. SCROPHULARIOIDES	EXT	HAW			
LAMIACEAE	STENOGYNE SCROPHULARIOIDES VAR. SKOTTSBERGII	END	HAW			
LAMIACEAE	STENOGYNE SESSILIS VAR. HEXANTHA	END	HAW			
LAMIACEAE	STENOGYNE SESSILIS VAR. LANAIENSIS	END				LAN
LAMIACEAE	STENOGYNE SESSILIS VAR. WILKESII	EXT	HAW			
LAMIACEAE	STENOGYNE SHERFFII	END	OAH			
LAMIACEAE	STENOGYNE SORORIA	EXT				

ENDANGERED, THREATENED, AND RECENTLY EXTINCT SPECIES OF HAWAII

FAMILY	SPECIES	STATUS	RANGE
LAMIACEAE	STENOGYNE VAGANS	EXT	MAU
LAMIACEAE	STENOGYNE VIRIDIS	EXT	MAU
LAURACEAE	CRYPTOCARYA OAHUENSIS	END	OAH
LILIACEAE	ASTELIA VERATROIDES VAR. GRACILIS	THR	OAH
LILIACEAE	ASTELIA VERATROIDES VAR. VERATROIDES	END	OAH
LILIACEAE	ASTELIA VERATROIDES SSP. MACROSPERMA	END	KAU
LILIACEAE	DRACAENA AUREA	END	HAW OAH KAU
LILIACEAE	DRACAENA HAWAIIENSIS	END	HAW
LILIACEAE	PLEOMELE (DRACAENA) FORBESII	END	OAH
LILIACEAE	SMILAX MELASTOMIFOLIA VAR. MELASTOMIFOLIA	EXT	OAH
LOGANIACEAE	LABORDIA BAILLONII	EXT	HAW
LOGANIACEAE	LABORDIA CYRTANDRAE VAR. NAHIKUANA	THR	MAU
LOGANIACEAE	LABORDIA DECURRENS VAR. DECURRENS	EXT	OAH
LOGANIACEAE	LABORDIA FAGRAEOIDEA VAR. FAGRAEOIDEA	END	OAH
LOGANIACEAE	LABORDIA FAGRAEOIDEA VAR. LONGISEPALA	END	OAH
LOGANIACEAE	LABORDIA FAGRAEOIDEA VAR. SAINT-JOHNIANA	THR	OAH
LOGANIACEAE	LABORDIA FAGRAEOIDEA VAR. WAIANAEANA	END	OAH
LOGANIACEAE	LABORDIA GLABRA VAR. GLABRA	EXT	MAU
LOGANIACEAE	LABORDIA GLABRA VAR. LATISEPALA	END	OAH
LOGANIACEAE	LABORDIA GLABRA VAR. ORIENTALIS	END	MAU
LOGANIACEAE	LABORDIA HEDYOSMIFOLIA VAR. KILAUEANA	END	HAW
LOGANIACEAE	LABORDIA HEDYOSMIFOLIA VAR. MAGNIFOLIA	END	HAW
LOGANIACEAE	LABORDIA HEDYOSMIFOLIA VAR. ROBUSTA	END	HAW
LOGANIACEAE	LABORDIA HEDYOSMIFOLIA VAR. ROCKII	END	MAU
LOGANIACEAE	LABORDIA HEDYOSMIFOLIA VAR. SKOTTSBERGII	END	HAW
LOGANIACEAE	LABORDIA HIRTELLA VAR. IMBRICATA	END	HAW
LOGANIACEAE	LABORDIA HIRTELLA VAR. LAEVIS	END	MOL
LOGANIACEAE	LABORDIA HIRTELLA VAR. LAEVISEPALA	END	MOL
LOGANIACEAE	LABORDIA HIRTELLA VAR. MICROCALYX	EXT	HAW
LOGANIACEAE	LABORDIA HIRTELLA VAR. MICROPHYLLA	EXT	MAU
LOGANIACEAE	LABORDIA KAALAE VAR. BRACHYPODA	END	OAH
LOGANIACEAE	LABORDIA KAALAE VAR. FOSBERGII	END	OAH
LOGANIACEAE	LABORDIA KAALAE VAR. KAUAIENSIS	END	KAU
LOGANIACEAE	LABORDIA KAALAE VAR. MENDAX	END	OAH
LOGANIACEAE	LABORDIA LYDGATEI	THR	KAU
LOGANIACEAE	LABORDIA MEMBRANACEA VAR. EXIGUA	END	MAU
LOGANIACEAE	LABORDIA MEMBRANACEA VAR. MEMBRANACEA	END	OAH
LOGANIACEAE	LABORDIA MOLOKAIANA VAR. MOLOKAIANA	END	MOL
LOGANIACEAE	LABORDIA MOLOKAIANA VAR. MUNROI	END	LAN
LOGANIACEAE	LABORDIA MOLOKAIANA VAR. SETOSA	END	MOL
LOGANIACEAE	LABORDIA NELSONII	EXT	HAW
LOGANIACEAE	LABORDIA OLYMPIANA	END	OAH
LOGANIACEAE	LABORDIA PALLIDA VAR. HISPIDULA	END	KAU

ENDANGERED, THREATENED, AND RECENTLY EXTINCT SPECIES OF HAWAII

FAMILY	SPECIES	STATUS	RANGE							
LOGANIACEAE	LABORDIA PALLIDA VAR. PALLIDA	END				KAU				
LOGANIACEAE	LABORDIA PEDUNCULATA	EXT	MAU?							
LOGANIACEAE	LABORDIA TINIFOLIA VAR. EUPHORBIOIDEA	THR	MAU							
LOGANIACEAE	LABORDIA TINIFOLIA VAR. FORBESII	END					MOL			
LOGANIACEAE	LABORDIA TINIFOLIA VAR. HONOLULUENSIS	EXT			OAH					
LOGANIACEAE	LABORDIA TINIFOLIA VAR. LANAIENSIS	THR	MAU HAW					LAN		
LOGANIACEAE	LABORDIA TINIFOLIA VAR. MICROGYNA	END				KAU				
LOGANIACEAE	LABORDIA TINIFOLIA VAR. PARVIFOLIA	EXT					MOL			
LOGANIACEAE	LABORDIA TINIFOLIA VAR. TENUIFOLIA	END					MOL			
LOGANIACEAE	LABORDIA TRIFLORA	EXT					MOL			
LOGANIACEAE	LABORDIA WAWRANA	THR				KAU				
LYCOPODIACEAE	LYCOPODIUM HALEAKALAE	THR	MAU							
LYCOPODIACEAE	LYCOPODIUM MANNII	EXT	MAU HAW							
LYCOPODIACEAE	LYCOPODIUM NUTANS	THR	MAU	OAH						
MALVACEAE	ABUTILON EREMITOPETALUM	EXT						LAN		
MALVACEAE	ABUTILON MENZIESII	EXT		HAW				LAN		
MALVACEAE	ABUTILON SANDWICENSE VAR. SANDWICENSE	END		OAH						
MALVACEAE	GOSSYPIUM TOMENTOSUM	THR	MAU HAW	OAH	KAU	MOL	LAN	NII	KAH	
MALVACEAE	HIBISCADELPHUS BOMBYCINUS	EXT		HAW						
MALVACEAE	HIBISCADELPHUS DISTANS	END				KAU				
MALVACEAE	HIBISCADELPHUS GIFFARDIANUS	END		HAW						
MALVACEAE	HIBISCADELPHUS HUALALAIENSIS	END		HAW						
MALVACEAE	HIBISCADELPHUS WILDERIANUS	EXT	MAU							
MALVACEAE	HIBISCUS BRACKENRIDGEI VAR. BRACKENRIDGEI	END	MAU					LAN		
MALVACEAE	HIBISCUS BRACKENRIDGEI VAR. MOKULEIANA	END		OAH						
MALVACEAE	HIBISCUS BRACKENRIDGEI VAR. MOLOKAIANUS	EXT					MOL			
MALVACEAE	HIBISCUS BRACKENRIDGEI VAR. (FROM HAWAII)	END	HAW							
MALVACEAE	HIBISCUS CLAYI	END				KAU				
MALVACEAE	HIBISCUS IMMACULATUS	END					MOL			
MALVACEAE	HIBISCUS KAHILII	END				KAU				
MALVACEAE	HIBISCUS KOKIO VAR. KOKIO	END		OAH	KAU?					
MALVACEAE	HIBISCUS KOKIO VAR. PUKOONIS	END					MOL			
MALVACEAE	HIBISCUS NEWHOUSEI	END				KAU				
MALVACEAE	HIBISCUS ROEATAE	END				KAU				
MALVACEAE	HIBISCUS SAINTJOHNIANUS	END				KAU				
MALVACEAE	HIBISCUS WAIMEAE	END				KAU				
MALVACEAE	KOKIA COOKEI	EXT					MOL			
MALVACEAE	KOKIA DRYNARIOIDES	END		HAW						
MALVACEAE	KOKIA KAUAIENSIS	END				KAU				
MALVACEAE	KOKIA LANCEOLATA	EXT		OAH						
MALVACEAE	SIDA LEDYARDII	EXT		HAW						
MALVACEAE	SIDA NELSONII	EXT		HAW						
MARSILEACEAE	MARSILEA VILLOSA	END			OAH		MOL			

ENDANGERED, THREATENED, AND RECENTLY EXTINCT SPECIES OF HAWAII

FAMILY	SPECIES	STATUS	MAU	HAW	OAH	KAU	MOL	LAN	KAH
MENISPERMACEAE	COCCULUS INTEGER	THR						LAN	
MENISPERMACEAE	COCCULUS LONCHOPHYLLUS	THR	MAU						
MENISPERMACEAE	COCCULUS VIRGATUS	THR				KAU	MOL	LAN	
MYRSINACEAE	MYRSINE FERNSEEI	END			OAH	KAU?			
MYRSINACEAE	MYRSINE LANAIENSIS VAR. OAHUENSIS	END			OAH				
MYRSINACEAE	MYRSINE LINEARIFOLIA VAR. LINEARIFOLIA	END				KAU			
MYRSINACEAE	MYRSINE MEZII	END				KAU			
MYRSINACEAE	MYRSINE PETIOLATA	THR				KAU			
MYRSINACEAE	MYRSINE ST.-JOHNII	END				KAU			
MYRTACEAE	EUGENIA MOLOKAIANA	EXT					MOL		
MYRTACEAE	METROSIDEROS COLLINA VAR. NEWELLII	END	MAU	HAW					
OPHIOGLOSSACEAE	BOTRYCHIUM SUBBIFOLIATUM	EXT	MAU	HAW	OAH	KAU	MOL	LAN	
OPHIOGLOSSACEAE	OPHIOGLOSSUM CONCINNUM	END	MAU	HAW	OAH		MOL	LAN	
ORCHIDACEAE	HABENARIA HOLOCHILA	THR	MAU		OAH	KAU	MOL		
PAPAVERACEAE	ARGEMONE GLAUCA VAR. INERMIS	END		HAW					KAH
PIPERACEAE	PEPEROMIA COOKIANA VAR. MINUTILIMBA	THR		HAW					
PIPERACEAE	PEPEROMIA CORNIFOLIA	END		HAW					
PIPERACEAE	PEPEROMIA DEGENERI	END					MOL		
PIPERACEAE	PEPEROMIA EXPALLESCENS VAR. BREVIPILOSA	THR	MAU						
PIPERACEAE	PEPEROMIA FAURIEI	END					MOL		
PIPERACEAE	PEPEROMIA FORBESII	END					MOL		
PIPERACEAE	PEPEROMIA HAUPUENSIS	THR				KAU			
PIPERACEAE	PEPEROMIA HELLERI VAR. KNUDSENII	THR				KAU			
PIPERACEAE	PEPEROMIA KULENSIS	END	MAU						
PIPERACEAE	PEPEROMIA LILIIFOLIA VAR. OBTUSATA	THR		HAW			MOL		
PIPERACEAE	PEPEROMIA MAUNAKEANA	END		HAW					
PIPERACEAE	PEPEROMIA OAHUENSIS VAR. ST.-JOHNII	END			OAH				
PIPERACEAE	PEPEROMIA PLINERVATA	EXT		HAW					
PIPERACEAE	PEPEROMIA RIGIDILIMBA	THR		HAW					
PIPERACEAE	PEPEROMIA SUBPETIOLATA	THR	MAU						
PIPERACEAE	PEPEROMIA TRELEASEI	END					MOL		
PIPERACEAE	PEPEROMIA WAIKAMOIANA	THR	MAU						
PITTOSPORACEAE	PITTOSPORUM ACUMINATUM VAR. LEPTOPODUM	END				KAU			
PITTOSPORACEAE	PITTOSPORUM ACUMINATUM VAR. MAGNIFOLIUM	END				KAU			
PITTOSPORACEAE	PITTOSPORUM ACUMINATUM VAR. WAIMEANUM	END				KAU			
PITTOSPORACEAE	PITTOSPORUM AMPLECTENS	THR		HAW					
PITTOSPORACEAE	PITTOSPORUM ARGENTIFOLIUM VAR. ARGENTIFOLIUM	THR	MAU						
PITTOSPORACEAE	PITTOSPORUM ARGENTIFOLIUM VAR. SESSILE	END	MAU						
PITTOSPORACEAE	PITTOSPORUM CAULIFLORUM VAR. CAULIFLORUM	EXT			OAH				
PITTOSPORACEAE	PITTOSPORUM CAULIFLORUM VAR. CLADANTHOIDES	END			OAH				
PITTOSPORACEAE	PITTOSPORUM CAULIFLORUM VAR. PEDICELLATUM	END			OAH				
PITTOSPORACEAE	PITTOSPORUM CLADANTHUM VAR. GRACILIPES	THR						LAN	
PITTOSPORACEAE	PITTOSPORUM CONFERTIFLORUM VAR. LONGIPES	THR	MAU						

ENDANGERED, THREATENED, AND RECENTLY EXTINCT SPECIES OF HAWAII

FAMILY	SPECIES	STATUS	RANGE						
			MAU	HAW	OAH	KAU	MOL	LAN	NII/LEEWARD
PITTOSPORACEAE	PITTOSPORUM CONFERTIFLORUM VAR. MICROPHYLLUM	THR						LAN	
PITTOSPORACEAE	PITTOSPORUM GLABRUM VAR. GLOMERATUM	EXT			OAH				
PITTOSPORACEAE	PITTOSPORUM GLABRUM VAR. INTERMEDIUM	END			OAH				
PITTOSPORACEAE	PITTOSPORUM GLABRUM VAR. TINIFOLIUM	THR	MAU						
PITTOSPORACEAE	PITTOSPORUM HALOPHILOIDES	THR						LAN	
PITTOSPORACEAE	PITTOSPORUM HALOPHILUM	THR					MOL		
PITTOSPORACEAE	PITTOSPORUM HELLERI	END				KAU			
PITTOSPORACEAE	PITTOSPORUM HOSMERI VAR. HOSMERI	END		HAW					
PITTOSPORACEAE	PITTOSPORUM HOSMERI VAR. SAINT-JOHNII	EXT		HAW					
PITTOSPORACEAE	PITTOSPORUM INSIGNE VAR. MICRANTHUM	END	MAU						
PITTOSPORACEAE	PITTOSPORUM KAHANANUM	THR			OAH				
PITTOSPORACEAE	PITTOSPORUM KAUAIENSE VAR. REPENS	EXT				KAU			
PITTOSPORACEAE	PITTOSPORUM TERMINALIOIDES VAR. LANAIENSE	END						LAN	
PITTOSPORACEAE	PITTOSPORUM TERMINALIOIDES VAR. MACROPUS	THR		HAW					
PITTOSPORACEAE	PITTOSPORUM TERMINALIOIDES VAR. MAUIENSE	END	MAU						
PLANTAGINACEAE	PLANTAGO PRINCEPS VAR. ACAULIS	THR			OAH				
PLANTAGINACEAE	PLANTAGO PRINCEPS VAR. DENTICULATA	THR					MOL		
PLANTAGINACEAE	PLANTAGO PRINCEPS VAR. ELATA	END			OAH				
PLANTAGINACEAE	PLANTAGO PRINCEPS VAR. LAXIFOLIA	END	MAU	HAW		KAU	MOL?		
PLANTAGINACEAE	PLANTAGO PRINCEPS VAR. PRINCEPS	END			OAH				
PLANTAGINACEAE	PLANTAGO PRINCEPS VAR. QUELENIANA	THR			OAH				
POACEAE	CENCHRUS AGRIMONIOIDES VAR. AGRIMONIOIDES	END	MAU		OAH			LAN	
POACEAE	CENCHRUS AGRIMONIOIDES VAR. LAYSANENSIS	END							LEEWARD
POACEAE	CENCHRUS PEDUNCULATUS	END			OAH				
POACEAE	DISSOCHONDRUS BIFLORUS	THR	MAU?	HAW?	OAH	KAU?	MOL	LAN	
POACEAE	ERAGROSTIS FOSBERGII	END			OAH				
POACEAE	ERAGROSTIS MAUIENSIS	END	MAU						
POACEAE	ERAGROSTIS NIIHAUENSIS	THR							NII
POACEAE	ERAGROSTIS PAUPERA	EXT			OAH				
POACEAE	ISCHAEMUM BYRONE	END	MAU	HAW	OAH		MOL		
POACEAE	PANICUM ALAKAIENSE	END				KAU			
POACEAE	PANICUM CARTERI	END			OAH				
POACEAE	PANICUM FAURIEI	END		HAW	OAH		MOL		
POACEAE	PANICUM LAMIATILE	END	MAU						
POACEAE	PANICUM LUSTRIALE	END	MAU						
POACEAE	PANICUM NIIHAUENSE	THR							NII
POACEAE	POA MANNII	EXT				KAU			
POACEAE	POA SANDVICENSIS	END				KAU			
POACEAE	POA SIPHONOGLOSSA	EXT				KAU			
POLYPODIACEAE	ADENOPHORUS PERIENS	END	MAU	HAW	OAH	KAU	MOL	LAN	
POLYPODIACEAE	ASPLENIUM FRAGILE VAR. INSULARIS	END	MAU	HAW					
POLYPODIACEAE	ASPLENIUM LEUCOSTEGIOIDES	EXT	MAU						
POLYPODIACEAE	CTENITIS SQUAMIGERA	EXT	MAU		OAH	KAU	MOL	LAN	

ENDANGERED, THREATENED, AND RECENTLY EXTINCT SPECIES OF HAWAII

FAMILY	SPECIES	STATUS	MAU	HAW	OAH	KAU	MOL	LAN	KAH
POLYPODIACEAE	DIELLIA ERECTA	END	MAU	HAW	OAH	KAU	MOL	LAN	
POLYPODIACEAE	DIELLIA FALCATA	END			OAH				
POLYPODIACEAE	DIELLIA LACINIATA	THR				KAU			
POLYPODIACEAE	DIELLIA MANNII	EXT				KAU			
POLYPODIACEAE	DIELLIA UNISORA	THR			OAH				
POLYPODIACEAE	DIPLAZIUM MOLOKAIENSE	END	MAU		OAH	KAU	MOL		
POLYPODIACEAE	PTERIS LYDGATEI	END	MAU		OAH		MOL		
PORTULACACEAE	PORTULACA HAWAIIENSIS	END		HAW					
PORTULACACEAE	PORTULACA SCLEROCARPA	END	MAU	HAW				LAN	KAH
PRIMULACEAE	LYSIMACHIA FILIFOLIA	THR			OAH?KAU				
PRIMULACEAE	LYSIMACHIA HILLEBRANDII VAR. HILLEBRANDII	END	MAU?		OAH		MOL?		
PRIMULACEAE	LYSIMACHIA KALALAUENSIS	END				KAU			
PRIMULACEAE	LYSIMACHIA OVATA	THR			OAH				
PRIMULACEAE	LYSIMACHIA SP. (FROM MAUI)	END	MAU						
RHAMNACEAE	ALPHITONIA PONDEROSA	END	MAU?HAW		OAH?KAU		MOL	LAN	
RHAMNACEAE	COLUBRINA OPPOSITIFOLIA	END		HAW	OAH				
RHAMNACEAE	GOUANIA BISHOPII	EXT	MAU						
RHAMNACEAE	GOUANIA CUCULLATA	EXT							KAH
RHAMNACEAE	GOUANIA FAURIEI	END					MOL		
RHAMNACEAE	GOUANIA GAGNEI	END			OAH				
RHAMNACEAE	GOUANIA HAWAIIENSIS	EXT		HAW					
RHAMNACEAE	GOUANIA HILLEBRANDI	END	MAU						
RHAMNACEAE	GOUANIA LYDGATEI	EXT	MAU						
RHAMNACEAE	GOUANIA MANNII	EXT						LAN	
RHAMNACEAE	GOUANIA MEYENII	EXT			OAH				
RHAMNACEAE	GOUANIA OLIVERI	EXT			OAH				
RHAMNACEAE	GOUANIA PILATA	EXT	MAU						
RHAMNACEAE	GOUANIA REMYI	EXT							KAH
RHAMNACEAE	GOUANIA SANDWICHIANA	EXT		HAW					
RHAMNACEAE	GOUANIA THINOPHILA	EXT	MAU						
RHAMNACEAE	GOUANIA VITIFOLIA	EXT			OAH				
ROSACEAE	ACAENA EXIGUA VAR. EXIGUA	THR				KAU			
ROSACEAE	ACAENA EXIGUA VAR. GLABERRIMA	END	MAU						
ROSACEAE	ACAENA EXIGUA VAR. GLABRIUSCULA	THR				KAU			
ROSACEAE	ACAENA EXIGUA VAR. SUBTUSSTRIGULOSA	THR				KAU			
RUBIACEAE	BOBEA SANDWICENSIS	END	MAU		OAH?		MOL	LAN	
RUBIACEAE	BOBEA TIMONIOIDES	END		HAW					
RUBIACEAE	COPROSMA FAURIEI VAR. LANAIENSIS	END						LAN	
RUBIACEAE	COPROSMA MONTANA VAR. ORBICULARIS	END		HAW					
RUBIACEAE	COPROSMA OCHRACEA VAR. KAALAE	END			OAH				
RUBIACEAE	COPROSMA PUBENS VAR. SESSILIFLORA	END						LAN	
RUBIACEAE	COPROSMA SERRATA	END		HAW					
RUBIACEAE	GARDENIA BRIGHAMII	END	MAU	HAW	OAH		MOL	LAN	

ENDANGERED, THREATENED, AND RECENTLY EXTINCT SPECIES OF HAWAII

FAMILY	SPECIES	STATUS	RANGE
RUBIACEAE	GOULDIA SP. (FROM KAUAI)	THR	KAU
RUBIACEAE	GOULDIA ST.-JOHNII VAR. MUNROI	THR	LAN
RUBIACEAE	GOULDIA TERMINALIS VAR. BOBEOIDES	END	LAN
RUBIACEAE	GOULDIA TERMINALIS VAR. CONGESTA	END	HAW
RUBIACEAE	GOULDIA TERMINALIS VAR. CRASSICAULIS	END	MAU
RUBIACEAE	GOULDIA TERMINALIS VAR. DEGENERI	END	OAH
RUBIACEAE	GOULDIA TERMINALIS VAR. LANAI	END	LAN
RUBIACEAE	GOULDIA TERMINALIS VAR. PARVIFOLIA	THR	MAU
RUBIACEAE	GOULDIA TERMINALIS VAR. PSEUDODICHOTOMA	END	LAN
RUBIACEAE	GOULDIA TERMINALIS VAR. PUBESCENS	END	MAU
RUBIACEAE	GOULDIA TERMINALIS VAR. QUADRANGULARIS	END	HAW
RUBIACEAE	GOULDIA TERMINALIS VAR. ROTUNDIFOLIA	END	MOL
RUBIACEAE	GOULDIA TERMINALIS VAR. SUBCORDATA	END	LAN
RUBIACEAE	HEDYOTIS ANGUSTA VAR. ANGUSTA	EXT	OAH
RUBIACEAE	HEDYOTIS ANGUSTA VAR. UMBROSA	END	OAH
RUBIACEAE	HEDYOTIS COOKIANA	EXT	HAW OAH
RUBIACEAE	HEDYOTIS CORIACEA	EXT	MAU HAW OAH
RUBIACEAE	HEDYOTIS DEGENERI VAR. COPROSMIFOLIA	END	OAH
RUBIACEAE	HEDYOTIS DEGENERI VAR. DEGENERI	END	OAH
RUBIACEAE	HEDYOTIS ELATIOR VAR. ELATIOR	THR	OAH KAU MOL
RUBIACEAE	HEDYOTIS ELATIOR VAR. HERBACEA	END	MAU
RUBIACEAE	HEDYOTIS FLUVIATILIS VAR. KAUAIENSIS	THR	KAU
RUBIACEAE	HEDYOTIS FOLIOSA	EXT	MAU
RUBIACEAE	HEDYOTIS FORMOSA	THR	MAU
RUBIACEAE	HEDYOTIS GLAUCIFOLIA VAR. HELLERI	THR	KAU
RUBIACEAE	HEDYOTIS LITTORALIS	END	MAU HAW OAH KAU MOL
RUBIACEAE	HEDYOTIS MANNII VAR. CUSPIDATA	THR	LAN
RUBIACEAE	HEDYOTIS MANNII VAR. MANNII	THR	MAU
RUBIACEAE	HEDYOTIS MANNII VAR. MUNROI	THR	LAN
RUBIACEAE	HEDYOTIS MANNII VAR. SCAPOSA	EXT	HAW
RUBIACEAE	HEDYOTIS PARVULA	END	OAH
RUBIACEAE	HEDYOTIS SCHLECHTENDAHLIANA VAR. NUTTALLI	EXT	OAH
RUBIACEAE	HEDYOTIS SCHLECHTENDAHLIANA VAR. PLANA	END	OAH MOL?
RUBIACEAE	HEDYOTIS SCHLECHTENDAHLIANA VAR. RETICULATA	END	MOL
RUBIACEAE	HEDYOTIS ST.-JOHNII	END	KAU
RUBIACEAE	HEDYOTIS THYRSOIDEA VAR. HILLEBRANDII	EXT	MOL
RUBIACEAE	HEDYOTIS THYRSOIDEA VAR. THYRSOIDEA	END	MOL
RUBIACEAE	MORINDA SANDWICENSIS	END	OAH
RUBIACEAE	MORINDA TRIMERA	THR	MAU OAH?
RUBIACEAE	PSYCHOTRIA GRANDIFLORA	END	KAU
RUBIACEAE	PSYCHOTRIA INSULARUM VAR. PARADISII	EXT	
RUTACEAE	PELEA ANISATA VAR. HAUPUANA	END	KAU
RUTACEAE	PELEA BALLOUI	END	MAU

ENDANGERED, THREATENED, AND RECENTLY EXTINCT SPECIES OF HAWAII

FAMILY	SPECIES	STATUS	RANGE
RUTACEAE	PELEA CHRISTOPHERSENII	END	OAH
RUTACEAE	PELEA CINEREA VAR. CINEREA	END	OAH
RUTACEAE	PELEA CINEREA VAR. MAUIANA	EXT	MAU
RUTACEAE	PELEA CINEREA VAR. SKOTTSBERGII	END	OAH
RUTACEAE	PELEA CINEREOPS	END	OAH
RUTACEAE	PELEA CLUSIIFOLIA VAR. PICKERINGII	END	HAW?
RUTACEAE	PELEA DEGENERI	END	KAU
RUTACEAE	PELEA DESCENDENS	END	OAH
RUTACEAE	PELEA ELLIPTICA VAR. MAUIENSIS	EXT	MAU
RUTACEAE	PELEA GLABRA	END	KAU
RUTACEAE	PELEA GRANDIFOLIA VAR. LIANOIDES	END	LAN
RUTACEAE	PELEA GRANDIFOLIA VAR. MONTANA	END	MAU
RUTACEAE	PELEA GRANDIFOLIA VAR. TERMINALIS	END	MAU HAW
RUTACEAE	PELEA HAUPUENSIS	END	KAU
RUTACEAE	PELEA HAWAIENSIS VAR. BRIGHAMII	END	MAU
RUTACEAE	PELEA HAWAIENSIS VAR. HAWAIENSIS	END	MAU HAW
RUTACEAE	PELEA HAWAIENSIS VAR. MOLOKAIANA	EXT	MOL
RUTACEAE	PELEA HAWAIENSIS VAR. PILOSA	EXT	MAU LAN
RUTACEAE	PELEA HAWAIENSIS VAR. RACEMIFLORA	EXT	MAU
RUTACEAE	PELEA HAWAIENSIS VAR. REMYANA	EXT	HAW
RUTACEAE	PELEA HAWAIENSIS VAR. RUBRA	END	MAU HAW
RUTACEAE	PELEA HAWAIENSIS VAR. SULFUREA	END	LAN
RUTACEAE	PELEA HIIAKAE	END	OAH
RUTACEAE	PELEA HOSAKAE	END	OAH
RUTACEAE	PELEA KAUAENSIS	END	OAH
RUTACEAE	PELEA KAVAIENSIS	END	KAU
RUTACEAE	PELEA KNUDSENII	EXT	KAU
RUTACEAE	PELEA LAKAE	END	OAH
RUTACEAE	PELEA LANCEOLATA	END	HAW
RUTACEAE	PELEA LEVEILLEI	END	KAU
RUTACEAE	PELEA LYDGATEI	END	OAH
RUTACEAE	PELEA MACROPUS	EXT	KAU
RUTACEAE	PELEA MAKAHAE	END	OAH
RUTACEAE	PELEA MUCRONULATA	EXT	MAU
RUTACEAE	PELEA MULTIFLORA	END	MAU
RUTACEAE	PELEA MUNROI	EXT	LAN
RUTACEAE	PELEA NEALAE	END	KAU
RUTACEAE	PELEA OBLONGIFOLIA VAR. MANUKAENSIS	END	HAW
RUTACEAE	PELEA OBLONGIFOLIA VAR. OBLONGIFOLIA	END	HAW MOL?
RUTACEAE	PELEA OBOVATA	EXT	MAU? LAN?
RUTACEAE	PELEA OLOWALUENSIS	END	MAU
RUTACEAE	PELEA ORBICULARIS VAR. ORBICULARIS	END	MAU
RUTACEAE	PELEA ORBICULARIS VAR. TONSA	END	MAU

ENDANGERED, THREATENED, AND RECENTLY EXTINCT SPECIES OF HAWAII

FAMILY	SPECIES	STATUS	MAU	HAW	OAH	KAU	MOL	LAN	LAY
RUTACEAE	PELEA OVALIS	END	MAU						
RUTACEAE	PELEA OVATA	END				KAU			
RUTACEAE	PELEA PALLIDA	END			OAH				
RUTACEAE	PELEA PANICULATA	END				KAU			
RUTACEAE	PELEA PARVIFOLIA VAR. APODA	END		HAW					
RUTACEAE	PELEA PARVIFOLIA VAR. SESSILIS	END					MOL		
RUTACEAE	PELEA PEDUNCULARIS VAR. CORDATA	END			OAH				
RUTACEAE	PELEA PEDUNCULARIS VAR. NIUENSIS	END			OAH				
RUTACEAE	PELEA PEDUNCULARIS VAR. NUMMULARIA	END			OAH				
RUTACEAE	PELEA PLUVIALIS	END				KAU			
RUTACEAE	PELEA QUADRANGULARIS	END				KAU			
RUTACEAE	PELEA RECURVATA	END				KAU			
RUTACEAE	PELEA REFLEXA	END					MOL		
RUTACEAE	PELEA SAINT-JOHNII VAR. ELONGATA	END			OAH				
RUTACEAE	PELEA SAINT-JOHNII VAR. SAINT-JOHNII	END			OAH				
RUTACEAE	PELEA SANDWICENSIS	EXT			OAH		MOL		
RUTACEAE	PELEA STOREYANA	EXT			OAH				
RUTACEAE	PELEA TOMENTOSA	EXT	MAU						
RUTACEAE	PELEA VOLCANICA VAR. KOHALAE	END		HAW					
RUTACEAE	PELEA WAHIAWAENSIS	END				KAU			
RUTACEAE	PELEA WAIMEAENSIS	EXT				KAU			
RUTACEAE	PELEA ZAHLBRUCKNERI	END		HAW					
RUTACEAE	PLATYDESMA REMYI	END		HAW					
RUTACEAE	ZANTHOXYLUM BLUETTIANUM	THR		HAW					
RUTACEAE	ZANTHOXYLUM DIPETALUM VAR. DEGENERI	THR			OAH				
RUTACEAE	ZANTHOXYLUM DIPETALUM VAR. DIPETALUM	THR			OAH				
RUTACEAE	ZANTHOXYLUM DIPETALUM VAR. GEMINICARPUM	END		HAW					
RUTACEAE	ZANTHOXYLUM DIPETALUM VAR. TOMENTOSUM	THR		HAW					
RUTACEAE	ZANTHOXYLUM GLANDULOSUM	THR	MAU						
RUTACEAE	ZANTHOXYLUM HAWAIIENSE VAR. CITRIODORUM	END						LAN	
RUTACEAE	ZANTHOXYLUM HAWAIIENSE VAR. HAWAIIENSE	THR		HAW					
RUTACEAE	ZANTHOXYLUM HAWAIIENSE VAR. VELUTINOSUM	THR		HAW					
RUTACEAE	ZANTHOXYLUM KAUAENSE VAR. KOHUANA	THR				KAU			
RUTACEAE	ZANTHOXYLUM MAVIENSE VAR. ANCEPS	THR		HAW					
RUTACEAE	ZANTHOXYLUM MAVIENSE VAR. CRANWELLIAE	THR					MOL		
RUTACEAE	ZANTHOXYLUM MAVIENSE VAR. MAVIENSE	THR	MAU					LAN	
RUTACEAE	ZANTHOXYLUM MAVIENSE VAR. RIGIDUM	THR	MAU						
RUTACEAE	ZANTHOXYLUM SEMIARTICULATUM VAR. SEMIARTICULATUM	THR			OAH				
RUTACEAE	ZANTHOXYLUM SEMIARTICULATUM VAR. SESSILE	END			OAH				
RUTACEAE	ZANTHOXYLUM SKOTTSBERGII	THR			OAH				
SANTALACEAE	EXOCARPUS GAUDICHAUDII	END	MAU	HAW	OAH	KAU	MOL	LAN	
SANTALACEAE	EXOCARPUS LUTEOLUS	END				KAU			
SANTALACEAE	SANTALUM ELLIPTICUM VAR. LITTORALE	END			OAH				LAY

ENDANGERED, THREATENED, AND RECENTLY EXTINCT SPECIES OF HAWAII

FAMILY	SPECIES	STATUS	RANGE
SANTALACEAE	SANTALUM LANAIENSE	END	LAN
SANTALACEAE	SANTALUM SALICIFOLIUM	END	
SAPINDACEAE	ALECTRYON MACROCOCCUM	END	MAU KAU MOL
SAPINDACEAE	ALECTRYON MAHOE	END	OAH
SAPINDACEAE	DODONAEA ERIOCARPA VAR. CONFERTIOR	END	HAW
SAPINDACEAE	DODONAEA ERIOCARPA VAR. COSTULATA	THR	LAN
SAPINDACEAE	DODONAEA ERIOCARPA VAR. FORBESII	END	HAW LAN
SAPINDACEAE	DODONAEA ERIOCARPA VAR. LANAIENSIS	END	LAN
SAPINDACEAE	DODONAEA ERIOCARPA VAR. MOLOKAIENSIS	END	MOL
SAPINDACEAE	DODONAEA ERIOCARPA VAR. OBLONGA	END	LAN
SAPINDACEAE	DODONAEA ERIOCARPA VAR. PALLIDA	END	MOL
SAPINDACEAE	DODONAEA ERIOCARPA VAR. SKOTTSBERGII	END	HAW
SAPINDACEAE	DODONAEA ERIOCARPA VAR. VARIANS	THR	LAN
SAPINDACEAE	DODONAEA SANDWICENSIS VAR. LATIFOLIA	THR	LAN
SAPINDACEAE	DODONAEA SANDWICENSIS VAR. SIMULANS	END	MOL
SAPINDACEAE	DODONAEA STENOPTERA VAR. FAURIEI	END	OAH
SAPINDACEAE	DODONAEA STENOPTERA VAR. STENOPTERA	END	MOL
SAPOTACEAE	POUTERIA AUAHIENSIS	END	MAU
SAPOTACEAE	POUTERIA RHYNCHOSPERMA	END	MAU
SOLANACEAE	NOTHOCESTRUM BREVIFLORUM VAR. BREVIFLORUM	END	HAW
SOLANACEAE	NOTHOCESTRUM BREVIFLORUM VAR. LONGIPES	EXT	
SOLANACEAE	NOTHOCESTRUM LATIFOLIUM	END	MAU HAW? KAU MOL LAN
SOLANACEAE	NOTHOCESTRUM LONGIFOLIUM VAR. RUFIPILOSUM	END	HAW
SOLANACEAE	NOTHOCESTRUM PELTATUM	END	KAU
SOLANACEAE	NOTHOCESTRUM SUBCORDATUM	EXT	OAH
SOLANACEAE	SOLANUM HALEAKALAENSE	EXT	MAU
SOLANACEAE	SOLANUM HILLEBRANDII	EXT	HAW
SOLANACEAE	SOLANUM INCOMPLETUM VAR. GLABRATUM	END	MAU LAN
SOLANACEAE	SOLANUM INCOMPLETUM VAR. INCOMPLETUM	END	HAW
SOLANACEAE	SOLANUM INCOMPLETUM VAR. MAUIENSIS	END	MAU
SOLANACEAE	SOLANUM KAUAIENSE	THR	KAU
SOLANACEAE	SOLANUM NELSONI VAR. NELSONI	END	MAU MOL NII LEEWARD
SOLANACEAE	SOLANUM NELSONI VAR. THOMASIAEFOLIUM	END	KAU
SOLANACEAE	SOLANUM SANDWICENSE	END	OAH KAU
STERCULIACEAE	WALTHERIA PYROLAEFOLIA	EXT	MAU
THEACEAE	EURYA SANDWICENSIS VAR. GRANDIFOLIA	END	OAH KAU
THYMELAEACEAE	WIKSTROEMIA BASICORDATA	END	OAH
THYMELAEACEAE	WIKSTROEMIA HANALEI	THR	KAU
THYMELAEACEAE	WIKSTROEMIA ISAE	THR	OAH
THYMELAEACEAE	WIKSTROEMIA LEPTANTHA	END	OAH
THYMELAEACEAE	WIKSTROEMIA MONTICOLA VAR. OCCIDENTALIS	END	MAU
THYMELAEACEAE	WIKSTROEMIA PERDITA	END	HAW
THYMELAEACEAE	WIKSTROEMIA SKOTTSBERGIANA	END	KAU

ENDANGERED, THREATENED, AND RECENTLY EXTINCT SPECIES OF HAWAII

FAMILY	SPECIES	STATUS	RANGE
THYMELAEACEAE	WIKSTROEMIA VILLOSA	END	MAU
URTICACEAE	HESPEROCNIDE SANDWICENSIS	END	HAW
URTICACEAE	NERAUDIA ANGULATA VAR. ANGULATA	END	OAH
URTICACEAE	NERAUDIA ANGULATA VAR. DENTATA	END	OAH
URTICACEAE	NERAUDIA COOKII	EXT	HAW
URTICACEAE	NERAUDIA KAHOOLAWENSIS	EXT	KAH
URTICACEAE	NERAUDIA KAUAIENSIS VAR. HELLERI	EXT	KAU
URTICACEAE	NERAUDIA KAUAIENSIS VAR. KAUAIENSIS	END	KAU
URTICACEAE	NERAUDIA MELASTOMIFOLIA VAR. GAUDICHAUDII	END	OAH
URTICACEAE	NERAUDIA MELASTOMIFOLIA VAR. MELASTOMIFOLIA	END	OAH
URTICACEAE	NERAUDIA MELASTOMIFOLIA VAR. PALLIDA	END	MAU
URTICACEAE	NERAUDIA MELASTOMIFOLIA VAR. PARVIFOLIA	END	OAH
URTICACEAE	NERAUDIA MELASTOMIFOLIA VAR. PUBESCENS	END	KAU
URTICACEAE	NERAUDIA MELASTOMIFOLIA VAR. UNCINATA	END	OAH
URTICACEAE	NERAUDIA OVATA	END	HAW
URTICACEAE	NERAUDIA SERICEA	END	MAU MOL LAN
URTICACEAE	URERA KAALAE	END	OAH
URTICACEAE	URERA KONAENSIS	EXT	HAW
VIOLACEAE	ISODENDRION FORBESII	END	KAU
VIOLACEAE	ISODENDRION HAWAIIENSE	EXT	HAW
VIOLACEAE	ISODENDRION HILLEBRANDII	EXT	OAH
VIOLACEAE	ISODENDRION HOSAKAE	EXT	HAW
VIOLACEAE	ISODENDRION LANAIENSE	EXT	LAN
VIOLACEAE	ISODENDRION LAURIFOLIUM	END	OAH
VIOLACEAE	ISODENDRION LONGIFOLIUM	END	OAH
VIOLACEAE	ISODENDRION LYDGATEI	EXT	OAH
VIOLACEAE	ISODENDRION MACULATUM	END	KAU
VIOLACEAE	ISODENDRION MOLOKAIENSE	EXT	MOL
VIOLACEAE	ISODENDRION PYRIFOLIUM	EXT	OAH
VIOLACEAE	ISODENDRION REMYI	EXT	NII
VIOLACEAE	ISODENDRION SUBSESSILIFOLIUM	END	KAU
VIOLACEAE	ISODENDRION WAIANAEENSE	EXT	OAH
VIOLACEAE	VIOLA CHAMISSONIANA	END	OAH
VIOLACEAE	VIOLA HELENA VAR. HELENA	END	KAU
VIOLACEAE	VIOLA HELENA VAR. LANAIENSIS	END	LAN
VIOLACEAE	VIOLA KAUAENSIS VAR. WAHIAWAENSIS	END	KAU
VIOLACEAE	VIOLA OAHUENSIS	END	OAH
VIOLACEAE	VIOLA ROBUSTA	END	MOL

Endangered, Threatened, and Recently Extinct Species of Puerto Rico and the Virgin Islands

With regard to the Endangered Species Act of 1973, the term "United States," when used in a geographical context, includes the Commonwealth of Puerto Rico and the U. S. Virgin Islands.

In this list of endangered and threatened species, those of Puerto Rico itself are indicated, as well as those of the adjacent islets under its administration: Desecheo, Mona, Muertos, Icacos, Culebra, Culebrita, Pineros, Vieques, and several smaller isles. The United States Virgin Islands are St. Thomas, St. John and St. Croix.

The British Virgin Islands, which lie in very close proximity to the U. S. Virgin Islands, are included in the lists because Puerto Rico and the U. S. and British Virgin Islands are treated together as a geographical, geological, and biological province in published floras, and the data are available for them as a unit. The British Virgin Islands comprise Anegada, Tortola, Virgin Gorda, Jost Van Dyke, and several smaller isles.

The first step in developing the present contribution was to make a list of endemic species, based on N. L. Britton and P. Wilson's flora (*Scientific Survey of Porto Rico and the Virgin Islands*, 1923-1940). Recent revisions of families and genera were then checked for changes in taxonomy of the groups since the time of Britton and Wilson, and the data was entered onto file cards. Papers dealing with new additions and changes in the flora were consulted; these are listed below (excluding the taxonomic revisions).

Two excellent volumes on trees of Puerto Rico and the Virgin Islands by E. L. Little, Jr. et al. (1964, 1974) have been especially useful, because rare or endangered endemic species are so designated therein.

The list of endemics was then circulated to various botanists having specialized knowledge of the status of the flora as a whole or of certain plant groups. The botanists are listed below; their comments are most gratefully acknowledged. The list was later compared with a further valuable reference: Rare and Endangered Plants of Puerto Rico - A Committee Report (1975). Mr. Roy O. Woodbury, botanist of the Puerto Rico Department of Natural Resources and the University of Puerto Rico, is the principal author of that report, which is published by the U. S. Department of Agriculture, Soil Conservation Service in cooperation with the Puerto Rico Department of Natural Resources.

One species is thought to be extinct, Solanum conocarpum from St. John. In the event of rediscovery, it is recommended that it be officially listed as endangered.

The total of native and introduced plant species in Puerto Rico and the Virgin Island is in excess of 2590. The present lists indicate that 102 species, or approximately 3.9 percent of the flora of Puerto Rico and the Virgin Islands, is endangered or threatened.

Contributors of information used for the lists are Dr. F. R. Fosberg, Dr. D. B. Lellinger, Dr. D. H. Nicolson, Miss Dulcie Powell, Dr. R. W. Read, Dr. T. R. Soderstrom, Dr. J.J. Wurdack, all in the Department of Botany, Smithsonian Institution; Dr. W. G. D'Arcy, Missouri Botanical Garden, St. Louis; Dr. R. A. Howard, Arnold Arboretum of Harvard University, Jamaica Plain, Massachusetts; Dr. A. H. Liogier, Jardin Botanico Dr. Rafael M. Moscoso, Santo Domingo, Republica Dominica; Dr. E. L. Little, Jr., U. S. Forest Service, Washington, D.C.; Mr. R. O. Woodbury, Santurce, Puerto Rico; Mr. J. A. Yntema, Bureau of Fish and Wildlife, Frederiksted, St. Croix.

ENDANGERED, THREATENED, AND RECENTLY EXTINCT SPECIES OF PUERTO RICO AND THE VIRGIN ISLANDS

FAMILY	SPECIES	STATUS	RANGE
ACANTHACEAE	DICLIPTERA KRUGII	ENDANGERED	PUERTO RICO
ACANTHACEAE	JUSTICIA BORINQUENSIS	ENDANGERED	PUERTO RICO
ACANTHACEAE	JUSTICIA CULEBRITAE	ENDANGERED	VIRGIN GORDA, CULEBRITA
AQUIFOLIACEAE	ILEX COOKII	ENDANGERED	PUERTO RICO
ARECACEAE	CALYPTRONOMA RIVALIS	ENDANGERED	PUERTO RICO
ASCLEPIADACEAE	CYNANCHUM MONENSE	ENDANGERED	MONA
ASCLEPIADACEAE	MARSDENIA ELLIPTICA	ENDANGERED	PUERTO RICO
ASTERACEAE	EUPATORIUM BORINQUENSE	THREATENED	PUERTO RICO
ASTERACEAE	EUPATORIUM DROSEROLEPIS	ENDANGERED	PUERTO RICO
ASTERACEAE	EUPATORIUM OTEROI	THREATENED	MONA
ASTERACEAE	MIKANIA STEVENSIANA	ENDANGERED	PUERTO RICO
ASTERACEAE	VERNONIA BORINQUENSIS	THREATENED	PUERTO RICO
BIGNONIACEAE	CRESCENTIA PORTORICENSIS	ENDANGERED	PUERTO RICO
BORAGINACEAE	CORDIA BELLONIS	ENDANGERED	PUERTO RICO
BORAGINACEAE	CORDIA RUPICOLA	THREATENED	ANEGADA, PUERTO RICO
BORAGINACEAE	CORDIA WAGNERORUM	ENDANGERED	PUERTO RICO
BORAGINACEAE	HELIOTROPIUM GUANICENSE	ENDANGERED	PUERTO RICO
BROMELIACEAE	TILLANDSIA LINEATISPICA	ENDANGERED	VIEQUES, CULEBRA, ST. JOHN
BUXACEAE	BUXUS VAHLII	ENDANGERED	PUERTO RICO, ST. CROIX
CACTACEAE	HARRISIA PORTORICENSIS	ENDANGERED	MONA, DESECHEO
CACTACEAE	LEPTOCEREUS QUADRICOSTATUS	THREATENED	PUERTO RICO
CACTACEAE	OPUNTIA BORINQUENSIS	ENDANGERED	PUERTO RICO
CANELLACEAE	PLEODENDRON MACRANTHUM	THREATENED	PUERTO RICO
CANNACEAE	CANNA PERTUSA	ENDANGERED	PUERTO RICO
CELASTRACEAE	MAYTENUS CYMOSA	THREATENED	PUERTO RICO, VIEQUES, ST. CROIX, ST. THOMAS, VIRGIN GORDA, PINEROS
CELASTRACEAE	MAYTENUS ELONGATA	THREATENED	PUERTO RICO
CELASTRACEAE	MAYTENUS PONCEANA	THREATENED	PUERTO RICO
CONVOLVULACEAE	IPOMOEA KRUGII	ENDANGERED	PUERTO RICO
CONVOLVULACEAE	OPERCULINA TRIQUETRA	ENDANGERED	ST. CROIX, ST. THOMAS
CUCURBITACEAE	ANGURIA COOKIANA	ENDANGERED	PUERTO RICO
CYATHEACEAE	CYATHEA DRYOPTEROIDES	ENDANGERED	PUERTO RICO
CYPERACEAE	CYPERUS URBANII	ENDANGERED	PUERTO RICO, VIEQUES
CYPERACEAE	SCLERIA DORADOENSIS	ENDANGERED	PUERTO RICO
EUPHORBIACEAE	CROTON FISHLOCKII	THREATENED	VIRGIN GORDA, ANEGADA, TORTOLA
EUPHORBIACEAE	CROTON IMPRESSUS	ENDANGERED	PUERTO RICO
FABACEAE	CAESALPINIA CULEBRAE	ENDANGERED	CULEBRA
FABACEAE	CAESALPINIA MONENSIS	THREATENED	MONA
FABACEAE	CAESALPINIA PORTORICENSIS	ENDANGERED	PUERTO RICO
FABACEAE	CASSIA EXUNGUIS	ENDANGERED	PUERTO RICO
FABACEAE	CASSIA MIRABILIS	ENDANGERED	PUERTO RICO
FABACEAE	GALACTIA EGGERSII	THREATENED	ST. JOHN, ST. THOMAS, TORTOLA
FABACEAE	SCHRANKIA PORTORICENSIS	ENDANGERED	PUERTO RICO

ENDANGERED, THREATENED, AND RECENTLY EXTINCT SPECIES OF PUERTO RICO AND THE VIRGIN ISLANDS

FAMILY	SPECIES	STATUS	RANGE
FABACEAE	STAHLIA MONOSPERMA	THREATENED	PUERTO RICO, VIEQUES
FLACOURTIACEAE	BANARA VANDERBILTII	ENDANGERED	PUERTO RICO
GESNERIACEAE	GESNERIA PAUCIFLORA	ENDANGERED	PUERTO RICO
ICACINACEAE	OTTOSCHULZIA RHODOXYLON	ENDANGERED	PUERTO RICO
LILIACEAE	AGAVE EGGERSIANA	ENDANGERED	ST. CROIX, ST. THOMAS
LORANTHACEAE	DENDROPEMON SINTENISII	ENDANGERED	PUERTO RICO
LYCOPODIACEAE	LYCOPODIUM PORTORICENSIS	ENDANGERED	PUERTO RICO
MALPIGHIACEAE	MALPIGHIA PALLENS	ENDANGERED	ST. CROIX
MALVACEAE	ABUTILON COMMUTATUM	ENDANGERED	PUERTO RICO
MALVACEAE	SIDA EGGERSII	ENDANGERED	CULEBRA, TORTOLA, JOST VAN DYKE
MELIACEAE	TRICHILIA TRIACANTHA	ENDANGERED	PUERTO RICO
MYRTACEAE	CALYPTRANTHES LUQUILLENSIS	THREATENED	PUERTO RICO
MYRTACEAE	CALYPTRANTHES PEDUNCULARIS	ENDANGERED	PUERTO RICO
MYRTACEAE	CALYPTRANTHES THOMASIANA	ENDANGERED	VIEQUES, ST. THOMAS
MYRTACEAE	CALYPTRANTHES TRIFLORUM	ENDANGERED	PUERTO RICO
MYRTACEAE	EUGENIA HAEMATOCARPA	THREATENED	PUERTO RICO
MYRTACEAE	EUGENIA MARGARETTAE	ENDANGERED	PUERTO RICO
MYRTACEAE	EUGENIA UNDERWOODII	ENDANGERED	PUERTO RICO
MYRTACEAE	MARLIEREA SINTENISII	THREATENED	PUERTO RICO
MYRTACEAE	MYRCIA PAGANII	THREATENED	PUERTO RICO
MYRTACEAE	PSIDIUM SINTENISII	THREATENED	PUERTO RICO
OLACACEAE	SCHOEPFIA ARENARIA	ENDANGERED	PUERTO RICO
ORCHIDACEAE	BRACHIONIDIUM CILIOLATUM	ENDANGERED	PUERTO RICO
ORCHIDACEAE	EPIDENDRUM BRITTONIANUM	THREATENED	MONA
ORCHIDACEAE	EPIDENDRUM KRUGII	ENDANGERED	PUERTO RICO
ORCHIDACEAE	EPIDENDRUM LACERUM	ENDANGERED	PUERTO RICO
ORCHIDACEAE	EPIDENDRUM SINTENISII	ENDANGERED	PUERTO RICO
ORCHIDACEAE	LEPANTHES DODIANA	ENDANGERED	PUERTO RICO
ORCHIDACEAE	LEPANTHES ELTOROENSIS	ENDANGERED	PUERTO RICO
PIPERACEAE	PEPEROMIA WHEELERI	ENDANGERED	CULEBRA
POACEAE	ARISTIDA PORTORICENSIS	ENDANGERED	PUERTO RICO
POACEAE	PANICUM STEVENSIANUM	ENDANGERED	PUERTO RICO
POLYGALACEAE	POLYGALA COWELLII	THREATENED	PUERTO RICO
PORTULACACEAE	PORTULACA CAULERPOIDES	THREATENED	MONA, MUERTOS
RUBIACEAE	ERITHALIS REVOLUTA	ENDANGERED	PUERTO RICO
RUBIACEAE	MITRACARPUS MAXWELLIAE	ENDANGERED	PUERTO RICO
RUBIACEAE	MITRACARPUS POLYCLADUS	ENDANGERED	PUERTO RICO
RUBIACEAE	RANDIA PORTORICENSIS	ENDANGERED	PUERTO RICO
RUTACEAE	RAVENIA URBANII	THREATENED	PUERTO RICO
RUTACEAE	ZANTHOXYLUM THOMASIANUM	ENDANGERED	PUERTO RICO, ST. JOHN, ST. THOMAS
SOLANACEAE	BRUNFELSIA PORTORICENSIS	THREATENED	PUERTO RICO
SOLANACEAE	GOETZEA ELEGANS	ENDANGERED	PUERTO RICO
SOLANACEAE	SOLANUM CONOCARPUM	EXTINCT	ST. JOHN

ENDANGERED, THREATENED, AND RECENTLY EXTINCT SPECIES OF PUERTO RICO AND THE VIRGIN ISLANDS

FAMILY	SPECIES	STATUS	RANGE
SOLANACEAE	SOLANUM DRYMOPHILUM	THREATENED	PUERTO RICO
SOLANACEAE	SOLANUM MUCRONATUM	ENDANGERED	PUERTO RICO
SOLANACEAE	SOLANUM WOODBURYI	ENDANGERED	PUERTO RICO
STYRACACEAE	STYRAX PORTORICENSIS	THREATENED	PUERTO RICO
THEACEAE	LAPLACEA PORTORICENSIS	THREATENED	PUERTO RICO
THEACEAE	TERNSTROEMIA LUQUILLENSIS	THREATENED	PUERTO RICO
THEACEAE	TERNSTROEMIA SUBSESSILIS	THREATENED	PUERTO RICO
THEOPHRASTACEAE	JACQUINIA UMBELLATA	ENDANGERED	PUERTO RICO
THYMELAEACEAE	DAPHNOPSIS HELLERIANA	ENDANGERED	PUERTO RICO
URTICACEAE	PILEA LEPTOPHYLLA	ENDANGERED	PUERTO RICO
URTICACEAE	PILEA MULTICAULIS	ENDANGERED	PUERTO RICO
URTICACEAE	PILEA RICHARDII	ENDANGERED	PUERTO RICO, ST. THOMAS
URTICACEAE	PILEA YUNQUENSIS	THREATENED	PUERTO RICO
VERBENACEAE	CALLICARPA AMPLA	THREATENED	PUERTO RICO
VERBENACEAE	CORNUTIA OBOVATA	THREATENED	PUERTO RICO
VERBENACEAE	PRIVA PORTORICENSIS	ENDANGERED	PUERTO RICO
ZANNICHELLIACEAE	RUPPIA ANOMALA	ENDANGERED	PUERTO RICO

References Used in Compiling Lists for Puerto Rico and the Virgin Islands

Britton, N. L., and P. Wilson. Scientific Survey of Porto Rico and the Virgin Islands. Volume 5(1923/1924), 626 pp. Volume 6(1925/1930), 636 pp. New York: New York Academy of Science.

D'Arcy, W. G. Annotated Checklist of the Dicotyledons of Tortola, Virgin Islands. Rhodora 69: 385-450 (1967).

D'Arcy, W. G. The Island of Anegada and its Flora. Atoll Research Bulletin (Smithsonian Institution) 139: 1-21 (1971).

D'Arcy, W. G. Anegada Island: Vegetation and Flora. Atoll Research Bulletin (Smithsonian Institution) 188: 1-40 (1975).

Fosberg, F. R. Revisions in the Flora of St. Croix, U. S. Virgin Islands. Rhodora 78(813): 79-119 (1976).

Howard, R. A. Notes on Some Plants of Puerto Rico. Journal of Arnold Arboretum 47: 137-146 (1966).

Howard, R. A. The Vegetation of the Antilles. In A. Graham, editor, Vegetation and Vegetational History of Northern Latin America, pp. 1-38. (1973).

Liogier, Alain H. Novitates Antillanae, I. Bulletin of Torrey Botanical Club 90(2): 186-192 (1963).

Liogier, Alain H. Nomenclatural Changes and Additions to Britton and Wilson's "Flora of Porto Rico and the Virgin Islands." Rhodora 67: 315-361 (1965).

Liogier, Alain H. Novitates Antillanae, II. Bulletin of Torrey Botanical Club 92(4): 288-304 (1965).

Liogier, Alain H. Further Changes and Additions to the Flora of Porto Rico and the Virgin Islands. Rhodora 69: 372-376 (1967).

Little, E. L., Jr. Trees of Mona Island. Caribbean Forester 16(1-2): 36-53 (January-June 1955).

Little, E. L., Jr. Trees of Jost Van Dyke (British Virgin Islands) . Forest Service Research Paper ITF-9, 12 pp. Institute of Tropical Forestry, Rio Piedras, Puerto Rico and U. S. Dept. of Agriculture, Forest Service. (August 1969)

Little, E. L., Jr. Relationships of Trees of the Luquillo Experimental Forest. Chapter B-3 (pp. B-47 to B-58) in H. T. Odum and R. F. Pigeon, editors, A Tropical Rain Forest: A Study of Irradiation and Ecology at El Verde, Puerto Rico. U. S. Atomic Energy Commission, Washington, D. C. (1970)

Little, E. L., Jr., and F. H. Wadsworth. Common Trees of Puerto Rico and the Virgin Islands. Agriculture Handbook No. 249; 548 pp. U. S. Dept. of Agriculture, Forest Service. (July 1964)

Little, E. L., Jr., and R. O. Woodbury. Trees of the Caribbean National Forest, Puerto Rico. Forest Service Research Paper ITF-20, 27 pp. U. S. Dept. of Agriculture, Forest Service (September 1976).

Little, E. L., Jr., R. O. Woodbury, and F. H. Wadsworth. Trees of Puerto Rico and the Virgin Islands - Second Volume. Agriculture Handbook No. 449, 1024 pp. U. S. Dept. of Agriculture, Forest Service. (September 1974)

Little, E. L., Jr., R. O. Woodbury and F. H. Wadsworth. Flora of Virgin Gorda (British Virgin Islands). Forest Service Research Paper ITF-21, 36 pp. Institute of Tropical Forestry, Rio Piedras, Puerto Rico and U. S. Dept. of Agriculture, Forest Service. (September 1976).

232

Woodbury, R. O. Rare and Endangered Plants of Puerto Rico: A Committee Report. 85 pp. U. S. Dept. of Agriculture, Soil Conservation Service and Dept. of Natural Resources, Commonwealth of Puerto Rico. (1975)

Woodbury, R. O., and E. L. Little, Jr. Flora of Buck Island Reef National Monument (U. S. Virgin Islands). Forest Service Research Paper ITF-19, 27 pp. Institute of Tropical Forestry, Rio Piedras, Puerto Rico and U. S. Dept. of Agriculture, Forest Service. (August 1976)

Woodbury, R. O., L. F. Martorell, and J. C. Garcia Tuduri. The Flora of Desecheo Island, Puerto Rico. Journal of Agriculture of the University of Puerto Rico 55(4): 478-505 (October 1971).

COORDINATION

The U. S. Department of Agriculture has been active with respect to endangered and threatened plants. For several years the Forest Service has listed rare and endangered plants that occur in the national forests. These lists have been made available to the Smithsonian Institution. The Forest Service also has made a detailed analysis of the rare and endangered trees of the United States and has determined their exact distribution and status. Investigations of sensitive species on Forest Service lands are being conducted in some regional offices, and the results are being made available to the Smithsonian on a reciprocal basis.

The National Arboretum has an interest in rare and endangered plants, especially ornamentals and other species of direct value to man. The Soil Conservation Service also has contributed to the program in the Department of Agriculture.

Canada has developed a list of endangered and rare plants. At the Smithsonian Workshop, the Chairman of the Canadian Committee on Rare and Endangered Plants reviewed the progress on the list. Canadian and American efforts will be coordinated.

Data sheets on United States plants will be submitted by the Smithsonian for inclusion in the *Red Data Book, Volume 5: Angiospermae,* published by the International Union for the Conservation of Nature and Natural Resources, under the aegis of its Survival Service Commission, which has established a Threatened Plants Committee. This Committee is developing a world list of endangered and vulnerable flowering plants that is being issued at intervals in the *Red Data Book.*

The methods developed by the United States, including computer lists of species and information sheets, should prove to be of great value in developing this world list. Thus, the United States will be able to assist countries throughout the world in preserving valuable plant species--hopefully before more extinction occurs.

CONTRIBUTORS OF DATA USED IN REVISING THE LISTS
FOR THE UNITED STATES

Name	*Organization*	*Data*
Amerson, P. A.	Espey, Huston & Associates, Austin, Texas	*Physostegia*
Barneby, R. C.	New York Botanical Garden	Fabaceae
Benson, L.	Pomona College	Cactaceae
Brohn, A.	Missouri Department of Conservation	Missouri
Browne, E. T.	Memphis State University	*Apios*
Canfield, J. E.	University of Washington	*Douglasia*
Churchill, J. A.	Birmingham, Michigan	General
Clausen, R. T.	Cornell University	*Sedum*
Cooperrider, T. S.	Kent State University	Ohio
Cronquist, A.	New York Botanical Garden	Asteraceae
Cummings, E.	San Francisco Bay National Wildlife Refuge	*Lasthenia*
D'Arcy, W. G.	Missouri Botanical Garden	Solanaceae
Dorn, R. D.	Bureau of Land Management, Rawlins, Wyoming	Montana, Wyoming
Drapalik, D. J.	Georgia Southern College	*Matelea*
Duncan, W. H.	University of Georgia	Georgia
Ettman, J.	Kentucky Department of Parks	Orchidaceae
Evans, A. M.	University of Tennessee	Tennessee
Feddema, C.	U.S. Forest Service	Colorado, Wyoming
Fisher, T. R.	Bowling Green State University	*Silphium*
Folkerts, G. W.	Auburn University	Carnivorous
Gentry, H. S.	Desert Botanical Garden, Phoenix	*Agave*

Name	*Organization*	*Data*
Gibson, T. C.	University of Arizona	*Sarracenia*
Gunn, C. R.	U.S. Department of Agriculture	*Vicia*
Hardin, J. W.	North Carolina State University	North Carolina
Heckard, L. R.	University of California, Berkeley	*Cordylanthus*
Herbst, D.	H. L. Lyon Arboretum	Hawaii
Hermann, F. J.	U.S. Department of Agriculture	*Carex*
Howard, A. Q.	University of Califronia, Berkeley	*Lupinus*
Iltis, H. H.	University of Wisconsin	Capparidaceae
Isley, D.	Iowa State University	Fabaceae
Johnston, M. C.	University of Texas, Austin	Texas
Koelling, A. C.	Illinois State Museum	Illinois
Kral, R.	Vanderbilt University	Southeast
Kruckeberg, A. R.	University of Washington	Washington
Lamoureux, C. B.	University of Hawaii	Hawaii
Lazell, J. D.	Massachusetts Audubon Society	Mississippi
Lehto, E.	Arizona State University	Arizona
Lyons, G.	Huntington Botanical Gardens	Cactaceae, Liliaceae
Mayes, R.	Virginia Polytechnic Institute	*Pyrrocoma*
McDaniel, S.	Mississippi State University	Mississippi
McGregor, R. L.	University of Kansas	Kansas
McWilliam, A. L.	Mena, Arkansas	Arkansas
Mahler, W. F.	Southern Methodist University	*Physostegia*
Mathias, M. E.	University of California	Apiaceae
Meijer, W.	University of Kentucky	Kentucky
Mitchell, R.	New York State Museum	Polygonaceae

Name	Organization	Data
Moran, R.	San Diego Natural History Museum	Crassulaceae
Musselman, L. J.	Old Dominion University	*Schwalbea*
Newland, K. C.	Boyce Thompson Southwestern Arboretum	Succulents
Philbrick, R. N.	Santa Barbara Botanic Garden	Santa Barbara Is.
Piehl, M.	Waterville, Ohio	Louisiana
Pierce, P.	Albuquerque, New Mexico	Cactaceae
Pittillo, J. D.	Western Carolina University	North Carolina
Popenoe, J.	Fairchild Tropical Garden	*Asimina*
Powell, W. R.	University of California, Davis	California
Quarterman, E.	Vanderbilt University	Tennessee
Radford, A. E.	University of North Carolina	North Carolina
Raven, P. H.	Missouri Botanical Garden	Onagraceae
Reed, C. F.	Baltimore, Maryland	General
Reeder, C. G.	University of Wyoming	Wyoming
Reveal, J. L.	University of Maryland	*Eriogonum*, Utah
Richardson, J.	Southern Illinois University	Cactaceae
Riskind, D. H.	Texas Parks and Wildlife Department	Texas
Rogers, C. M.	Wayne State University	*Linum*
Rollins, R. C.	Harvard University	Brassicaceae
Schnell, D. E.	Statesville, North Carolina	Carnivorous
Schuyler, A. E.	Academy of Natural Sciences, Philadelphia	*Scirpus*
Sherman, H. L.	Mississippi University for Women	*Schoenolirion*
Siddall, J. L.	Oregon Nature Conservancy	Oregon
Skinner, H. T.	U.S. National Arboretum	*Rhododendron*
Smith, E. B.	University of Arkansas	*Coreopsis*

Name	*Organization*	*Data*
Smith, L. C.	Birmingham, Alabama	*Leptogramma*
Smith, S. J.	New York State Museum	New York
Soule, L. T.	U.S. Forest Service	Oregon, Washington
Spellenberg, R. W.	New Mexico State University	New Mexico
Stuckey, R. L.	Ohio State University	Ohio
Taylor, C. & R. J.	Southeastern Oklahoma State University	Oklahoma
Terrell, E. E.	U.S. Department of Agriculture	*Zizania, Hedyotis*
Thomas, J. L.	University of North Carolina	Alabama
Thorne, R. F.	Rancho Santa Ana Botanic Garden	California
Todsen, T. K.	Las Cruces, New Mexico	*Cirsium*
Treubig, R. J.	Louisiana Forestry Commission	Louisiana
Tucker, G.	Arkansas Polytechnic College	*Castanea*, Arkansas
Voigt, L. P.	Wisconsin Department of Natural Resources	Wisconsin
Wagner, W. L.	University of New Mexico	New Mexico
Ware, D. M.	College of William and Mary	*Valerianella*
Watson, J. R.	Mississippi State University	Mississippi
Weber, W. A.	University of Colorado Museum	Colorado
Welsh, S. L.	Brigham Young University	Alaska, Utah
Wendt, T.	University of Texas, Austin	Texas
Wherry, E. T.	University of Pennsylvania	Pennsylvania
Williams, M. J.	Northern Nevada Native Plant Society	Nevada

INVITED PARTICIPANTS IN WORKSHOP ON ENDANGERED AND
THREATENED HIGHER PLANTS OF THE UNITED STATES (1974)

Dr. Edward S. Ayensu (Convener)
Chairman
Department of Botany
Smithsonian Institution
Washington, D. C. 20560

Dr. Ralph I. Blouch
Senior Wildlife Executive
Department of Natural Resources
Stevens T. Mason Building
Lansing, Michigan 48926

Ms. Joan E. Canfield
Museum Technician
Department of Botany
Smithsonian Institution
Washington, D. C. 20560

Dr. David Challinor
Assistant Secretary for Science
Smithsonian Institution
Washington, D. C. 20560

Mr. Thomas L. Cobb
Administrative Assistant, Forestry
National Parks and Conservation
Association
1701 18th Street, N. W.
Washington, D. C. 20009

Mr. John E. Cooper
Director of Research and Collections
N.C. State Museum of Natural History
Jones and Halifax Streets
Raleigh, North Carolina 27611

Dr. Charles T. Cushwa
Staff, Range/Wildlife Scientist
Division of Forest Environment
Research
Forest Service
U.S. Department of Agriculture
Washington, D. C. 20250

Dr. O. Keister Evans
Executive Director
American Horticultural Society
Mt. Vernon, Virginia 22121

Dr. F. Raymond Fosberg
Curator, Department of Botany
Smithsonian Institution
Washington, D. C. 20560

Dr. Jerry F. Franklin
Program Director
Ecosystem Analysis
Division of Biology and Medicine
National Science Foundation
Washington, D. C. 20550

Ms. Margaret C. Gaynor
Special Assistant to the Secretary
Smithsonian Institution
Washington, D. C. 20560

Mr. Michael R. Huxley
Deputy Assistant Secretary for Science
Smithsonian Institution
Washington, D. C. 20560

Dr. Howard S. Irwin
President
New York Botanical Garden
Bronx, New York 10458

Dr. Dale W. Jenkins
Consultant to Endangered Flora Project
Department of Botany
Smithsonian Institution
Washington, D. C. 20560

Dr. Robert E. Jenkins
Vice President for Science
The Nature Conservancy
Arlington, Virginia 22209

Dr. Marshall C. Johnston
Director, Rare Plant Study Center
Department of Botany
University of Texas
Austin, Texas 78712

Dr. Porter M. Kier
Director
National Museum of Natural History
Smithsonian Institution
Washington, D.C. 20560

Dr. Robert Kral
Curator of Herbarium
Department of Biology
Vanderbilt University
Nashville, Tennessee 37203

Dr. Arthur R. Kruckeberg
Professor and Chairman
Department of Botany
University of Washington
Seattle, Washington 98195

Dr. Gerald A. Lieberman
Ecologist
The Nature Conservancy
Arlington, Virginia 22209

Dr. Elbert L. Little, Jr.
Chief Dendrologist
Forest Service
U.S. Department of Agriculture
Washington, D.C. 20250

Dr. Ronald L. McGregor
Director, State Biological Survey
  of Kansas
University of Kansas
Lawrence, Kansas 66044

Dr. James F. Mello
Assistant Director
National Museum of Natural History
Smithsonian Institution
Washington, D.C. 20560

Dr. Frederick G. Meyer
Taxonomist
National Arboretum
U.S. Department of Agriculture
Washington, D.C. 20002

Dr. J. K. Morton
Professor and Chairman
Department of Botany
University of Waterloo
Waterloo, Ontario, Canada

Dr. Robert Ornduff
Director of Botanical Garden
University of California
Berkeley, California 94720

Mr. Frank Potter
Majority Counsel
Office of the House Merchant
  Marine Committee
1334 Longworth Building
Washington, D.C. 20555

Dr. W. Robert Powell
Director, CNPS Rare
  Plant Project
Department of Agronomy and Range Science
University of California
Davis, California 95616

Dr. James L. Reveal
Department of Botany
University of Maryland
College Park, Maryland   20901

Dr. Jack E. Schmautz
Staff Specialist of Range Ecology
Division of Range Management
Forest Service
U.S. Department of Agriculture
Washington, D. C.   20250

Dr. Thomas N. Shiflet
Chief Range Conservationist
Soil Conservation Service
U.S. Department of Agriculture
Washington, D. C.   20250

Dr. Ronald O. Skoog
Chief, Office of Endangered Species and
   International Activities
Fish and Wildlife Service
U.S. Department of the Interior
Washington, D. C.   20240

Dr. Dixie R. Smith
Principal Range/Widlife Scientist
Division of Forest Environment Research
Forest Service
U.S. Department of Agriculture
Washington, D. C.   20250

Major James Stevenson
Chief Naturalist
Division of Recreation and Parks
Department of Natural Resources
Tallahassee, Florida   32304

Dr. Lee Talbot
Senior Scientist
Council on Environmental Quality
Washington, D.C.   20006

Dr. Warren H. Wagner, Jr.
Professor and Curator
Department of Botany
University of Michigan
Ann Arbor, Michigan   48104

Dr. Daniel B. Ward
Curator, The Herbarium
Department of Botany
University of Florida
Gainesville, Florida   32601

Dr. Caroll E. Wood, Jr.
Professor of Biology and Curator
Arnold Arboretum of
   Harvard University
Cambridge, Massachusetts   02138

Dr. John J. Wurdack
Curator, Department of Botany
Smithsonian Institution
Washington, D. C.   20560

BIBLIOGRAPHIES

State Lists of Endangered and Threatened Plant Species

ALABAMA

Endangered and Threatened Plants and Animals of Alabama. 1976.
J. L. Thomas, in *Bulletin of the Alabama Museum of Natural History*,
No. 2. Alabama Museum of Natural History, University of Alabama,
Box 5897, University 35496.

ARIZONA

Cacti and Succulents and Their Status as Endangered Species. 1974.
W. Hubert Earle, Desert Botanical Garden, P. O. Box 5415, Phoenix
85010. Unpublished manuscript.

Rare and Endangered Plants of Arizona. 1973. Anonymous.

ARKANSAS

Threatened Native Plants of Arkansas. 1974. G. E. Tucker, in
Arkansas Natural Area Plan. Arkansas Department of Planning,
400 Train Station Square, Victory at Markham, Little Rock 72201.

CALIFORNIA

Inventory of Rare and Endangered Vascular Plants of California.
1974. W. R. Powell, editor. California Native Plant Society,
Room 317, 2490 Channing Way, Berkeley 94704.

COLORADO

List of Endangered and Threatened Plants in Colorado. 1975.
W. A. Weber, University of Colorado Museum, Boulder 80302.
Unpublished manuscript.

244

FLORIDA

Rare and Endangered Florida Plants. 1974. *Florida Flora Newsletter*
No. 19. D. B. Ward, Botany Department, University of Florida,
Gainesville 32611.

Inventory of Rare and Endangered Biota of Florida. 1976.
Florida Audubon Society, P. O. Drawer 7, Maitland 32751. Microfiche.

GEORGIA

Rare and Endangered Vascular Plants of Georgia. 1973. W. H. Duncan,
Botany Department, University of Georgia, Athens 30602.
Unpublished manuscript.

Endangered Species of Georgia: Proceedings of the 1974 Conference,
Sponsored by the Georgia Department of Natural Resources. 1974.
J. L. McCollum, editor. Natural Areas Unit, Office of Planning
and Research, Georgia Department of Natural Resources, Atlanta
30334.

Endangered, Threatened and Unusual Plants-Protected Plants List.
1976. J. L. McCollum, Outdoors in Georgia 5(9): 27-31 (September).

HAWAII

Rare and Endangered Species of Hawaiian Vascular Plants. 1975.
F. R. Fosberg and D. Herbst, Allertonia 1(1): 1-72 (March).

ILLINOIS

Interim List of Extinct, Endangered, Threatened, Vulnerable, Rare and
Restricted Plant Species in Illinois. 1976. G. A. Paulson and
J. Schwegman, Illinois Nature Preserves Commission and Illinois
Department of Conservation, 819 North Main Street, Rockford 61103.
Unpublished manuscript.

INDIANA

Rare and Endangered Plants in Indiana. 1975. W. B. Barnes,
Division of Nature Preserves, Indiana Department of Natural
Resources, 616 State Office Building, Indianapolis 46204.
Unpublished manuscript.

KANSAS

Rare and Endangered Plants in Kansas. 1973. J. E. Bare and R. L.
McGregor, Division of Biological Sciences, University of Kansas,
Lawrence 66044. Unpublished manuscript.

LOUISIANA

Rare and Threatened Vascular Plants in Louisiana. 1975. M. G.
Curry, VTN Louisiana, 2701 Independence Street, Metairie
70002. Unpublished manuscript.

MICHIGAN

Michigan Endangered Plants. 1976. W. H. Wagner, Jr., in *Michigan's
Endangered and Threatened Species Program*. Michigan Department
of Natural Resources, Stevens T. Mason Building, Lansing 48926.

MINNESOTA

Rare or Endangered Plants of Minnesota. 1972. T. Morley, Botany
Department, University of Minnesota, Minneapolis 55455.
Unpublished manuscript.

MISSISSIPPI

Rare and Endangered Plant Species in Mississippi. 1975. T. M.
Pullen, Biology Department, University of Mississippi, University
38677. Unpublished manuscript.

MISSOURI

Rare and Endangered Species of Missouri. 1974. Missouri Department of Conservation, 2901 North Ten Mile Drive, Jefferson City 65101, and U.S. Department of Agriculture, Soil Conservation Service.

NEBRASKA

Threatened and Endangered Species of Vascular Plants in Nebraska. 1975. U.S. Department of Agriculture, Soil Conservation Service, Lincoln, Nebraska 68508. Unpublished manuscript.

NEVADA

Endangered Plant Species of the Nevada Test Site, Ash Meadows, and Central-Southern Nevada. 1977. J. C. Beatley. U.S. Energy Research and Development Administration Contract E (11-1)-2307. Biological Sciences Department, University of Cincinnati, Cincinnati 45221.

Status of Endangered and Threatened Plant Species on Nevada Test Site, Part 1 - Endangered Species. 1977. W. A. Rhoads and M. P. Williams. E, G, and G, Inc., 130 Robin Hill Road, Goleta, California 93017.

NEW HAMPSHIRE

Endangered Plants of New Hampshire: A Selected List of Endangered Species. 1973. A. R. Hodgdon, Forest Notes, pages 2-6. Society for the Protection of New Hampshire Forests, 5 South State Street, Concord 03301.

NEW JERSEY

Rare or Endangered Vascular Plants of New Jersey. 1973. D. E. Fairbrothers and M. Y. Hough, Science Notes No. 14, New Jersey State Museum, Cultural Center, Trenton 08625.

NEW MEXICO

Endangered and Threatened Cacti of New Mexico. 1975. Prince Pierce, Alburquerque, New Mexico. Unpublished manuscript.

NEW YORK

Our Wild Flowers and a Program to Protect Them. 1975. J. W. Aldrich, The Conservationist, pp. 23-29, April-May.

NORTH CAROLINA

List of Endangered, Threatened, Extinct Plants in North Carolina. 1975. J. W. Hardin, Botany Department, North Carolina State University, Raleigh 27607. Unpublished manuscript.

Preliminary List of Endangered Plant and Animal Species in North Carolina. 1973. Assistant Secretary for Resource Management, North Carolina Department of Natural and Economic Resources, P. O. Box 27687, Raleigh 27611.

Rare Vascular Plants of Western North Carolina. 1974. J. D. Pittillo, Department of Biological Sciences, Western Carolina University, Cullowhee 28723. Unpublished manuscript.

NORTH DAKOTA

Rare and Endangered Plant Species in North Dakota. 1972. U.S. Department of Agriculture, Soil Conservation Service, Bismarck, North Dakota 58501. Unpublished manuscript.

OKLAHOMA

Rare and Endangered Vertebrates and Plants of Oklahoma. 1975. Rare and Endangered Species of Oklahoma Committee and U.S. Department of Agriculture, Soil Conservation Service, Stillwater 74074.

248

OREGON

Lists of Endangered and Threatened Plants in Oregon. 1975.

J. L. Siddall, The Nature Conservancy, Oregon Chapter, 535
Atwater Road, Lake Oswego, Oregon 97034. Unpublished manuscript.

PENNSYLVANIA

Lists of Endangered and Threatened Plants in Pennsylvania. 1975.

E. T. Wherry, Morris Arboretum, 1414 Meadowbrook Avenue,
Philadelphia 19118. Unpublished manuscript.

Rare Plants of Southeastern Pennsylvania. 1975-1976.

E. T. Wherry, Bartonia 44: 22-26.

SOUTH DAKOTA

Endangered Plants of South Dakota-Preliminary Draft. 1975. C. M.
Schumacher, South Dakota Endangered Species Plants Committee, U.S.
Department of Agriculture, Soil Conservation Service, Huron
57350. Unpublished manuscript.

TENNESSEE

Endangered Species of Plants in Middle Tennessee. 1973. E. Quarterman,
Biology Department, Vanderbilt University, Nashville 37235.
Unpublished manuscript.

Rare Plants of Tennessee. 1974. A. J. Sharp, The Tennessee
Conservationist 40(7): 20-31, July.

The Tennessee List of Possibly Extinct, Endangered, Threatened and
Special Concern Vascular Plants. 1976. B. E. Wofford, R. Kral,
A. M. Evans, H. R. DeSelm, and J. L. Collins (Tennessee Committee
for Rare Plants). Botany Department, University of Tennessee,
Knoxville 37916. Unpublished manuscript.

TEXAS

Rare and Endangered Plants Native to Texas. 1974. M. C. Johnston, Rare Plant Study Center, University of Texas at Austin, P. O. Box 8495, Austin 78712.

Provisional List of Texas' Threatened and Endangered Plant Species. 1975. D. Riskind, Texas Organization of Threatened and Endangered Species, Texas Parks and Wildlife Department, J. H. Reagan Building, Austin 78701.

UTAH

Endangered, Threatened, Extinct, Endemic, and Rare or Restricted Utah Vascular Plants. 1975. S. L. Welsh, N. C. Atwood and J. L. Reveal, The Great Basin Naturalist 35(4): 327-376 (December).

VERMONT

Rare and Endangered Vascular Plant Species in Vermont: Part 1, Monocotyledoneae. 1972. W. D. Countryman, Biology Department, Norwich University, Northfield, Vermont 05663. Unpublished manuscript.

VIRGINIA

Lists of Rare, Endangered, Endemic (Local), and Depleted Plants. 1973. L. J. Uttal. Biology Department, Virginia Polytechnic Institute and State University, Blacksburg 24061. Unpublished manuscript.

WASHINGTON

Partial Inventory of Rare, Threatened, and Unique Plants of Washington. 1974. A. R. Kruckeberg, in C. T. Dyrness, et al., *Research Natural Area Needs in the Pacific Northwest--A Contribution to Land-Use Planning*. Report on Natural Area Needs held November 29- December 1, 1973 at Wemme, Oregon. pages 311-319. Review Draft.

WEST VIRGINIA

Rare and Endangered Plant Species in West Virginia. Undated.

E. L. Core. West Virginia Department of Agriculture, Charleston
25305.

WISCONSIN

Endangered and Threatened Vascular Plants in Wisconsin. 1976.

R. H. Read. *Technical Bulletin* No. 92. Scientific Areas
Preservation Council, Department of Natural Resources, Madison
53701.

Endangered and Threatened Plants of the United States

Alcorn, S. M.

    1966. The Saguaro Cactus in Arizona. American Horticultural

        Magazine 45(3): 286-295.

Alfieri, S. A., A. P. Martinez, and C. Wehlburg

    1967. Stem and Needle Blight of Florida Torreya, Torreya taxifolia

        Arn. Proceedings of Florida State Horticultural Society

        80: 428-431.

American Institute of Biological Sciences

    1968. Papers from Symposium on The Role of the Biologist in

        Preservation of the Biotic Environment. BioScience 18(5):

        383-424.

Ayensu, E. S.

    1975. Endangered and Threatened Orchids of the United States.

        American Orchid Society Bulletin 44(5): 384-394.

    1977. Evaluating Impacts on Endangered and Threatened Flora. In

        Biological Evaluation of Environmental Impact. Proceedings

        of a Symposium sponsored by the American Institute of Biological

        Sciences, Ecological Society of America, and the Council

        on Environmental Quality, Tulane University, New Orleans,

        Louisiana.

Baetsen, R. H.

    1977. The Impact of Snowmobiling on Ground Layer Vegetation near

        Sage Lake, Ogemaw County, Michigan. The Michigan Botanist

        16(1): 19-25 (January).

252

Benedict, R. C.

 1928. How Shall We Save Rare Plant Species from Extinction? *Wild*

  Flower 5: 45-46.

Benson, L.

 1975. Cacti-Bizarre, Beautiful, but in Danger. *National Parks &*

  Conservation Magazine 49(7): 17-21.

 1976. Endangered Species - Heads in the Clouds or the Sand?

  Cactus and Succulent Journal 48(5): 207-212.

Branson, B. A.

 1974. Stripping the Appalachians. *Natural History* 83(9): 52-61.

Bryan, J. Y.

 1975. Organ Pipe: Trouble Along the Devil's Road. *National Parks*

  & Conservation Magazine 49(12): 4-9.

Burt, D. E.

 1973. The Geography of Extinct and Endangered Species in the United

  States. *The Explorer* 15(3): 4-10.

Case, F. W., and R. B. Case

 1974. *Sarracenia alabamensis*, a Newly Recognized Species from

  Central Alabama. *Rhodora* 76: 650-666.

Cathey, H. M.

 1976. Endangered Plants. *American Horticulturist* 55(6): 2-3.

Coblentz, B. E.

 1976. Wild Goats of Santa Catalina. *Natural History* 85(6): 71-77.

Connolly, L.

 1974. Chaparral Mismanagement. *Sierra Club Bulletin* 59(5): 10-11,

  30-31.

Cory, V. L.

    1952. The Disappearance of Plant Species from the Range in Texas.
        Wild Flower 28: 51-68.

Cowan, B. D.

    1975. Protecting and Restoring Native Dune Plants. Fremontia 3(2):
        3-7.

    1976. The Menace of Pampas Grass. Fremontia 4(2): 14-16.

Cox, C.

    1976. Our Unknown Vanishing Breeds. The Sciences 16(2): 21-24.

Crampton, B.

    1976. Rare Grasses in a Vanishing Habitat. Fremontia 4(3): 22-23.

Cranston, A.

    1976. The Battle for Death Valley. National Parks & Conservation
        Magazine 50(1): 4-9.

DeFilipps, R. A.

    1976 a. Conservation Action for Carnivorous Plants. Carnivorous
        Plant Newsletter 5(1): 8.

    1976 b. Endangered and Threatened Plants of Maryland. Presented at
        Maryland Endangered Species Symposium, Sponsored by the
        Chesapeake Audubon Society, at the University of Maryland-
        Baltimore County Campus, October 30, 1976.

Dolan, R., P. J. Godfrey, and W. E. Odum

    1973. Man's Impact on the Barrier Islands of North Carolina.
        American Scientist 61(2): 152-162.

Dolan, R., A. Howard and A. Gallenson

    1974. Man's Impact on the Colorado River in the Grand Canyon.
        American Scientist 62(4): 392-401.

254

Douglas, E.

    1975. Cactus Rustlers Beware: Arizona Means Business. Christian Science Monitor. page 26 (17 November).

DuMond, D. M.

    1973. A Guide for the Selection of Rare, Unique and Endangered Plants. Castanea 38: 387-395.

Duncan, W. H.

    1970. Endangered, Rare and Uncommon Wildflowers Found on the Southern National Forests. U.S. Department of Agriculture, Forest Service, Southern Region.

Editor.

    1975. How to Save a Wildflower. National Parks & Conservation Magazine 49(4): 10-14.

Elias, T. S.

    1975. Vascular Plants. In Proceedings of the Symposium on Endangered and Threatened Species of North America, Washington, D.C., 1974, pages 88-93. St. Louis, Missouri: Wild Canid Survival and Research Center.

    1976. Extinction is Forever: What Can Be Done to Save Endangered Species? Garden Journal 26(2): 52-55.

Emerson, B. H.

    1976. Botanical Will-O-The-Wisp: Franklinia. The Green Scene (Pennsylvania Horticultural Society) 4(4): 2-6

Faust, M. E.

    1969. Conservation of the Hart's-Tongue Fern in North America. Biological Conservation 1(3): 256-257.

Fay, R. C., et al.

1973. Southern California's Deteriorating Marine Environment.
Center for California Public Affairs, Claremont. (Kelp)

Federal Committee on Ecological Reserves.

1977. A Directory of Research Natural Areas on Federal Lands of
the United States of America. Washington, D. C.

Fernald, M. L.

1940. The Problem of Conserving Rare Native Plants. Smithsonian
Report for 1939, pages 375-391.

Fialka, J.

1975. Running Wild. National Wildlife 13(2): 36-41. (Off-road
vehicles).

Fosberg, F. R.

1971. Endangered Island Plants. Bulletin of the Pacific Tropical
Botanical Garden 1(3): 1-7.

1972. Our Native Plants, a Look to the Future. National Parks &
Conservation Magazine 46(11): 17-21.

1975. The Deflowering of Hawaii. National Parks & Conservation
Magazine 49(10): 4-10.

1976. Local Floras in Relation to Conservation. Jeffersonia 10(1):
1-2.

1977. Expert on Hawaiian Plants Responds. National Parks & Conservation
Magazine 51(2): 28-29

Franklin, J. F.

1977. The Biosphere Reserve Program in the United States.
Science 194(4275): 262-267.

Gagne, W. C.

    1975. Hawaii's Tragic Dismemberment. Defenders 50(6): 461-469.

Ghiselin, J.

    1973-1974. Wilderness and the Survival of Species: Declining
        Populations Lose Options in Recessive Genes. The Living
        Wilderness, winter issue, pages 22-27.

Gleason, P. J., editor

    1974. Environments of South Florida: Present and Past. 452 pages.
        Memoir 2: Miami Geological Survey.

Godfrey, R. K., and H. Kurz

    1962. The Florida Torreya Destined for Extinction. Science 136
        (3519): 900, 902; American Horticultural Magazine 42(1):
        65-66 (1963).

Goldstein, J.

    1976. How Gardeners Can Help Save Endangered Plants. Organic
        Gardening and Farming 23(2): 110-112.

Gore, R.

    1976. Florida, Noah's Ark for Exotic Newcomers. National Geographic
        150(4): 538-559. (Introduced Hydrilla, Eichhornia, Melaleuca)

Gosnell, M.

    1976. Please Don't Pick the Butterworts. National Wildlife
        14(3): 32-37.

Graves, J.

    1974. Redwood National Park: Controversy & Compromise. National
        Parks & Conservation Magazine 48(10): 14-19.

Greenburg, W.

    1976. Showdown in Death Valley. Sierra Club Bulletin 61(2): 41-46.

Gwynne, P., and M. Gosnell

    1975. Fading Flowers. Newsweek 86(2): 72 (July 14).

Harcombe, P. A., and P. L. Marks

    1976. Species Preservation. Science 194(4263): 383.

Herrman, J.

    1975-1976. Who's Protecting Endangered Plants? Brooklyn Botanic
        Garden Record: Plants and Gardens 31(4): 43-45.

Hillson, C. J.

    1976. Codium Invades Virginia Waters. Bulletin of the Torrey
        Botanical Club 102(6): 266-267.

Hirano, R. T.

    1973. Preservation of the Hawaiian Flora. Arboretum and Botanical
        Gardens Bulletin 7(1): 10-11.

Hood, L.

    1976. What's Next for Endangered Plants. Fremontia 4(1): 14-16.

Howe, S. E.

    1976. An Orchid Conservation Review for South Florida. American
        Orchid Society Bulletin 45(5): 423-425.

Iltis, H. H.

    1969. A Requiem for the Prairie. Prairie Naturalist 1(4): 51-57.

Irwin, H. S.

    1976. The Future of our North American Plant Heritage. Garden
        Journal 26(4): 140.

Jenkins, D. W.

    1975. At Last--A Brighter Outlook for Endangered Plants.
        National Parks & Conservation Magazine 49(1): 13-17.

258

Jenkins, D. W., and E. S. Ayensu

 1975. One-tenth of Our Plant Species May Not Survive. <u>Smithsonian</u> 5(10): 92-96.

Jenkins, R. E.

 1975. Endangered Plant Species: A Soluble Ecological Problem. The <u>Nature Conservancy News</u> 25(4): 20-21.

Johnson, R. R., Carothers, S. W., Dolan, R., Hayden, B. P., and A. Howard

 1977. Man's Impact on the Colorado River in the Grand Canyon. <u>National Parks & Conservation Magazine</u> 51(3): 13-16.

Jones, R. A.

 1976. Wholesale Plant Thefts Imperil Desert Environment. <u>Los Angeles Times</u>, part 2, pages 1,5 (19 December).

Josephy, A. M., Jr.

 1976. Kaiparowits: The Ultimate Obscenity. <u>Audubon</u> 78(2): 64-90.

Kilburn, P. D.

 1961. Endangered Relic Trees. <u>Natural History</u> 70(10): 56-63.

 1962. Endangered Relic Trees: The West's Rare Trees are Becoming Rarer. <u>Plants and Gardens</u> 17: 31-37.

 1976. Environmental Implications of Oil-Shale Development. <u>Environmental Conservation</u> 3(2): 101-115.

Kinkead, E.

 1976. The Search for <u>Betula uber</u>. The <u>New Yorker</u>, pages 58-69 (January 12).

Lamoureux, C. H.

 1973. Conservation Problems in Hawaii. In A. B. Costin and R. H. Groves, editors, Nature Conservation in the Pacific. <u>IUCN Publications</u>, new series, 25: 315-319.

Layne, E. N.

    1973. Who will Save the Cacti? Audubon 75(4): 4.

Lehr, J. H.

    1974. The Torrey Symposium on Endangered Plant Species. Bulletin
        of the Torrey Botanical Club 101(4): 213.

Little, E. L., Jr.

    1975 a. Our Rare and Endangered Trees. American Forests 81(7):
        16-21, 55-57.

    1975 b. Rare and Local Conifers in the United States. Conservation
        Research Report Number 19. 25 pages. U. S. Dept. of Agriculture,
        Forest Service (January).

    1976. Rare Tropical Trees of South Florida. Conservation Research
        Report Number 20. 20 pages. U. S. Dept. of Agriculture, Forest
        Service (December).

Love, D.

    1971. Reflections Around a Mutilated Tree. Biological Conservation
        3(4): 274-278. (Pinus aristata)

Luckenbach, R.

    1975. What the ORV's Are Doing to the Desert. Fremontia 2(4): 3-11.

Lynch, J. J.

    1970. Conservation of America's Native Wild Orchids. American Orchid
        Society Bulletin 39(12): 1060-1065.

Lyons, G.

    1972. Conservation: A Waste of Time? Cactus and Succulent Journal
        44(4): 173-177.

    1974. Some New Perspectives in Conservation. In D. R. Hunt, editor,
        Succulents in Peril: Supplement to International Organization
        for Succulent Plant Study Bulletin 3(3), pages 26-28.

260

1976. Some New Developments in Conservation. *Cactus and Succulent Journal* 48: 155-162.

McCluney, W. R., editor

1971. *The Environmental Destruction of South Florida.* 134 pages. Coral Gables, Florida: University of Miami.

Medders, S.

1975. California's Channel Islands. *National Parks & Conservation Magazine* 49(10): 11-15.

Moore, G.

1977. The Deflowering of the Endangered Species Act. *Horticulture* 55(5): 36-39.

Moser, D.

1976. Dig They Must, the Army Engineers, Securing Allies and Acquiring Enemies. *Smithsonian* 7(9): 40-51.

1977. Mangrove Island is Reprieved by Army Engineers. *Smithsonian* 7(10): 68-77.

Mueller-Dombois, D., and G. Spatz

1972. *The Influence of Feral Goats on the Lowland Vegetation in Hawaii Volcanoes National Park*. Island Ecosystems IRP/IBP Hawaii, Technical Report Number 13, 46 pages.

Murchie, D. R.

1973. What Trackless Desert? *Sierra Club Bulletin* 58(3): 25-28. (California off-road vehicles)

Niering, W. A., R. H. Whittaker, and C. H. Lowe

1963. The Saguaro: A Population in Relation to Environment. *Science* 142(3588): 15-23.

Nisbet, I. C. T.

1976. Pacific Follies, or the Ravishing of Hawaii. *Technology Review* 78: 8-9.

Ogle, D. W., and P. M. Mazzeo

    1976. *Betula uber*, the Virginia Round-Leaf Birch, Rediscovered in Southwest Virginia. *Castanea* 41(3): 248-256.

Parsons, D. J.

    1976. The Role of Fire in Natural Communities: An Example from the Southern Sierra Nevada, California. *Environmental Conservation* 3(2): 91-99.

Pickoff, L. J.

    1975. Our Role in Conservation. *Cactus and Succulent Journal* 47(1): 20-22.

Preston, D. J.

    1975. Endangered Plants. *American Forests* 81(4): 8-11, 46-47.

    1976. The Rediscovery of *Betula uber*. *American Forests* 82(8): 16-20.

Raskin, M.

    1975. Smog Alert for Our Southwestern National Parks. *National Parks & Conservation Magazine* 49(7): 9-15.

Reed, C. F.

    1975. *Betula uber* (Ashe) Fernald Rediscovered in Virginia. *Phytologia* 32(4): 305-311 (1975).

Rowley, G.

    1973. Save the Succulents!--A Practical Step to Aid Conservation. *Cactus and Succulent Journal* 45(1): 8-11.

Sayers, W. B.

    1974. The King is Dead: Long Live the King. *National Parks & Conservation Magazine* 49(9): 8-13. (*Castanea dentata*)

262

Smithsonian Institution

    1975. Report on Endangered and Threatened Plant Species of the
       United States. House Document Number 94-51, Serial Number
       94-A. Washington, D.C.: Government Printing Office.

Soucie, G.

    1973. Subdividing and Conquering the Desert. Audubon 75(4): 26-35.

Spatz, G., and D. Mueller-Dombois

    1973. The Influence of Feral Goats on Koa Tree Reproduction in
       Hawaii Volcanoes National Park. Ecology 54(4): 870-876.

Stansbery, D. H.

    1972. The Responsibility of the Professional Ecologist in the
       Preservation of Natural Areas. Ohio Journal of Science
       72: 1-3.

Stebbins, R. C.

    1974 a. Off-Road Vehicles and the Fragile Desert, Part I. The American
       Biology Teacher 36(4): 203-208, 220 .

    1974 b. Off-Road Vehicles and the Fragile Desert, Part II. The American
       Biology Teacher 36(5): 294-304

Stebbins, R. C., and N. W. Cohen

    1976. Off-Road Menace--A Survey of Damage in California. Sierra
       Club Bulletin 61(7): 33-37.

Stevens, L.

    1976. King Kong Kudzu, Menace to South. Smithsonian 7(9): 93-94,
       96, 98-99. (Introduced Pueraria)

Styron, C. E., editor

    1976. Symposium on Rare, Endangered and Threatened Biota of the
       Southeast. ASB Bulletin 23(3): 137-167.

Summer, F. B.

    1920. The Need for a More Serious Effort To Rescue a Few Fragments of Vanishing Nature. Scientific Monthly 10: 236-248.

    1921. The Responsibility of the Biologist in the Matter of Preserving Natural Conditions. Science 54: 39-43.

Sumner, D., and C. Johnson

    1974 a. The Great Shale Robbery. Sierra Club Bulletin 59(3): 10-14, 16.

    1974 b. 700,000,000,000 Barrels of Soot. Sierra Club Bulletin 59(4): 25-30. (Oil shale)

Toner, M. F.

    1976. Farming the Everglades. National Parks & Conservation Magazine 50(8): 4-9.

    1977. Putting the Big Cypress Together Again. National Parks & Conservation Magazine 51(3): 4-9.

Twisselmann, E. C.

    1969. Status of the Rare Plants of Kern County. California Native Plant Society Newsletter 5(3): 1-7.

Van Dersal, W. R.

    1972. Why Living Organisms Should Not Be Exterminated. Atlantic Naturalist 27(1): 7-10.

Voss, E. G.

    1972. The State of Things. The Michigan Academician 5(1): 1-7.

Watkins, T. H.

    1975. Pastures of Hell. Sierra Club Bulletin 60(5): 4-8, 20, 28. (The Great Plains)

Weinberg, J. H.

    1975. Botanocrats and the Fading Flora. Science News 102(6): 92-95
        (August 9).

Willard, B. E., and J. W. Marr

    1971. Recovery of Alpine Tundra Under Protection After Damage by
        Human Activities in the Rocky Mountains of Colorado.
        Biological Conservation 3(3): 181-190.

Zieman, J. C.

    1976. The Ecological Effects of Physical Damage from Motor Boats
        on Turtle Grass Beds in Southern Florida. Aquatic Botany
        2(2): 127-139. (Thalassia)

Zimmerman, J. H., and H. H. Iltis

    1961. Conservation of Rare Plants and Animals. Wisconsin Academy
        Review, pages 7-11 (Winter).

Endangered and Threatened Flora of the World

Act of Parliament (United Kingdom)

    1976. Endangered Species (Import and Export) Act 1976; 1976 Chapter

        72. Enacted 22 November 1976; in force February 3, 1977.

Adjanohoun, E.

    1971. Les problemes souleves par la conservation de la flore en Cote

        d'Ivoire. Bulletin du Jardin Botanique National de Belgique

        41(1): 107-133. (Ivory Coast)

Allen, R.

    1977. Margin of Life. International Wildlife 7(2): 20-29

        (Wetlands)

Alon, A.

    1973. Saving the Wild Flowers of Israel. Biological Conservation

        5: 150.

Alphonso, A. G.

    1966. The Need for Conservation of Malaysian Orchid Species. In

        Proceedings of the Fifth World Orchid Conference, Long Beach,

        California, pages 125-132.

Anderson, W. R., R. McVaugh, and E. G. Voss

    1975. Resolutions Accepted at the Closing Plenary Session of the 12th

        International Botanical Congress, 10 July 1975, in Leningrad.

        Taxon 24(5-6): 701-703.

Andrews, N.

    1973. Tropical Forestry: The Timber Industry Finds a New Last

        Stand. Sierra Club Bulletin 58(4): 4-9.

266

Anonymous.

    1974. Appello per la protezione della flora italiana. <u>Webbia</u>

        29(1): 361-363.

Augustyn, O. P. H.

    1975. Vanishing Bulbs of the Veld, IV: <u>Herschelia lugens</u> (Bol.)

        Kraenzl. Orchidaceae. <u>Indigenous Bulb Growers Association</u>

        <u>of South Africa Bulletin</u> 25: 5-6 (September).

Ayensu, E. S.

    1975. Threatened or Endangered Plants of Sri Lanka. In <u>Natural</u>

        <u>Products for Sri Lanka's Future</u>, <u>National Science Council</u>,

        <u>Colombo</u>, pages 45-46.

Aymonin, G. G.

    1974. Listes preliminaires des especes endemiques et des especes

        menacees en France. <u>Convention-Societe Botanique de France</u>

        <u>Rapport</u>, 2: 1-53.

Bandurski, B., and M. Buchinger

    1969. More Recreation: Implications for the Tropical Ecosystem. In

        J. M. Idrobo, editor, <u>II Simposio y Foro de Biologia Tropical</u>

        <u>Amazonica</u>, pages 197-267. Asociacion pro Biologia Tropical,

        Bogota.

Barnard, T. T.

    1974. Vanishing Bulbs of the Veld, III: <u>Gladiolus aureus</u> Baker.

        <u>Indigenous Bulb Growers Association of South Africa Bulletin</u>

        24: 4-5 (December).

Baum, P.

    1975. European List of Endemic and Endangered Plants. <u>Environmental</u>

        <u>Conservation</u> 2(3): 204.

Beaumont, T.

    1975. Saving our Wild Flowers. The Illustrated London News
        263(6922): 49 (May). (Great Britain)

Belville, L. S.

    1971. They are Destroying Brazil's Paradise. International Wildlife
        1(5): 36-40.

Bernardi, L.

    1974. Problemes de conservation de la nature dans les iles de
        l'Ocean Indien, 1: Meditation a propos de Madagascar.
        Saussurea 5: 37-47.

Berwick, S.

    1976. The Gir Forest: An Endangered Ecosystem. American Scientist
        64(1): 28-40. (India)

Bieloussova, L. S., and L. V. Denissova

    1970. Protection de la Flore en URSS. Biological Conservation
        2(3): 218-220.

Birkmane, K.

    1974. Protected Plants in Latvia. 58 pages. Riga: Zinatne.

Borromeo, C. R.

    1976. The Urgent Need for the Conservation of Philippine Orchid
        Species. In Proceedings of the Eighth World Orchid Conference,
        Palmengarten, Frankfurt, 10-17 April, 1975, pages 341-342.

Boudet, G.

    1972. Desertification de l'Afrique tropicale seche. Adansonia,
        series 2, 12: 505-524.

Carlozzi, C. A., and A. A. Carlozzi

    1968. Conservation and Caribbean Regional Progress. 151 pages. The
        Antioch Press.

Challinor, D., and D. B. Wingate

    1971. The Struggle for Survival of the Bermuda Cedar. Biological
Conservation 3(3): 220-222.

Chew, W. L.

    1968. Conservation of Habitats: Conservation in Tropical South
East Asia. IUCN Publications, new series, 10: 337-339.
(Malay Peninsula)

Chiariello, N.

    1976. Plant Endangerment and Ecological Stability. Conservation
News 41(13): 10-13.

Cloudsley-Thompson, J. L.

    1974. The Expanding Sahara. Environmental Conservation 1(1): 5-13.

Copyk, V. I.

    1970. Scientific Grounds for Protection of Rare Species of
Ukrainian Flora. Ukrajinskyi Botanicnyii Zhurnal 27(6): 693-704.

Corbett, R.

    1975. Notes on some Rare Vanishing Orchid Species found in Perak.
Bulletin of the Orchid Society of South East Asia 3: 8.

Cronquist, A.

    1971. Adapt or Die! Bulletin du Jardin Botanique National de
Belgique 41(1): 135-144.

Crusz, H.

    1973. Nature Conservation in Sri Lanka (Ceylon). Biological
Conservation 5: 199-208.

Curry-Lindahl, K.

    1972. Conservation for Survival--An Ecological Strategy. 335 pages.
New York: William Morrow & Company, Inc.

Dafni, A., and M. Agami

1976. Extinct Plants of Israel. Biological Conservation 10(1): 49-52.

D'Arcy, W. G.

1976. Near Extinct Plant in Climatron. Missouri Botanical Garden Bulletin 64(3) (March). (Lebronnecia - Tahiti)

Daugherty, H. E.

1973. The Montecristo Cloud Forest of El Salvador--A Chance for Protection. Biological Conservation 5: 228-230.

Decker, B. G.

1975. Unique Dry-Island Biota Under Official Protection in Northwestern Marquesas Islands (Isles Marquises). Biological Conservation 5(1): 66-67.

Delpierre, G. R.

1974. Vanishing Bulbs of the Veld, II: Gladiolus bullatus Thunb. ex Lewis. Iridaceae. Indigenous Bulb Growers Association of South Africa Bulletin 23: 3-4 (June).

Deneven, W. M.

1973. Development and the Imminent Demise of the Amazon Rain Forest. Professional Geographer 25: 130-135.

Denisova, L. V., and L. S. Belousova

1974. Rare and Vanishing Plants of the USSR. 150 pages. Moscow.

Department of Nature Conservation of the Cape Provincial Administration

1967. Some Protected Wildflowers of the Cape Province. 224 plates. Cape Town.

Dillon, G. W.

1976. A.O.S. Response to Endangered Species Proposals. American Orchid Society Bulletin 45(9): 814-816.

Dodson, C. H.

    1968. Conservation of Orchids. Proceedings of the Latin American Conference on the Conservation of Renewable Natural Resources. IUCN Publications, new series, 13: 170.

Douglas, I.

    1975. Pressures on Australian Rain-forests. Environmental Conservation 2(2): 109-119.

Dourojeanni, M. J.

    1968. Estado actual de la conservacion de la flora y la fauna en el Peru. Ciencia Interamericana 9(1-6): 51-64.

Dring, M. J., and L. C. Frost

    1972. Studies of Ranunculus ophioglossifolius in Relation to Its Conservation at the Badgeworth Nature Reserve, Gloucestershire, England. Biological Conservation 4: 48-56.

Dunsterville, G. C. K.

    1968. Please Don't Strip the Forests--Even From Your Armchair. American Orchid Society Bulletin 37(5): 397-401.

    1974. La conservation de las Orquideas en Venezuela. In Proceedings of the Seventh World Orchid Conference, Medellin, Colombia, pages 93-105.

    1975. A Letter to Orchid Conservationists. American Orchid Society Bulletin 44(10): 882-885.

Dunsterville, G. C. K., and E. Dunsterville

    1976. Bifrenaria maguirei, Zygosepalum tatei and Odontoglossum arminii--Three Fine Orchids Safe in Venezuela's Hinterland. American Orchid Society Bulletin 45(9): 783-787.

Eckholm, E. P.

    1975. The Firewood Crisis. Natural History 84(8): 6-22.

Eden, M. J.

    1975. Last Stand of the Tropical Forest. Geographical Magazine
    47(9): 578-582.

Editor.

    1974. Two Botanists in a Search to Track Down Plants in Danger.
    African Wildlife 28(2): 31-32. (South Africa)

Edmonson, J.

    1977. Plant Conservation in the USSR. Notes from the Royal Botanic
    Garden Edinburgh 35(2): 246. Book review.

Elbert, G. A.

    1969. On the Conservation of Orchids. American Orchid Society
    Bulletin 38(6): 497-498.

Eliasson, U.

    1968. On the Influence of Introduced Animals on the Natural
    Vegetation of the Galapagos Islands. Notiser Galapagos
    11: 19-21.

Elliott, H. F. I., editor

    1973. Conservation and Development. Papers and Proceedings of IUCN
    12th Technical Meeting, Banff, 1972. IUCN Publications, new
    series, 28: 1-383.

Emmelin, L.

    1975. Protection of Fauna and Flora in Sweden. Current Sweden
    58: 1-3 (November).

Everard, B.

    1974. Vanishing Flowers of the World. Bartholomew World Pictorial
    Map Series. John Bartholomew & Son Ltd., Duncan Street,
    Edinburgh EH9 1TA, Scotland (May).

272

Farnworth, E. G., and F. B. Golley

    1974. Fragile Ecosystems: Evaluation of Research and Application
        in the Neotropics. 258 pages. The Institute of Ecology. New
        York: Springer.

Fitch, C. M.

    1969. Orchid Conservation in the Tropics. American Orchid Society
        Bulletin 38(6): 476-485.

Fosberg, F. R.

    1968. Scientific Need for Parks and Reserves. Conservation in
        Tropical South East Asia. IUCN Publications, new series,
        10: 380-381.

    1969. Sabah (North Borneo), Malaysia: Sepilok Forest Endangered.
        Biological Conservation 1: 339-340.

    1972 a. The Value of Systematics in the Environmental Crisis.
        Taxon 21(5-6): 631-634.

    1972 b. On the Importance of Preserving Diversity in Ecosystems.
        In Comptes Rendus de la Conference Internationale sur la
        Conservation de la Nature et de ses Ressources a Madagascar,
        Tananarive, Madagascar, 7-11 Octobre 1970. IUCN Publications,
        new series, supplementary paper 36: 81-84.

    1973 a. On Present Condition and Conservation of Forest in Micronesia.
        Pacific Science Association Standing Committee on Pacific
        Botany Symposium: Planned Utilization of the Lowland Tropical
        Forest, Bogor, Indonesia, 1971: 165-171.

    1973 b. Temperate Zone Influence on Tropical Forest Land Use: A
        Plea for Sanity. In B. J. Meggers, E. S. Ayensu, and W. D.
        Duckworth, editors, Tropical Forest Ecosystems in Africa
        and South America: A Comparative Review, pages 345-350.
        Washington, D.C.: Smithsonian Institution Press.

1974. Studies in American Rubiaceae, 2: Ayuque, Balmea stormae, and Endangered Mexican Species. Sida 5(4): 268-270.

Fosberg, F. R. and M. -H. Sachet

1972. Status of Floras of Western Indian Ocean Islands. In Comptes Rendus de la Conference Internationale sur la Conservation de la Nature et de ses Ressources a Madagascar, Tananarive, Madagascar, 7-11 Octobre 1970. IUCN Publications, new series, supplementary paper 36: 152-155.

Galiano, E. F.

1971. Problemes de la conservation de la vegetation et de la flore en Espagne. Boissiera 19: 81-86.

Gardner, S.

1976. Some Thoughts on the Future of Bromeliads in the Wild. Journal of the Bromeliad Society 26(2): 47-48, 65.

Germishuizen, G.

1975. Some Endangered Succulents of the Transvaal. Fauna Flora (Pretoria) 26: 5-7.

Gilbert, J. L.

1975. A History of Plant Conservation in Britain. Country-Side 22(11): 542-546.

Golden, F.

1977. Shifting Sands. International Wildlife 7(1): 20-27. (World Deserts)

Goldsmith, F. B.

1974. An Assessment of the Nature Conservation Value of Majorca. Biological Conservation 6(2): 79-83.

Gomez-Pompa, A., C. Vazquez-Yanes, and S. Guevera

    1972. The Tropical Rain Forest: A Nonrenewable Resource. _Science_
        177: 762-765 (September 1).

Goodland, R. J. A., and H. S. Irwin

    1975. _Amazon Jungle: Green Hell to Red Desert?_ 155 pages. New
        York: Elsevier Scientific Publishing Company.

Gorman, M. L., and S. Siwatibau

    1975. The Status of _Neoveitchia storckii_ (Wendl.): A Species of
        Palm Tree Endemic to the Fijian Island of Viti Levu.
        Biological Conservation 8(1): 73-76.

Gosnell, M.

    1976. The Island Dilemma. _International Wildlife_ 6(5): 24-35.

Guillarmod, A. J.

    1975. Point of No Return? _African Wildlife_ 29(4): 28-29, 31.
        (_Aloe_)

Gwynne, P.

    1976. Doomed Jungles? _International Wildlife_ 6(4): 36-47.

Haber, W.

    1976. Orchids and Conservation. In _Proceedings of the Eighth World_
        Orchid Conference, Palmengarten, Frankfurt, 10-17 April,
        _1975,_ pages 73-79.

Hagsater, E.

    1976 a. Can there be a Different View on Orchids and Conservation?
        American Orchid Society Bulletin 45(1): 18-21.

    1976 b. Orchids and Conservation in Mexico. _The Orchid Review_
        84(992): 39-42.

    1976 c. Our View on Orchids and Conservation. _Orquidea_ 6(2): 54-67.

Hamilton, B.

    1976. Forest in Isolation. Australian Natural History 18(10):
        358-361. (New Zealand)

Hamilton, L. S.

    1976. Whither the Tropical Rainforest? Sierra Club Bulletin
        61(4): 9-11.

Hartmann, W.

    1974. At Last, Something Positive--Orchid Conservation. American
        Orchid Society Bulletin 43(9): 802-803. (Mexico)

Hedberg, I., and O. Hedberg

    1968. Conservation of Vegetation in Africa South of the Sahara.
        Proceedings of a Symposium Held at the 6th Plenary Meeting of
        the Association pour l'Etude Taxonomique de la Flore d'Afrique
        Tropicale, Uppsala, September 12th-16th, 1966. Acta
        Phytogeographica Suecica 54.

Hepper, F. N.

    1969. The Conservation of Rare and Vanishing Species of Plants.
        In J. Fisher, N. Simon, and J. Vincent, Wildlife in Danger,
        pages 353-360. Viking Press.

Hernandez, M. O.

    1968. The Disappearance of Valuable Native Orchids in Latin
        America. Proceedings of the Latin American Conference on
        the Conservation of Renewable Natural Resources. IUCN
        Publications, new series, 13: 168-169. Heywood, V. H.

    1971. Preservation of the European Flora. The Taxonomist's Role.
        Bulletin du Jardin Botanique National de Belgique 41(1): 153-166.

276

HRH The Prince of the Netherlands

    1976. Plants and the Future of Man. Environmental Conservation 3(1): 23-26.

Hunt, D. R., editor

    1974. Succulents in Peril. Supplement to International Organization for Succulent Plant Study Bulletin 3(3): 1-32.

Hunt, P. F.

    1968. Conservation of Orchids. The Orchid Review 76(905): 320-327.

Iltis, H. H.

    1967. To the Taxonomist and Ecologist-Whose Fight is the Preservation of Nature? BioScience 17(12): 886-890.

    1970. Man First? Man Last? The Paradox of Human Ecology. BioScience 20(14): 820.

    1972 a. Conservation, Contraception and Catholicism, A 20th Century Trinity. The Biologist 54(1): 35-47.

    1972 b. The Extinction of Species and the Destruction of Ecosystems. The American Biology Teacher 34(4): 201-205, 221.

    1973. Can One Love a Plastic Tree? Bulletin of the Ecological Society of America 54(4): 5-7, 9.

    1974. Flowers and Human Ecology. In C. Selmes, editor, New Movements in the Study and Teaching of Biology, pages 289-317. London: Maurice Temple Smith.

Jackson, W.

    1976. Toward an Ecological Ethic. Association of Southeastern Biologists Bulletin 23(3): 123-132.

Jain, S. K., and P. K. Hajra

    1976. Orchids in Some Protected Habitats in Assam in Eastern India. American Orchid Society Bulletin 45(12): 1103-1109.

Janzen, D. H.

    1972. The Uncertain Future of the Tropics. Natural History 81(9): 80-94.

    1974. The Deflowering of Central America. Natural History 83(4): 48-53. (Pollinators)

Jenkins, R. E.

    1973. The Use of Natural Areas to Establish Environmental Baselines. Biological Conservation 5: 168-174.

Johnson, A.

    1968. Rare Plants and the Community in South East Asia: Conservation in Tropical South East Asia. IUCN Publications, new series, 10: 340-343.

Jordan, C. F.

    1971. A World Pattern in Plant Energetics. American Scientist 59: 425-433.

Kammer, F.

    1976. The Influence of Man on the Vegetation of the Island of Hierro (Canary Islands). In G. Kunkel, editor, Biogeography and Ecology in the Canary Islands (Monographiae Biologicae, volume 30), pages 327-346. The Hague: W. Junk Publishers.

Kataki, S. K.

    1976. Indian Orchids--A Note on Conservation. American Orchid Society Bulletin 45(10): 912-914.

Katende, A. B.

    1976. The Problem of Plant Conservation and the Endangered Plant Species in Uganda. Boissiera 24b: 451-456.

278

Kennedy, G. C.

1975. Orchids and Conservation--A Different View. American Orchid
Society Bulletin 44(5): 401-405.

Keraudren-Aymonin, M.

1972. Un Patrimoine Naturel de Valeur Universelle a Preserver:
Les Fourres a Didieracees du Sud de Madagascar. In Comptes
Rendus de la Conference Internationale sur la Conservation
de la Nature et de ses Ressources a Madagascar, Tananarive,
Madagascar, 7-11 Octobre 1970. IUCN Publications, new series,
supplementary paper 36: 145-151.

Keraudren-Aymonin, M., and G. Aymonin

1971. La flore malgache: Un joyau a sauvegarder. Science et
Nature 106: 5-14. (July-August). (Madagascar)

Kershaw, L., J. K. Morton, and J. M. Venn

1976. A List of Rare or Endangered Species in the Canadian Flora--
Vascular Plants. 38 pages. Ontario: Department of Biology,
University of Waterloo.

Kevan, P. G.

1975. Pollination and Environmental Conservation. Environmental
Conservation 2(4): 293-298.

Kimberley, M. J.

1975. Plant Protection Legislation in Rhodesia. Excelsa 5: 3-16.

Kornas, J.

1971. Changements recents de la Flore polonaise. Biological
Conservation 4(1): 43-47.

Kornas, J., and A. Medwecka-Kornas

1967. The Status of Introduced Plants in the Natural Vegetation of
Poland. IUCN Publications, new series, 9: 38-45.

Kostermans, A. J. G. H.

    1960. The Influence of Man on the Vegetation of the Humid Tropics.
In _Symposium on the Impact of Man on Humid Tropics Vegetation_,
pages 332-338. Goroka.

Krieger, M. H.

    1973. What's Wrong with Plastic Trees? _Science_ 179: 446-455
(February 2).

Kunkel, G.

    1972. Gran Canaria. Plantas en peligro. In _Aves y Plantas de
Gran Canaria en Peligro de Extincion_. A.S.C.A.N. Las
Palmas de Gran Canaria. (Canary Islands)

    1975. _Aves y Plantas de Fuerteventura en Peligro de Extincion_
A.S.C.A.N. Las Palmas de Gran Canaria. (Canary Islands)

Lamb, K. P., and J. L. Gressitt, editors

    1976. _Ecology and Conservation in Papua New Guinea_. 153 pages.
Wau Ecology Institute, Pamphlet No. 2.

Lampe, D.

    1975. The Benevolent Hedgerows of Britain. _International
Wildlife_ 5(4): 36-41.

Landolt, E.

    1970. _Plantes protegees de Suisse (Sauvons la Flore Suisse)_.
215 pages. Ligue Suisse pour la Protection de la Nature.

Lapin, P. L.

    1975. Our Endangered Environment: The Russian View. _Garden
Journal_ 25(6): 171-175.

Lawalree, A.

    1971 a. L'appauvrissement de la flore en Belgique depuis 1850.
Boissiera 19: 65-72.

1971 b. L'appauvrissement de la flore belge. Bulletin du Jardin
Botanique National de Belgique 41(1): 167-171.

Lawder, M.

1973. Vanishing Bulbs of the Veld, I: Hyacinthus corymbosus L.
Liliaceae. Indigenous Bulb Growers Association of South
Africa Newsletter 22: 5-6 (November).

Lawson, G. W.

1972. The Case for Conservation in Ghana. Biological Conservation
4(4): 292-300.

Lebrun, J.

1971. La conservation du peuplement vegetal au centre de l'Afrique,
specialement au Congo. Bulletin du Jardin Botanique National
de Belgique 41(1): 203-218.

Lemieux, G.

1972. Une prospective de la conservation des especes vegetales au
Quebec. Detoute Urgence 3(3): 17-30.

Lionnet, J. F. G.

1968. Striking Plants of the Seychelles. Journal of the Seychelles
Society (Mahe) 6: 32-42.

Lothian, T. R. N.

1969. The Importance of Preserving our Native Plants. Education
Gazette (Australia), pages 1-7.

Loubser, J. W.

1976. Vanishing Bulbs of the Veld: Moraea loubseri Goldblatt.
Indigenous Bulb Growers Association of South Africa
Bulletin 26: 3-4 (October).

Lovejoy, T.

   1976. We Must Decide Which Species Will Go Forever. *Smithsonian* 7(4): 52-59.

Lucas, G. L.

   1975. Problems and Approaches in the Conservation of Threatened Plants and Plant Genetic Resources. Presented at IUCN 13th Technical Meeting, Kinshasa, Zaire.

Lucas, G. L., and S. M. Walters

   1976. *List of Rare, Threatened and Endemic Plants for the Countries of Europe.* 166 pages. Kew: IUCN Threatened Plants Committee, Royal Botanic Gardens.

Lyons, G.

   1974. In Search of Dragons, or: The Plant that Roared. *Cactus and Succulent Journal* 46(6): 267-282. (Dracaena draco)

MacFarlane, C. I.

   1971. The Importance of Conservation of the Seaweed Resource. In *Proceedings, Meeting on the Canadian Atlantic Seaweeds Industry.* Sponsored by the Federal-Provincial Atlantic Fisheries Committee, Charlottetown, Prince Edward Island, October 5-6, 1971. Pages 43-58.

Maheshwari, J. K.

   1970. The Need for Conservation of Flora and Floral Provinces in South East Asia. *IUCN Publications,* new series, 18(2): 89-94.

   1972 a. Plant Wildlife and Conservation in Shivpuri National Park, India. *Biological Conservation* 4(3): 214-219.

   1972 b. The Baobab Tree: Disjunctive Distribution and Conservation. *Biological Conservation* 4(1): 57-60.

282

Malato-Beliz, J.

    1975. The Mediterranean Flora must be Saved. *Naturopa* 22: 3-6.

Marshall, A. G.

    1973. Conservation in West Malaysia: The Potential for International

        Cooperation. *Biological Conservation* 5(2): 133-140.

Marshall, A. J., editor

    1966. The Great Extermination. 221 pages. London: William

        Heinemann Ltd. (Australia)

Martin, W.

    1941. Preservation of Native Plants. *Journal of the Royal New*

        Zealand Institute of Horticulture 11: 43-46.

Mathews, W. H.

    1971. *Man's Impact on Terrestrial and Oceanic Ecosystems.* 540 pages.

        Cambridge: Massachusetts Institute of Technology Press.

McClintock, D.

    1968. Britain's Vanished Orchid. *Country Life* 144(3723): 73.

        (*Spiranthes aestivalis*)

McLean, A.

    1976. Protection of Vegetation in Ecological Reserves in Canada.

        Canadian Field-Naturalist 90: 144-148.

McMaster, C.

    1976. *Protea simplex*: Is It an Endangered Species? *Veld & Flora*

        62(2): 21-22.

Meijer, W.

    1973 a. Endangered Plant Life. *Biological Conservation* 5(3): 163-167.

    1973 b. Endangered Plant Life in Brazil. *Biological Conservation*

        5: 147.

1973 c. Devastation and Regeneration of Lowland Dipterocarp Forests
in Southeast Asia. BioScience 23(9): 528-533.

Melville, R.

1969. The Endemics of Phillip Island. Biological Conservation
1(2): 170-172.

1970 a. Endangered Plants and Conservation in the Islands of the
Indian Ocean. IUCN Publications, new series, 18(2): 103-107.

1970 b. Plant Conservation and the Red Book. Biological Conservation
2(3): 185-188.

1970 c. Red Data Book, Volume 5: Angiospermae. Morges, Switzerland:
International Union for Conservation of Nature and Natural
Resources.

1971. Endangered Angiosperms and Conservation in Australia.
Bulletin du Jardin Botanique National de Belgique 41: 145-152.

1972. The Significance of the Madagascan Flora Among the Floras of
the World. In Comptes Rendus de la Conference Internationale
sur Conservation de la Nature et de ses Ressources a Madagascar,
Tananarive, Madagascar, 7-11 Octobre 1970. IUCN Publications,
new series, supplementary paper 36: 139-142.

1973. Relict Plants in the Australian Flora and their Conservation.
In A. B. Costin and R. H. Groves, editors, Nature Conservation
in the Pacific. IUCN Publications, new series, 25: 83-90.

Mennema, J.

1975. Threatened and Protected Plants in the Netherlands. Naturopa
22: 10-13.

Merwe, P. van der

1975. Impossible to Save the Marsh Rose Protea? Veld & Flora
61(1): 4-5.

284

Miege, J.

    1971. L'importance de la sauvegarde du patrimoine floristique. Bulletin du Jardin Botanique National de Belgique 41(1): 93-106.

Miller, R. S., and D. B. Botkin

    1974. Endangered Species: Models and Predictions. American Scientist 62: 172-181.

Milne-Redhead, E.

    1971. Botanical Conservation in Britain, Past, Present and Future. Watsonia 8: 195-203.

Ministerie van Cultuur, Recreatie en Maatschappelijk Werk

    1975. Beschermde Planten en Dieren. 92 pages. Gravenhage. (The Netherlands)

Misra, R.

    1970. Save the Tropical Ecosystems. INTECOL Bulletin 2: 29-32.

Moir, W. H.

    1972. Natural Areas. While Harboring Valuable Species, Natural Areas also Serve as Bench Marks in Evaluating Landscape Change. Science 177: 396-400 (August 4).

Moir, W. W. G.

    1968. Conservation and Orchids. The Orchid Review 76(905): 327-331.

Morton, J. K., editor

    1976. Man's Impact on the Canadian Flora: Proceedings of a Symposium. Supplement to Canadian Botanical Association Bulletin 9(1): 1-30.

Mudd, J. B., and T. T. Kozlowski, editors

    1975. Responses of Plants to Air Pollution. 392 pages. New York: Academic Press.

Mueller-Dombois, D.

    1973. Natural Area System Development for the Pacific Region: A Concept and Symposium. Island Ecosystems IRP/IBP Hawaii, Technical Report 26: 1-55.

Munoz Pizarro, C.

    1973. Chile: Plantas en Extincion. 248 pages. Santiago de Chile: Editorial Universitaria.

Myers, N.

    1976. An Expanded Approach to the Problem of Disappearing Species. Science 193: 198-202 (July 16).

New York Botanical Garden

    1976. Resolutions from the Endangered Species Symposium, May 1976. Brittonia 28(3): 379-380.

Ngan, P. T.

    1968. The Status of Conservation in South Vietnam. Conservation in Tropical South East Asia. IUCN Publications, new series, 10: 519-522. (Gymnosperms)

Nicot, J.

    1973. Les champignons dans la destruction de la nature: Les mycologues et la protection de l'environment. Revue Mycologique 37(1-2): 96-99.

Oza, G. M.

    1974. Indian Doum Palm Faces Extinction. Biological Conservation 6(1): 65-67. (Hyphaene indica)

Patter, T. I.

    1897. The Destructive Collection of Orchids. The Orchid Review 5(54): 204-205. (Trinidad)

Perring, F. H.

 1971 a. The Biolgical Records Centre--A Data Centre. *Biological Journal of the Linnean Society* 3(3): 237-243.

 1971 b. Rare Plant Recording and Conservation in Great Britain. *Boissiera* 19: 73-79.

 1974. The Flora of a Changing Britain. *Botanical Society of the British Isles Conference Report*, Number 11.

 1975 a. Plant Conservation--Without Botanists? *New Scientist* 67: 194-195 (July 24).

 1975 b. Problems of Conserving the Flora of Britain. *Botanical Society of the British Isles News* 9: 18-24.

Perring, F. H., and S. M. Walters.

 1971. Conserving Rare Plants in Britain. *Nature* 229(5284): 375-377 (February 5).

 1972. Europe's Flora Threatened. *Nature in Focus* 12: 12-13 (Summer).

Petter, J. J., and M. Pariente

 1971. The Rational Use and Conservation of Nature. *Nature and Resources* 7(3): 2-8.

Polunin, N.

 1970. Botanical Conservation in the Arctic. *Biological Conservation* 2(3): 197-205.

Poore, D.

 1976. The Value of Tropical Moist Forest Ecosystems and the Environmental Consequences of Their Removal. *Unasylva* 28(112-113): 127-143, 145-146.

Pradhan, G. M.

 1974. Orchid Conservation in India. *American Orchid Society Bulletin* 43(2): 135-139.

1976. Habitat Destruction of Himalayan Orchid Jungles. In Proceedings of the Eighth World Orchid Conference, Palmengarten, Frankfurt, 10-17 April, 1975, pages 331-334.

Pradhan, U. C.

1975. Conservation of Eastern Himalayan Orchids: Problems and Prospects: Part 1. The Orchid Review 83(987): 314-317.

1976. Conservation of Eastern Himalayan Orchids: Problems and Prospects. In Proceedings of the Eighth World Orchid Conference, Palmengarten, Frankfurt, 10-17 April, 1975, pages 335-340.

1977. Conserving Indian Orchids. American Orchid Society Bulletin 46(2): 117-121.

Prance, G. T., and T. S. Elias, editors

1977. Extinction Is Forever. New York: New York Botanical Garden.

Pritchard, N. M.

1972. Where Have All the Gentians Gone? Transactions of the Botanical Society of Edinburgh 41: 279-291.

Quaintance, C. W.

1971. The Vanishing Flora of Colombia. Biological Conservation 3(2): 145-147.

Quisumbing, E.

1960. The Vanishing Species of Plants in the Philippines. In Symposium on the Impact of Man on Humid Tropics Vegetation, pages 344-348. Goroka.

1967. Philippine Species of Plants Facing Extinction. ARENETA Journal of Agriculture 14(3): 135-162.

288

Qureshi, I. M., and O. N. Kaul

    1970. Some Endangered Plants and Threatened Habitats in South

        East Asia. IUCN Publications, new series, 18(2): 115-126.

Rapp, A., H. N. Le Houerou, and B. Lundholm, editors

    1976. Can Desert Encroachment be Stopped? Swedish Natural Science

        Research Council Ecological Bulletins, 24: 1-241.

Raven, P. H., and T. Engelhorn

    1971. A Plea for the Collection of Common Plants. New Zealand

        Journal of Botany 9: 217-222.

Raven, P. H., R. F. Evert, and H. Curtis

    1976. The Conservation of Plants. Biology of Plants, Second

        edition, page 585. New York: Worth Publishers, Inc.

Reecher, H.

    1976. An Ecologist's View: The Failure of the National Parks

        System. Australian Natural History 18(11): 398-405.

Ricciuti, E. R.

    1976. Mountains Besieged. International Wildlife 6(6): 24-34.

Richards, A. J.

    1972. The Code of Conduct: A List of Rare Plants. Watsonia 9:

        67-72. (Great Britain)

Richards, P. W.

    1971. Some Problems of Nature Conservation in the Tropics.

        Bulletin du Jardin Botanique National de Belgique 41(1):

        173-187.

Richardson, D. H. S.

    1974. The Vanishing Lichens. New York: Hafner Press.

Riedl, H.

    1975. Plant Species Conservation in the Alps: Possibilities and
        Problems. Naturopa 22: 6-9.

Roche, L.

    1971. The Conservation of Forest Gene Resources in Canada. The
        Forestry Chronicle 47(4): 215-217.

Rogaly, J. M.

    1976. Ecology and Conservation of Sub-Tropical Orchids of Southern
        Africa--Mainly Natal Province. In Proceedings of the Eighth
        World Orchid Conference, Palmengarten, Frankfurt, 10-17
        April, 1975, pages 326-330.

Sahni, K. C.

    1970. Protection of Rare and Endangered Plants in the Indian Flora.
        IUCN Publications, new series, 18(2): 95-102.

Sanford, W. W.

    1969. Conservation of West African Orchids, I: Nigeria.
        Biological Conservation 1(2): 148-150.

    1970 a. Conservation des Orchidees en Afrique Occidentale, II:
        La Republique du Cameroun. Biological Conservation 3(1): 47-50.

    1970 b. Practical Conservation of Orchids in Nigeria. Nigerian
        Journal of Science 4: 49-57.

Santapau, H.

    1970. Endangered Plant Species and Their Habitats. IUCN Publications,
        new series, 18(2): 83-88. (India and South East Asia)

Schofield, E. K.

    1973 a. Galapagos Flora: The Threat of Introduced Plants.
        Biological Conservation 5(1): 48-51.

1973 b. A Unique and Threatened Flora. Garden Journal 23(3): 68-73. (Galapagos)

Shelyag-Sosonko, Y. R., and G. S. Kukovitsya

1974. New and Rare Species of Western Podolya Flora and Their Protection. Ukrajinskyi Botanicnyii Zhurnal 31(4): 522-525.

Sherbrooke, W. C., and P. Paylore

1973. World Desertification, Cause and Effect: A Literature Review and Annotated Bibliography. 168 pages. Tucson: University of Arizona, Office of Arid Land Studies.

Simmons, J. B., R. I. Beyer, P. E. Brandham, G. L. Lucas, and V. T. H. Parry, editors

1976. Conservation of Threatened Plants. NATO Conference Series, 1 Ecology, Volume 1. New York and London: Plenum Press.

Sjögren, E. A.

1973. Conservation of Natural Plant Communities on Madeira and in the Azores. Monographiae Biologicae Canariensis 4: 148-153.

Skvortsov, A. K.

1976. "Red Book" of the Native Flora. Priroda 8: 121-128. (Soviet Union)

Smith, R. F.

1972. The Impact of the Green Revolution on Plant Protection in Tropical and Sub-tropical Areas. Bulletin of the Entomological Society of America 18: 7-14.

Smith, R. L.

1976. Ecological Genesis of Endangered Species: The Philosophy of Preservation. Annual Review of Ecology and Systematics 7: 33-55.

Snook, L. C.

    1977. Unrecognized Evils in Lantana (L. camara). Tigerpaper
        4(1): 8-9 (January)

Societa botanica italiana, Florence. Gruppo di lavoro per la
conservazioni della natura

    1961. Censimento dei biotopi di rilevante interesse vegetazionale
        meritenoli di conservazione in Italia. 668 pages. Camerino:
        Savini-Mercuri.

Specht, R. L.

    1961. Flora Conservation in South Australia: The Preservation
        of Plant Formations and Associations Recorded in South
        Australia. Transactions of the Royal Society of South
        Australia 85: 177-196.

Specht, R. L., E. M. Roe, and V. H. Boughton, editors

    1974. Conservation of Major Plant Communities in Australia and
        Papua New Guinea. Australian Journal of Botany, supplementary
        series 7: 1-667.

Srinivasan, K. S.

    1959. Protection of Wild (Plant) Life. Bulletin of the Botanical
        Survey of India 1(1): 85-89.

Stacy, H. P.

    1976. Hakea bakerana--A Species in Need of Preservation?
        Australian Plants 8(67): 295-296.

Stanev, S. T.

    1975. Stars in the Mountains are Vanishing: Stories about Our
        Rare Plants. 129 pages. Sofia: Zemizdot.

Stewart, P.

    1969. Cupressus duprezianus, Threatened Conifer of the Sahara.
        Biological Conservation 2(1): 10-12.

292

Stoop van de Kasteele, F. S. C.

    1974. Conservation of Wild Lilium Species. Biological

        Conservation 6(1): 26-31.

Straatmans, W.

    1964. Dynamics of Some Pacific Island Forest Communities in

        Relation to the Survival of the Endemic Flora. Micronesica

        1: 113-122. (Tonga Islands)

Subramanyam, K., and C. P. Sreemadhavan

    1970. Endangered Plant Species and Their Habitats--A Review of

        the Indian Situation. IUCN Publications, new series, 18(2):

        108-114.

Sukopp, H.

    1974. "Rote Liste" der in der Bundesrepublik Deutschland gefahrdeten

        Arten von Farn--und Blutenflanzen (1. Fassung). Natur und

        Landschaft 49(12): 315-322.

Sutton, D. M.

    1973. The Conservation and Use of Endangered Islands.

        Monographiae Biologicae Canariensis 4: 154-157.

Sutton, M.

    1976. Conservation of Fragile Ecosystems in the Canary Islands.

        In G. Kunkel, editor, Biogeography and Ecology in the Canary

        Islands (Monographiae Biologicae, volume 30), pages 479-483.

        The Hague: W. Junk Publishers.

Swabey, C.

    1970. The Endemic Flora of the Seychelle Islands and Its

        Conservation. Biological Conservation 2(3): 171-177.

Takhtajan, A., editor

    1975. Red Book: Native Plant Species to be Protected in the U.S.S.R. 204 pages. Leningrad: Nauka Publishers.

Tannowa, T., A. Yoshida, and K. R. Woolliams

    1976. Tentative List of Rare and Endangered Plants of the Ogasawara Islands. Notes From Waimea Arboretum 3(2): 10-12 (December).

Tem, S.

    1968. Some Rare and Vanishing Plants of Thailand. Conservation in Tropical South East Asia. IUCN Publications, new series, 10: 344-346.

Thayer, G. W., D. A. Wolfe, and R. B. Williams

    1975. The Impact of Man on Seagrass Systems. American Scientist 63(3): 288-296.

The Nature Conservancy

    1975. The Preservation of Natural Diversity: A Survey and Recommendations. 212 pages + appendices. Arlington, Virginia: The Nature Conservancy.

Thiem, K.

    1968. General Information on Conservation Activities and Problems in Thailand: Conservation in Tropical South East Asia. IUCN Publications, new series, 10: 508-518.

Thomas, W. L.

    1965. Man's Role in Changing the Face of the Earth. 1193 pages. Chicago: University of Chicago Press.

Turner, J. S., C. N. Smithers, and R. D. Hoogland

    1968. The Conservation of Norfolk Island. Australian Conservation Foundation, Inc. Special Publication, 1: 1-41.

294

Tyler, P. A.

    1976. Lagoon of Islands, Tasmania--Death Knell for a Unique Ecosystem? _Biological Conservation_ 9(1): 1-11.

Ulychna, K. O., and L. Y. Partyka

    1972. Rare Species of Bryo Flora in the Ukraine and Necessity of Their Protection. _Ukrajinskyi Botanicnyii Zhurnal_ 29: 581-585.

UNESCO

    1970. _Use and Conservation of the Biosphere_. Proceedings of the intergovernmental conference of experts on the scientific basis for rational use and conservation of the resources of the biosphere, Paris, 1968. 272 pages. UNESCO Natural Resources Research Series, Volume 10.

Van Steenis, C. G. G. J.

    1971. Plant Conservation in Malesia. _Bulletin du Jardin Botanique National de Belgique_ 41(1): 189-202.

Veblen, T. T.

    1976. The Urgent Need for Forest Conservation in Highland Guatemala. _Biological Conservation_ 9(2): 141-154.

Vermah, J. C. and K. C. Sahni

    1976. Rare Orchids of the Northeastern Region and Their Conservation. _Indian Forester_ 102: 424-431.

Vesey-FitzGerald, D.

    1974. Rare Plants and the Environment. _Environmental Conservation_ 1(4): 249-250.

Vietmeyer, N. D.

  1975. The Beautiful Blue Devil. *Natural History* 84(9): 65-73.
  (*Eichhornia crassipes*)

de Vogel, E. F.

  1976. Tropical Orchids as an Endangered Plant Group. *Flora
  Malesiana Bulletin* 29: 2602-2604; *Tigerpaper* 3(4): 21-23
  (October 1976).

Vorster, P., and R. Watmough

  1974. Wanton Destruction of an old Cycad. *Excelsa* 4: 47-49.
  (*Encephalartos*)

Walters, S. M.

  1971 a. Index to the Rare Endemic Vascular Plants of Europe.
  Boissiera 19: 87-89.

  1971 b. Taxonomic and Floristic Aspects of Plant Conservation. In
  P. H. Davis, P. C. Harper, and I. C. Hedge, editors, *Plant
  Life of South-West Asia*, pages 293-296; discussion, pages
  306-310. Botanical Society of Edinburgh.

  1976. The Conservation of Threatened Vascular Plants in Europe.
  Biological Conservation 10(1): 31-41.

WATCH

  1976. Help us Save the Threatened Wild Flowers of our Roadside
  Verges. *The Sunday Times Magazine* (London), pages 14-15
  (May 23).

Weber, D.

  1971. Pinta, Galapagos: Une Ile a sauver. *Biological Conservation*
  4(1): 8-12.

296

Wendelbo, P.

    1976. Endangered Flora and Vegetation, With Notes on Some
Results of Protection. Ecological Guidelines for the Use of
Natural Resources in the Middle East and South West Asia.
IUCN Publications, new series, 34: 189-195. (Iran)

Westing, A. H.

    1971. Ecological Effects of Military Defoliation in the Forests
of South Vietnam. BioScience 21(17): 893-898.

Whitmore, T. C.

    1977. Conservation of Tropical Rain Forests, I. Tigerpaper 4(1): 20-23
(January).

Wisniewski, N.

    1976. Ecological Aspects of Orchid Protection in the GDR. In
Proceedings of the Eighth World Orchid Conference, Palmengarten,
Frankfurt, 10-17 April, 1975, pages 307-322.

Woods, D.

    1974. The Disappearing Cape. African Wildlife 28(4): 26-29.

Yoshida, A. and T. Tannowa

    1976. Endangered Plant Species of the Ogasawara Islands. Notes
From Waimea Arboretum 3(2): 8-9 (December).

The Needed Diversity of Plant Resources

American Society of Pharmacognosy

    1962. Search for New Medicinal Agents from Plants. *Lloydia*
        25(4): 241-335.

Bernard, J.

    1968. Treatment of Leukemias, Hodgkin's Disease and Allied Diseases
        by Natural Products: Symposium on Natural Products and
        Cancer Chemotherapy. *Lloydia* 30(4): 291-323.

Burley, J., and B. T. Styles, editors

    1976. *Tropical Trees--Variation, Breeding and Conservation.*
        Linnean Society Symposium Series Number 2. 262 pages.
        New York: Academic Press.

Cannon, H. L.

    1971. The Use of Plant Indicators in Ground Water Surveys,
        Geologic Mapping, and Mineral Prospecting. *Taxon* 20(2/3):
        227-256.

Clark, J. A.

    1956. Collection, Preservation and Utilization of Indigenous
        Strains of Maize. *Economic Botany* 10(2): 194-200.

Committee on Jojoba Utilization, National Research Council

    1975. *Products from Jojoba--A Promising New Crop for Arid Lands.*
        Washington, D.C.: National Academy of Sciences.

Correll, D. S.

    1975. The Search for Plant Precursors of Cortisone. *Economic
        Botany* 9(4): 307-375.

298

Creech, J. L., and Q. Jones

   1969.   Symposium on Centers of Plant Diversity and the Conservation
           of Crop Germ Plasm.   Economic Botany 23(4): 297-338.

Cruse, R. T.

   1959.   Recent Highlights in the Chemurgy of Xerophytic Plants.
           Economic Botany 13(3): 243-260.

Dolan, D. D.

   1957.   New Germ Plasm--The Merits and Uses of Some Plant Intro-
           ductions.   Economic Botany 11(3): 244-248.

Drury, W. H.

   1974.   Rare Species.   Biological Conservation 6(3): 162-169.

Duke, J. A.

   1976.   Economic Appraisal of Endangered Plant Species.   Phytologia
           34(1): 21-27.

Editor.

   1972.   It's a Sperm Whale, it's Superbean!   Smithsonian 3(1):
           54-55.   (Jojoba)

   1975.   A Blooming Desert Project.   Science News 107(21): 335.
           (Jojoba)

Ehrenfeld, D. W.

   1976.   The Conservation of Non-Resources.   American Scientist
           64(6): 648-656.

Evans, H. E.

   1973.   Taxonomists' Curiosity may Help Save the World.
           Smithsonian 4(6): 36-45.

Frankel, O. H.

   1970.   Genetic Conservation of Plants Useful to Man.   Biological
           Conservation 2(3): 162-169.

Frankel, O. H., and E. Bennett, editors

    1970. Genetic Resources in Plants--Their Exploration and Conservation.

        IBP Handbook, number 11.

Frankel, O. H., and J. G. Hawkes, editors

    1975. Crop Genetic Resources for Today and Tomorrow. IBP Handbook,

        number 2. Cambridge University Press.

Frei, E., W. W. Sutow, and J. K. Luce

    1967. Plant Products and the Treatment of Cancer in Man:

        Symposium on Natural Products and Cancer Chemotherapy.

        Lloydia 30(4): 324-331.

Gade, D. W.

    1969. Vanishing Crops of Traditional Agriculture: The Case of Tarwi

        (Lupinus mutabilis) in the Andes. Proceedings of the Association

        of American Geographers 1: 47-51.

Gentry, H. S.

    1958. The Natural History of Jojoba (Simmondsia chinensis)

        and its Cultural Aspects. Economic Botany 12(3): 261-295.

Gentry, H. S., and R. W. Miller

    1965. The Search for New Industrial Crops, IV: Prospectus of

        Limnanthes. Economic Botany 19(1): 25-32.

Hartwell, J. L.

    1971. Plants Used Against Cancer. Lloydia 34(4): 427-438. [index.]

Hawkes, J. G.

    1971. Conservation of Plant Resources. Outlook on Agriculture

        6: 248-253.

Heslop-Harrison, J.

    1975. Man and the Endangered Plant. The International Yearbook and

        Statesmen's Who's Who. pages xiii-xvi. Surrey: Kelly's

        Directories Ltd.

300

Hodge, W. H.

    1955. Some New or Noteworthy Industrial Raw Materials of Plant

        Origin. Economic Botany 9(2): 99-107.

Hoffman, D. B.

    1975. Schistosomiasis Research, The Strategic Plan. New York City:

        The Edna McConnell Clark Foundation. (Molluscicide from

        Phytolacca)

Iltis, H. H.

    1974. Freezing the Genetic Landscape--The Preservation of Diversity

        in Cultivated Plants as an Urgent Social Responsibility of

        the Plant Geneticist and Plant Taxonomist. Maize Genetics

        Cooperation News Letter 48: 199-200.

Jacobson, M.

    1975. Insecticides from Plants, a Review of the Literature, 1954-

        1971. United States Department of Agriculture, Agriculture

        Handbook, number 461.

Jones, Q., and I. A. Wolff

    1960. The Search for New Industrial Crops. Economic Botany 14(1):

        56-68.

    1961. Using Germ Plasm for New Products. In R. E. Hodgson,

        Germ Plasm Resources, xii + 381 pages. Washington, D.C.

Krochmal, A., S. Paur, and P. Duisberg

    1954. Useful Native Plants in the American Southwestern Deserts.

        Economic Botany 8(1): 3-20.

Levin, D. A.

    1976. Alkaloid--Bearing Plants: An Ecogeographic Perspective.

        The American Naturalist 110(972): 261-284.

Malato-Beliz, J.

>   1976. Conservacion de la Naturaleza y Recursos Geneticos. Botanica
>   Macaronesica 1: 67-82. (Macaronesia)

Meyer, W.

>   1974. Podophyllum peltatum--May Apple: A Potential New Cash-Crop
>   Plant. Economic Botany 28(1): 68-72.

Morgan, D.

>   1976. Seeds for Posterity. Washington Post, April 14.

National Academy of Sciences

>   1975. Underexploited Tropical Plants with Promising Economic Value.
>   Washington, D.C.

Raffauf, R. F.

>   1960. Plants as Sources of New Drugs. Economic Botany 14(4):
>   276-279.

>   1970. Some Notes on the Distribution of Alkaloids in the Plant
>   Kingdom. Economic Botany 24(1): 34-38.

Ratsimamanga, R.

>   1972. L'Exploitation des Plantes Medicinales et la Protection de la
>   Nature. In comptes Rendus de la Conference Internationale sur
>   la Conservation de la Nature et de ses Ressources a Madagascar,
>   Tananarive, Madagascar, 7-11 Octobre 1970. IUCN Publications,
>   new series, supplementary paper 36: 143-144.

Roark, R. C.

>   1947. Some Promising Insecticidal Plants. Economic Botany 1(4):
>   437-445.

Sharp, W. C.

>   1970. New Plants for Conservation. Economic Botany 24(1): 53-54.

302

Stewart, C.

    1975. Saving Endangered Vegetables. The Christian Science Monitor, July 1, page 2.

Stuhr, E. T.

    1947. The Distribution, Abundance and Uses of Wild Drug Plants in Oregon and Southern California. Economic Botany 1(1): 57-68.

UNESCO

    1973. Programme on Man and the Biosphere, Expert Panel on Project 8: Conservation of Natural Areas and of the Genetic Material they Contain. MAB Report Series Number 12. Morges, Switzerland.

Vietmeyer, N. D.

    1975. Can a Whale Find Life in the Desert? Audubon 77(5): 101-105. (Jojoba)

Willaman, J. J., and Hui-Lin Li.

    1970. Alkaloid-Bearing Plants and Their Contained Alkaolids. Lloydia 33 (3A-supplement): 1-286.

Winters, H. F.

    1970. Our Hardy Hibiscus Species as Ornamentals. Economic Botany 24(2): 155-164.

Woodward, E. F.

    1947. Botanical Drugs: A Brief Review of the Industry with Comments on Recent Developments. Economic Botany 1(4): 402-414.

Yermanos, D. M.

    1974. Agronomic Survey of Jojoba in California. Economic Botany 28(2): 160-174.

The Role of Botanical Gardens in the

Conservation of Endangered Plants

Alphonso, A. G.

    1976. The Role of the Botanic Gardens in the Conservation of Orchid
        Species. In Proceedings of the Eighth World Orchid
        Conference, Palmengarten, Frankfurt, 10-17 April, 1975, pages
        323-325.

Anonymous.

    1975. Role of Botanical Gardens in the U.S.S.R. in Plant Protection.
        Byulleten Glavnogo Botanicheskogo Sada (Moscow) 95: 1-99.

Ayensu, E. S.

    1976. International Co-operation Conservation-Orientated
        Botanical Gardens and Institutions. In J. B. Simmons, et al.,
        editors, Conservation of Threatened Plants. NATO Conference
        Series, I Ecology, Volume 1, pages 259-269. New York and
        London: Plenum Press.

Breckon, W.

    1974. The Other Side of Kew. The Illustrated London News, May, pages
        53-59.

Brown, R.

    1973. Plants in the Computer. American Horticulturist 52(4): 36-42
        (Winter).

Budowski, G.

    1976. The Global Problems of Conservation and the Potential Role
        of Living Collections. In J. B. Simmons, et al., editors,
        Conservation of Threatened Plants. NATO Conference Series,

I Ecology, Volume 1, pages 9-13. New York and London:
Plenum Press.

Burkill, H. M.

1959. The Botanic Gardens and Conservation in Malaya. The
Gardens' Bulletin (Singapore) 17: 201-205.

Editor.

1976. The Plant Dig-a New Family Outing. Southern Living, March,
page 182.

Esser, K.

1976. Genetic Factors to be Considered in Maintaining Living Plant
Collections. In J. B. Simmons, et al., editors, Conservation
of Threatened Plants. NATO Conference Series, I Ecology,
Volume 1, pages 185-198. New York and London: Plenum Press.

Farmer, R. E., editor.

1975. Rare-Endangered-Threatened Flora Propagation Newsletter. Fall,
4 pages. Division of Forestry, Fisheries, and Wildlife
Development, Tennessee Valley Authority, Norris, Tennessee
37828.

Favarger, C.

1971. Heurs et malheurs d'un jardin experimental. Bulletin du
Jardin Botanique National de Belgique 41(1): 27-41.

Frankel, O.

1976. The Time Scale of Concern. In J. B. Simmons, et al.,
editors. Conservation of Threatened Plants. NATO Conference
Series, I Ecology, Volume 1, pages 245-248. New York and
London: Plenum Press.

Gomez-Campo, C.

    1972. Preservation of West Mediterranean Members of the Cruciferous
            Tribe Brassiceae. Biological Conservation 4(5): 355-360.

    1973. Hacia un Banco de Germoplasma de Endemismos Vegetales
            Ibericos Macaronesicos. Monographiae Biologicae Canariensis
            4: 143-147.

Gyer, J. O.

    1974. Rescue: The Aim is Rescue, Not Just Removal. The Green
            Scene (Pennsylvania Horticultural Society), pages 2-6 (September).

Heslop-Harrison, J.

    1973. The Plant Kingdom: An Exhaustible Resource? Transactions
            of the Botanical Society of Edinburgh 42: 1-15.

    1974. Genetic Resource Conservation: The End and the Means.
            Journal of the Royal Society of Arts, pages 157-169 (February).

Heywood, V. H.

    1976. The Role of Seed Lists in Botanic Gardens Today. In J. B.
            Simmons, et al., editors, Conservation of Threatened Plants.
            NATO Conference Series, I Ecology, Volume 1, pages 225-231.
            New York and London: Plenum Press.

Hondelmann, W.

    1976. Seed Banks. In J. B. Simmons, et al., editors, Conservation
            of Threatened Plants. NATO Conference Series, I Ecology,
            Volume 1, pages 213-224. New York and London: Plenum Press.

Hunt, D. R.

    1974. The Role of Reserve Collections. In D. R. Hunt, editor,
            Succulents in Peril. Supplement to International Organization
            for Succulent Plant Study Bulletin 3(3): 17-20.

306

Huxley, A.

    1974. The Ethics of Plant Collecting. Journal of the Royal
        Horticultural Society 99: 242-249 (1974); Fremontia 4(2):
        17-21 (July 1976).

Lucas, G. L.

    1976. Conservation: Recent Developments in International Co-operation
        and Legislation. In J. B. Simmons, et al., editors,
        Conservation of Threatened Plants. NATO Conference Series,
        I Ecology, Volume 1, pages 271-277. New York and London:
        Plenum Press.

Melville, R.

    1970. Plant Conservation in Relation to Horticulture. Journal
        of the Royal Horticultural Society 95(11): 473-480.

Moore, J. K., and C. R. Bell

    1974. The North Carolina Botanical Garden--A Natural Garden of
        Native Plants. American Horticulturist 53(5): 23-29.

Morton, J. K.

    1972. The Role of Botanic Gardens in Conservation of Species and
        Genic Material. In P. F. Rice, editor, Proceedings of the
        Symposium on A National Botanical Garden System for Canada.
        Royal Botanical Gardens Technical Bulletin 6: 46-54.

Polunin, N.

    1968. Conservational Significance of Botanical Gardens. Biological
        Conservation 1(1): 104-105.

Raven, P. H.

    1976. Ethics and Attitudes. In J. B. Simmons, et al., editors,
        Conservation of Threatened Plants. NATO Conference Series,

I Ecology, Volume 1, pages 155-179. New York and London:
Plenum Press.

Shaw, R. L.

    1976. Future: Integrated International Policies. In J. B. Simmons,
et al., editors, Conservation of Threatened Plants. NATO
Conference Series, I Ecology, Volume 1, pages 39-47. New York
and London: Plenum Press.

Simmons, J. B.

    1976. Present: The Resource Potential of Existing Living Plant
Collections. In J. B. Simmons, et al., editors, Conservation
of Threatened Plants . NATO Conference Series, I Ecology,
Volume 1, pages 27-38. New York and London: Plenum Press.

Simmons, J. B., R. I. Beyer, P. E. Brandham, G. L. Lucas, and V. T. H.
Parry, editors.

    1976. Conservation of Threatened Plants. NATO Conference Series,
I Ecology, Volume 1. New York and London: Plenum Press.

Thompson, P. A.

    1970. Seed Banks as a Means of Improving the Quality of Seed.
Taxon 19(1): 59-62.

    1974. The Use of Seed-Banks for Conservation of Populations of
Species and Ecotypes. Biological Conservation 6(1): 15-19
(January).

    1975 a. The Collection, Maintenance, and Environmental Importance
of the Genetic Resources of Wild Plants. Environmental
Conservation 2(3): 223-228.

1975 b.   Should Botanic Gardens Save Rare Plants? New Scientist

    68: 636-638 (December 11).

1976.   Factors Involved in the Selection of Plant Resources for

    Conservation as Seed in Gene Banks. Biological Conservation

    10(3): 159-167.

Walters, S. M.

1973.   The Role of Botanic Gardens in Conservation. Journal of

    the Royal Horticultural Society 98(7): 311-315.

Woolliams, K. R.

1975.   The Propagation of Hawaiian Endangered Species. Hawaiian

    Botanical Society Newsletter 14(4): 59-68 (October).

1976.   Propagation of Hawaiian Endangered Species. In J. B. Simmons,

    et al., editors, Conservation of Threatened Plants. NATO

    Conference Series, I Ecology, Volume 1, pages 73-83. New York

    and London: Plenum Press.

EXAMPLES OF COMPUTER-PRINTED INFORMATION

The Smithsonian computer file on endangered, threatened, and extinct plants currently includes only the name of the taxon, its family, the states in which it occurs, and its status designation. More detailed information is stored on file cards.

To facilitate the storage of detailed information on the taxonomy, localities, population levels, habitats, threats, and reference sources for each species, and to be easily able to retrieve that information, an expanded computer file is being developed using the Smithsonian's SELGEM system.

The file will be designed to encompass all pertinent information required by botanists, government agencies, and conservation groups. This system, which is still in the planning stages, will thus be capable of printing complete data reports for each species, or listing each species whose characteristics answer to a given question asked of the computer.

Several preliminary reports are shown to illustrate the extent of the file envisioned. The data elements in the new file are not yet arranged in final form, and thus do not necessarily reflect the retrievability of each element.

For a pilot set of species, the commercially and privately exploited ones, latitudes and longitudes (coordinates) of the known localities have been made available to the computer, outside of the SELGEM system.

Computer-drawn maps of two species produced from those data are shown. The locations of exploited species of the southeastern states have been similarly mapped by computer, in order to help identify the areas where different species are found aggregated together in the same habitat. The resultant map is shown as an example of what could be similarly prepared for the entire country.

The species to which the symbols refer on the map of southeastern exploited plants are listed below. It should be noted that the range of Sarracenia rubra (M), which was listed as threatened in the 1975 Report, was plotted before the species was removed from the present lists due to its abundance in the field.

A	Cereus eriophorus var. fragrans
B	Cereus gracilis var. aboriginum
C	Cereus gracilis var. simpsonii
D	Cereus robinii var. deeringii
E	Cereus robinii var. robinii
F	Cladrastis lutea
G	Dionaea muscipula
H	Encyclia boothiana var. erythronioides
I	Rhapidophyllum hystrix
J	Rhododendron vaseyi
K	Sarracenia oreophila
M	Sarracenia rubra
N	Sedum nevii
Q	Zamia integrifolia
R	Zephyranthes simpsonii
S	Zephyranthes treatiae
T	Sarracenia jonesii

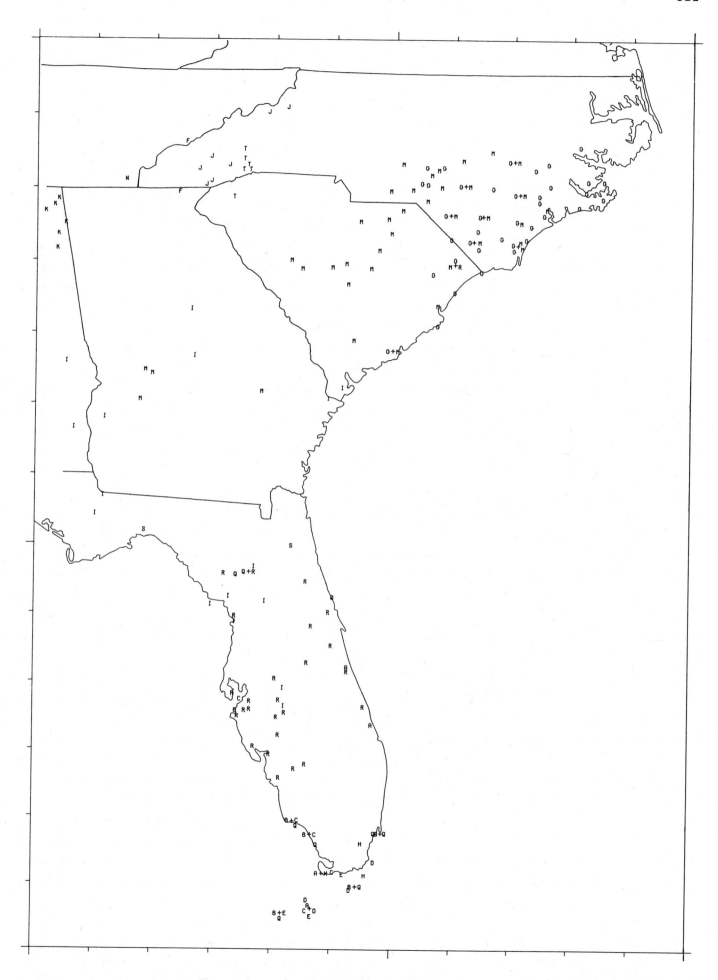

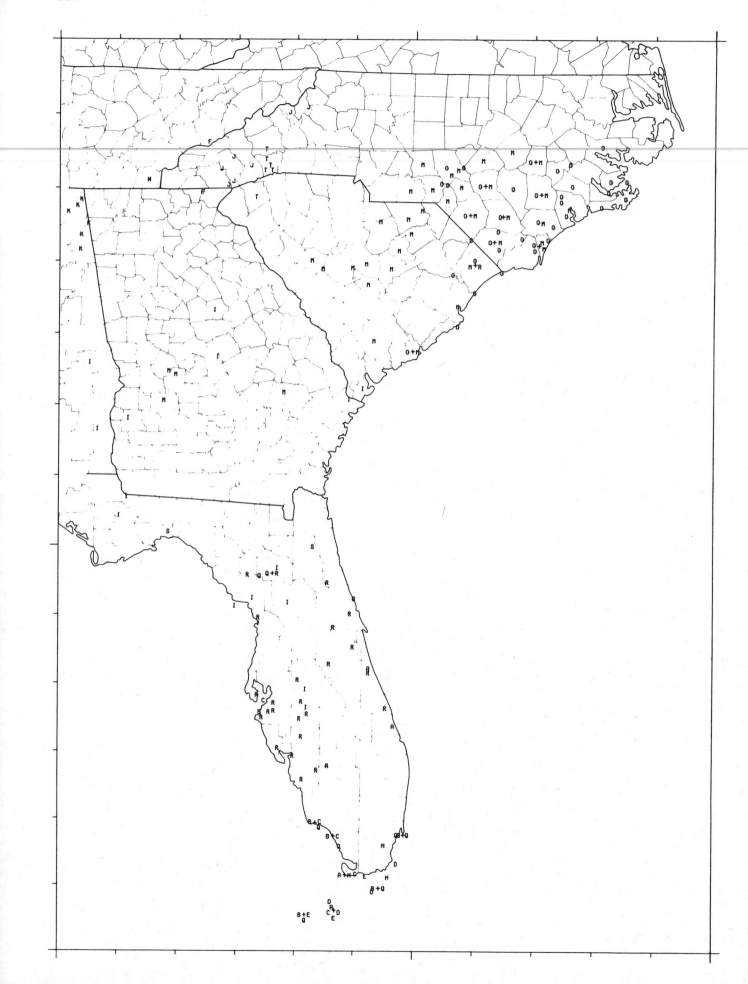

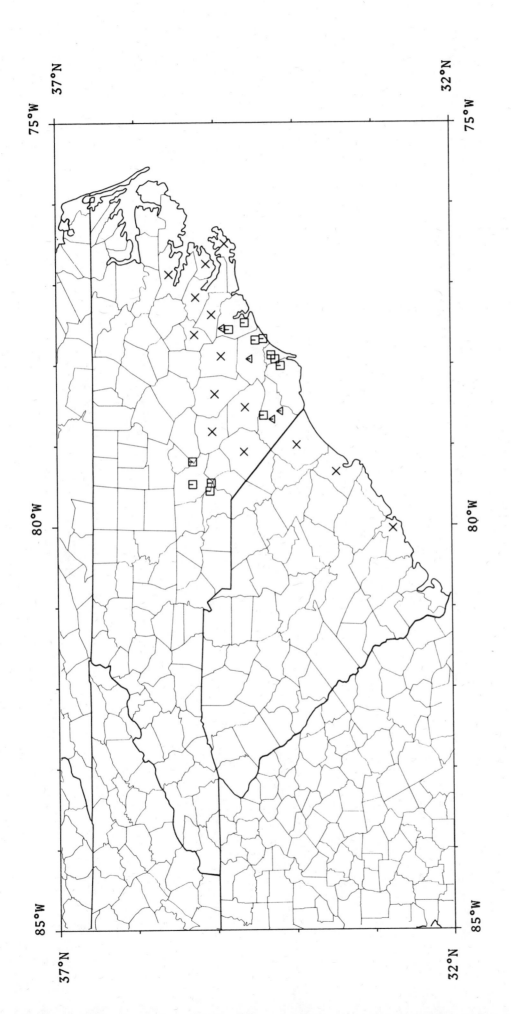

Distribution of *Dionaea muscipula* Ellis

△ accurate locations
□ approximate locations
✕ county records

314

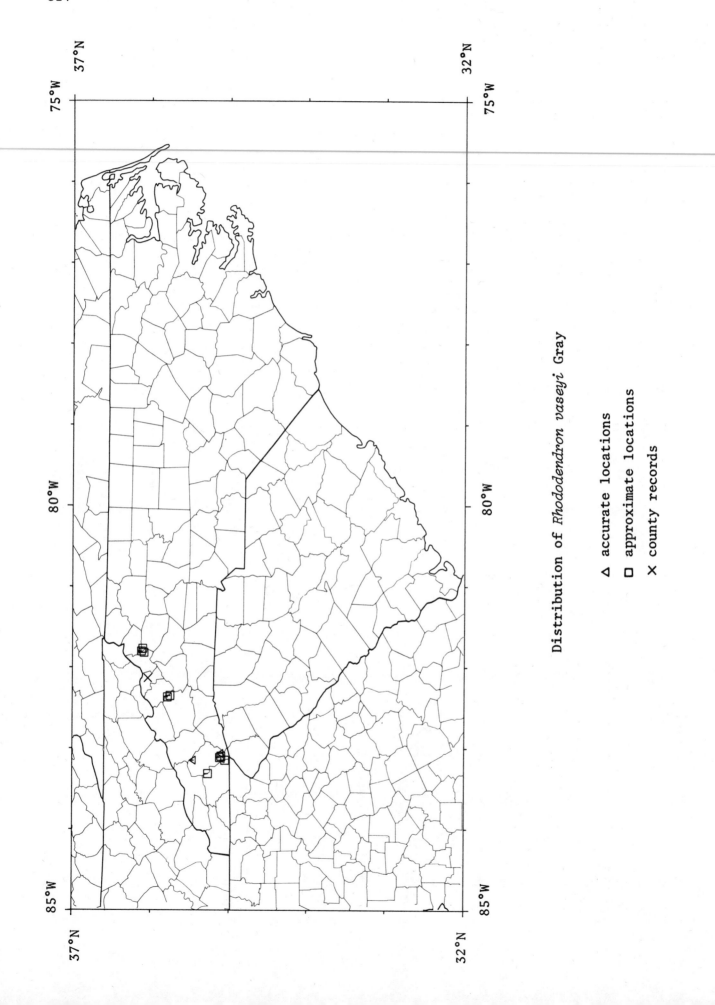

Distribution of *Rhododendron vaseyi* Gray

△ accurate locations

☐ approximate locations

✕ county records

DATE: 04/18/77

EXAMPLE OF COMPUTER-PRINTED DATA SHEETS
MASTER LIST

PAGE NO: 1

SERIAL	CATEG	LINE	
			....2....V....3....V....4....V....5....V....6....V....7....V....
00000000	010	01	FAMILY
	020	01	NAME
	030	01	AUTHOR, PLACE OF PUBLICATION
	040	01	SYNONYMS
	050	01	COMMON NAME
	060	01	TYPE
	070	01	GEOGRAPHICAL SUMMARY
	160	01	PHYSIOGRAPHIC PROVINCE
	170	01	NATURAL REGION
	180	01	ASSOCIATION
	190	01	ASSOCIATES
	200	01	HABITAT
	210	01	ELEVATION
	220	01	S. I. STATUS
	230	01	DATE LAST RECORDED
	240	01	EXPLOITATION
	250	01	LEGAL STATUS
	260	01	OTHER LISTS
	270	01	POPULATION DATA
	280	01	THREATS
	290	01	NATURAL PRESERVES
	300	01	CULTIVATION
	310	01	BIOLOGY
	320	01	POTENTIAL VALUE
	330	01	REFERENCES
	340	01	DATE OF REPORT
	400	01	RANGE
	410	01	COUNTRY
	420	01	STATE
	430	01	COUNTY
	440	01	ISLAND
	450	01	DISTRICT
	460	01	SITE
	470	01	COORDINATES

EXAMPLE OF COMPUTER-PRINTED DATA SHEETS
MASTER LIST

DATE: 04/18/77   PAGE NO: 2

...2....V....3....V....4....V....5....V....6....V....7....V....

SERIAL	CATEG	LINE	CAT-DEFINITION	
00000110	010	01	FAMILY:	BURMANNIACEAE
	011	01	NAME:	THISMIA AMERICANA
	020	01	AUTHOR, PLACE OF PUBLICATION:	PFEIFFER, N. E./ 1914/ BOT. GAZ./ 57:123.
	021	01		
	030	01	SYNONYMS:	SARCOSIPHON AMERICANUS (PFEIFFER) SCHLECHTER
	031	01		
	040	01	TYPE:	CHICAGO, ILLINOIS/ PFEIFFER, N. E./ 1912.
	041	01		
	060	01	GEOGRAPHICAL SUMMARY:	ILLINOIS: TYPE LOCALITY ONLY, EXTINCT
	061	01		
	070	01	ASSOCIATES:	AMONG SOLIDAGO SEROTINA, SOLIDAGO TENUIFOLIA, RUDBECKIA HIRTA,
	071	01		EUPATORIUM PERFOLIATUM, ASCLEPIAS INCARNATA, IRIS VERSICOLOR,
	190	01		ACORUS CALAMUS, AND AGROSTIS ALBA VAR. VULGATA./ SELAGINELLA
		02		APUS, ANEURA PINGUIS, AND HYPNUM "ON THE SOIL ITSELF."
		03		USUALLY IN SPOTS WHERE SOIL NOT CLOSELY COVERED BY ANEURA AND
		04		SELAGINELLA; OCCASIONALLY AMONG MOSS.
		05		
		06		
	191	01	HABITAT:	LOW WET PRAIRIES
	200	01		
	220	01	S. I. STATUS:	EXTINCT
	221	01		
	230	01	DATE LAST RECORDED:	1913
	231	01		
	250	01	LEGAL STATUS:	NOTICE OF REVIEW, FEDERAL REGISTER 40: 27918. 1 JUL 1975.
	251	01		
	270	01	POPULATION DATA:	REPEATED UNSUCCESSFUL SEARCHES OF ORIGINAL AREA AND SIMILAR
		02		SITES BY SWINK; WAGNER AND CHURCHILL
	271	01		
	280	01	THREATS:	AREA NOW COVERED BY INDUSTRIAL SITES
	281	01		
	300	01	CULTIVATION:	ATTEMPTS AT SEED GERMINATION FAILED
	301	01		
	310	01	BIOLOGY:	NEAREST RELATIVE IN NEW ZEALAND AND TASMANIA
	311	01		
	330	01	REFERENCES:	JONES, G. N. & FULLER, G. D./ 1955/ VASC. PL. ILLINOIS/ 142.
		02		MOHLENBROCK, R. H./ 1970/ ILLUS. FL. ILLINOIS: LILIES TO
		03		ORCHIDS/ 178, T.88. (DESCRIP., ILLUS.)
		04		PFEIFFER, N. E./ 1914/ BOT. GAZ./ 57: 123-135, T.7-11.
		05		(DESCRIP., ILLUS.)
	331	01	DATE OF REPORT:	APR 1976
	340	01		
00000111	410	01	COUNTRY:	U. S. A.
	420	01	STATE:	ILLINOIS
	430	01	COUNTY:	COOK
00000112	460	01	SITE:	CHICAGO: ALONG MARGIN OF A GRASS FIELD, BOTTOM PRAIRIE SWALE ON
		02		EAST SIDE OF LAKE CALUMET, BETWEEN TORRENCE AVE. (AT ABOUT 11900
		03		SOUTH) AND NICKEL PLATE RR, BETWEEN FORD PLANT AND SOLWAY COKE

...2....V....3....V....4....V....5....V....6....V....7....V....

SERIAL	CATEG	LINE	CAT-DEFINITION	
00000112	460	04		PLANT.(EXTIRPATED)
	470	01	COORDINATES:	41 41 30 N X 087 33 30 W
00000113	500	01		*
		02		*
		03		*
		04		*
00000210	010	01	FAMILY:	DROSERACEAE
	011	01		
	020	01	NAME:	DIONAEA MUSCIPULA
	021	01		
	030	01	AUTHOR, PLACE OF PUBLICATION:	ELLIS EX LINNAEUS/ 1771/ MANTISSA PLANTARUM/ 2: 2388
	031	01		
	050	01	COMMON NAME:	VENUS' FLY TRAP
	051	01		
	060	01	TYPE:	CAROLINA, IN BOG/ DANIÉL SOLANDER/ NO DATE
	061	01		
	070	01	GEOGRAPHICAL SUMMARY:	COASTAL PLAIN OF NORTH AND SOUTH CAROLINA
	071	01		
	160	01	PHYSIOGRAPHIC PROVINCE:	COASTAL PLAIN
	161	01		
	180	01	ASSOCIATION:	SAVANNAHS
	181	01		
	190	01	ASSOCIATES:	SARRACENIA SPP., DROSERA SPP., PINGUICULA SPP., ILEX GLABRA,
	190	02		MYRICA CERIFERA, PINUS SEROTINA
	191	01		
	200	01	HABITAT:	MARGINS OF POCOSINS OR BOGS
	200	02		WET SANDY DITCHES
	200	03		USUALLY ACID SOIL
	201	01		
	220	01	S. I. STATUS:	THREATENED
	221	01		
	240	01	EXPLOITATION:	COMMERCIALLY EXPLOITED, PRIVATELY EXPLOITED
	241	01		
	250	01	LEGAL STATUS:	NOTICE OF REVIEW, FED. REG. 40:27866, 27872. 1 JUL 1975 1975
	250	02		IN NORTH CAROLINA, THE PLANT IS PROTECTED BY LAW, GENERAL
	250	03		STATUTE, SECT. 14-129.1
	251	01		
	260	01	OTHER LISTS:	THREATENED, ENDEMIC, EXPLOITED -N. C. DEPT. OF NAT. & ECON.
	260	02		RES., ENDANGERED SPECIES COMMITTEE/ DEC 1975, UPDATED LIST
	261	01		
	280	01	THREATS:	EXPLOITATION
	281	01		
	300	01	CULTIVATION:	WIDELY CULTIVATED AS A HOUSEPLANT
	300	02		AM. HORT. SOC. PLANT SERVICE DATA CENTER.
	300	04		NEW JERSEY, ATLANTIC CO., OCEANVILLE, SEEN BY JOHN CHURCHILL
	300	05		IN 1956. GOOD STAND IN BOG WITH SCHIZAEA TRANSPLANTED THERE
	300	06		BY SOMEONE AND SPREADING. NOT SEEN IN 1969 BECAUSE OF HIGH
	300	07		WATER LEVEL, POSSIBLY FLOODED OUT. (CHURCHILL)
	301	01		
	310	01	BIOLOGY:	INSECTIVOROUS
	310	02		MONOTYPIC GENUS

318

EXAMPLE OF COMPUTER-PRINTED DATA SHEETS
MASTER LIST

DATE: 04/18/77   PAGE NO: 4

```
          ....2....v....3....v....4....v....5....v....6....v....7....v....
                                       *         *              *
SERIAL    CATEG  LINE  CAT-DEFINITION

00000210  311    01    SCIENTIFIC
          320    01    POTENTIAL VALUE:
          321    01
          330    01    REFERENCES:     CHURCHILL, J./ DEC 1974/ LETTER.
          330    02                    RADFORD, A. E., AHLES, H. E., & BELL, C. R./ 1968/ MANUAL OF
          330    03                    VASC. FLORA OF CAROLINAS/ 518. (ILLUS.)
          330    04                    UNITED STATES NATIONAL HERBARIUM
          331    01    DATE OF REPORT: APR 1975
          340    01

00000211  410    01    COUNTRY:        U. S. A.
          420    01    STATE:          NORTH CAROLINA
          430    01    COUNTY:         COLUMBUS

00000212  460    01    SITE:           10 MI NW OF LAKE WACCAMAW (USNH)
          470    01    COORDINATES:    34 23 -- N X 078 38 -- W

00000213  460    01    SITE:           OLD DOCK (USNH)
          470    01    COORDINATES:    34 10 -- N X 078 35 -- W

00000214  460    01    SITE:           10 MI N OF OLD DOCK, ON ROAD TO WHITEVILLE (USNH)
          470    01    COORDINATES:    34 16 -- N X 078 41 -- W

00000215  410    01    COUNTRY:        U. S. A.
          420    01    STATE:          NORTH CAROLINA
          430    01    COUNTY:         MOORE

00000216  460    01    SITE:           SE SECTION (USNH)
          470    01    COORDINATES:    35 03 -- N X 079 28 -- W
          460    02                    35 04 -- N X 079 34 -- W
          470    03                    35 17 -- N X 079 12 -- W
          470    04                    35 17 -- N X 079 29 -- W

00000217  410    01    COUNTRY:        U. S. A.
          420    01    STATE:          NORTH CAROLINA
          430    01    COUNTY:         NEW HANOVER

00000218  460    01    SITE:           US HWY 17 AND ATLANTIC COAST LINE RAILROAD, 10 MI NE OF
          460    02                    WILMINGTON (USNH)
          470    01    COORDINATES:    34 17 -- N X 077 53 -- W

00000219  460    01    SITE:           WILMINGTON (USNH)
          470    01    COORDINATES:    34 14 -- N X 077 56 -- W

00000220  410    01    COUNTRY:        U. S. A.
          420    01    STATE:          NORTH CAROLINA
          430    01    COUNTY:         PENDER

00000221  460    01    SITE:           8 MI N OF HAMSTEAD (USNH)
          470    01    COORDINATES:    34 29 -- N X 077 42 -- W

00000222  460    01    SITE:           BURGAW (USNH)
          470    01    COORDINATES:    34 33 -- N X 077 56 -- W
```

EXAMPLE OF COMPUTER-PRINTED DATA SHEETS
MASTER LIST

....2....V....3....V....4....V....5....V....6....V....7....V....

SERIAL	CATEG	LINE	CAT-DEFINITION	
00000223	410	01	COUNTRY:	U. S. A.
	420	01	STATE:	NORTH CAROLINA
	430	01	COUNTY:	BRUNSWICK
00000224	460	01	SITE:	7 MI SW OF WILMINGTON (USNH)
	470	01	COORDINATES:	34 10 -- N X 078 01 -- W
00000225	410	01	COUNTRY:	U. S. A.
	420	01	STATE:	NORTH CAROLINA
	430	01	COUNTY:	ONSLOW
00000226	460	01	SITE:	NEAR LAKE CATHERINE (USNH)
	470	01	COORDINATES:	34 49 -- N X 077 34 -- W
00000227	460	01	SITE:	RICHLANDS (USNH)
	470	01	COORDINATES:	34 54 -- N X 077 33 -- W
00000228	460	01	SITE:	10 MI S OF JACKSONVILLE (USNH)
	470	01	COORDINATES:	34 37 -- N X 077 29 -- W
00000229	460	01	SITE:	30 MI S OF JACKSONVILLE (USNH)
	470	01	COORDINATES:	34 23 -- N X 077 41 -- W
00000230	410	01	COUNTRY:	U. S. A.
	420	01	STATE:	NORTH CAROLINA
	430	01	COUNTY:	BLADEN/ RADFORD ET AL.
	470	01	COORDINATES:	34 30 -- N X 078 30 -- W
00000231	410	01	COUNTRY:	U. S. A.
	420	01	STATE:	NORTH CAROLINA
	430	01	COUNTY:	ROBESON/ RADFORD ET AL.
	470	01	COORDINATES:	34 30 -- N X 079 00 -- W
00000232	410	01	COUNTRY:	U. S. A.
	420	01	STATE:	NORTH CAROLINA
	430	01	COUNTY:	CUMBERLAND/ RADFORD ET AL.
	470	01	COORDINATES:	35 00 -- N X 078 45 -- W
00000233	410	01	COUNTRY:	U. S. A.
	420	01	STATE:	NORTH CAROLINA
	430	01	COUNTY:	SAMPSON/ RADFORD ET AL.
	470	01	COORDINATES:	35 00 -- N X 078 15 -- W
00000234	410	01	COUNTRY:	U. S. A.
	420	01	STATE:	NORTH CAROLINA
	430	01	COUNTY:	DUPLIN/ RADFORD ET AL.
	470	01	COORDINATES:	35 00 -- N X 078 00 -- W
00000235	410	01	COUNTRY:	U. S. A.
	420	01	STATE:	NORTH CAROLINA
	430	01	COUNTY:	LENOIR/ RADFORD ET AL.
	470	01	COORDINATES:	35 15 -- N X 077 45 -- W

EXAMPLE OF COMPUTER-PRINTED DATA SHEETS
MASTER LIST

```
....2....V....3....V....4....V....5....V....6....V....7....V....
```

SERIAL	CATEG	LINE	CAT-DEFINITION	
00000236	410	01	COUNTRY:	U. S. A.
	420	01	STATE:	NORTH CAROLINA
	430	01	COUNTY:	JONES/ RADFORD ET AL.
	470	01	COORDINATES:	35 00 -- N X 077 30 -- W
00000237	410	01	COUNTRY:	U. S. A.
	420	01	STATE:	NORTH CAROLINA
	430	01	COUNTY:	CRAVEN/ RADFORD ET AL.
	470	01	COORDINATES:	35 00 -- N X 077 00 -- W
00000238	410	01	COUNTRY:	U. S. A.
	420	01	STATE:	NORTH CAROLINA
	430	01	COUNTY:	CARTERET/ RADFORD ET AL.
	470	01	COORDINATES:	34 45 -- N X 077 00 -- W
00000239	410	01	COUNTRY:	U. S. A.
	420	01	STATE:	NORTH CAROLINA
	430	01	COUNTY:	PAMLICO/ RADFORD ET AL.
	470	01	COORDINATES:	35 00 -- N X 076 45 -- W
00000240	410	01	COUNTRY:	U. S. A.
	420	01	STATE:	NORTH CAROLINA
	430	01	COUNTY:	BEAUFORT/ RADFORD ET AL.
	470	01	COORDINATES:	35 30 -- N X 077 00 -- W
00000241	410	01	COUNTRY:	U. S. A.
	420	01	STATE:	SOUTH CAROLINA
	430	01	COUNTY:	HORRY/ RADFORD ET AL.
	470	01	COORDINATES:	34 00 -- N X 079 00 -- W
00000242	410	01	COUNTRY:	U. S. A.
	420	01	STATE:	SOUTH CAROLINA
	430	01	COUNTY:	GEORGETOWN/ RADFORD ET AL.
	470	01	COORDINATES:	33 30 -- N X 079 15 -- W
00000243	410	01	COUNTRY:	U. S. A.
	420	01	STATE:	SOUTH CAROLINA
	430	01	COUNTY:	CHARLES/ RADFORD ET AL.
	470	01	COORDINATES:	32 45 -- N X 080 00 -- W
00000244	500	01		*
		02		*
		03		*
		04		*
00000310	010	01	FAMILY:	ERICACEAE
	011	01		
	020	01	NAME:	RHODODENDRON VASEYI
	021	01		
	030	01	AUTHOR, PLACE OF PUBLICATION:	GRAY, A./ 1879/ PROC. AM. ACAD./ 15: 48.
	031	01		
	040	01	SYNONYMS:	AZALEA VASEYI REHDER.
		02		BILTIA VASEYI (GRAY) SMALL

EXAMPLE OF COMPUTER-PRINTED DATA SHEETS
MASTER LIST

DATE: 04/18/77

...2....V....3....V....4....V....5....V....6....V....7....V....

SERIAL	CATEG	LINE	CAT-DEFINITION	
00000310	041	01		PINK-SHELL AZALEA
	050	01	COMMON NAME:	
	051	01	TYPE:	MTS. OF JACKSON CO., NORTH CAROLINA, 7 MI SW OF WEBSTER/ VASEY,
	060	02		G. R./ 3 JUN 1878.
	061	01	GEOGRAPHICAL SUMMARY:	MOUNTAINS OF NORTH CAROLINA
	070	01		
	071	01	PHYSIOGRAPHIC PROVINCE:	BLUE RIDGE
	160	01		
	161	01	NATURAL REGION:	BLUE RIDGE MTS
	170	01		
	171	01	ASSOCIATION:	SPRUCE FOREST
	180	01		
	181	01	HABITAT:	FOREST
	200	02		BOG
	201	01	ELEVATION:	5000 + FT
	210	01		
	211	01	S. I. STATUS:	THREATENED
	220	01		
	221	01	EXPLOITATION:	COMMERCIALLY EXPLOITED
	240	01		
	241	01	LEGAL STATUS:	NOTICE OF REVIEW, FED. REG. 40(127):27866. 1 JUL 1975
	250	01		
	251	01	OTHER LISTS:	THREATENED, ENDEMIC -N. C. DEPT. OF NAT. & ECON. RES.,
	260	02		ENDANGERED SPECIES COMMITTEE/ DEC 1975, UPDATED LIST
	261	01	THREATS:	OFTEN CULTIVATED, OFTEN SOLD BY NURSERIES
	280	01		
	281	01	NATURAL PRESERVES:	PISGAH NATIONAL FOREST
	290	02		NANTAHALA NATIONAL FOREST
	291	01	CULTIVATION:	AM. HORT. SOC. PLANT SERVICE DATA CENTER
	300	01		
	301	01	BIOLOGY:	SHRUB
	310	01		
	311	01	POTENTIAL VALUE:	SHOWY FLOWERS
	320	01		
	321	01	REFERENCES:	GRAY, A./ 1879/ PROC. AM. ACAD./ 15: 48 (DESCRIP.)
	330	02		RADFORD, A. E., AHLES, H. E., & BELL, C. R./ 1968/ MANUAL OF
		03		VASCULAR FLORA OF CAROLINAS/ 798-799 (ILLUS., DESCRIP.)
		04		UNITED STATES NATIONAL HERBARIUM
		05		WOOD, C./ PERSONAL COMMUNICATION.
	331	01	DATE OF REPORT:	APR 1975
	340	01		
00000311	410	01	COUNTRY:	U. S. A.
	420	01	STATE:	NORTH CAROLINA
	430	01	COUNTY:	WATAUGA
00000312	460	01	SITE:	GRANDFATHER MT. (USNH)
	470	01	COORDINATES:	36 07 -- N X 081 48 -- W

EXAMPLE OF COMPUTER-PRINTED DATA SHEETS
MASTER LIST

DATE: 04/18/77

```
....2....V....3....V....4....V....5....V....6....V....7....V....

SERIAL    CATEG LINE    CAT-DEFINITION

00000313    410  01     COUNTRY:    U. S. A.
            420  01     STATE:      NORTH CAROLINA
            430  01     COUNTY:     AVERY

00000314    460  01     SITE:        S-SE SLOPES OF GRANDFATHER MT. (USNH)
            470  01     COORDINATES: 36 05 -- N X 081 49 -- W
                 02                  36 06 -- N X 081 46 -- W

00000315    410  01     COUNTRY:    U. S. A.
            420  01     STATE:      NORTH CAROLINA
            430  01     COUNTY:     MACON

00000316    460  01     SITE:        3 MI E OF HIGHLANDS (USNH)
            470  01     COORDINATES: 35 03 -- N X 083 09 -- W

00000317    410  01     COUNTRY:    U. S. A.
            420  01     STATE:      NORTH CAROLINA
            430  01     COUNTY:     JACKSON

00000318    460  01     SITE:        7 MI SW OF WEBSTER (USNH)
            470  01     COORDINATES: 35 16 -- N X 083 19 -- W.

00000319    460  01     SITE:        SUMMIT OF PLOTT BALSAMS (USNH)
            470  01     COORDINATES: 35 27 -- N X 083 09 -- W

00000320    460  01     SITE:        CASHIER'S VALLEY (USNH)
            470  01     COORDINATES: 35 06 -- N X 083 06 -- W
                 02                  35 07 -- N X 083 07 -- W

00000321    460  01     SITE:        CHIMNEY TOP GAP (USNH)
            470  01     COORDINATES: 35 06 -- N X 083 04 -- W

00000322    410  01     COUNTRY:    U. S. A.
            420  01     STATE:      NORTH CAROLINA
            430  01     COUNTY:     TRANSYLVANIA

00000323    460  01     SITE:        CLUB GAP TRAIL, PISGAH NATIONAL FOREST (USNH)
            470  01     COORDINATES: 35 20 -- N X 082 46 -- W

00000324    410  01     COUNTRY:    U. S. A.
            420  01     STATE:      NORTH CAROLINA
            430  01     COUNTY:     MITCHELL/ RADFORD ET AL.

00000325    470  01     COORDINATES: 35 55 -- N X 082 09 -- W

00000326    500  01                  *
                 02                  *
                 03                  **
                 04                  *

00001000    010  01     FAMILY:     ASTERACEAE
            011  01
            020  01     NAME:       ARGYROXIPHIUM KAUENSE
```

DATE: 04/18/77   PAGE NO: 9

EXAMPLE OF COMPUTER-PRINTED DATA SHEETS
MASTER LIST

```
....2....v....3....v....4....v....5....v....6....v....7....v....
```

SERIAL	CATEG	LINE	CAT-DEFINITION
00001000	021	01	**AUTHOR, PLACE OF PUBLICATION:**
	030	01	(ROCK & NEAL) DEGENER & DEGENER/ 27 DEC 1957/ FL. HAWAIIENSIS.
	031	01	**SYNONYMS:**
	040	01	ARGYROXIPHIUM SANDWICENSE VAR. KAUENSE ROCK & NEAL
	041	01	**COMMON NAME:**
	050	01	KAU SILVERSWORD
	051	01	**TYPE:**
	060	01	HAWAII: KAHUKU, ABOVE KAU FOREST RESERVE AT CHARLIE STONE/
	060	02	BRYAN, L. W./ 1956.
	061	01	**GEOGRAPHICAL SUMMARY:**
	070	01	HAWAII: TYPE LOCALITY ONLY
	071	01	**NATURAL REGION:**
	170	01	MAUNA LOA
	171	01	**ASSOCIATION:**
	180	01	METROSIDEROS WOODLAND
	181	01	**ASSOCIATES:**
	190	01	GERANIUM CUNEATUM VAR. HYPOLEUCUM, VACCINIUM RETICULATUM,
	190	02	STYPHELIA, OREOBOLUS
	191	01	**HABITAT:**
	200	01	OPEN, SCRUBBY RAIN FOREST IN FOG BELT, ON WET HUMUS, AA LAVA
	201	01	**ELEVATION:**
	210	01	6000 FT.
	211	01	**S. I. STATUS:**
	220	01	ENDANGERED
	221	01	**LEGAL STATUS:**
	250	01	NOTICE OF REVIEW, FEDERAL REGISTER 40: 27893. 1 JUL 1975.
	251	01	**OTHER LISTS:**
	260	01	VL, R, EN - FOSBERG & HERBST/ 1975/ ALLERTONIA 1(1): 19.
	261	01	**POPULATION DATA:**
	270	01	50 - 70 PLANTS, 1966 (CARLQUIST).
	270	02	1000 PLANTS OVER 20-30 ACRES, 1974 (DEGENERS).
	271	01	**THREATS:**
	280	01	LAVA FLOWS.
	280	02	INTRODUCED INSECTS, EXOTIC WEEDS, BROWSING ANIMALS.
	280	03	GRAZING BY MOUFLON INTRODUCED IN 1974.
	280	04	EATEN BY PIGS.
	281	01	**CULTIVATION:**
	300	01	HILO
	300	02	HAWAII VOLCANOES NATIONAL PARK
	301	01	**BIOLOGY:**
	310	01	GROWS 5-10 YEARS, FLOWERS, MATURES SEED, DIES.
	311	01	**REFERENCES:**
	330	01	CARLQUIST, S./ PERSONAL COMMUNICATION.
	330	02	DEGENER, O./ 1957/ FL. HAWAIIENSIS/ V.5
	330	03	DEGENER, O. & DEGENER, I./ 1975/ SILVERSWORDS & BLUE DATA BOOK/
	330	04	NOTES FROM WAIMEA ARBORETUM/2(1): 3-6. (ILLUS.)
	330	05	DEGENER, DEGENER, SUNADA & SUNADA/ 1976/ PHYTOLOGIA/
	330	06	33(3): 173-177. (DESCRIP., ILLUS.)
	330	07	FOSBERG, F. R./ DATA CARDS.
	331	01	**DATE OF REPORT:**
	340	01	DEC 1976
00001001	410	01	**COUNTRY:** U. S. A.

DATE: 04/18/77

....2....v....3....v....4....v....5....v....6....v....7....v....

SERIAL	CATEG	LINE	CAT-DEFINITION
00001001	420	01	STATE: HAWAII
	430	01	COUNTY: HAWAII
	440	01	ISLAND: HAWAII
	450	01	DISTRICT: KAU
00001002	460	01	SITE: KAHUKU, SW RIFT ZONE OF MAUNA LOA, ABOVE KAU FOREST RESERVE AT
	460	02	CHARLIE STONE
	470	01	COORDINATES: 19 12 30 N X 155 42 30 W
00001003	500	01	*
	500	02	*
	500	03	*
	500	04	*
00001010	010	01	FAMILY: ASTERACEAE *
	011	01	NAME: ARGYROXIPHIUM MACROCEPHALUM *
	020	01	
	021	01	AUTHOR, PLACE OF PUBLICATION: GRAY, A./ 1852/ PROC. AM. ACAD./ 2: 160. *
	030	01	
	031	01	SYNONYMS: ARGYROXIPHIUM SANDWICENSE VAR. MACROCEPHALUM HILLEBRAND *
	040	01	
	041	01	COMMON NAME: HALEAKALA SILVERSWORD, AHINAHINA, HINAHINA, POHINAHINA *
	050	01	
	051	01	TYPE: HAWAII: AT THE BASE OF MT. HALEAKALA, MAUI/ WILKES EXPEDITION *
	060	01	
	061	01	GEOGRAPHICAL SUMMARY: HAWAII: MAUI
	070	01	
	071	01	NATURAL REGION: HALEAKALA; MAUNA KEA; MAUNA LOA *
	170	01	
	171	01	HABITAT: BARREN, LAVA-STREWN PEAKS, ON CINDERS AND LAVA LEDGES OF ALPINE
	200	01	REGION
	200	02	ELEVATION: 8,000-10,000 FT. *
	201	01	
	210	01	S. I. STATUS: ENDANGERED *
	211	01	
	220	01	EXPLOITATION: SHIPPED TO ORIENT AS ORNAMENTS AS LATE AS 1915. *
	221	01	
	240	01	POPULATION DATA: FORMERLY ABUNDANT ON SLOPES OF MAUNA KEA, HUALALAI, AND MAUNA
	241	01	LOA, HAWAII, AND HALEAKALA, MAUI.
	270	01	VERY COMMON WITHIN CRATER AND ON RIM OF HALEAKALA UNTIL 1900.
	270	02	IN 1927, BARELY 100 SPECIMENS WITHIN HALEAKALA CRATER (DEGENER,
	270	03	1930).
	270	04	TOTAL PRESENT ESTIMATED POPULATION MORE THAN 20,000 (HUBBARD &
	270	05	BENDER, 1960).
	270	06	HAWAII POPULATION VERY RARE, ONE SMALL POPULATION OF ABOUT 30
	270	07	PLANTS (LAMOUREUX, 1975). *
	270	08	
	270	09	
	271	01	THREATS: UPROOTED BY MAN.
	280	01	FLOWER BUDS EATEN BY GOATS.
	280	02	ATTACKED BY 6 SPECIES OF INJURIOUS INSECTS, INCLUDING: ILBURNIA
	280	03	ARGYROXYPHII, TEPHRITIS CRATERICOLA, AND A PHYCITID MOTH.
	280	04	

DATE: 04/18/77

EXAMPLE OF COMPUTER-PRINTED DATA SHEETS
MASTER LIST

PAGE NO: 11

```
....2....v....3....v....4....v....5....v....6....v....7....v....
```

SERIAL	CATEG	LINE	CAT-DEFINITION	
00001010	281	01	NATURAL PRESERVES:	HALEAKALA NATIONAL PARK *
	290	01		HAWAII VOLCANOES NATIONAL PARK *
		02		KAUPULEHU FOREST RESERVE *
		03		
	291	01	BIOLOGY:	DIES AFTER FRUITING
	310	01		
	311	01	REFERENCES:	DEGENER, O./ 1930/ PL. HAWAII NATL. PK./ 306-308.
	330	01		GRAY, A./ 1852/ PROC. AM. ACAD./ 160. (DESCRIP.)
		02		HERBST, D./ 3 DEC 1975/ PERSONAL COMMUNICATION.
		03		HILLEBRAND, W./ 1888/ FL. HAWAIIAN IS./ 219. (DESCRIP.)
		04		KECK, D./ 1936/ HAW. SILVERSWORDS/ OCC. PAP. B. P. BISHOP MUS./
		05		11(19): 16-17.
		06		LAMOUREUX, C. H./ 11 SEP 1975/ PERSONAL COMMUNICATION. *
		07		
	331	01	DATE OF REPORT:	DEC 1976
	340	01		
00001011	410	01	COUNTRY:	U. S. A.
	420	01	STATE:	HAWAII
	430	01	COUNTY:	MAUI
	440	01	ISLAND:	MAUI
00001012	460	01	SITE:	HALEAKALA CRATER, SLIDING SAND TRAIL & LAVA BEDS AT FOOT
00001013	410	01	COUNTRY:	U. S. A.
	420	01	STATE:	HAWAII
	430	01	COUNTY:	HAWAII
	440	01	ISLAND:	HAWAII
	450	01	DISTRICT:	HAMAKUA
00001014	460	01	SITE:	MAUNA KEA, SOURCE OF WAILUKU RIVER
	470	01	COORDINATES:	19 49 00 N X 155 25 30 W
00001015	460	01	SITE:	MAUNA KEA, KUKAIAU RANCH
00001016	410	01	COUNTRY:	U. S. A.
	420	01	STATE:	HAWAII
	430	01	COUNTY:	HAWAII
	440	01	ISLAND:	HAWAII
	450	01	DISTRICT:	KAU
00001017	460	01	SITE:	KAA, POHINA
00001018	460	01	SITE:	ABOVE KAPAPALA
	470	01	COORDINATES:	19 20 -- N X 155 27 -- W
00001019	500	01		*
		02		*
		03		*
		04		*
00001020	010	01	FAMILY:	ASTERACEAE
	011	01		

EXAMPLE OF COMPUTER-PRINTED DATA SHEETS
MASTER LIST

```
....2....v....3....v....4....v....5....v....6....v....7....v....
```

SERIAL	CATEG	LINE	CAT-DEFINITION	
00001020	020	01	NAME:	ARGYROXIPHIUM VIRESCENS VAR. VIRESCENS
	021	01		
	030	01	AUTHOR, PLACE OF PUBLICATION:	HILLEBRAND/ 1888/ FL. HAWAIIAN IS./ 219.
	031	01		
	050	01	COMMON NAME:	GREENSWORD
	051	01		
	060	01	TYPE:	HAWAII: KOOLAU GAP, N. SLOPE OF HALEAKALA, MAUI
	061	01		HAWAII: MAUI
	070	01	GEOGRAPHICAL SUMMARY:	ALPINE REGIONS OF HALEAKALA
	071	01		
	170	01	NATURAL REGION:	
	171	01		
	200	01	HABITAT:	FOG SWEPT, ALPINE SCRUB, MOSS-COVERED, ERODED LAVA
	201	01		
	210	01	ELEVATION:	6000 - 10000 FT.
	211	01		
	220	01	S. I. STATUS:	EXTINCT
	221	01		
	230	01	DATE LAST RECORDED:	1915
	231	01		
	250	01	LEGAL STATUS:	NOTICE OF REVIEW, FEDERAL REGISTER 40: 27893. 1 JUL 1975.
	251	01		
	260	01	OTHER LISTS:	D, VR, U, POEX - FOSBERG & HERBST/ 1975/ ALLERTONIA/ 1(1): 19.
	261	01		
	290	01	NATURAL PRESERVES:	HALEAKALA NATIONAL PARK
	291	01		
	310	01	BIOLOGY:	NOT EXPOSED TO INTENSE SUNLIGHT OR EXTREMELY DRY CONDITIONS,
	310	02		POSSESSES SPARSER HAIR COVERING THAN CONGENERS.
	310	03		DIES AFTER FRUITING.
	311	01	REFERENCES:	CARLQUIST, S./ 1 APR 1970/ PERSONAL COMMUNICATION.
	330	01		DEGENER, O./ 1930/ PL. HAWAII NATL. PK./ 310, T.95. (ILLUS.)
	330	02		FOSBERG, F. R./ DATA CARDS.
	330	03		KECK, D./ 1936/ HAWAIIAN SILVERSWORDS/ OCC. PAP. B. P. BISHOP
	330	04		MUS./ 11(19): 17-18.
	330	05		HILLEBRAND, W./ 1888/ FL. HAWAIIAN IS./ 219. (DESCRIP.)
	330	06		NEAL, M. C./ 1965/ IN GARDENS OF HAWAII/ 847. (DESCRIP.)
	330	07		
	331	01	DATE OF REPORT:	DEC 1976
	340	01		
00001021	410	01	COUNTRY:	U. S. A.
	420	01	STATE:	HAWAII
	430	01	COUNTY:	MAUI
	440	01	ISLAND:	MAUI
	450	01	DISTRICT:	MAKAWAO
00001022	460	01	SITE:	KOOLAU GAP, N. SLOPE OF HALEAKALA
	470	01	COORDINATES:	20 44 -- N X 156 12 30 W
00001023	460	01	SITE:	HALEAKALA CRATER
	470	01	COORDINATES:	20 43 00 N X 156 12 30 W
00001024	460	01	SITE:	PUU NIANIAU

EXAMPLE OF COMPUTER-PRINTED DATA SHEETS
MASTER LIST

DATE: 04/18/77   PAGE NO: 13

...2....V....3....V....4....V....5....V....6....V....7....V....

SERIAL	CATEG	LINE	CAT-DEFINITION
00001024	470	01	COORDINATES: 20 46 30 N X 156 12 30 W
00001025	460	01	SITE: UKULELE, EAST & SOUTHEAST OF
	470	01	COORDINATES: 20 47 00 N X 156 15 00 W
00001026	410	01	COUNTRY: U. S. A.
	420	01	STATE: HAWAII
	430	01	COUNTY: MAUI
	440	01	ISLAND: MAUI
	450	01	DISTRICT: HANA
00001027	460	01	SITE: EDGE OF KIPAHULA
	470	01	COORDINATES: 20 43 -- N X 156 07 -- W
00001028	460	01	SITE: KEANAE GAP
	470	01	COORDINATES: 20 48 00 N X 156 11 30 W
00001029	500	01	*
		02	*
		03	*
		04	*
00001100	010	01	FAMILY: CAMPANULACEAE
	011	01	
	020	01	NAME: ROLLANDIA ST.-JOHNII
	021	01	AUTHOR, PLACE OF PUBLICATION: HOSAKA/ 1935/ OCC. PAP. B. P. BISHOP MUS./ 11(13): 15.
	030	01	
	031	01	COMMON NAME: ST. JOHN ROLLANDIA
	050	01	
	051	01	TYPE: HAWAII: KIPAPA GULCH, WAIPIO, KOOLAU RANGE, OAHU/ HOSAKA/ 1933.
	060	01	
	061	01	GEOGRAPHICAL SUMMARY: HAWAII: OAHU
	070	01	
	071	01	NATURAL REGION: KOOLAU RANGE
	170	01	
	171	01	ASSOCIATES: LABORDIAS, GOULDIAS
	190	01	
	191	01	HABITAT: WET WINDSWEPT SCRUB ON MAIN DIVIDE, RIDGES AND RAVINES
	200	01	
	201	01	ELEVATION: 2800 FT., 900 M.
	210	01	
	211	01	S. I. STATUS: ENDANGERED
	220	01	
	221	01	LEGAL STATUS: NOTICE OF REVIEW, FEDERAL REGISTER 40: 27899. 1 JUL 1975.
	250	01	
	251	01	OTHER LISTS: VR, DE, EN - FOSBERG \ HERBST/ 1975/ ALLERTONIA/ 1(1): 16.
	260	01	
	261	01	THREATS: TRAFFIC ON SUMMIT TRAIL
	280	01	MILITARY MANEUVERS.
	281	02	
	290	01	NATURAL PRESERVES: EWA FOREST RESERVE
	291	01	

EXAMPLE OF COMPUTER-PRINTED DATA SHEETS
MASTER LIST

```
....2....v....3....v....4....v....5....v....6....v....7....v....
```

SERIAL	CATEG	LINE	CAT-DEFINITION	
00001100	330	01	REFERENCES:	CARLQUIST, S./ 1 APR 1970/ PERSONAL COMMUNICATION.
		02		DEGENER, O./ 1938/ FL. HAWAIIENSIS. (DESCRIP., ILLUS.)
		03		FOSBERG, F. R./ DATA CARDS.
		04		ST. JOHN, H. & HOSAKA, E. Y./ 1935/ OCC. PAP. B. P. BISHOP MUS./
		05		11(13): 15. (DESCRIP., ILLUS.)
	331	01	DATE OF REPORT:	
	340	01		DEC 1976
00001101	410	01	COUNTRY:	U. S. A.
	420	01	STATE:	HAWAII
	430	01	COUNTY:	HONOLULU
	440	01	ISLAND:	OAHU
	450	01	DISTRICT:	EWA
00001102	460	01	SITE:	SUMMIT OF KOOLAU RANGE, NE OF KAIWIPOO & MAUKA OF AIEA
	470	01	COORDINATES:	21 25 00 N X 157 51 30 W
00001103	460	01	SITE:	KIPAPA GULCH, WAIPIO, KOOLAU RANGE
	470	Q1	COORDINATES:	21 29 -- N X 157 54 00 W
00001104	500	01		*
		02		*
		03		*
		04		*
00001200	010	01	FAMILY:	FABACEAE
	011	01	NAME:	SESBANIA TOMENTOSA VAR. TOMENTOSA
	020	02		FORMA TOMENTOSA
		03		FORMA ARBOREA ROCK
	021	01	AUTHOR, PLACE OF PUBLICATION:	HOOKER, W. J. & ARNOTT, G. A. W./ 1838/ BOTANY OF CAPT.
	030	01		BEECHEY'S VOYAGE/ 286.
		02		
	031	01	COMMON NAME:	OAHU SESBANIA, OHAI
	050	01		
	051	01	TYPE:	(INCORRECTLY GIVEN AS ACAPULCO, MEXICO)
	060	01	GEOGRAPHICAL SUMMARY:	HAWAII
	061	01		
	070	01	ASSOCIATES:	JACQUEMONTIA, HELIOTROPIUM
	071	01		
	190	01	HABITAT:	JACQUEMONTIA AND HELIOTROPIUM COVERED LIMESTONES AND CLAYS,
	191	01		COASTAL SAND DUNES, WHITE CORAL SANDS
	200	01		
		02		
	201	01	ELEVATION:	NEAR THE SEA
	210	01		
	211	01		
	220	01	S. I. STATUS:	ENDANGERED
	221	01		
	250	01	LEGAL STATUS:	NOTICE OF REVIEW, FEDERAL REGISTER 40: 27901. 1 JUL 1975.
	251	01		
	260	01	OTHER LISTS:	D, R, EN, P, C - FOSBERG & HERBST/ 1975/ ALLERTONIA/ 1(1): 46.
	261	01		

328

EXAMPLE OF COMPUTER-PRINTED DATA SHEETS
MASTER LIST

DATE: 04/18/77   PAGE NO: 15

```
....2....v....3....v....4....v....5....v....6....v....7....v....
   *         *         *       *   *                    *
```

SERIAL	CATEG	LINE	CAT-DEFINITION
00001200	270	01	POPULATION DATA: LESS THAN 40 PLANTS (DEGENER, 1946).
	271	01	
	280	02	THREATS: ROAD BUILDING AROUND KAENA PT. (DEGENER)
	281	01	INFESTED BY WOOLY COCCUS (HILLEBRAND).
	300	02	CULTIVATION: DIFFICULT (HILLEBRAND)
	301	01	DEGENER PLANTED SEEDS AT WAIMANALO, OAHU.
	310	02	BIOLOGY: PROSTRATE SHRUB
	311	01	FORMA ARBOREA - A TREE TO 12-15 FT.
	320	01	POTENTIAL VALUE: ORNAMENTAL
	321	01	REFERENCES: DEGENER, O./ 1946/ FL. HAWAIIENSIS.
	330	02	FOSBERG, F. R./ DATA CARDS.
		03	HILLEBRAND, W./ 1888/ FL. HAWAIIAN IS./ 95.
		04	LAMOUREUX, C. H./ OCT 1975/ PERSONAL COMMUNICATION.
		05	ROCK, J. F./ 1920/ LEGUMINOUS PLANTS OF HAWAII/ 155-156.
		06	ST. JOHN, H./ 1973/ LIST FL. PL. HAWAII/ 193.
	331	01	DATE OF REPORT:
	340	01	DEC 1976
00001201	410	01	COUNTRY: U. S. A.
	420	01	STATE: HAWAII
	430	01	COUNTY: HAWAII
	440	01	ISLAND: HAWAII
	450	01	DISTRICT: PUNA
00001202	460	01	SITE: COAST OF
00001203	410	01	COUNTRY: U. S. A.
	420	01	STATE: HAWAII
	430	01	COUNTY: MAUI
	440	01	ISLAND: MOLOKAI
	450	01	DISTRICT: MOLOKAI
00001204	460	01	SITE: SOUTHERN SHORE
00001205	460	01	SITE: ON SAND DUNES AT MOOMOMI, WEST END OF ISLAND
	470	01	COORDINATES: 21 12 00 N X 157 09 30 W
00001206	460	01	SITE: MAHANA
	470	01	COORDINATES: 21 09 30 N X 157 09 00 W
00001207	460	01	SITE: KALANI
	470	01	COORDINATES: 21 12 00 N X 157 10 30 W
00001208	410	01	COUNTRY: U. S. A.
	420	01	STATE: HAWAII
	430	01	COUNTY: MAUI
	440	01	ISLAND: LANAI
	450	01	DISTRICT: LANAI

EXAMPLE OF COMPUTER-PRINTED DATA SHEETS
MASTER LIST

DATE: 04/18/77

```
....2....V....3....V....4....V....5....V....6....V....7....V....
```

SERIAL	CATEG	LINE	CAT-DEFINITION
00001209	460	01	SITE: KAUIOKU
00001210	410	01	COUNTRY: U.S.A.
	420	01	STATE: HAWAII
	430	01	COUNTY: HONOLULU
	440	01	ISLAND: OAHU
	450	01	DISTRICT: WAIANAE
00001211	460	01	SITE: BETWEEN RR TRACKS & OCEAN, EXTENDING FROM KAENA PT. ABOUT 2 MIS.
	470	02	TOWARD KAWAIHAPAI
		01	COORDINATES: 21 34 30 N X 158 17 00 W
00001212	410	01	COUNTRY: U.S.A.
	420	01	STATE: HAWAII
	430	01	COUNTY: HONOLULU
	440	01	ISLAND: OAHU
	450	01	DISTRICT: KOOLAUPOKO
00001213	460	01	SITE: MOKAPU CRATER
	470	01	COORDINATES: 21 26 -- N X 178 45 -- W
00001214	460	01	SITE: KAOHIAIPU ISLET, WAIMANALO BAY
	470	01	COORDINATES: 21 19 30 N X 157 39 30 W
00001215	410	01	COUNTRY: U.S.A.
	420	01	STATE: HAWAII
	430	01	COUNTY: HONOLULU
	440	01	ISLAND: NECKER
00001216	460	01	SITE: SUMMIT HILL
00001217	460	01	SITE: ANNEXATION HILL
00001218	460	01	SITE: FLAGPOLE HILL
00001219	410	01	COUNTRY: U.S.A.
	420	01	STATE: HAWAII
	430	01	COUNTY: HONOLULU
	440	01	ISLAND: NIHOA
00001220	410	01	COUNTRY: U.S.A.
	420	01	STATE: HAWAII
	430	01	COUNTY: KAUAI
	440	01	ISLAND: KAUAI
	450	01	DISTRICT: WAIMEA
00001221	460	01	SITE: NEAR MANA
	470	01	COORDINATES: 22 02 30 N X 159 46 30 W
00001222	500	01	*
		02	*
		03	*
		04	*

EXAMPLE OF COMPUTER-PRINTED DATA SHEETS
MASTER LIST

DATE: 04/18/77   PAGE NO: 17

....2....v....3....v....4....v....5....v....6....v....7....v....

SERIAL	CAT-DEFINITION	CATEG	LINE	
00001300	FAMILY:	010	01	GOODENIACEAE *
		011	01	*
	NAME:	020	01	SCAEVOLA CORIACEA *
	AUTHOR, PLACE OF PUBLICATION:	021	01	*
		030	01	NUTTALL/ 1843/ TRANS. AM. PHIL. SOC./ N. S. 8: 253. *
		031	01	*
	COMMON NAME:	050	01	FALSE JADETREE, NAUPAKA *
	TYPE:	051	01	*
		060	01	HAWAII: KAUAI, NEAR THE SEA/ NELSON/ COOK EXPEDITION *
	GEOGRAPHICAL SUMMARY:	061	01	*
		070	01	HAWAII: AT ONE TIME PROBABLY ON ALL OF LARGER ISLANDS, NOW
		070	02	EXTANT ONLY AT WAIHEE, MAUI. *
	HABITAT:	071	01	*
		200	01	DRY, CONSOLIDATED, LITHIFIED, COASTAL SAND, ARID LOWLANDS, LAVA
	ELEVATION:	201	01	.
		210	01	NEAR SEA LEVEL
	S. I. STATUS:	211	01	*
		220	01	ENDANGERED *
	LEGAL STATUS:	221	01	*
		250	01	NOTICE OF REVIEW, FEDERAL REGISTER 40: 27904. 1 JUL 1975. *
	OTHER LISTS:	251	01	*
		260	01	VL, VR, EN, C - FOSBERG & HERBST/ 1975/ ALLERTONIA/ 1(1): 38.
	POPULATION DATA:	261	01	
		270	01	100 PLANTS AT WAIHEE, MAUI, 1968 (LAMOUREUX).
		270	02	LESS THAN 500 WILD PLANTS IN ONLY ONE AREA.
	THREATS:	271	01	*
		280	01	AREA GRAZED BY CATTLE, SUBJECT TO DEVELOPMENT IN NEAR FUTURE. *
	BIOLOGY:	281	01	*
		310	01	TRAILING SHRUB *
	REFERENCES:	311	01	
		330	01	CARLQUIST, S./ 1970/ HAWAII: NATURAL HISTORY/ 155. (ILLUS.)
		330	02	CARLQUIST, S./ 1 APR 1970/ PERSONAL COMMUNICATION.
		330	03	DEGENER, O./ 1950/ FL. HAWAIIENSIS. (DESCRIP., ILLUS.)
		330	04	FOSBERG, F. R./ DATA CARDS.
		330	05	HILLEBRAND, W./ 1888/ FL. HAWAIIAN ISLANDS/ 266. (DESCRIP.)
		330	06	LAMOUREUX, C. H./ 28 APR 1970/ PERSONAL COMMUNICATION.
		330	07	LAMOUREUX, C. H./ SEP 1975/ PERSONAL COMMUNICATION.
		330	08	ST. JOHN, H./ 1973/ LIST FL. PL. HAWAII/ 346.
	DATE OF REPORT:	331	01	
		340	01	DEC 1976 *
00001301	COUNTRY:	410	01	U. S. A.
	STATE:	420	01	HAWAII
	COUNTY:	430	01	HAWAII
	ISLAND:	440	01	HAWAII
00001302	COUNTRY:	410	01	U. S. A.
	STATE:	420	01	HAWAII
	COUNTY:	430	01	MAUI
	ISLAND:	440	01	MAUI
00001303	SITE:	460	01	"ISTHMUS" (DEGENER)

EXAMPLE OF COMPUTER-PRINTED DATA SHEETS
MASTER LIST

DATE: 04/18/77

...2....V....3....V....4....V....5....V....6....V....7....V....

SERIAL	CATEG	LINE	CAT-DEFINITION	
00001304	410	01	COUNTRY:	U. S. A.
	420	01	STATE:	HAWAII
	430	01	COUNTY:	MAUI
	440	01	ISLAND:	MAUI
	450	01	DISTRICT:	WAILUKU
00001305	460	01	SITE:	WAIHEE, WAIHEE GOLF COURSE
	470	01	COORDINATES:	20 55 30 N X 156 31 00 W
00001306	460	01	SITE:	KALEPOLEPO
	470	01	COORDINATES:	20 45 00 N X 156 27 30 W
00001307	460	01	SITE:	WAILUKU TO WAIHEE PT.
00001308	410	01	COUNTRY:	U. S. A.
	420	01	STATE:	HAWAII
	430	01	COUNTY:	MAUI
	440	01	ISLAND:	LANAI
	450	01	DISTRICT:	LANAI
00001309	460	01	SITE:	KEONOHAU
00001310	410	01	COUNTRY:	U. S. A.
	420	01	STATE:	HAWAII
	430	01	COUNTY:	MAUI
	440	01	ISLAND:	LANAI
00001311	460	01	SITE:	LAE WAHIA, MAHANA
00001312	410	01	COUNTRY:	U. S. A.
	420	01	STATE:	HAWAII
	430	01	COUNTY:	HONOLULU
	440	01	ISLAND:	OAHU
	450	01	DISTRICT:	WAIANAE
00001313	460	01	SITE:	KAENA PT.
	470	01	COORDINATES:	21 34 15 N X 158 16 45 W
00001314	410	01	COUNTRY:	U. S. A.
	420	01	STATE:	HAWAII
	430	01	COUNTY:	HONOLULU
	440	01	ISLAND:	OAHU
	450	01	DISTRICT:	EWA
00001315	460	01	SITE:	BARBER'S PT.
	470	01	COORDINATES:	21 17 45 N X 158 06 30 W
00001316	410	01	COUNTRY:	U. S. A.
	420	01	STATE:	HAWAII
	430	01	COUNTY:	KAUAI
	440	01	ISLAND:	KAUAI
00001317	410	01	COUNTRY:	U. S. A.

EXAMPLE OF COMPUTER-PRINTED DATA SHEETS
MASTER LIST

....2....V....3....V....4....V....5....V....6....V....7....V....

SERIAL	CATEG	LINE	CAT-DEFINITION	
00001317	420	01	STATE:	HAWAII
	430	01	COUNTY:	KAUAI
	440	01	ISLAND:	NIIHAU
	450	01	DISTRICT:	WAIMEA
00001318	500	01		*
		02		*
		03		*
		04		*
00001410	010	01	FAMILY:	MALVACEAE
	011	01		*
	020	01	NAME:	HIBISCADELPHUS BOMBYCINUS
	021	01		*
	030	01	AUTHOR, PLACE OF PUBLICATION:	FORBES, C. N./ 1920/ OCCAS. PAP. B. P. BISHOP MUS./ 7(3): T.3.
	031	01		*
	050	01	COMMON NAME:	KAWAIHAE HIBISCADELPHUS, HUA KUAHIWI
	051	01		*
	060	01	TYPE:	HAWAII: KAWAIHAE-UKA, KOHALA RANGE, HAWAII/ HILLEBRAND/ BEFORE
		02		1868.
	061	01		*
	070	01	GEOGRAPHICAL SUMMARY:	HAWAII: TYPE COLLECTION ONLY, EXTINCT.
	071	01		*
	170	01	NATURAL REGION:	KOHALA MOUNTAIN RANGE
	171	01		*
	200	01	HABITAT:	DRY KIPUKA FOREST
	201	01		*
	220	01	S. I. STATUS:	EXTINCT
	221	01		*
	230	01	DATE LAST RECORDED:	1868
	231	01		*
	250	01	LEGAL STATUS:	NOTICE OF REVIEW, FEDERAL REGISTER 40: 27908. 1 JUL 1975
	251	01		*
	260	01	OTHER LISTS:	PREX - FOSBERG & HERBST/ 1975/ ALLERTONIA/ 1(1): 49.
		02		EXTINCT - I.U.C.N./ JAN 1970/ RED DATA SHEET.
	261	01		*
	270	01	POPULATION DATA:	KNOWN ONLY FROM HERBARIUM SPECIMENS
	271	01		*
	280	01	THREATS:	LAVA FLOWS.
		02		GRAZING AND DESTRUCTION OF HABITAT BY CATTLE.
		03		HABITAT DISPLACEMENT BY ALGAROBA, CACTI, GRASSES, AND INTRODUCED
		04		WEEDS.
	281	01		*
	310	01	BIOLOGY:	TREE.
		02		PROBABLY POLLINATED BY ENDEMIC NECTAR FEEDING DREPANIIDAE BIRDS.
	311	01		*
	320	01	POTENTIAL VALUE:	SCIENTIFIC INTEREST IN CO-EVOLUTION OF ENDEMIC ANIMAL AND PLANT
		02		SPECIES.
	321	01		*
	330	01	REFERENCES:	CARLQUIST, S./ PERSONAL COMMUNICATION.
		02		DEGENER, O./ 1946/ FLORA HAWAIIENSIS. (DESCRIP., ILLUS.)
		03		FOSBERG, F. R./ DATA CARDS.
		04		I.U.C.N./ JAN 1970/ RED DATA SHEET. (DESCRIP.)

334

EXAMPLE OF COMPUTER-PRINTED DATA SHEETS
MASTER LIST

DATE: 04/18/77

....2....v....3....v....4....v....5....v....6....v....7....v....

SERIAL	CATEG	LINE	CAT-DEFINITION
00001410	330	05	
	331	06	
	340	01	DATE OF REPORT:
00001411	410	01	COUNTRY:
	420	01	STATE:
	430	01	COUNTY:
	440	01	ISLAND:
	450	01	DISTRICT:
00001412	460	01	SITE:
	470	01	COORDINATES:
00001413	500	01	
		02	
		03	
		04	
00001420	010	01	FAMILY:
	011	01	NAME:
	020	01	
	021	01	AUTHOR, PLACE OF PUBLICATION:
	030	01	
	031	01	COMMON NAME:
	050	01	TYPE:
	051	01	
	060	01	
		02	
	061	01	GEOGRAPHICAL SUMMARY:
	070	01	
	071	01	NATURAL REGION:
	170	01	
	171	01	ASSOCIATES:
	190	01	
		02	
	191	01	HABITAT:
	200	01	
	201	01	ELEVATION:
	210	01	
	211	01	S. I. STATUS:
	220	01	
	221	01	LEGAL STATUS:
	250	01	
	251	01	OTHER LISTS:
	260	01	
		02	
		03	
	261	01	POPULATION DATA:
	270	01	
		02	
		03	
		04	

LAMOUREUX, C. H./ 28 APR 1970/ PERSONAL COMMUNICATION.
ST. JOHN, H./ 1973/ LIST FL. PL. HAWAIIAN IS./ 229.   *

APR 1976

U. S. A.
HAWAII
HAWAII
HAWAII
SOUTH KOHALA

KAWAIHAE-UKA
20 05 00 N X 155 45 30 W

*
*
*
*

MALVACEAE   *

HIBISCADELPHUS GIFFARDIANUS   *

ROCK, J. F./ 1911/ BULL. HAWAIIAN BOARD AGRIC. & FOR./ 1: 10.   *

GIFFARD HIBISCADELPHUS, HUA KUAHIWI   *

HAWAII: EAST SLOPE OF MAUNA LOA, KIPUKA PUAULU, KEAUHOU, A FEW
MILES FROM KILAUEA VOLCANO, HAWAII/ ROCK/ 1911.   *

HAWAII: TYPE LOCALITY ONLY, EXTIRPATED, REESTABLISHED FROM CULT.   *

MAUNA LOA   *

SAPINDUS SAPONARIA, PELEA, XANTHOXYLUM, URERA, STRAUSSIA, AND
OCHROSIA.   *

OPEN MESOPHYTIC FOREST, ON WEATHERED LAVA   *

4300 FT.   *

ENDANGERED   *

NOTICE OF REVIEW, FEDERAL REGISTER 40: 27908. 1 JUL 1975.   *

VR, EN, C - FOSBERG & HERBST/ 1975/ ALLERTONIA/ 1(1): 49.
EXTINCT IN ITS NATIVE STATION - I.U.C.N./ JAN 1970/ RED DATA
SHEET.   *

ONE TREE DISCOVERED IN WILD, 1911.
SEEDS PLANTED AT GIFFARD HOME NEAR TYPE LOCALITY, KILAUEA, TWO
SURVIVED UNTIL 1946.
EXTINCT IN WILD, 1930.

EXAMPLE OF COMPUTER-PRINTED DATA SHEETS
MASTER LIST

....2....V....3....V....4....V....5....V....6....V....7....V....

SERIAL	CATEG	LINE	CAT-DEFINITION
00001420	270	05	SEEDLINGS REESTABLISHED IN KIPUKA KI AND KIPUKA PUAULU,
		06	KILAUEA AREA, HAWAII, 10 HEALTHY TREES IN 1968.
		07	APPROXIMATELY 60 TREES IN KILAUEA AREA (FOSBERG, 1969).
	271	01	THREATS: LAVA FLOWS.
	280	01	CATTLE GRAZING IN NATIVE FORESTS.
		02	EXTINCTION OF POLLINATORS.
	281	03	
	290	01	NATURAL PRESERVES: HAWAII VOLCANOES NATIONAL PARK: KIPUKA KI, KIPUKA PUANLU
	291	01	CULTIVATION: NOW KNOWN IN CULTIVATION ONLY.
	300	01	FLEMING ARBORETUM, MAUI.
		02	REESTABLISHED IN HAWAII VOLCANOES NATIONAL PARK.
		03	ELSEWHERE IN THE ISLANDS.
		04	
	301	01	BIOLOGY: TREE, 7 M. HIGH
	310	01	SPECIMENS REESTABLISHED IN WILD FORMING FERTILE HYBRIDS WITH
		02	H. HUALALAIENSIS
		03	PROBABLY POLLINATED BY ENDEMIC NECTAR FEEDING DREPANIIDAE BIRDS.
		04	PROPAGATED BY CUTTINGS.
		05	
	311	01	POTENTIAL VALUE: SCIENTIFIC INTEREST IN CO-EVOLUTION OF ENDEMIC ANIMAL AND PLANT
	320	01	SPECIES.
		02	
	321	01	REFERENCES: DEGENER, O./ 1946/ FLORA HAWAIIENSIS. (DESCRIP., ILLUS.)
	330	01	FOSBERG, F. R./ DATA CARDS.
		02	I.U.C.N./ JAN 1970/ RED DATA SHEET. (DESCRIP.)
		03	LAMOUREUX, C. H./ 28 APR 1970/ PERSONAL COMMUNICATION.
		04	ROCK, J. F./ 1913/ INDIG. TREES HAWAIIAN IS./ 297, 299, T.117.
		05	(DESCRIP, ILLUS.)
		06	ST. JOHN, H./ 1973/ LIST FL. PL. HAWAIIAN IS./ 229.
		07	
	331	01	DATE OF REPORT: APR 1976
	340	01	
00001421	410	01	COUNTRY: U. S. A.
	420	01	STATE: HAWAII
	430	01	COUNTY: HAWAII
	440	01	ISLAND: HAWAII
	450	01	DISTRICT: KAU
00001422	460	01	SITE: EASTERN SLOPE OF MAUNA LOA, KIPUKA PUAULU, KEAUHOU, A FEW MILES
		02	FROM KILAUEA VOLCANO. (EXTIRPATED)
	470	01	COORDINATES: 19 27 00 N X 155 18 30 W
00001423	500	01	
		02	
		03	
		04	
00001430	010	01	FAMILY: MALVACEAE
	011	01	
	020	01	NAME: HIBISCADELPHUS HUALALAIENSIS
	021	01	

EXAMPLE OF COMPUTER-PRINTED DATA SHEETS
MASTER LIST

DATE: 04/18/77   PAGE NO: 22

```
SERIAL      CATEG  LINE  CAT-DEFINITION      ....2....V....3....V....4....V....5....V....6....V....7....V....

00001430    030    01    AUTHOR, PLACE OF PUBLICATION:   ROCK, J. F./ 1911/ BULL. HAWAII BOARD AGRIC. & FOR./ 1: 14.   *
            031    01
            050    01    COMMON NAME:                    HUALALAI HIBISCADELPHUS, HAU KUAHIWI                          *
            051    01
            060    01    TYPE:                           HAWAII: MT. HUALAI, NORTH KONA, HAWAII/ ROCK/ 1909.           *
            061    01
            070    01    GEOGRAPHICAL SUMMARY:           HAWAII                                                       *
            071    01
            170    01    NATURAL REGION:                 MT. HUALALAI                                                 *
            171    01
            200    01    HABITAT:                        LAVA FIELDS
            200    02                                    DRY FORESTS                                                  *
            201    01
            210    01    ELEVATION:                      900 M., 2700 FT.                                             *
            211    01
            220    01    S. I. STATUS:                   ENDANGERED                                                   *
            221    01
            250    01    LEGAL STATUS:                   NOTICE OF REVIEW, FEDERAL REGISTER 40: 27908. 1 JUL 1975.    *
            251    01
            260    01    OTHER LISTS:                    VR, EN, C - FOSBERG & HERBST/ 1975/ ALLERTONIA/ 1(1): 50.
            260    02                                    ENDANGERED - I.U.C.N./ JAN 1970/ RED DATA SHEET.             *
            261    01
            270    01    POPULATION DATA:                LESS THAN 20 TREES IN FOREST OF WAIHOU (DEGENER, 1946).
            270    02                                    FORMERLY MORE WIDESPREAD ON LAVA FIELDS OF HUALALAI, LESS
            270    03                                    THAN 10 WILD PLANTS (LAMOUREUX, 1975).
            270    04                                    "ONE TREE LEFT," WAIHOU (ROCK, 1955 - FIDE LAMOUREUX).
            270    05                                    AT LEAST 3 WILD TREES PERSIST ON SLOPE OF HUALALAI AT 1300 M,
            270    06                                    DECLINING IN VIGOR (BISHOP & HERBST, 1973).                  *
            271    01
            280    01    THREATS:                        LAVA FLOWS.
            280    02                                    GRAZING AND DESTRUCTION OF HABITAT BY CATTLE.
            280    03                                    LEAVES AND CAPSULES ATTACKED BY SEVERAL SPECIES OF MOTH.
            280    04                                    PROPOSED RELEASE OF AXIS DEER (BISHOP & HERBST, 1973).       *
            281    01
            290    01    NATURAL PRESERVES:              IN CULTIVATION AT KIPUKA PUAULU, HAWAII VOLCANOES NATL. PARK.
            290    02                                    KAUPULEHU FOREST RESERVE, NORTH KONA, HAWAII ISLAND.         *
            291    01
            300    01    CULTIVATION:                    KIPUKA PUAULU, HAWAII VOLCANOES NATL. PARK, HAWAII.
            300    02                                    FLEMING ARBORETUM, PUU MAHOE, MAUI.
            300    03                                    PERHAPS ON OAHU (LAMOUREUX).
            300    04                                    PLANTS RAISED FROM SEED BY G. WILDER.                        *
            301    01
            310    01    BIOLOGY:                        TREE, 5-7 M. HIGH.
            310    02                                    PROBABLY POLLINATED BY ENDEMIC NECTAR FEEDING DREPANIIDAE BIRDS.
            310    03                                    LARVAE OF ENDEMIC MOTH FEED ON LEAVES AND MATURE CAPSULES.
            310    04                                    SPECIMENS PLANTED IN WILD FORMING FERTILE HYBRIDS.           *
            311    01
            320    01    POTENTIAL VALUE:                SCIENTIFIC INTEREST IN CO-EVOLUTION OF ENDEMIC ANIMAL AND PLANT
            320    02                                    SPECIES.                                                     *
            321    01
            330    01    REFERENCES:                     BISHOP & HERBST/ 1973/ BRITTONIA/ 25: 290-293.
            330    02                                    DEGENER, O./ 1946/ FLORA HAWAIIENSIS. (DESCRIP., ILLUS.)
            330    03                                    FOSBERG, F. R./ DATA CARDS.                                  *
```

EXAMPLE OF COMPUTER-PRINTED DATA SHEETS
MASTER LIST

DATE: 04/18/77   PAGE NO: 23

....2....V....3....V....4....V....5....V....6....V....7....V....

SERIAL	CATEG	LINE	CAT-DEFINITION	
00001430	330	04		I.U.C.N./ JAN 1970/ RED DATA SHEET. (DESCRIP.)
		05		LAMOUREUX, C. H./ 28 APR 1970/ PERSONAL COMMUNICATION.
		06		ROCK, J. F./ 1913/ INDIG. TREES HAWAII IS./ 301-302, T.119.
		07		(DESCRIP., ILLUS.)
		08		ST. JOHN, H./ 1973/ LIST FL. PL. HAWAIIAN IS./ 229.
	331	01		
	340	01	DATE OF REPORT:	APR 1976
00001431	410	01	COUNTRY:	U. S. A.
	420	01	STATE:	HAWAII
	430	01	COUNTY:	HAWAII
	440	01	ISLAND:	HAWAII
	450	01	DISTRICT:	NORTH KONA
00001432	460	01	SITE:	LAVA FIELDS OF MT. HUALALAI
	470	01	COORDINATES:	19 41 30 N X 155 52 00 W
00001433	460	01	SITE:	FOREST OF WAIHOU
	470	01	COORDINATES:	19 32 30 N X 155 54 30 W
00001434	500	01		*
		02		*
		03		*
		04		*
00001440	010	01	FAMILY:	MALVACEAE
	011	01	NAME:	HIBISCADELPHUS DISTANS
	020	01		
	021	01	AUTHOR, PLACE OF PUBLICATION:	BISHOP & HERBST/ 1973/ BRITTONIA/ 25: 290-293.
	030	01	TYPE:	HAWAII: BLUFF ABOVE KOAIE STREAM, WAIMEA CANYON, KAUAI/ BISHOP &
	031	01		HERBST/ 1972.
	060	02		
	061	01	GEOGRAPHICAL SUMMARY:	HAWAII: KAUAI, TYPE LOCALITY ONLY
	070	01	ASSOCIATES:	ERYTHRINA SANDWICENSIS, SAPINDUS OAHUENSIS
	071	01		
	190	01	HABITAT:	REMNANT NATIVE FOREST, ROCKY BLUFFS
	191	01		
	200	01	ELEVATION:	350 M.
	210	01		
	211	01	S. I. STATUS:	ENDANGERED
	220	01		
	221	01	LEGAL STATUS:	NOTICE OF REVIEW, FEDERAL REGISTER 40: 27908. 1 JUL 1975.
	250	01		
	251	01	OTHER LISTS:	VL, VR, EN, C - FOSBERG & HERBST/ 1975/ ALLERTONIA/ 1(1): 49.
	260	01		
	261	01	POPULATION DATA:	ONE KNOWN COLONY IMMINENTLY THREATENED (BISHOP & HERBST, 1973)
	270	01		
	271	01	THREATS:	GRAZING AND DESTRUCTION OF HABITAT BY GOATS.
	280	01		
	281	01	NATURAL PRESERVES:	PUU KA PELE FOREST RESERVE
	290	01		

338

EXAMPLE OF COMPUTER-PRINTED DATA SHEETS
MASTER LIST

...2....V....3....V....4....V....5....V....6....V....7....V....

SERIAL	CATEG	LINE	CAT-DEFINITION	
00001440	291	01		*
	300	01	CULTIVATION:	PACIFIC TROPICAL BOTANIC GARDEN, KOLOA, HAWAII
	301	01		*
	310	01	BIOLOGY:	TREE, 5 M. HIGH.
				PROBABLY POLLINATED BY ENDEMIC NECTAR FEEDING DREPANIIDAE BIRDS.
	311	02		*
	320	01	POTENTIAL VALUE:	SCIENTIFIC INTEREST IN CO-EVOLUTION OF ENDEMIC ANIMAL AND PLANT
				SPECIES.
	321	02		*
	330	01	REFERENCES:	BISHOP & HERBST/ 1973/ BRITTONIA/ 25: 290-293 (DESCRIP., ILLUS.)
		02		FOSBERG, F. R./ DATA CARDS.
		03		LAMOUREUX, C. H./ 1975/ PERSONAL COMMUNICATION.
	331	01		*
	340	01	DATE OF REPORT:	DEC 1976
00001441	410	01	COUNTRY:	U. S. A.
	420	01	STATE:	HAWAII
	430	01	COUNTY:	KAUAI
	440	01	ISLAND:	KAUAI
	450	01	DISTRICT:	WAIMEA
00001442	460	01	SITE:	BLUFF ABOVE KOAIE STREAM, WAIMEA CANYON
	470	01	COORDINATES:	25 05 00 N X 159 39 -- W
00001443	500	01		*
		02		*
		03		*
		04		*
00001450	010	01	FAMILY:	MALVACEAE
	011	01		*
	020	01	NAME:	HIBISCADELPHUS WILDERIANUS
	021	01		*
	030	01	AUTHOR, PLACE OF PUBLICATION:	ROCK, J. R./ 1911/ BULL. HAWAIIAN BOARD AGR. & FOR./ 1: 12.
	031	01		*
	050	01	COMMON NAME:	HAU KUAHIWI
	051	01		*
	060	01	TYPE:	HAWAII: LAVA FIELD OF AUAHI, S. SLOPE OF MT. HALEAKALA, MAUI/
		02		ROCK/ 1910.
	061	01		*
	070	01	GEOGRAPHICAL SUMMARY:	HAWAII: KNOWN FROM ONLY ONE WILD TREE AND ONE CULTIVATED
		02		SEEDLING, BOTH NOW DEAD.
	071	01		*
	170	01	NATURAL REGION:	MT. HALEAKALA
	171	01		*
	190	01	ASSOCIATES:	USNEA
	191	01		*
	200	01	HABITAT:	DRY FOREST ON ANCIENT LAVA FLOWS
	201	01		*
	210	01	ELEVATION:	2500-2600 FT.
	211	01		*
	220	01	S. I. STATUS:	EXTINCT
	221	01		*

EXAMPLE OF COMPUTER-PRINTED DATA SHEETS
MASTER LIST

```
....2....V....3....V....4....V....5....V....6....V....7....V....
```

SERIAL	CATEG	LINE	CAT-DEFINITION	
00001450	230	01	CAT-DEFINITION	1912
	231	01		
	250	01	DATE LAST RECORDED:	
	251	01	LEGAL STATUS:	NOTICE OF REVIEW, FEDERAL REGISTER 40: 27908. 1 JUL 1975.
	260	01	OTHER LISTS:	EX - FOSBERG & HERBST/ 1975/ ALLERTONIA/ 1(1): 50. *
	261	02		EXTINCT IN NATIVE HABITAT - I.U.C.N./ JAN 1970/ RED DATA SHEET. *
	270	01	POPULATION DATA:	SINGLE DYING PLANT DISCOVERED IN 1910.
		02		REPEATED UNSUCCESSFUL SEARCHES MADE OF TYPE LOCALITY. *
	271	01	THREATS:	LAVA FLOWS.
	280	02		DESTRUCTION OF HABITAT BY GRAZING CATTLE.
		03		EXTINCTION OF POLLINATORS.
	281	01	CULTIVATION:	WILDER RAISED ONE PLANT FROM SEED, FATE UNKNOWN. *
	300	01	BIOLOGY:	TREE, 5 M. HIGH.
	301	01		PROBABLY POLLINATED BY ENDEMIC NECTAR FEEDING DREPANIIDAE BIRDS. *
	310	01	POTENTIAL VALUE:	SCIENTIFIC INTEREST IN CO-EVOLUTION OF ENDEMIC ANIMAL AND PLANT
		02		SPECIES. *
	311	01	REFERENCES:	FOSBERG, F. R./ DATA CARDS.
	320	01		LAMOUREUX, C. H./ 28 APR 1970/ PERSONAL COMMUNICATION.
		02		ROCK, J. F./ 1911/ BULL. HAWAIIAN BOARD AGR. & FOR./ 1: 12, T.5.
	321	01		ROCK, J. F./ 1913/ INDIG. TREES HAWAIIAN IS./ 299, T. 301. *
	330	01	DATE OF REPORT:	DEC 1976
		02		
		03		
		04		
	331	01		
	340	01		
00001451	410	01	COUNTRY:	U. S. A.
	420	01	STATE:	HAWAII
	430	01	COUNTY:	MAUI
	440	01	ISLAND:	MAUI
00001452	460	01	SITE:	LAVA FIELDS OF AUAHI, ON LAND OF KAHIKINUI, S. SLOPE OF HALEAKAL
00001453	500	01		*
		02		*
		03		*
		04		*
00001500	010	01	FAMILY:	RUTACEAE
	011	01		
	020	01	NAME:	PLATYDESMA REMYI
	021	01		*
	030	01	AUTHOR, PLACE OF PUBLICATION:	(SHERFF) DEGENERS, SHERFF, & STONE/ 28 DEC 1960/FL. HAWAIIENSIS. *
	031	01		
	040	01	SYNONYMS:	PLATYDESMA AURICULAEFOLIA (GRAY) HILLEBRAND SENSU HILLEBRAND
		02		PELEA AURICULAEFOLIA GRAY SENSU GRAY
		03		CLAOXYLON INSIGNE BAILLON
		04		PLATYDESMA CAMPANULATUM VAR. SESSILIFOLIUM ROCK
		05		CLAOXYLON REMYI SHERFF
	041	01		*

EXAMPLE OF COMPUTER-PRINTED DATA SHEETS
MASTER LIST

DATE: 04/18/77

....2....v....3....v....4....v....5....v....6....v....7....v....

SERIAL	CATEG	LINE	CAT-DEFINITION	
00001500	050	01	COMMON NAME:	REMY PLATYDESMA, PILOKEA
	051	01		
	060	01	TYPE:	HAWAII: HAWAII/ REMY/ 1851-1855.
	061	01		
	070	01	GEOGRAPHICAL SUMMARY:	HAWAII: HAWAII
	071	01		
	170	01	NATURAL REGION:	MAUNA KEA
	170	02		KOHALA MTS. (EXTIRPATED)
	171	01	ASSOCIATES:	METROSIDEROS & ACACIA HOA
	190	01		
	191	01	HABITAT:	WET FOREST
	200	01		
	201	01	ELEVATION:	2500-3000 FT.
	210	01		
	211	01	S. I. STATUS:	ENDANGERED
	220	01		
	221	01	LEGAL STATUS:	NOTICE OF REVIEW, FEDERAL REGISTER 40: 27914. 1 JUL 1975.
	250	01		
	251	01	OTHER LISTS:	VL, R, EN - FOSBERG & HERBST/ 1975/ ALLERTONIA/ 1(1): 65.
	260	01		
	261	01	POPULATION DATA:	TYPE LOCALITY (KOHALA MTS.) POPULATION PROBABLY EXTINCT.
	270	01		6 PLANTS AT MAUNA KEA, 1970.
	270	02		
	271	01	THREATS:	HABITAT DETERIORATION.
	280	01		LOGGING ROAD RECENTLY BULLDOZED THROUGH MAUNA KEA AREA
	280	02		(FOSBERG):
	280	03		
	281	01	NATURAL PRESERVES:	MAUNA KEA SITE PROPOSED AS A NATURAL AREA.
	290	01		
	291	01	REFERENCES:	DEGENER, O./ 1960/ FL. HAWAIIENSIS. (DESCRIP., ILLUS.)
	330	01		FOSBERG, F. R./ DATA CARDS.
	330	02		HILLEBRAND/ 1888/ FL. HAWAIIAN.IS./ 72-73. (DESCRIP.)
	330	03		LAMOUREUX, C. H./ 28 APR 1970/ PERSONAL COMMUNICATION.
	330	04		
	331	01	DATE OF REPORT:	DEC 1976
	340	01		
00001501	410	01	COUNTRY:	U. S. A.
	420	01	STATE:	HAWAII
	430	01	COUNTY:	HAWAII
	440	01	ISLAND:	HAWAII
	450	01	DISTRICT:	HAMAKUA
00001502	460	01	SITE:	E SLOPES OF MAUNA KEA, AHUPUAA OF LAUPAHOEHOE
00001503	460	01	SITE:	INLAND SIDE OF WAIPIO VALLEY, KOHALA MTS.
	470	01	COORDINATES:	20 03 -- N X 155 39 -- W
00001504	410	01	COUNTRY:	U. S. A.
	420	01	STATE:	HAWAII
	430	01	COUNTY:	HAWAII
	440	01	ISLAND:	HAWAII
	450	01	DISTRICT:	SOUTH KOHALA

DATE: 04/18/77   PAGE NO: 27

EXAMPLE OF COMPUTER-PRINTED DATA SHEETS
            MASTER LIST

SERIAL   CATEG LINE   CAT-DEFINITION

00001505 460   01   SITE:

....2....V....3....V....4....V....5....V....6....V....7....V....

UPPER HAMAKUA DITCH TRAIL ABOVE KOIAWE VALLEY, KOHALA MTS.

ENDANGERED FLORA PROJECT--SMITHSONIAN INSTITUTION

NAME	FAMILY
Viguiera ludens (Shinners) Johnston	Asteraceae

ORIGINAL LOCALITY	TYPE OF HABITAT
TEXAS: Culberson Co., Lobo Flat (1953)	ditch

MAXIMUM KNOWN RANGE
as above

PROBABLE PRESENT RANGE
as above

WHERE AND WHEN LAST SEEN OR COLLECTED AND BY WHOM
Known only from type collection in ditch beside cotton field

ENDANGERED

PROBABLE PRESENT STATUS (Underline)   INCREASING   DECREASING   STATIONARY   RARE
VERY RARE   UNCERTAIN   PROBABLY EXTINCT

REMARKS (Use Back If Necessary)
Johnston, Southwestern Naturalist 13:250(1968); Shinners, Sida 1:377 (1964); Heiser, et al., Mem. Torrey Bot. Club 22(3):81-84(1969); Correll & Johnston, Texas Flora 1646(1970). Status: Johnston list (1974).

REPORT BY:

SI-3058
11-22-74

ENDANGERED FLORA PROJECT--SMITHSONIAN INSTITUTION

NAME	FAMILY
Rorippa subumbellata Rollins	Brassicaceae

ORIGINAL LOCALITY CALIFORNIA: El Dorado Co., Lake Tahoe (1919)	TYPE OF HABITAT sandy shores

MAXIMUM KNOWN RANGE
CALIFORNIA: Lakes Tahoe, Truckee and Tallac

PROBABLE PRESENT RANGE

WHERE AND WHEN LAST SEEN OR COLLECTED AND BY WHOM
Bliss Memorial Park in 1963 by R.L. Stuckey, in shifting sand where much trampled by swimmers and sunbathers.   THREATENED

PROBABLE PRESENT STATUS (Underline)   INCREASING   DECREASING   STATIONARY   RARE
VERY RARE   UNCERTAIN   PROBABLY EXTINCT

REMARKS (Use Back If Necessary)
Contrib. Dudley Herb. 3:177(1941).
Stuckey, Sida 4(4):296-297(1972); Stuckey letter 10/74.

REPORT BY: CNPS Inventory 1974:2-2-2-3

SI-3058
11-22-74

ENDANGERED FLORA PROJECT--SMITHSONIAN INSTITUTION

NAME
Coryphantha ramillosa Cutak

FAMILY
Cactaceae

ORIGINAL LOCALITY

TYPE OF HABITAT

MAXIMUM KNOWN RANGE
Texas: restricted to Big Bend area;   Rio Grande in

PROBABLE PRESENT RANGE Brewster and s Terrell Cos.; Mexico in Coah.

WHERE AND WHEN LAST SEEN OR COLLECTED AND BY WHOM   N & E of Big Bend Nat'l. Pk.
(Benson, 12/72)

PROBABLE PRESENT STATUS (Underline)   INCREASING   DECREASING   STATIONARY   RARE
VERY RARE   UNCERTAIN   PROBABLY EXTINCT

REMARKS (Use Back If Necessary)
Commercial exploitation and abuse.
Being collected in great quantities by commercial dealers for
direct sale from the field (Benson, 12/72)

REPORT BY:

SI-3058
11-22-74

ENDANGERED FLORA PROJECT--SMITHSONIAN INSTITUTION

NAME
Cyperus grayioides Mohlenbrock

FAMILY
Cyperaceae

ORIGINAL LOCALITY
ILLINOIS: Mason Co. (1954)

TYPE OF HABITAT
sand prairies and blowouts

MAXIMUM KNOWN RANGE
ILLINOIS: Mason and Whiteside Cos.

PROBABLE PRESENT RANGE
as above

WHERE AND WHEN LAST SEEN OR COLLECTED AND BY WHOM

ENDANGERED

PROBABLE PRESENT STATUS (Underline)   INCREASING   DECREASING   STATIONARY   RARE
VERY RARE   UNCERTAIN   PROBABLY EXTINCT

REMARKS (Use Back If Necessary)
Brittonia 11(4): 255(1959). Status: Myers, Annot. Cat. 45(1972).
Endangered due to habitat destruction (Fell letter 9/75).
Endangered (Dean letter 8/75).

REPORT BY:
R.A. DeFilipps

SI-3058
11-22-74

344

ENDANGERED FLORA PROJECT--SMITHSONIAN INSTITUTION

NAME
Astragalus lentiginosus var. micans

FAMILY
Fabaceae

ORIGINAL LOCALITY   CALIFORNIA: Inyo Co.,
Eureka Valley (1955)

TYPE OF HABITAT
sand dunes

MAXIMUM KNOWN RANGE
as above

PROBABLE PRESENT RANGE
as above

WHERE AND WHEN LAST SEEN OR COLLECTED AND BY WHOM

ENDANGERED

PROBABLE PRESENT STATUS (Underline)   INCREASING   DECREASING   STATIONARY   RARE
VERY RARE   UNCERTAIN   PROBABLY EXTINCT

REMARKS (Use Back If Necessary)
Barneby, Leaflets Western Botany 8:22(1956).
Barneby, Atlas Astragalus, Mem. New York Bot. Garden 13(2):952-953∧   (1964).
Endangered by off-road vehicles (Henry 11/75, Thorne 11/75 letters).

REPORT BY:
CNPS Inventory 1974: 3-2-1-3

SI-3058
11-22-74

ENDANGERED FLORA PROJECT--SMITHSONIAN INSTITUTION

NAME   Yucca toftiae Welsh

FAMILY   Liliaceae

ORIGINAL LOCALITY
UTAH: San Juan Co., Lake Powell (1973)

TYPE OF HABITAT
sandy alluvium and outcrops

MAXIMUM KNOWN RANGE
UTAH: San Juan and Kane Cos., Lake Powell.

PROBABLE PRESENT RANGE
as above

WHERE AND WHEN LAST SEEN OR COLLECTED AND BY WHOM

THREATENED

PROBABLE PRESENT STATUS (Underline)   INCREASING   DECREASING   STATIONARY   RARE
VERY RARE   UNCERTAIN   PROBABLY EXTINCT

REMARKS (Use Back If Necessary)
Great Basin Naturalist 34(4): 308(1974).   Status: Welsh Utah list
(1975).   Much of known range inundated by Lake Powell.

REPORT BY:
J.L. Reveal (1976).

SI-3058
11-22-74

ENDANGERED FLORA PROJECT--SMITHSONIAN INSTITUTION

NAME	FAMILY
<u>Mentzelia</u> packardiae Glad	Loasaceae

ORIGINAL LOCALITY	TYPE OF HABITAT
OREGON: Malheur Co. (1974)	volcanic ash on talus slopes

MAXIMUM KNOWN RANGE
OREGON: Malheur Co., Leslie Gulch

PROBABLE PRESENT RANGE
as above

WHERE AND WHEN LAST SEEN OR COLLECTED AND BY WHOM
June 1975 by J.L. Reveal

ENDANGERED

PROBABLE PRESENT STATUS (Underline)   INCREASING   DECREASING   STATIONARY   <u>RARE</u>
VERY RARE   UNCERTAIN   PROBABLY EXTINCT

REMARKS (Use Back If Necessary)
<u>Madrono</u> 23(5): 289(1976). Status: Reveal (1976, pers. comm.).
Area just opened up to recreational use by new road to Lake Owyhee.
Heavy cattle grazing in area (Reveal).

REPORT BY:
J.L. Reveal (1976).

SI-3058
11-22-74

ENDANGERED FLORA PROJECT--SMITHSONIAN INSTITUTION

NAME	FAMILY
<u>Boerhaavia</u> mathisiana F.B. Jones	Nyctaginaceae

ORIGINAL LOCALITY	TYPE OF HABITAT
TEXAS: San Patricio Co. (1968)	caliche slope

MAXIMUM KNOWN RANGE
TEXAS: San Patricio and Live Oak Cos.

PROBABLE PRESENT RANGE
as above

WHERE AND WHEN LAST SEEN OR COLLECTED AND BY WHOM

THREATENED

PROBABLE PRESENT STATUS (Underline)   INCREASING   DECREASING   STATIONARY   <u>RARE</u>
VERY RARE   UNCERTAIN   PROBABLY EXTINCT

REMARKS (Use Back If Necessary)
Flora of the Texas Coastal Bend 231(1975). Only a few scattered
individuals now remain, due to removal of caliche for road
construction.

REPORT BY: Johnston letter 9/75.

SI-3058
11-22-74

346

ENDANGERED FLORA PROJECT--SMITHSONIAN INSTITUTION

NAME	FAMILY
Eriogonum ephedroides Reveal	Polygonaceae

ORIGINAL LOCALITY	TYPE OF HABITAT
UTAH: Uintah Co., s. of Bonanza(1965)	white shale outcrops

MAXIMUM KNOWN RANGE
UTAH: Uintah Co. and COLORADO: Rio Blanco Co.

PROBABLE PRESENT RANGE

WHERE AND WHEN LAST SEEN OR COLLECTED AND BY WHOM

ENDANGERED

PROBABLE PRESENT STATUS (Underline)   INCREASING   DECREASING   STATIONARY   <u>RARE</u>
VERY RARE   UNCERTAIN   PROBABLY EXTINCT

REMARKS (Use Back If Necessary)
<u>Madrono</u> 19(8):295(1968). Status: Welsh Utah list (1975). On oil shale, the 3 known locations subject to immediate mining; also highway construction in area (Reveal).

REPORT BY:
J.L. Reveal

SI-3058
11-22-74

ENDANGERED FLORA PROJECT--SMITHSONIAN INSTITUTION

NAME	FAMILY
Prunus gravesii Small	Rosaceae

ORIGINAL LOCALITY	TYPE OF HABITAT low shrubby
CONNECTICUT: New London Co.	thickets on gravelly sand ridges near Long Island Sound

MAXIMUM KNOWN RANGE
CONNECTICUT: New London Co.: Groton: Esker Pt. Park, Noank

PROBABLE PRESENT RANGE

WHERE AND WHEN LAST SEEN OR COLLECTED AND BY WHOM

ENDANGERED

Place of Publication: <u>Bull. Torrey Bot. Club</u> 24:45,pl.292(1897).

PROBABLE PRESENT STATUS (Underline)   INCREASING   DECREASING   STATIONARY   RARE
VERY RARE   UNCERTAIN   PROBABLY EXTINCT

REMARKS (Use Back If Necessary)   One of rarest plants in Conn.. Immediately endangered. 15 individuals in 50 sq. meters. Moderately disturbed site. Adjacent to heavily used recreation area. Individuals healthy. No seedlings. Limited vegetative reproduction.

REPORT BY:
Status Report by J.J. Dowhan (see Bampton letter 8-75)

SI-3058
11-22-74

DATA SHEETS ON PLANTS LISTED IN
APPENDIX I TO THE 1973 CONVENTION

The Endangered Species Act of 1973 notes that the United States has pledged itself as a sovereign state in the international community to conserve to the extent practicable the various species of plants and animals facing extinction, pursuant to the Convention on International Trade in Endangered Species of Wild Fauna and Flora. The United States had signed the Convention on March 3, 1973; it later ratified and became a party to the Convention as of July 1, 1975.

The Convention provides for listings of plants in three appendices. Appendix I, the most critical, is designed to include all species of plants threatened with extinction which are or may be affected by international trade. Trade in specimens of these species must be subject to particularly strict regulation in order not to endanger further their survival and must only be authorized in exceptional circumstances. In order for trade in a species listed in Appendix I to be legal, both an export permit from the country of origin and an import permit from the country of destination are required. Certain findings must be made by the Management Authority and Scientific Authority in each country.

On April 13, 1976, the President signed Executive Order Number 11911, which names the Secretary of the Interior as the Management Authority under the Convention for the United States. That Order made it possible for the United States to begin the implementation of the Convention, and the U.S. Fish and Wildlife Service published, in the Federal Register of June 16, 1976, a proposed implementation including proposed regulations for permits to trade in Appendix I plants. Further, the Service proposed endangered status for the currently listed Appendix I plants, in the Federal Register of September 26, 1975. On February 22, 1977 the Fish and Wildlife Service published a rulemaking in the Federal Register which officially implements the Convention for the United States, authorizing the Service to issue permits and certificates for trade in plants listed in the Appendices to the Convention. Meanwhile, the Service will examine the species on the Appendices on an individual basis, to determine if they qualify for listing as endangered or threatened under the Endangered Species Act of 1973.

A meeting of the Conference of the Parties to the Convention took place in Berne, Switzerland, on November 2-6, 1976, at which it was recognized that certain species presently listed on the Appendices should be reviewed to determine if they are indeed qualified for listing. A special working session of the Convention will review the Appendices during 1977 and make recommendations for action at the next Conference of the Parties in 1978.

The meeting decided that, in determining the appropriate Appendix into which a species should be placed, the biological status and trade status of the species should be evaluated together with respect to the following guidelines:

Appendix I

1. Biological status. To qualify for Appendix I, a species must be currently threatened with extinction. Information of any one of the following types should be required, in order of preference: (a) scientific reports on the population size or geographic range of the species over a number of years, (b) scientific reports on the population size or geographic range of the species based on single surveys, (c) reports by reliable observers other than scientists on the population size or geographic

range of the species over a number of years, or (d) reports from various
sources on habitat destruction, heavy trade or other potential causes of
extinction. Genera should be listed if most of their species are
threatened with extinction and if identification of individual species
within the genus is difficult.

2. <u>Trade status</u>. Species meeting the biological criteria should be
listed in Appendix I if they are or may be affected by international
trade. This should include any species that might be expected to be
traded for any purpose, scientific or otherwise. Particular attention
should be given to any species for which trade might, over a period
of time, involve numbers of specimens constituting a significant portion
of the total population size necessary for the continued survival of the
species. When biological data show a species to be declining seriously,
there need be only a probability of trade. When trade is known to occur,
information on the biological status need not be as complete. This
principle especially applies to groups of related species, where trade
can readily shift from one species that is well-known to another for
which there is little biological information.

Appendix II

1. <u>Biological status</u>. To qualify for Appendix II, species need not
currently be threatened with extinction, but there should be some indication that they might become so. Such an indication might be a decreasing
or very limited population size or geographic range of distribution.
Information on biological status should be one of the types required for
Appendix I species.

2. <u>Trade status</u>. Species meeting the biological criteria should be listed
if they presently are subject to trade or are likely to become subject to

trade. The amount of trade that a species can sustain without threat of extinction generally will be greater for species in Appendix II than for those in Appendix I, so there should be evidence of actual or expected trade in such volume as to constitute a potential threat to survival of the species.

There are no plants native to the United States listed in Appendix I to the Convention, which comprises 44 species in various families plus all species of *Encephalartos*, a cycad genus having approximately 44 recognized species. Data on the present Appendix I plants have been compiled by the Endangered Flora Project of the Smithsonian, and data on selected species is included below. The species are in alphabetical order by genus; only one of the *Encephalartos* species is included. The extent to which a number of the species figure in international trade is questionable. Many seem to be narrow endemics or less localized species not in trade but otherwise endangered by habitat destruction; others may not be truly endangered at all. The current list of Appendix I plants will undoubtedly be modified in the future.

The 1976 Conference of the Parties to the Convention has determined that criteria for deletion, or transfer of a species from Appendix I to Appendix II, should require positive scientific evidence that the plant can withstand the exploitation resulting from the removal of protection. Such evidence should include at least a well-documented population survey, an indication of the population trend of the species, showing recovery sufficient to justify deletion, and an analysis of the potential for commercial trade in the species or population.

Abies guatemalensis Rehder, Journal of the Arnold Arboretum 20: 285 (1939).

Guatemalan fir, Pinabete

PINACEAE

STATUS. In imminent danger of disappearing from the forests of Guatemala. One of the most common trees in the western highlands in the 19th century, still locally abundant in the 1940s, now extremely rare and the most endangered conifer in Guatemala. Excessive cutting of saplings for Christmas trees, destruction of seedlings by sheep and other livestock, and infrequent production of cones will probably result in its elimination from Guatemala (3).

Local, abundant in some localities, sometimes forming dense forests, sometimes isolated. Does not reproduce freely in native habitat. Grazing by sheep destroys seedlings (2).

DISTRIBUTION. Mexico and Guatemala. Mexico: Chiapas: Coapila; Copainala; Male, Porvenir; Tacana. Oaxaca: Ixtepeji; Rancho Tablas (Racho Benito Juarez), near Lachatao; Sierra de Juarez; Siltepec. Guerrero: Cerro de Tecolote. Jalisco: mountains of Oscuro, Rincon de La Mesa, Verdura de Javiel, Cumbres de Santas Marias, Cuale. Guatemala: Quiche: Cerro Maria Tecun. Totonicapan: Cumbre del Aire; Region of Desconsuelo and between San Francisco El Alto and Momostenago; Cerro Calel, Sierra Madre Mts.; Barranco Buena Vista, Cuesta El Caracol, Sierra Madre Mts., about 5 km NW of San Juan Ostuncalco; Volcan de Zunzil; Mts. SE of Palestina.

Huehuetenagno: Las Cumbres del Aire; Sierra de los Cuchumatanes, at
Chancol, region of San Mateo Ixtatlan; Cerca de Chantla; Sierra de los
Cuchumatanes, between km 324 and 325 on Ruta Nacional 9N (between Chemal
and San Juan Ixcoy). San Marcos: Volcanoes of Tajumulco and Tacana;
region of Serchil.

HABITAT AND ECOLOGY. Moist or wet forests of the high mountains, at
1800-4000 m. Mixed coniferous forest of Pinus, Abies, and Cupressus (1, 2).
It rarely descends to an elevation of 1800 m in Mexico, where it occurs
associated with oaks (Quercus spp.) and madrones (Arbutus spp.) (1).

CONSERVATION MEASURES TAKEN. Cutting prohibited on national land in
Guatemala (2).

BIOLOGY AND POTENTIAL VALUE. Wood in demand for making Indian hand looms.
Lumber is not available due to scarcity. Branches are cut for temporary
shelters. Primarily used as decoration in churches and dwellings, including
as Christmas trees (2).

CULTIVATION. Rarely planted in parks and fincas.

BOTANY. Tree to 45 m high, 1 m or more in diameter. Branches dark or
grayish brown. Young twigs purplish red, sparsely hirtellous near apex.
Leaves 10-45 X 1-2 mm, linear, appearing 2 ranked, spreading-ascending or
almost divaricate, obtuse or emarginate, lustrous and dark or light
green above, usually silvery beneath, the upper surface mostly sulcate,

costa elevated beneath; margin recurved, stomata conspicuous beneath.
Resin canals 2, subepidermal; hypoderm well developed, interrupted.
Cones subsessile, 8.5-11.5 X 4.5-5 cm, broadly rounded or truncate and
erose-denticulate at apex, the cusp usually exserted. Bracts cuneate-
obovate or oblanceolate, half as long as or somewhat exceeding the scales.
Scales 1.5-2.2 X 2.7-3 cm, broadly cuneate-obovate or transverse-oblong,
the margin hirtellous-puberulent. Seed 8-10 mm, cuneate-obovoid, pale
brown, the wings 10-15 X 1.4-1.5 mm, obovate.

ILLUSTRATIONS. See references (1) and (2).

REFERENCES. (1) Liu, Tang-Shui. (1971). A Mongraph of the Genus
Abies, pages 270-272, 362-363; plate 33; map 32. Taiwan.

(2) Standley, P. C., and J. A. Steyermark. (1958). Flora of
Guatemala. Fieldiana, Botany 24(1): 37-40, figure 7.

(3) Veblen, T. T. (1976). The Urgent Need for Forest
Conservation in Highlands Guatemala. Biological
Conservation 9(2): 141-154.

Aloe polyphylla Schonland ex Pillans, South African Gardening and Country
Life 24: 267 (1934).

Spiral Aloe, Lekhala Kharatsa

LILIACEAE

STATUS. This plant has been almost eradicated from its native habitat. It
is estimated that there are fewer than 500 plants in nature (4). The
plants are heavily collected for sale as horticultural items despite the
fact that this has been an illegal practice for 40 years. Overgrazing
on the mountain slopes where it grows is also a factor leading to its
decline (4, 8).

DISTRIBUTION. Lesotho (Western Basutoland). Near Butha Buthe in the north;
Phurumela Mountain near Nyakoesuba and near Makhaleng, also on another
mountain slope 2 mi (3 km) further east; NW of Phurumela at foot of Mount
Machache; 50 mi (80 km) E of Mafeteng, near the source of the Kubake
River; Kubung in the south (12).

HABITAT AND ECOLOGY. This aloe grows at an altitude of about 2500 m, and
is usually under snow in the winter (12). Grows on well-drained hill
slopes that are abundantly supplied with clean, pure rainwater seeping out
of rock crevices (4). Once commonly grew in colonies of a dozen or more
individuals before the species numbers were depleted (4, 12).

CONSERVATION MEASURES TAKEN. Since 1938 there has been an ordinance prohibiting the removal, export, sale, or destruction of certain plant species,
including Aloe polyphylla. Subsequent proclamations have upheld this (4).

CONSERVATION MEASURES PROPOSED. The International Union for Conservation
of Nature and Natural Resources has recommended the creation of a national
park from which domestic and feral grazing animals are excluded (8).

BIOLOGY AND POTENTIAL VALUE. The flowering period is from August to
December and plants are in full bloom during the months of September and
October (10).

The long-beaked Malachite Sunbird (Nectarinia famosa) has been observed
acting as pollinator, but seed set is poor. This failure of cross-pollination
may be due to the increasing impoverishment of the ecosystem resulting from
overgrazing, roadbuilding, and other activites. Since the many genera of
flowering plants attractive to the Sunbirds are becoming increasingly
rarer in this region, there is not sufficient nectar to support the Sunbird
population at its former high level (4, 8).

Seedlings are rarely found in the natural habitat. It has been
suggested that the frost heaving, which occurs on exposed soil during the
winter, uproots and kills the seedlings. Trampling by domesticated animals
is also likely to play a part (4). The larvae of a beetle also destroys
seeds and seedlings (5, 8). Young plants have been found to be infected
with a gall mite (Eriophyces), which can destroy the plant (5).

This species is unique in the striking spiral arrangement of the
leaves (8).

356

CULTIVATION. Transplants, whether young or old, almost invariably die (4, 12). Plants can be raised from seed and occasionally from offsets (8). Soil pH may be an important consideration; pH in the field was found to be 6.2 and 6.4. Four plants are in cultivation and thriving at the National Botanic Gardens in Pretoria (5).

BOTANY. Succulent perennial with a rosette of leaves up to 90 cm across, arranged in 5 spiral rows, which may run clockwise or anticlockwise. Leaves 15-30 X 7-13 cm, very fleshy, oblong-lanceolate, acute, nearly flat above and eccentrically keeled below, the margin with rather soft, white teeth. Flowering shoot 50-80 cm, branching from near the base, with 4-11 branches bearing the flowers crowded at the tips. Flowers with a narrow, triangular, violet-tinged bract; corolla 4-5 cm, cylindrical, salmon to yellow. Capsule 3 cm (8).

TYPE. Collected by Reynolds in 1934 on the western slopes of Phurumela Mountains (12).

ILLUSTRATIONS. See references (2), (3), (4), (5), (6), (10), (11), and (12).

REFERENCES. (1) Basutoland Government. (1938). Resident Commisioner's Notice, 20 September 1938.
(2) Bornmann, H., and D. Hardy. (1971). Aloes of the South African Veld, page 69, plate 35. Johannesburg.
(3) Coombs, S. V. (1935). South Africa--A Land of Many Flowers. Journal of New York Botanic Garden 36: 49-53.

(4) Guillarmod, A. J. (1975). Point of No Return? *African Wildlife* 29(4): 28-29, 31. Summer.

(5) Hardy, D. (1968). The Spiral *Aloe* from the Maluti Mountains. *Cactus and Succulent Journal* 40: 49-51.

(6) Judd, E. (1967). *What Aloe is That* ?, page 8, plate 2. Cape Town/ Johannesburg: Purnell.

(7) Kofler, L. (1966). Biology and Cultivation of *A. polyphylla*, the Spiral Aloe. *National Cactus and Succulent Society Journal* 21: 16-19.

(8) Melville, R. (1971). *Aloe polyphylla*. *Red Data Book, Volume 5: Angiospermae*. Morges, Switzerland: Survival Service Commission, International Union for Conservation of Nature and Natural Resources.

(9) Pillans, N. S. (1934). *South African Gardening and Country Life* 24: 267.

(10) Pole, Evans, I. B. (1935). *The Flowering Plants of South Africa*, page 15, plate 571.

(11) Reynolds, G. W. (1934). The Quest of *Aloe polyphylla*. *Journal of Botanical Society of South Africa* 20: 11-12, plate 2, 3.

(12) Reynolds, G. W, (1974). *The Aloes of South Africa*, pages 194-196, figures 187-190. Cape Town/Rotterdam: A. A. Balkema.

358

Balmea stormae Martinez, Bulletin of the Torrey Botanical Club 69: 438 (1942).

Ayuque

RUBIACEAE

STATUS. Due to its conical habit with pointed top when young, it is commonly cut and sold in markets in the Uruapan area as a Christmas tree. Although this tree may be commoner than suspected in lesser known parts of Mexico, it is likely that this species will soon become very rare in any place where it may be discovered. This use of Balmea arose when laws were enforced making it illegal to cut conifer saplings for this purpose (1).

DISTRIBUTION. Mexico and Guatemala. Mexico. Temascaltepec: Nanchititla; Michoacan: Palo Verde, near Uruapan; Jicalan, near Uruapan; Cañon del Mal Pais; NW Aguililla; Uruapan. Guatemala: Huehuetenango: NW of Cuilco, above Carrizal. Jalapa: Potrero Carrillo, NE of Jalapa. Zacapa: San Lorenzo, Sierra de las Minas (1).

HABITAT AND ECOLOGY. Dry, stony areas (2).

CONSERVATION MEASURES TAKEN. Attempts are being made to bring this plant into cultivation (1).

BIOLOGY AND POTENTIAL VALUE. Due to the brilliant scarlet-red flowers, it has long been a favorite of the people of the region in which it grows. It has potential value as an ornamental (1, 2).

CULTIVATION. If it were brought into cultivation, it might do well in rather dry, warm temperate, and subtropical regions (1).

BOTANY. Shrub 4-7 m, pyramidal or conic when young, irregular when older. Leaves 9-13 cm, suborbicular, somewhat broader distally, acuminate at apex, cordate to subcordate at base, clustered at the ends of the branches. Flowers in terminal pendent cymes; corolla tubular, deep red to dark maroon, the tube 20-27 mm (1, 2).

TYPE. Collected in 1941 by Martinez in Palo Verde (2).

ILLUSTRATIONS. See references (2) and (3).

REFERENCES. (1) Fosberg, F. R. (1974). Studies in American Rubiaceae, 2: Ayuque, Balmea stormae, an Endangered Mexican Species. Sida 5(4): 268-270.

(2) Martinez, M. (1942). A New Genus of Rubiaceae from Mexico. Bulletin of the Torrey Botanical Club 69(6): 438-441, figures 1-11.

(3) Standley, P. C., and L. O. Williams. (1975). Flora of Guatemala: Rubiaceae. Fieldiana, Botany 24 (11), nos. 1-3: 1-274.

Encephalartos barteri Carruthers ex Miquel, Archives Neerlandaises des
Sciences Exactes et Naturalles   3: 243 (1868).

West African Cycad, Ghost Palm, Pardi Attar

ZAMIACEAE

STATUS.   Limited distribution. The extremely slow growth rate observed
in this species has been suggested as responsible for its inability to
colonize wide areas. During the last million years, the West African
environment has changed from moist forest to dry savanna, but E. barteri
has remained in areas of shallow, rocky soil where formerly it had been
free of competition from dense forest (4).

DISTRIBUTION.   Ghana, Togo, Dahomey, Nigeria, Mali, Sudan, and Uganda.

HABITAT AND ECOLOGY.   The guinea savanna of Ghana (4). Found on dry stony
hill slopes, at an elevation of approximately 900 m (7). The outer third
of the thickness of the stem consists of a fireproof armour of fleshy,
tightly overlapping, persistent leaf-bases (4).

BIOLOGY AND POTENTIAL VALUE.   Special side-roots growing upwards contain
colonies of Anabaena, a blue-green alga. The alga fixes nitrogen and
supplies it to the host (2). The pith of some other species of Encephalartos is known to be edible, but there is no record of E. barteri being
eaten regularly. Branching of the subterranean stem may be important in

the propagation of the species, as young seedlings seem to be very scarce. The leaves wither and fall during the dry season. The cones develop at this time, followed by the flushing of new leaves. It is unlikely that every plant produces cones every year (4).

CULTIVATION. Members of this genus may be propagated by seeds, also by offsets or suckers. Most species prefer a sunny, tropical greenhouse and they are slow-growing except in very warm greenhouses. They do best in a strong, loamy soil, and while making new growth they need plenty of water. Encephalartos seldom produces seed in conservatories, and plants are, therefore, usually imported. When transplanted, they often remain dormant for a year or more (1).

BOTANY. Dioecious; resembles a young oil palm. Stem subterranean. Leaves arise at ground level, may reach 5 ft (1.5 m) in length. Leaflets about 3-5 in (10 cm) long, with slightly irregular margins bearing a few sharp points. Male cone 9 X 2-3 in (23 X 6 cm) on a long stalk about 7 in (18 cm) long; scales green with brownish tips. Female cone 12-14 X 5-6 in (33 X 14 cm), on a shorter stalk than that of the male; scales T-shaped when viewed from above, with a large seed, scarlet when ripe, suspended from each arm (2).

TYPE. Nigeria, Jebba, collected by Barter (7).

ILLUSTRATIONS. See references (4) and (8).

REFERENCES. (1) Bailey, L. H. (1906). Cyclopedia of American Horticulture
2: 530. 4th edition. New York.

(2) Berrie, A., and G. K. Berrie. (1956). The West African
Cycad. Nigerian Field 21(1): 36-41.

(3) Greguss, P. (1968). Xylotomy of the Living Cycads.
260 pages. Budapest.

(4) Hall, J. B., and J. Jenik. (1967). Observations on the
West African Cycad in Ghana. Nigerian Field 32: 75-81.

(5) Hutchinson, J., and J. M. Dalziel. (1927). Flora of West
Tropical Africa 1: 45. London.

(6) Melville, R. (1957). Encephalartos in Central Africa.
Kew Bulletin 12: 237-257, map on page 248.

(7) Melville, R. (1958). Cycadaceae. In W. B. Turrill and
E. Milne-Redhead, editors, Flora of Tropical East Africa
Gymnospermae, pages 1-10. London.

(8) Prain, D. (1909). Curtis's Botanical Magazine, volume 135,
plate 8232.

Fitzroya cupressoides (Molina) Johnston, Contributions from the Gray
Herbarium 70: 91 (1924).

Alerce, lahuen

TAXACEAE

STATUS. Exploited throughout its range. According to the Comite Nacional
pro Defensa de la Fauna y Flora, Casilla 3675, Santiago, Chile, the
species will be extinct in ten years unless the present rate of felling
is abated; a leaflet campaign to save this species, aimed particularly
at architects and builders, has been launched by the Comite (5).

DISTRIBUTION. Chile and Argentina. Chile : Range begins near the city
of Valdivia (39°45') and ends near the Rio Futalelfu (43°29') (4);
Neuquen, Rio Negro, Chubut (1); Cautin (3). Middle of Chiloe Island and
on the "out-jutting" bit of land from just south of Osorno to Puerto
Montt, occuring to the north, west, and south of Lake Llanquihue, also
on the mainland just east of the Gulf of Ancud (in Chile). Argentina:
Directly to the west of Lake Nahuel Huapi, and also around the two small
lakes in Argentina located due east of the center of Chiloe Island (2).

HABITAT AND ECOLOGY. Typically a species of low swamps but grows at higher
elevations on the Isla de Chiloe and in the Territory of Aysen in Patagonia.
It is unique among the Chilean conifers in forming dense, nearly pure
forests over thousands of acres (4). At heights of 600 m above sea level

there are about 13,000 hectares which contain this species, with 220 trees
per hectare (3).

BIOLOGY AND POTENTIAL VALUE. The tree bears considerable resemblance to
the California Redwood (Sequoia sempervirens). On Chiloe and in the
Cordillera de Piuchue the average height at maturity is about 100 ft
(30 m), but near Puerto Montt the old trees are mostly 130-150 ft (40-46 m).
One tree measured 240 ft (73 m). The famous "Silla del Presidente," which
formerly stood near Puerto Montt, was 15 ft(4.5 m) in diameter; others
in the same region reach a diameter (breast high) of 9 ft (2.7 m). On
Chiloe the average diameter is about 4 ft (1 m). The tree grows slowly
and attains an age of 1000 to 3000 years (1, 4).

Young trees have a conic-pyramidal form and very dense, dark green
foliage; old trees have a small crown at the top of the long straight bole;
so many of the branches are dry that the forest, seen from a distance,
appears to be dead. The bark is several centimeters thick, corky, and,
upon incision, exudes a resin of agreeable odor.

Value: The inner bark, "estopa de Alerce," is harvested in the
summer and used in caulking boats and ships. The resin is also collected
to burn as incense. The wood is of fine and uniform texture, straight
and fine grain, low density, and is strong and elastic for its weight. The
sapwood which is thin and white is not utilized; the heartwood is red and
long-lived under exposure. The timber is the finest produced in Chile and
ranks with the best and most useful in the world. It is used for carpentry
and light and durable construction of all kinds. Temporary native dwellings
in the forests of Cordillera are built entirely of rived Alerce. The

wood may also be used for honey barrels, musical instruments (violins, **guitars,** piano), small tiles, and small ships. Limited quantities are exported for making pencils and cigar boxes. Lumbermen make and sell troughs noted for their durability (1, 4).

BOTANY. Leaves ericoid, pointed, limited to the young twigs. Male cones 3-6 X 2-2.5 mm, axillary, cylindrical. Female cones 8 mm in diameter, subglobose, the terminal with 6 scales and 9 winged seeds 3-4.5 mm long. Cones ripen in the spring (1, 4).

ILLUSTRATIONS. See references (1) and (3).

REFERENCES. (1) Arboles Forestales Argentinos. (undated). Ministerio de Agricultura y Ganaderia. Administracion Nacional de Bosques. Direccion de Investigaciones Forestales. (see "Alerce")

(2) Hueck, K. (1972). Vegetationskarte von Südamerika (map). In Hueck, K., and P. Seibert, Vegetationskarte von Südamerika. Stuttgart: Gustav Fischer Verlag.

(3) Pizarro, C. M. (1973). Chile: plantas en extincion, page 27. Editorial Universitaria, Chile.

(4) Record, S. J., and R. W. Hess. (1943). Timbers of the New World, pages 7-8. New Haven: Yale University Press.

(5) Comite Nacional pro Defensa de la Fauna y Flora, Santiago. (1976). Save Chile's Larch and Native Forests. IUCN Bulletin, new series, 7(11): 60 (November).

Laelia jongheana Reichenbach fil., Gardeners Chronicle, page 425 (1872).

ORCHIDACEAE

STATUS. Extinguished from Minas Gerais. Many years ago the trees in its limited habitat were cut down to make charcoal (3). Extremely rare in nature, threatened with extinction; never particularly plentiful (5).

DISTRIBUTION. Minas Gerais, Brazil.

HABITAT AND ECOLOGY. Dry regions (1). In areas of savanna or prarie character with long periods of drought and extreme temperature fluctuation between hot days and cool nights (4).

BIOLOGY AND POTENTIAL VALUE. Ornamental.

CULTIVATION. Introduced to cultivation in 1834 (7). Very rare in collections. (3). Rarely offered commercially. Plants in trade mostly the results of selfings or sibling crosses (5).

BOTANY. Terrestrial. Pseudobulbs up to more than 5 cm, subfusiform to ovoid-oblong, compressed, arising from a stout, creeping rhizome. Leaves 8-16 cm, solitary, ovate-oblong, rigidly leathery, erect. Peduncles shorter than leaves. Spathe absent. Inflorescence 1-(2-)flowered. Flowers 10-13 cm in diameter, heavy-textured, long-lasting, spreading out flat, soft rose-purple, except for the lip which has a yellow disc, in front of which is a white blotch. Sepals lanceolate, acute. Petals broader than sepals, elliptic-oblong, obtuse. Lip 3-lobed, ovate-oblong, the lateral lobes triangular, the mid-lobe obtuse, emarginate; all segments crisped and somewhat toothed marginally. Disc with 7 ridges. Column slender, pale rose-purple above, triquetrous. Flowering in February-April.

ILLUSTRATIONS. See references (2), (4), and (6).

REFERENCES. (1) Hawkes, A. D. (1965). Encyclopaedia of Cultivated Orchids, page 260. London: Faber & Faber.

(2) Hooker, J. D. (1872). Botanical Magazine, volume 99, plate 6038.

(3) Leinig, M. (1968). Distribution of Brazilian Laelias. American Orchid Society Bulletin 37(12): 1070.

(4) Pabst, G. F. J., and F. Dungs. (1975). Orchidaceae Brasiliensis 1: 40-41, 145, 211, 317. Hildesheim: Brücke-Verlag Kurt Schmersow.

(5) Peterson, R. (1976). Personal communication.

(6) Reichenbach, H. G. (1872). Gardeners' Chronicle, pages 425-426, figure 128.

(7) Richter, W. (1965). Orchid World, page 265. New York: Dutton.

(8) Veitch, J. & Sons (1887). Manual of Orchidaceous Plants 1: 73. London: Veitch.

Lycaste skinneri (Bateman ex Lindley) Lindley, Edwards's Botanical Register 29: 15-16 (1843).

Synonym: Lycaste virginalis (Scheidweiler) Linden, Lindenia 4: 22 (1888).

Monja blanca, White nun

ORCHIDACEAE

STATUS. Threatened by habitat destruction, native collecting, and selling (3). The white form is almost extinct (6).

DISTRIBUTION. Mexico, Guatemala, El Salvador, and Honduras. Mexico: Chiapas: NE Comitan near Santa Ana (7), Montelbello Lakes, vicinity of Tepancuapan (3), Forests of San Bartholo (3), Volcan Tacana not far from Tapachula (3), vicinity of Comitan (3). Guatemala: Alta Verapaz: Rio Tzimajil near Coban (3, 7), Coban (3, 7), mts. of Canton Coban, crest of Mt. Chicoye near Tactic (3), central slope Volcan Chucanep (3); Chiquimula: Montana Nonoja, 3-5 mi (5-8 km) E of Camotan (3). El Salvador: Volcan Chingo, 20 km NNW Chalchuapa- 89°45'W 14°10'N (3), Los Naranjos, 89°40'W 13°40'N (3), Santa Ana Volcano (6). Honduras: Santa Barbara: mountains (3), Francisco Morazon: Cerro Uyuca (3), Cordillera de Merendon (6), Sierra de Omoca (6).

HABITAT AND ECOLOGY. In the relatively dark interior of non-coniferous, humid and cool, montane cloud forests, usually on their crests but also

down to their lower limits from 1800-2200 m. On slender, heavily
moss-draped trees, 10-20 cm in diameter, 7 m above ground and up.
Occasionally in branch-forks, or on trunk or lower, heavier branches.
Rarely more than 2-3 on a tree, more frequent on slopes in vicinity of
creeks (3).

BIOLOGY AND POTENTIAL VALUE. National flower of Guatemala. Immense color
variation. Most important species of Lycaste in horticultural value and
for hybridizing (4).

CULTIVATION. Common in cultivation (R. Read). Easy to cultivate (8).
Cultivated since 1841, not now cultivated to previous extent. Requires
temperate, shady conditions (6).

BOTANY. Epiphyte. Pseudobulbs 10 X 6 cm, ovoid, 3 cm thick, compressed,
faintly ridged and furrowed, dark green. Leaves 2-3, 50-60 X 11 cm,
ribbed and plaited. Flowers 2-5, not fragrant, 12-14 cm in diameter;
sepals and petals cream white, variously shaded with different tones of
deep pink to lavender. Upper sepal 7.5 X 4.5 cm, elliptic-lanceolate,
inferiorly concave, the distal 1/3 convex, the apex apiculate, recurved;
lateral sepals 6.5 X 4.5 cm, recurved, anteriorly convex, with acuminate
apex. Petals 5 X 3.2 cm, broadly and asymmetrically elliptical, with
very erose dorsal margin, jutting forward with only the distal 1/10
gently recurved, with minutely acute apex. Labellum 4.8 X 4 cm,
prominently three-lobed, darker than the sepals and petals, usually highly
mottled with deep shades of dark purple-lavender and with a reddish callus.
Column 4 X 1.2 cm, 0.5 cm thick, gently incurved, deep red and velutinous
at the base, the distal 1/5 with a prominent anther bearing 4 pollinia.

ILLUSTRATIONS. See references (2), (3), (6), and (8).

REFERENCES. (1) Ames, O., and D. S. Correll. (1953). Orchids of Guatemala. Fieldiana, Botany 26: 556-557.

(2) Chickering, C. R. (1973). Flowers of Guatemala, page 80, plate 26. Norman: University of Oklahoma Press.

(3) Fowlie, J. A. (1970). The Genus Lycaste, pages 60-64. California: Azul Quinta Press.

(4) Gripp, P. (1976). The future of Lycastes. American Orchid Society Bulletin 45(9): 778-782.

(5) Hawkes, A. D. (1965). Encyclopedia of Cultivated Orchids, page 276. London: Faber & Faber.

(6) Richter, W. (1965). Orchid World, pages 60, 118, 152, 267, plate 22. New York: Dutton.

(7) United States National Herbarium Specimens.

(8) Veitch, J. & Sons. (1887-1894). Manual of Orchidaceous Plants, pages 93-95. London: Veitch.

Microcycas calocoma (Miquel) A. De Candolle, Prodromus 16(2): 538 (1868).

Palma corcho

ZAMIACEAE

STATUS. The rarest and most geographically restricted plant in Cuba (9).

DISTRIBUTION. Cuba. Pinar del Rio Province; occurs only between San Diego de los Banos on the east and Sumidero on the west (9). In 1907, Caldwell noted four locations in which it occurred, and in three of these the plants were few and distinctly local. In the fourth location, Cuchillo de Pinar, the plants were found one, two, or a half dozen together at infrequent intervals for a distance of 1 1/2 to 2 mi (3 km) (2).

HABITAT AND ECOLOGY. Arid sandy savannas (9). Does not appear to be confined to any particular exposure, but the species has a very restricted elevational range. It has been found growing on limestone, in places almost devoid of soil; in rocky soil; and in clay (2).

BIOLOGY AND POTENTIAL VALUE. Cones are produced in April or May. It has been suggested that the pollen is transported from the male to the female tree by ants. Pollination is not very effective, for often cones are found which contain only one or two fertile seeds (7). According to the folklore of the region, the roots are good for use as a rat poison (2).

372

BOTANY. Monotypic genus easily distinguished from the small Zamias in the region by its larger size, up to 3-10 m (4). Stem branched or unbranched. In all younger plants and in the shaded forms of the older trees, the stem is marked with conspicuous rings. The stem in cross-section shows a thick cortex, a single vascular cylinder, and pinkish, brittle, starch-bearing pith (2, 3). Leaves 0.6 - 1 m, 6 to 40 in the crown; petioles 10 cm, terete, with shield-like bases; leaflets 50-80 pairs, 8-12 cm, opposite or alternate, finely villous when young, glabrous and glistening when mature, bright green, acute. Staminate cones 25-30 X 5-8 cm, cylindrical. Ovulate cones 50-70 X 13-16 cm, cylindrical, slightly tapering from base to tip, obtuse. Ovules two, 3.25-3.5 X 1.25-1.75 cm, pink (3).

TYPE. Described from a specimen cultivated at Amsterdam.

ILLUSTRATIONS. See references (1), (2), (3), (4), (5), (7), and (9).

REFERENCES. (1) Barbour, T. (1946). A Naturalist in Cuba. Illustration facing page 18. Boston: Little, Brown and Company.

(2) Caldwell, O. W. (1907). Microcycas calocoma. Botanical Gazette 43: 330-335.

(3) Caldwell, O. W., and C. F. Baker. (1907). The Identity of Microcycas calocoma. Botanical Gazette 43: 330-335.

(4) Chamberlain, C. J. (1919). Living Cycads, pages 9-11. New York and London: Hafner Publishing Company.

(5) Graf, A. B. (1974). Exotica. Series 3, 7th edition New Jersey: Roehrs Co. Inc.

(6) Hutchinson, J. (1924). Contributions Toward a Phylogenetic Classification of Flowering Plants, III. Kew Bulletin 2: 49-66.

(7) Marie-Victorin, Frere, and Frere Leon. (1942). Itineraires Botanique Dans L'Ille De Cuba, series 1, pages 141-148, figures 81-86. Canada: University of Montreal.

(8) Miquel, F. A. K. (1851-1852). Sur une espece nouvelle de Zamia des Indes Occidentales, introduite dans l'establissement Van Houtte, a Gand. L. Van Houtte, editor, Flore des Serres et des Jardins de l'Europe 7: 141-142.

(9) Seifriz, W. (1943). The Plant Life of Cuba, pages 31-33, figures 39, 40. Reprinted from Ecological Monographs 13: 375-426.

Orothamnus zeyheri Pappe ex Hooker, Curtis's Botanical Magazine 74:
plate 4357 (1848).

Marsh Rose **Protea**

PROTEACEAE

STATUS.   In imminent danger of extinction (5).   In 1968, there was
estimated to be only 90 plants remaining (7).   It is a relict species
suffering a natural decline, which has been greatly accelerated by illegal
picking of the attractive blossoms, which are for sale in Cape Town (5).
Also, it has proved to be extremely susceptible to an exotic fungus
(Phytophthora cinnamomi) which has been accidentally introduced into the
soil of the Eddie Rubenstein Orothamnus Reserve.   This fungus could very
well destroy all Marsh Roses on the Reserve (8).

DISTRIBUTION.   South Africa.   Cape Province: Limited to three small
colonies on mountains near Caledon.   Formerly (1921), seven colonies were
known from the same area (5).

HABITAT AND ECOLOGY.   On south-facing peaty slopes in the mist belt on
mountains of Table Mountain Sandstone at about 3000 ft (900 m) (5).

CONSERVATION MEASURES TAKEN.   One population consisting of 50 plants on
the Eddie Rubenstein Orothamnus Reserve near Hermanus.   Here, the research
staff of the Department of Nature Conservation have been conducting
experiments into the biology and cultivation of the species (7, 8).   The

rest of the habitat is on state forest land and access is by permit only (4). The species is protected by law, consolidated by Ordinance no. 19 of 1974, schedule 3 of chapter VII (8).

BIOLOGY AND POTENTIAL VALUE. Branching, which is infrequent, usually takes place just below an inflorescence and picking by removing the only effective buds generally results in the death of a plant (4). In nature, regeneration takes place after long-interval summer fires. Seed set is poor and the germination rate is low (8). Orothamnus is a monotypic genus, and therefore of scientific value (7).

CULTIVATION. Can be successfully grafted onto Leucospermum conocarpodendron, one of the pin-cushion proteas. Softwood cuttings will root within eleven weeks when treated with commercial hormone powder and kept in a mist propagator. Branching can be induced by removing the apical bud in February or June. Seeds and seedlings have been planted on the Reserve (7, 8).

BOTANY. A single stemmed or sparingly branched shrub 1-3 m. Leaves 3-6 cm, elliptical, leathery. Flower heads nodding, 1-3 at the branch tips, 5-7 cm long with a number of oblong rose-red bracts enclosing the hairy lemon-yellow flowers.

TYPE. Collected by Zeyher in the Cape Province, Hottentot's Holland Mountains (5).

ILLUSTRATIONS. See references (1), (2), (3), (4), (6), (7), and (9).

REFERENCES. (1) Boucher, C., and G. McCann. (1975). The Orothamnus saga.
Veld & Flora 61(2): 2-5.

(2) Hooker, W. J. (1848). Curtis's Botanical Magazine,
volume 74, plate 4357.

(3) Marloth, R. (1913). The Flora of South Africa, volume 1,
plate 32. Cape Town & London: Darter Bros. & Co.

(4) Melville, R. (1970). Plant Conservation in Relation to
Horticulture. Journal of the Royal Horticultural Society
45(2): 473, figure 227.

(5) Melville, R. (1970). Orothamnus zeyheri. IUCN Red Data
Book 5: Angiospermae. Morges, Switzerland.

(6) Pole Evans, I. B. (1921). The Flowering Plants of South
Africa 1, plate 38.

(7) Van der Merwe, P. (1974). The Rarest Protea of the
Fairest Cape. African Wildlife 28(3): 28-29 (Spring).

(8) Van der Merwe, P. (1975). Impossible to Save the Marsh
Rose Protea? Veld & Flora 61(1): 4-5.

(9) Van Houtte, L. (1848). Flore des Serres et des Jardins
de l'Europe, series 1, volume 4: plate 338.

Peristeria elata Hooker, Curtis's Botanical Magazine, 58, plate 3116 (1831).

Holy Ghost, Dove Orchid, El Spirito Santo

ORCHIDACEAE

STATUS. Formerly very plentiful in the Panama Canal Zone. Commercial collecting has made it increasingly infrequent, restricting it to more inaccessible areas (1). Becoming extinct. Large areas of its previous range are now covered by waters of Lake Gatun. Reduced by overcollecting by local inhabitants for sale in markets (6).

DISTRIBUTION. Panama, Costa Rica, Venezuela, and Colombia. Panama: Hills near Juan Diaz (1); East of Panama City near Rio Tecumen (1); Chorrera (1); Cerro Campana (1); Colon: Cerro Santa Rita (1); Cocle: El Valle de Anton (1). Costa Rica: eastern (5). Venezuela: (1). Colombia: Antioquia: Central Cordillera (9); Tolima: Falan, region of Calamonte (9).

HABITAT AND ECOLOGY. Llanos and transitional zone between grass savannas and forest, also along river banks, grassy roadsides, and rocky outcrops (1). In unshaded or partly shaded areas with a great amount of rain throughout the year, at low to medium altitudes (7).

BIOLOGY AND POTENTIAL VALUE. Ornamental. The national flower of Panama.

CULTIVATION. Cultivated since 1826. Grown in well-drained, organic, fir bark-osmunda-sphagnum mixture, in strong light, high temperatures, and high humidity (2).

BOTANY. Terrestrial. Pseudobulbs 4-12 X 4-8 cm, stout, round, fleshy, subconic or broadly ovate, the broad bases enveloped in several closely imbricated papery bracts. Leaves 3-5, 30-100 X 6-12 cm, arising from apex of pseudobulb, broadly lanceolate, plicate, acuminate, deciduous. Inflorescence 80-130 cm, solitary, erect, of unbranched racemes, indeterminate, produced simultaneously beside the new growth at the base of the pseudobulb, developing concurrently with it. Flowering delayed until July-August, when pseudobulb has matured and leaves fallen. Flowers 10-15 or more, 5-7.5 cm in diameter, fleshy, subglobose, waxy-white, strongly fragrant; pedicels 4 cm. Sepals subequal, fleshy, broadly concave; dorsal sepal 2.5-3 X 2-2.5 cm, free, ovate, obtuse; lateral sepals 2.5-3 X 2.5-3 cm, somewhat connate at base, ovate or suborbicular, shortly acute. Petals 20-25 X 15-18 mm, elliptic-obovate, obtuse. Lip fleshy, the claw (hypochile) broad, continuous with the base of the column; lateral margin with ascending wings which are white, heavily spotted rose-red, the inner basal surface thickened into a fleshy lobule; apical lobe (epichile) white, articulated with the frontal margin of the hypochile, entire, subquadrate, retuse, nearly truncate, with a central glabrous, fleshy, pure white, ventricose or suborbicular crest. Column 9-11 mm, subconic, semiterete, pure white. Anther beaked, said to resemble a dove, pure white.

ILLUSTRATIONS. See references (2), (4), (7), and (8).

REFERENCES. (1) Allen, P. H. (1949). Orchidaceae. In Flora of Panama. Annals of the Missouri Botanical Garden 36(1): 48-50.

(2) Birk, L. (1971). Difficult Species: Peristeria elata and Its Flowering Requirements. Orchid Digest 35: 206-208.

(3) Hawkes, A. D. (1965). Encyclopaedia of Cultivated Orchids, page 363. London: Faber & Faber.

(4) Hooker, J. D. (1831). Curtis's Botanical Magazine, volume 58, plate 3116.

(5) Perez, A. B. (1942). Las Orquideas de Costa Rica, page 49. San Jose.

(6) Richter, W. (1965). Orchid World, pages 114, 117. New York: Dutton.

(7) Teuscher, H. (1975). Collector's Item. Peristeria elata and P. pendula. American Orchid Society Bulletin 44(12): 1056-1060.

(8) United States National Herbarium specimens.

(9) Veitch, J. & Sons. (1887-1894). Manual of Orchidaceous Plants 2: 128-129. London: Veitch.

Stangeria eriopus (Kunze) Nash, Journal of the New York Botanical Garden 10: 164 (1909).

Fern-leaved Stangeria (4). Finguane, Juma (S. Natal); Imfingo (Zulu); Umfingwani (Xhosa) (10).

STANGERIACEAE

STATUS. Many plants have been uprooted to meet the demand of scientific and educational institutions as well as private collectors throughout the world. The cultivation of pineapples and sugar cane has destroyed parts of the plants' habitat (10).

DISTRIBUTION. South Africa: Eastern Cape and Natal. Restricted to a relatively narrow strip of land along the coast, from about latitude 33° 30' S in the Bathurst district (in the eastern Cape) to 270° S, just south of the border between Mozambique and Natal (10).

HABITAT AND ECOLOGY. In coastal grassveld and inland forest (2, 3). It never occurs further than about 50 km from the sea, and is usually not closer than 2 or 3 km from the beach. When it is found very near the sea, it is only in places sheltered from the salt spray. The soil is mostly sandy, often derived from sandstone, although in the north it occurs on granite soil and in the Transkei, on heavy black clay. This distribution shows a close correlation with the "evergreen and deciduous bush and subtropical forest" vegetation type of Pole Evans (10). Various forms occur in different habitats (see BOTANY).

BIOLOGY AND POTENTIAL VALUE. Several species of weevils (family Curculionidae) have been collected from the cones, but there is as yet no proof that they play any part in pollinationa as has been suggested by Batten and Bokelmann. These insects use the pollen for food and breed in the female cones where they do great damage at times. Furthermore, the structure of the cones is typical of wind-pollinated gymnosperms. The red fleshy outer layer of the seeds is attractive to rodents and baboons (and possibly birds), which eat the flesh and discard the kernel, thereby spreading the seed. As in other cycads, Stangeria produces coralloid roots which grow upwards and contain a nitrogen-fixing alga (10).

CULTIVATION. Grows well under cultivation. If kept in light shade, it will produce more luxuriant foliage. If the underground parts have been damaged, the wounds should be dusted with sulfur or fungicide, and the underground part may be enclosed in a plastic bag. The soil should be sandy, slightly acid, very rich in humus and covered by a layer of dead leaves about 5 cm deep. The soil should be kept slightly moist and the plant must be protected from frost (10).

BOTANY. Monotypic. Strangeria differs from other living cycads in the venation of the leaflets: there is a strong midrib from which faint lateral veins branch out at angle of 10° to 45°, like a feather. In Cycas the leaflets contain a single prominent midrib, and in all the other cycads (Zamiaceae, of Johnson), the leaflets contain a number of equal parallel longitudinal veins (10).

Stem subterranean, sometimes branched. Root tuberous, elongated, swollen, and carrot-shaped. Leaves fern-like, pinnate, varying from 25-200 cm depending on environmental conditions; margin entire, serrate, or deeply fringed; petiole comprising about half the length of the leaf. Leaflets (pinnae) soft to slightly leathery, varying considerably in texture and size, arranged in 5-20 opposite or nearly opposite pairs. Male cone 10-25 X 3-4 cm, cylindrical, tapering to the apex, covered with short, silvery velvet hairs when young, yellow-brown when mature. Female cone up to 18 X 8 cm, elliptical or ovoid with a rounded tip, covered with silvery velvet hairs when young, becoming dark green at maturity. Seeds up to 3.5 X 2.5 cm, the dark red fleshy outer layer enveloping a hard kernel (10).

Stangeria varies considerably depending on the habitat. Grassland plants have shorter, much more compact, leathery leaves, compared to the more lax, fern-like leaves of forest plants. Grassland plants usually have pinnae with margins entire or at least with fewer teeth than those of forest plants. Grassland plants often branch into dense clumps a few meters across, while forest plants usually occur singly (3, 10).

TYPE. Collected in the eastern Cape, by Drege (2).

ILLUSTRATIONS. See references (1), (2), (3), (4), (7), (8), and (10).

REFERENCES. (1) Batten, A., and H. Bokelmann. (1966). Wild Flowers of the Eastern Cape Province, pages 2-3, plate 2, no. 2. Capetown: Books of Africa, Ltd.

(2) Dyer, R. A. (1966). Stangeriaceae. In Codd, L. E., De Winter, B. & H. B. Rycroft, editors, Flora of Southern Africa 1: 1-3, figure 1.

(3) Giddy, C. (1974). Cycads of South Africa, page 117, plate 29. Cape Town: Purnell and Sons.

(4) Hooker, W. J. (1859). Curtis's Botanical Magazine, volume 85, plate 5121.

(5) Hutchinson, J. (1924). Contributions Toward a Phylogenetic Classification of Flowering Plants, III. Kew Bulletin pages 49-66.

(6) Johnson, L. A. S. (1959). The Families of Cycads and the Zamiaceae of Australia. Proceedings of the Linnean Society of New South Wales 84: 64.

(7) Marloth, R. (1913). The Flora of South Africa 1: 98, plate 14. Cape Town and London: Darter Bros. & Co.

(8) Nash, G. V. (1909). A Rare Cycad. Journal of the New York Botanical Garden 10: 163-164, plate 62.

(9) Pole Evans, I. B. (1936). A Vegetation Map of South Africa. Botanical Survey of South Africa Memoire 15. Pretoria: Department of Agriculture and Forestry.

(10) Vorster, P., and E. Vorster. (1974). Stangeria eriopus. Excelsa 4: 79-89.

Welwitschia mirabilis Hooker fil., Gardeners' Chronicle , page 7 (1862).

Welwitschia

WELWITSCHIACEAE

STATUS. It is much commoner than was formerly realized. It is protected and in no danger of extinction (5).

DISTRIBUTION. Angola and Namibia (South West Africa). In the Namib Desert and on its eastern outskirts.

HABITAT AND ECOLOGY. There are two types of environments for Welwitschia, which are quite distinct on the basis of soils and rainfall. In Angola and in two localities in Namibia, a true desert habitat is encountered, which is nearly barren of vegetation except in abnormal years. These deserts are nearly level with a floor of sand and gravel and with exceedingly low rainfall. In these areas, Welwitschia is limited in its distribution mostly to within 50 mi (80 km) of the coast within the fog belt. It is thus possible that fog may be an important factor in its survival.

The second distinct environment is a semi-desert covered by bushes and low trees, and in an area with an annual rainy season. It is rough, hilly country and the plant occurs scattered over the dry hillsides and in dry river beds. Specimens here rarely attain a size over 30 cm in diameter, compared with the 1 m attained in the coastal locations. The factors that limit the size of the plants are unknown (8).

CONSERVATION MEASURES TAKEN. Protected by law in Namibia since 1916 (7); removal is absolutely forbidden (1). The largest concentration of Welwitschia occurs in the Namib Desert Park (11).

BIOLOGY AND POTENTIAL VALUE. Welwitschia has no close living relatives; it is now regarded as the only species of a natural order, the Welwitschiales (13). Individuals live for hundreds of years. A moderately sized specimen was carbon-dated at 700 years, and some of the largest may be as old as 5000 years (13). Since the chance of seed germination is slight, the long life of the mature plant is of primary importance in maintaing the species (1).

Associated with Welwitschia is a bug (Probergrothius sexpunctatis, Hemiptera). It is often assumed that this insect is the pollinating agent, but this is not likely since these sucking insects are not attracted to the male plants whose cones have little juice. It has been suggested that the plants are wind-pollinated (11).

BOTANY. Dioecious. Woody, with an obconic trunk of which a few centimeters rise above the soil (up to 1.5 m high in the largest specimens), looking like a round table top. Leaves 2, emerging from deep grooves along the edge of the crown, one leaf corresponding to each lobe of the trunk, flat, linear, very leathery, and in older plants split to the base, giving the impression of more than two leaves. Cones originate at base of leaves; female cones 7 X 5 cm, male cones 3 X 1 cm.

TYPE. Collected by F. Welwitsch, near Cape Negro, Angola, in 1860.

ILLUSTRATIONS. See references (1), (2), (5), (6), (7), (8), (10), (11), (12), and (13).

REFERENCES. (1) Benson, L. (1970). Welwitschia mirabilis in the Namib Desert, South West Africa. Cactus and Succulent Journal 42(5): 195-200.

(2) Bornman, C. H. (1972). Welwitschia mirabilis: Paradox of the Namib Desert. Endeavour 31(113): 95-99.

(3) Committee on Stabilization (1975). Report of the Standing Committee on Stabilization of Specific Names. Taxon 24(1): 176.

(4) Dyer, R. A., and I. C. Verdoorn. (1972). Science or Sentiment: The Welwitschia Problem. Taxon 21(4): 485-489.

(5) Hepper, F. N. (1969). The Conservation of Rare and Vanishing Species of Plants. In J. Fischer, N. Simon, and J. Vincent, Wildlife in Danger, pages 353-360.

(6) Hooker, J. D. (1863). On Welwitschia, a new Genus of Gnetaceae. Transactions of the Linnean Society, London 24(1): 1-48, plates 1-14.

(7) Kers, L. E. (1967). The Distribution of Welwitschia mirabilis. Svensk Botanisk Tidskrift 61(1): 97-125, figures 1-5, map.

(8) Rodin, R. J. (1953). Distribution of Welwitschia mirabilis. American Journal of Botany 40(4): 280-285.

(9)   Swinscow, T. D. V. (1972). Friedrich Welwitsch,
      1806-72: A Centennial Memoir. Biological Journal of the
      Linnean Society 4: 269-289.

(10) Tijmens, W. (1976). Die wonderplant Welwitschia mirabilis
      Hook. Veld & Flora 62(2): 4-7.

(11) Wager, V. A. (1976). Desert Antique. Wildlife 18(4).

(12) Wentzel, V. (1961). Unknown Africa. National Geographic,
      illustration, page 354. (September).

(13) Whellan, J. A. (1972). Welwitschia the Wonderful.
      Excelsa 2: 51-59.

Final Regulations for uses and permits for endangered and threatened plants under the Endangered Species Act of 1973 were published by the U. S. Fish and Wildlife Service on June 24. Federal Register 42(122): 32373-32381.

Interim Charter of the United States Endangered Species Scientific Authority for the Convention on International Trade in Endangered Species of Wild Fauna and Flora was published by the Executive Secretary of the Scientific Authority on July 11, Federal Register 42(132): 35799-35792.

Oversight Hearings on the Endangered Species Act of 1973 were held in Washington, D. C., on July 20-22, by the Senate Subcommittee on Resource Protection of the Committee on Environment and Public Works.

Notice of Special Working Session of parties to the Convention on International Trade, to be held in Geneva, Switzerland on October 17-28 to discuss matters regarding the Convention, including plants listed in the Appendices, was published by the U. S. Fish and Wildlife Service on August 10. Federal Register 42(154): 40459-40464.

Final Rulemaking of endangered status for four insular California plants, Lotus scoparius ssp. traskiae (San Clemente broom), Malacothamnus clementinus (San Clemente Island bushmallow), Delphinium kinkiense (San Clemente Island larkspur), and Castilleja grisea (San Clemente Island Indian paintbrush), was published by the U. S. Fish and Wildlife Service on August 11. Federal Register 42(155): 40682-40685.

Notice of Review of the status of American ginseng (Panax quinquefolius) was published by the U. S. Fish and Wildlife Service on August 11. Federal Register 42(155): 40821-40823.

Evaluation of biological status of American ginseng (Panax quinquefolius) was published by Endangered Species Scientific Authority on August 30. Federal Register 42(168): 43728-43767.

ADDENDA TO LISTS OF ENDANGERED,
THREATENED, AND EXTINCT SPECIES

p. 72.   Apiaceae - Cymopterus nivalis: add Idaho.
p. 72.   Apiaceae - Lomatium suksdorfii: add var. suksdorfii.
p. 76.   Brassicaceae - Arabis oxylobula: add Wyoming.
p. 95.   Apiaceae - Ligusticum porteri var. brevilobum: add Nevada.
p. 95.   Delete: Apiaceae - Lilaeopsis carolinensis.
p. 102. Brassicaceae - Arabis suffrutescens var. horizontalis: add
         California.
p. 102. Brassicaceae - Draba crassifolia var. nevadensis: add California.
p. 111. Gentianaceae - Frasera coloradensis: delete Oklahoma.
p. 113. Liliaceae - Lilium grayii: delete Maryland.
p. 115. Onagraceae - Gaura demareei: add Texas.
p. 116. Orchidaceae - Platanthera flava: add Oklahoma.
p. 116. Orchidaceae - Platanthera leucophaea: add Oklahoma.
p. 143. Add: California - Threatened - Brassicaceae - Arabis suffrutescens
         var. horizontalis.
p. 143. Add: California - Threatened - Brassicaceae - Draba crassifolia
         var. nevadensis.
p. 159. Add: Idaho - Endangered - Apiaceae - Cymopterus nivalis.
p. 162. Delete: Louisiana - Threatened - Apiaceae - Lilaeopsis
         carolinensis.
p. 163. Delete: Maryland - Threatened - Liliaceae - Lilium grayii.
p. 167. Add: Nevada - Threatened - Apiaceae - Ligusticum porteri
         var. brevilobum.
p. 172. Delete: North Carolina - Threatened - Apiaceae - Lilaeopsis
         carolinensis.
p. 174. Delete: Oklahoma - Threatened - Gentianaceae - Frasera
         coloradensis.
p. 174. Add: Oklahoma - Threatened - Orchidaceae - Platanthera flava.
p. 174. Add: Oklahoma - Threatened - Orchidaceae - Platanthera leucophaea.
p. 174. Oregon - Endangered - Apiaceae - Lomatium suksdorfii: add var.
         suksdorfii.
p. 179. Delete: South Carolina - Threatened - Apiaceae - Lilaeopsis
         carolinensis.
p. 186. Add: Texas - Threatened - Onagraceae - Gaura demareei.
p. 192. Delete: Virginia - Threatened - Apiaceae - Lilaeopsis carolinensis.
p. 193. Washington - Endangered - Apiaceae - Lomatium suksdorfii: add
         var. suksdorfii.
p. 196. Add: Wyoming - Endangered - Brassicaceae - Arabis oxylobula.
p. 201. Asteraceae - Argyroxiphium macrocephalum: read Thr - Mau.
p. 201. Add: Asteraceae - Argyroxiphium sandwicense - End - Haw.
p. 202. Asteraceae - Dubautia molokaiensis: add var. molokaiensis.
p. 202. Add: Asteraceae - Dubautia molokaiensis var. oppositifolia -
         End - Mol.
p. 202. Add: Asteraceae - Dubautia molokaiensis var. stipitata - End - Mau.
p. 206. Campanulaceae - Delissea parviflora - Ext: read Haw? Oah.
p. 206. Campanulaceae - Lobelia tortuosa: add var. tortuosa.
p. 210. Fabaceae - Sesbania tomentosa var. tomentosa - End: add Nii.

p. 216. Malvaceae - Hibiscus kokio var. kokio: delete Oah Kau?; add Mol.
p. 216. Add: Malvaceae - Hibiscus kokio var. pekeloi - End - Mol.
p. 216. Add: Malvaceae - Hibiscus oahuensis - End - Oah.
p. 216. Add: Malvaceae - Hibiscus ula - End - Mau.
p. 216. Malvaceae - Hibiscus waimeae: add var. waimeae.
p. 217. Myrsinaceae - Myrsine fernseei - End: read Oah Kau.
p. 227. Asteraceae - Eupatorium borinquense: read Chromolaena borinquensis.
p. 227. Asteraceae - Eupatorium droserolepis: read Koanophyllon droserolepis.
p. 227. Asteraceae - Eupatorium oteroi: read Chromolaena oteroi.

State Lists of Endangered and Threatened Plant Species

Page 243

ARKANSAS

Arkansas Plants Nominated for Listing as Endangered: A Fact Sheet.
1977. Arkansas Natural Heritage Commission, Suite 500,
Continental Building, Main and Markham, Little Rock 72201.

COLORADO

Natural History Inventory of Colorado 1. Vascular Plants, Lichens,
and Bryophytes. 1976. W. A. Weber and B. C. Johnston.
University of Colorado Museum, Boulder 80309.

CONNECTICUT

Rare and Endangered Species of Connecticut and Their Habitats. 1977.
J. J. Dowhan and R. J. Craig. Connecticut Geological and Natural
History Survey Report of Investigations No. 6. Connecticut
Department of Environmental Protection.

Page 244

IDAHO

Rare and Endangered Plants of Idaho. 1974. D. M. Henderson, in
C. A. Wellner and F. D. Johnson, Research Natural Area Needs in
Idaho--A First Estimate. Forest, Wildlife and Range Experiment
Station, University of Idaho.

Endangered and Threatened Plants of Idaho - A Summary of Current
Knowledge. 1977. D. M. Henderson, F. D. Johnson, P. Packard and
R. Steele. Bulletin Number 21, College of Forestry, Wildlife and
Range Sciences, University of Idaho, Moscow 83843.

KANSAS

Rare Native Vascular Plants of Kansas. 1977. R. L. McGregor. Technical Publications of the State Biological Survey of Kansas, No. 5. State Biological Survey of Kansas, 2045 Avenue A, Campus West, Lawrence 66044.

MICHIGAN

Endangered, Threatened, and Rare Vascular Plants in Michigan. 1977. W. H. Wagner, E. G. Voss, J. H. Beaman, E. A. Bourdo, F. W. Case, J. A. Churchill and P. W. Thompson, Michigan Botanist 16: 99-110.

Commentary on Endangered and Threatened Plants in Michigan. 1977. J. H. Beaman, Michigan Botanist 16: 110-122.

NEVADA

Threatened Plant Species of the Nevada Test Site, Ash Meadows, and Central-Southern Nevada. 1977. J. C. Beatley. U. S. Energy Research and Development Administration Contract E(11-1)-2307. Biological Sciences Department, University of Cincinnati, Cincinnati 45221.

Addendum to COO-2307-11 and COO-2307-12. (Endangered and Threatened Plant Species of the Nevada Test Site, Ash Meadows, and Central-Southern Nevada). 1977. J. C. Beatley. U. S. Energy Research and Development Administration Contract E(11-1)-2307. Biological Sciences Department, University of Cincinnati, Cincinnati 45221.

NEW JERSEY

Endangered, Threatened and Rare Vascular Plants of the Pine Barrens and Their Biogeography. 1978. D. E. Fairbrothers. In R.T.T. Forman, ed., The Pine Barrens of New Jersey. New York: Academic Press.

Page 247

NORTH CAROLINA

Endangered and Threatened Plants and Animals of North Carolina. 1977. Proceedings of a Symposium on Endangered and Threatened Biota of North Carolina 1. Biological Concerns. North Carolina State Museum of Natural History, P. O. Box 27647, Raleigh 27611.

OHIO

Rare and Endangered Aquatic Vascular Plants of Ohio: An Annotated List of the Imperiled Species. 1977. R. L. Stuckey and M. L. Roberts, Sida 7(1): 24-41.

Conservation of Wild Flowers. 1934. R. B. Gordon. In R. B. Gordon and V. H. Ries, Wild Flowers, pages 2-21. Agricultural Extension Service, The Ohio State University, Columbus, Bulletin 119.

Page 249

WASHINGTON

A Working List of Rare, Endangered, or Threatened Vascular Plant Taxa for Washington. 1977. M. Denton, B. Goldman, C. L. Hitchcock, A. R. Kruckeberg and M. Mueller. Department of Botany, University of Washington, Seattle 98195. Unpublished manuscript.

393

Page 251

Abrams, W. A.

1977. Cactus Rustlers Cause Thorny Problems. Brooklyn Botanic
Garden Record: Plants & Gardens 32(4): 46-48. (Condensed
from The Wall Street Journal, July 14, 1976)

Allen, G. E.

1974. Peperomia humilis, the Not Extinct Species. American
Horticulturist 53(5): 13-15. (Florida)

Austin, D. F.

1977. Endangered Species: Florida Plants. The Florida Naturalist
50(3): 15-21.

Baskin, J. M. and C. C. Baskin

1977. Leavenworthia torulosa Gray: An Endangered Plant Species in
Kentucky. Castanea 42(1): 15-17.

Page 252

Bean, M. J.

1977. The Endangered Species Act Under Fire. National Parks &
Conservation Magazine 51(6): 16-20.

Benson, L.

1977. How Do You Preserve an Ecosystem? Fremontia 5(2): 3-7.

Page 253

Crovello, T. J.

1975. Use of Computers to Determine the Endangered Status of Plant
Species. Proceedings of the Indiana Academy of Science 85: 352-353.
(Abstract, publ. 1976)

Editor.

    1977. Botanic Dilemma. The Herbarist 43: 56-57. (Pedicularis furbishiae)

Editor.

    1977. Crackdown on a Fabled Root. An Endangered Aphrodisiac? Time,

        page 34. (September 5) (Panax quinquefolius)

Emery, W. H. P.

    1967. The Decline and Threatened Extinction of Texas Wild Rice.

        Southwestern Naturalist 12: 203-204. (Zizania texana)

Fables, D.

    1962. 20 "Lost" Plants. Bartonia 31(1960-61): 7-10. (New Jersey)

Gentry, H. S.

    1977. Agave arizonica, a Rare Species. Saguaroland Bulletin

        40-42 (April).

Griffin, J. R.

    1976. Native Plant Reserves at Fort Ord. Fremontia 4(2): 25-28.

        (California)

Gruber, D.

    1977. The Mighty (Endangered) Furbish's Lousewort. Harvard Magazine

        79(7): 54-55. (Pedicularis furbishiae)

Herrman, J.

    1977. Who's Protecting Endangered Plants? Brooklyn Botanic Garden

        Record: Plants and Gardens 33(1): 15-18.

Irwin, H. S.

    1977. Miss Furbish's Lousewort Must Live. <u>Garden</u> 1(4): 6-11.

        (<u>Pedicularis</u> <u>furbishiae</u>)

Page 258

Kartesz, J. T., and R. Kartesz

    1977. <u>The</u> <u>Biota</u> <u>of</u> <u>North</u> <u>America</u>, <u>Part</u> <u>1</u>: <u>Vascular</u> <u>Plants</u>, <u>Volume</u> <u>I</u>:

        <u>Rare</u> <u>Plants</u>. 361 pages. Pittsburgh, Pennsylvania: Biota of

        North America Committee.

Kologiski, R. L., F. R. Hivick, C. W. Reed and D. W. Jenkins

    1975. Rare, Endangered, and Endemic Plants of the Chesapeake Bay

        Region. Appendix D (pages D1-D49), in <u>Compendium</u> <u>of</u>

        <u>Natural</u> <u>Features</u> <u>Information</u>, <u>Volume</u> <u>I</u>. Maryland Department

        of State Planning and Smithsonian Institution Center for

        Natural Areas.

Page 259

Little, E. L., Jr.

    1977. <u>Rare</u> <u>and</u> <u>Local</u> <u>Trees</u> <u>in</u> <u>the</u> <u>National</u> <u>Forests</u>. Conservation

        Research Report No. 21. 14 pages. U. S. Department of

        Agriculture, Forest Service (June).

Page 260

Moir, W. G.

    1977. Environments, Endangered Species and what we should do about

        them. <u>American</u> <u>Orchid</u> <u>Society</u> <u>Bulletin</u> 46(8): 725-726.

Morales, W. R.

    1977. The Endemic Orchid Species of Puerto Rico. <u>American</u> <u>Orchid</u>

        <u>Society</u> <u>Bulletin</u> 46(8): 727-730.

Morton, J. F.

    1976. Pestiferous Spread of Many Ornamental and Fruit Species in South Florida. _Proceedings of Florida State Horticultural Society_ 89: 348-353.

Page 261

Reveal, J. L.

    1977. Really, Who Gives a Damn? _Phytologia_ 35(5): 373-384.

Robertson, K. R.

    1977. _Cladrastis_: The Yellow-Woods. _Arnoldia_ 37(3): 137-150.

Root, G. W.

    1974. Protection of Rare and Endangered Species. _Soil Conservation_ 39(8): 4-5. (Pennsylvania)

Ross, L. T.

    1977. Conservation and Carnivorous Plants. _Carnivorous Plant Newsletter_ 6(2): 38-40.

Saltonstall, R.

    1977. Of Dams and Kate Furbish. _The Living Wilderness_ 40(136): 42-43. (_Pedicularis furbishiae_)

Smith, B. A.

    1977. Western Red Cedar: A Declining Giant. _Pacific Search_ 11(8): 14-15. (_Thuja plicata_)

Smith, D.

    1975. Texas Palms Rescued. _Principes_ 19(4): 146. (_Sabal mexicana_)

Page 262

Sparks, Z.

    1977. Saga of the Furbish Lousewort. Brooklyn Botanic Garden Record: Plants and Gardens 33(1): 13-14. (Pedicularis furbishiae)

Steinhart, P.

    1977. Mighty, Like a Furbish Lousewort. Audubon 79(3): 121-125. (Pedicularis furbishiae)

Page 263

Taylor, R. J., and C. E. Taylor

    1977. The Rare and Endangered Hypoxis longii Fern. (Amaryllidaceae) in Oklahoma. Bulletin of Torrey Botanical Club 104(3): 276.

USDA Forest Service

    1977. Conference on Endangered Plants in the Southeast, Proceedings. USDA Forest Service General Technical Report SE-11, 104 pages. Asheville, North Carolina: Southeastern Forest Experiment Station.

Page 264

Wester, L. L., and H. B. Wood

    1977. Koster's Curse (Clidemia hirta), a Weed Pest in Hawaiian Forests. Environmental Conservation 4(1): 35-41.

Wilson, D.

    1977. Endangered Species and the U. S. Army Corps of Engineers. Frontiers 41(4): 37-38. (Pedicularis furbishiae)

Page 267

Brunig, E. F.

    1977. The Tropical Rain Forest - A Wasted Asset or an Essential Biospheric Resource? Ambio 6(4): 187-191.

Devine, W. T.

    1977. A Programme to Exterminate Introduced Plants on Raoul Island. _Biological Conservation_ 11(3): 193-207.

Egler, F. E.

    1977. _The Nature of Vegetation, Its Management and Mismanagement._ 527 pages. Connecticut: Aton Forest, Norfolk and Connecticut Conservation Association, Bridgewater.

Ellis, E. A., F. Perring and R. E. Randall

    1977. Britain's Rarest Plants. Norwich: Jarrold & Sons Ltd.

Fosberg, F. R.

    1967. Some Ecological Effects of Wild and Semi-Wild Exotic Species of Vascular Plants. _Towards a New Relationship of Man and Nature in Temperate Lands._ Part III: _Changes Due to Introduced Species._ IUCN _Publications_, new series, 9: 98-109.

Guinea, E.

    1960. Some Considerations on the Future of the Spontaneous Vegetation of Europe. _Feddes Repertorium_ 63(2): 222-224.

Guppy, G. A.

    1977. Endangered Plants in British Columbia. _Davidsonia_ 8(2): 24-30.

Hall, A. V.

    1977. The Golden Gladiolus: Only 23 Plants to go to Extinction. _African Wildlife_ 31(3): 21. (_Gladiolus aureus_)

Page 276

Irwin, H. S.

> 1977. Coming to Terms With the Rain Forest. Garden 1(2): 28-33.
> (Brazil)

IUCN Threatened Plants Committee

> 1976. Amendments to the List of Rare, Threatened and Endemic Plants
> for the Countries of Europe. 22 Pages. Kew: Royal Botanic
> Gardens.

Jacobs, M.

> 1977. It is the Genera of Threatened Plants That Need Attention. Flora
> Malesiana Bulletin 30: 2828-830.

Page 278

Kolesnikov, B. P.

> 1976. The Problem of Vegetation Protection. Zurnal Obscej Biologii
> 37(5): 635-648.

Page 279

Laasimer, L.

> 1975. Rare Plant Communities and Their Conservation Problems,
> in Some Aspects of Botanical Research in the Estonian S.S.R.,
> pages 62-73. Tartu: Academy of Sciences of the Estonian S.S.R.

Page 280

Lodewick, K.

> 1976. How to Hunt a Penstemon. Penstemon Field Identifier, Part
> VII. Supplement to American Penstemon Society Bulletin 35.
> (P. cerrosensis, Baja California)

400

Page 281

Lovejoy, T.

1977. We Must Decide Which Species Will Go Forever. The Florida Naturalist
50(4): 4-10.

Page 282

Mattos, M. D. L. V. de, and C. C. L. V. de Mattos

1976. Palmito jucara--Euterpe edulis Mart. (Palmae)--uma especie
a plantar, manejar e proteger. Brasil Florestal 7(27): 9-20.

McMillan, A. J. S.

1969. How Rare Are Rare Cacti? National Cactus and Succulent Journal
(U.K.) 24(4): 84-85. (With comment by G. D. Rowley)

Medwecka-Kornas, A.

1977. Ecological Problems in the Conservation of Plant Communities,
With Special Reference to Central Europe. Environmental
Conservation 4(1): 27-33.

Page 284

Mosquin, T., and C. Suchal, editors

1977. Canada's Threatened Species and Habitats. 200 pages. Ottawa:
Canadian Nature Federation.

Page 285

Oliver, E. G. H.

1977. Orchids--or Houses? African Wildlife 31(2): 10-11.

Page 287

Queijo, J.

1977. Harvesting a Nuisance. Environment 19(2): 25-z 9. (Introduced
Eichhornia crassipes)

Page 289

Sachet, M.-H., P. A. Schafer and J. C. Thibault

    1975. Mohotani: une Ile Protegee aux Marquises. Bulletin de la

    Societe des Etudes Oceaniennes 16(6) (No. 193): 557-568.

Page 291

South Africa: National Programme of Environmental Sciences: Rare and
Endangered Plant Species Unit

    1976. An Interim Report on the Threatened, Rare and Endangered

    Plant Species on Table Mountain. 5 Pages. Cape Town:

    Bolus Herbarium.

Page 293

Temple, S. A.

    1977. Plant-Animal Mutualism: Coevolution with Dodo leads to Near

    Extinction of Plant. Science 197(4306): 885-886. (Calvaria

    major)

Page 295

Weber, B. E.

    1971. Is Dominica's Forest Doomed? American Forests 77(7): 12-15.

The Needed Diversity of Plant Resources

Page 297

Altschul, S. von R.

    1977. Exploring the Herbarium. Scientific American 236(5): 96-104.

Page 299

Title of J. G. Hawkes (1971) reads: Conservation of Plant Genetic

    Resources.

Lovejoy, T.

1977. We Must Decide Which Species Will Go Forever. The Florida Naturalist 50(4): 4-10.

Mattos, M. D. L. V. de, and C. C. L. V. de Mattos

1976. Palmito jucara--Euterpe edulis Mart. (Palmae)--uma especie a plantar, manejar e proteger. Brasil Florestal 7(27): 9-20.

McMillan, A. J. S.

1969. How Rare Are Rare Cacti? National Cactus and Succulent Journal (U.K.) 24(4): 84-85. (With comment by G. D. Rowley)

Medwecka-Kornas, A.

1977. Ecological Problems in the Conservation of Plant Communities, With Special Reference to Central Europe. Environmental Conservation 4(1): 27-33.

Mosquin, T., and C. Suchal, editors

1977. Canada's Threatened Species and Habitats. 200 pages. Ottawa: Canadian Nature Federation.

Oliver, E. G. H.

1977. Orchids--or Houses? African Wildlife 31(2): 10-11.

Queijo, J.

1977. Harvesting a Nuisance. Environment 19(2): 25-z 9. (Introduced Eichhornia crassipes)

Page 289

Sachet, M.-H., P. A. Schafer and J. C. Thibault

    1975. Mohotani: une Ile Protegee aux Marquises. *Bulletin de la*

        Societe des Etudes Oceaniennes 16(6) (No. 193): 557-568.

Page 291

South Africa: National Programme of Environmental Sciences: Rare and
Endangered Plant Species Unit

    1976. *An Interim Report on the Threatened, Rare and Endangered*

        Plant Species on Table Mountain. 5 Pages. Cape Town:

        Bolus Herbarium.

Page 293

Temple, S. A.

    1977. Plant-Animal Mutualism: Coevolution with Dodo leads to Near

        Extinction of Plant. *Science* 197(4306): 885-886. (*Calvaria*

        major)

Page 295

Weber, B. E.

    1971. Is Dominica's Forest Doomed? *American Forests* 77(7): 12-15.

The Needed Diversity of Plant Resources

Page 297

Altschul, S. von R.

    1977. Exploring the Herbarium. *Scientific American* 236(5): 96-104.

Page 299

Title of J. G. Hawkes (1971) reads: Conservation of Plant Genetic

Resources.

Page 281

Lovejoy, T.

    1977. We Must Decide Which Species Will Go Forever. The Florida Naturalist 50(4): 4-10.

Page 282

Mattos, M. D. L. V. de, and C. C. L. V. de Mattos

    1976. Palmito jucara--Euterpe edulis Mart. (Palmae)--uma especie a plantar, manejar e proteger. Brasil Florestal 7(27): 9-20.

McMillan, A. J. S.

    1969. How Rare Are Rare Cacti? National Cactus and Succulent Journal (U.K.) 24(4): 84-85. (With comment by G. D. Rowley)

Medwecka-Kornas, A.

    1977. Ecological Problems in the Conservation of Plant Communities, With Special Reference to Central Europe. Environmental Conservation 4(1): 27-33.

Page 284

Mosquin, T., and C. Suchal, editors

    1977. Canada's Threatened Species and Habitats. 200 pages. Ottawa: Canadian Nature Federation.

Page 285

Oliver, E. G. H.

    1977. Orchids--or Houses? African Wildlife 31(2): 10-11.

Page 287

Queijo, J.

    1977. Harvesting a Nuisance. Environment 19(2): 25-9. (Introduced Eichhornia crassipes)

Page 289

Sachet, M.-H., P. A. Schafer and J. C. Thibault

1975. Mohotani: une Ile Protegee aux Marquises. _Bulletin de la Societe des Etudes Oceaniennes_ 16(6) (No. 193): 557-568.

Page 291

South Africa: National Programme of Environmental Sciences: Rare and Endangered Plant Species Unit

1976. _An Interim Report on the Threatened, Rare and Endangered Plant Species on Table Mountain._ 5 Pages. Cape Town: Bolus Herbarium.

Page 293

Temple, S. A.

1977. Plant-Animal Mutualism: Coevolution with Dodo leads to Near Extinction of Plant. _Science_ 197(4306): 885-886. (_Calvaria major_)

Page 295

Weber, B. E.

1971. Is Dominica's Forest Doomed? _American Forests_ 77(7): 12-15.

The Needed Diversity of Plant Resources

Page 297

Altschul, S. von R.

1977. Exploring the Herbarium. _Scientific American_ 236(5): 96-104.

Page 299

Title of J. G. Hawkes (1971) reads: Conservation of Plant Genetic Resources.

JOURNAL OF THE TENNESSEE ACADEMY OF SCIENCE

VOLUME 53, NUMBER 4, OCTOBER, 1978

THE RARE VASCULAR PLANTS OF TENNESSEE

Committee for Tennessee Rare Plants*

ABSTRACT

This paper presents a list of rare Tennessee plants and their degree of endangerment. The list includes 52 possibly extirpated, 58 endangered, 108 threatened and 123 taxa of special concern. These data are based primarily on specimens deposited in the University of Tennessee Herbarium, Knoxville and the Herbarium of Vanderbilt University. The needs for understanding and protection of these taxa and management of the unique habitats in which they occur are discussed.

INTRODUCTION

It has become increasingly obvious that much of Tennessee's landscape has been severely altered or destroyed by man's activities. Few natural areas are adequately protected and without judicious land and water management in the future, most of the State's natural heritage will soon be destroyed. Recent national concern for the future quality of the earth's environment is reflected in the federal Endangered Species Act (Public Law 93-205), state, regional (Styron, 1976), federal plant and animal lists, and local environmental controversies. In Tennessee, concern for the production of a rare plant list has come from a variety of sources: interested individuals, garden clubs, lists from other states, progress on the national rare plant list, the scientific community, and those involved in the preparation of environmental impact statements. The list of Tennessee rare plants included herewith attempts to indicate their present state of endangerment and known county distribution. It was preceeded by those of Sharp (1948, 1974) and Goff, Stephenson, & Lewis (1975).

During the Spring of 1976 an *ad hoc* Committee for Tennessee Rare Plants began preparation of this list, using as reference sources the Herbarium of Vanderbilt University and the University of Tennessee Herbarium, Knoxville. Although by no means the only important collections of the State flora, these two herbaria represent by far the largest and most extensive collections. Our procedure was to catalog all vascular plants which are reported from three or fewer counties. Regional manuals, monographs, and other pertinent taxonomic papers were then consulted to determine the geographic distribution of each taxon. All introduced, cultivated, and escaped

*Alphabetical list of the Committee:

J. L. Collins, Div. of Forestry, Fisheries and Wildlife Dev., TVA, Norris, TN. 37828

H. R. DeSelm, Dept. of Botany and Grad. Program in Ecology, Univ. of Tennessee, Knoxville, TN. 37916

A.M. Evans, Dept. of Botany, Univ. of Tennessee, Knoxville, TN. 37916

R. Kral, Dept. of General Biology, Vanderbilt Univ., Nashville, TN. 37235

B. E. Wofford, Dept. of Botany, Univ. of Tennessee, Knoxville, TN. 37916 (Rapporteur for the Committee)

taxa were eliminated from the list. The remainder was then subjected to scrutiny by the committee and additional taxa were eliminated based on known weediness or other characteristics that would make their inclusion superfluous. In a few cases, we have included taxa known to occur in more than three counties. In general, these are taxa that are either commercially exploited (e.g., *Panax quinquefolium* and *Hydrastis canadensis*) or regionally endemic species with limited distributions and little or no possibility of further range extension (e.g., *Conradina verticillata*, Central Basin endemics, Southern Appalachian endemics). The resulting list is therefore a distillation of what is known of the county distribution and endangerment of our least frequently collected taxa, recognizing that working with these particular taxa represents a peculiar combination of both knowledge and ignorance. The final list includes some subjective judgments by the committee concerning inclusion or exclusion based on true rarity versus infrequent collection and the application of the criteria used below in assembling the list.

In September 1976, a letter was distributed to the State's botanists with the list appended. In November, a public workshop was held in conjunction with the Tennessee Academy of Science meeting and the list was examined and amended by approximately 40 participants. Since then we have had additional verbal and written communications and a few records have been added from other herbaria and from the literature.

The present list consists of four categories of taxa:

Possibly Extirpated—Species considered endangered but which have not been seen in Tennessee within the last 20 years.

Endangered—Species now in danger of becoming extinct in Tennessee because of:
   (a) their rarity throughout their range.
   (b) their rarity in Tennessee as a result of sensitive habitat or restricted area of distribution.

Threatened—Species likely to become endangered in the immediately foreseeable future as a result of rapid habitat destruction or commercial exploitation.

Special Concern—Species requiring particular attention because:
   (a) they are rare or distinctive in Tennessee because the State represents the limit or near-limit of their geographic range.
   (b) their status is undetermined because of insufficient information.

Of the 58 rare Tennessee plants in the Federal Register (U. S. Dept. of the Interior, 1975, 1976), we have excluded the following:

ENDANGERED

Clematis addisonii Britt.—does not occur in Tennessee (Dennis, 1976)
Clematis gattingeri Small—synonymous with *Clematis viorna* (Dennis, 1976)
Draba incisa—a *nomen nudum*
Elodea linearis (Rybd.) H. St. John—of doubtful validity, known only from the type
Geum geniculatum Michx.—no Tennessee specimens seen although it occurs on Roan Mountain in North Carolina (Radford, Ahles, & Bell, 1968)
Helianthus eggertii Small—not uncommon
Pycanathemum curvipes (Greene) Grant & Epling—no Tennessee specimens seen
Silphium integrifolium Michx, var. *gattingeri* Perry—of doubtful validity, known only from the type
Solidago spithamaea M. A. Curtis—no Tennessee specimens seen

THREATENED

Carex purpurifera Mackenzie—not uncommon
Platanthera flava (L.) Farwell—not uncommon
Platanthera peramoena Gray—not uncommon
Rhododendron bakeri (Lemmon & McKay) Hume—apparently conspecific with *Rhododendron cumberlandense* E. L. Braun
Saxifraga careyana Gray—not uncommon
Saxifraga caroliniana Gray—does not occur in Tennessee (Lord, 1960)
Synandra hispidula (Michx.) Baillon—not uncommon

The absence of cultivated plants on this list should not suggest that these taxa are of little value, as much of our native flora is already in cultivation (Bailey, 1961). Similarly, concern for rare plants need not be based solely on their uniqueness or esoteric value. The African species of *Rauwolfia* were rare botanical curiosities, but today they are cultivated for drugs such as the tranquilizer resperine. Also, the antihistamine Ephedrine is extracted from the slow *Ephedra* of the southwestern U. S. Several of our native plants are collected and sold as drug sources (Krochmal, 1968), thus depleting these natural resources. Several taxa in this text designated as commercially exploited are so indicated because of depletion by herb and nursery collectors, and wildflower gardeners.

The list in its present form represents approximately 10 per cent of the State's flora and should be considered tentative. New collections will uncover additional plants in need of protection while other species will be found to be more common than our current information indicates. Rare plants are known to occur in each of the State's major physiographic provinces with higher frequencies from the Central Basin, southeastern and northern Highland Rim, the Cumberland Plateau, and the Unaka Mountains. The patterns of distribution indicate that these taxa often occur together in unique landscape features which provide sound justification for permanent preservation of these unique habitats.

Interested persons across the State who find these plants are encouraged to notify staff of the University of Tennessee and/or the Herbarium of Vanderbilt University. Such information contributes not only to this project but also to State and regional projects underway. However, care should be taken to protect new populations, and collections should *not* be made from currently known localities.

It is also hoped that this publication will generate educational programs, necessary legislation, public awareness, and general concern for this significant portion of Tennessee's natural heritage.

POSSIBLY EXTIRPATED

ASTERACEAE
Aster praealtus Poir.—Davidson, Johnson
Eupatorium leucolepis (DC.) T & G.—Coffee
Helenium brevifolium (Nutt.) Wood—Morgan
Helianthus glaucophyllus D. M. Smith—Johnson
Ratibida columnifera (Nutt.) Woot. & Standl.—Davidson, Richards (1968)*
Rudbeckia subtomentosa Pursh—Montgomery
Tetragonotheca helianthoides L.—Knox

CAMPANULACEAE
Lobelia amoena Michx.—Polk

CAPRIFOLIACEAE
Linnaea borealis L.—Sevier

CARYOPHYLLACEAE
Paronychia argyrocoma (Michx.) Nutt.—Cumberland
Silene ovata Pursh—Cocke, Marion

CRASSULACEAE
Sedum rosea (L.) Scopoli—Carter, Gattinger (1901)

CYPERACEAE
Cladium mariscoides (Muhl.) Torr.—Middle Tennessee, Sharp, *et al.* (1956)
Dichromena latifolia Baldwin—Coffee, Gattinger (1901)
Rhynchospora capillacea Torr.—Campbell
Rhynchospora rariflora (Michx.) Ell.—Van Buren, Warren
Rhynchospora wrightiana Boeckler—Bledsoe
Scleria minor (Britt.) Stone—Bledsoe

ERICACEAE
Kalmia angustifolia L. var. *caroliniana* (Small) Fern.—Johnson

EUPHORBIACEAE
Croton alabamensis E. A. Smith—Coffee, Farmer & Thomas (1969)

FABACEAE
Apios priceana Robins.—Davidson, Montgomery
Crotalaria purshii DC.—Blount

HYPERICACEAE
Hypericum majus (Gray) Britt.—Morgan

ISOETACEAE
Isoetes melanopoda Gay & Durieu—Rutherford

LAMIACEAE
Pycnanthemum verticillatum (Michx.) Pers.—Fentress
Scutellaria montana Chapm.—Hamilton

LILIACEAE
Melanthium hybridum Walt.—Sevier, Unicoi
Melanthium virginicum L.—Lincoln
Zigadenus densus (Desr.) Fern.—Coffee

LYCOPODIACEAE
Lycopodium annotinum L.—Blount

MYRICACEAE
Comptonia peregrina (L.) Coulter—Scott

ORCHIDACEAE
Listera australis Lindl.—Coffee
Listera smallii Wieg.—Johnson, Sevier
Platanthera integra (Nutt.) Gray ex Beck—Coffee

POACEAE
Trisetum spicatum (L.) Richter var. *molle* (Michx.) Beal—Carter, Lamson-Scribner (1894)

POLYGONACEAE
Polygonum cilinode Michx.—Blount

PONTEDERIACEAE
Heteranthera limosa (Sw.) Willd.—Davidson, Montgomery

*Certain taxa for which we have seen no specimens are cited from the literature.

PORTULACACEAE
Talinum teretifolium Pursh—Rhea

RANUNCULACEAE
Anemone canadensis L.—Knox
Anemone caroliniana Walt.—Davidson, Rutherford
Ranunculus trichophyllus Chaix—Montgomery

RHAMNACEAE
Rhamnus alnifolius L'Hér.—Campbell

ROSACEAE
Crataegus harbisoni Beadle—Lawrence, Obion, Shelby
Spiraea virginiana Britton—Van Buren

SARRACENIACEAE
Sarracenia oreophila (Kearney) Wherry—Fentress, Sharp, et al. (1960)
Sarracenia purpurea L.—Mississippi, Tennessee & Duck Rivers, Gattinger (1901)

SCROPHULARIACEAE
Aureolaria patula (Chapm.) Pennell—Montgomery, Roane, Sevier
Gerardia gatesii (Benth.) Pennell—Polk
Gratiola floridana Nutt.—Franklin, Hamilton
Schwalbea americana L.—Coffee, Fentress, Pennell (1935)
Veronica comosa Richter—Rutherford
Veronica scutellata L.—Johnson

ENDANGERED

ANACARDIACEAE
Cotinus obovatus Raf.—Franklin, Marion

APIACEAE
Perideridia americana (Nutt.) Reichenb.—Davidson, Rutherford

ARACEAE
Symplocarpus foetidus (L.) Nutt.—Carter, Johnson

ASPLENIACEAE
Dryopteris spinulosa (Muell.) Watt—Johnson
Phyllitis scolopendrium (L.) Newman—Marion

ASTERACEAE
Cacalia rugelia (Chapm.) Barkley & Cronquist—Cocke, Sevier
Echinacea tennesseensis (Beadle) Small—Davidson, Rutherford
Heterotheca ruthii (Small) Harms—Polk
Liatris cylindracea Michx.—Rutherford
Marshallia grandiflora Beadle & Boynton—Cumberland, Roane
Senecio robbinsii Oakes ex Rusby—Carter
Silphium brachiatum Gattinger—Franklin, Polk
Solidago rupestris Raf.—Davidson, Montgomery

BORAGINACEAE
Onosmodium subsetosum Mackenzie & Bush—Rutherford

BRASSICACEAE
Arabis perstellata E. L. Br. var. ampla Rollins—Davidson
Arabis shortii (Fern.) Gl. var. phalacrocarpa (Hopkins) Steyerm.—Humphreys, Montgomery
Erysimum capitatum (Dougl.) Greene—DeKalb, Putnam, Smith
Leavenworthia exigua Rollins var. lutea Rollins—Maury
Lesquerella perforata Rollins—Smith, Wilson
Lesquerella stonensis Rollins—Rutherford

CARYOPHYLLACEAE
Arenaria fontinalis (Short & Peter) Shinners—Cheatham, Davidson, Williamson
Cerastium arvense L. var. oblongifolium (Torr.) Hollick & Britt.—Maury

CONVOLVULACEAE
Evolvulus pilosus Nutt.—Rutherford, Wilson

CRASSULACEAE
Sedum nevii Gray—Polk
Sedum smallii (Britt.) Ahles—Franklin, Grundy, Hamilton, Marion, Rhea

ERICACEAE
Pyrola rotundifolia L. var. americana (Sw.) Fern.—Sullivan

ERIOCAULACEAE
Lachnocaulon anceps (Walt.) Morong—Cumberland

FABACEAE
Thermopsis fraxinifolia M. A. Curtis—Hamilton

HAEMODORACEAE
Lachnanthes caroliniana (Lam.) Dandy—Coffee

HYMENOPHYLLACEAE
Trichomanes petersii Gray—Blount, Franklin, Monroe, Sevier, Shields (1939)

HYPERICACEAE
Hypericum ellipticum Hook.—Johnson

LAMIACEAE
Conradina verticillata Jennison—Cumberland, Fentress, Morgan, Scott, White

LILIACEAE
Allium stellatum Fraser—Davidson, Rutherford
Lilium grayi S. Wats.—Carter, Johnson; commercially exploited
Lilium philadelphicum L.—Grundy; commercially exploited
Trillium pusillum Michx.—Sumner
Veratrum woodii Robbins—Grundy

ONAGRACEAE
Oenothera missouriensis Sims—Rutherford

ORCHIDACEAE
Cypripedium reginae Walt.—Claiborne
Platanthera integrilabia (Correll) Luer—Cumberland, Fentress, Franklin, Grundy
Platanthera nivea (Nutt.) Luer—Coffee
Platanthera orbiculata (Pursh) Lindley—Carter, Unicoi

POACEAE
Calamagrostis cainii Hitchc.—Sevier
Calamovilfa arcuata K. E. Rogers—Cumberland
Glyceria nubigena W. A. Anderson—Blount, Sevier

POLYGONACEAE
Polygonella americana (F. & M.) Small—Morgan

RANUNCULACEAE
Caltha palustris L.—Carter, Johnson
Delphinium exaltatum Ait.—Anderson
Ranunculus longirostris Godr.—Montgomery, Robertson

ROSACEAE
Geum radiatum Gray—Blount, Carter, Sevier
Prunus pumila L.—Coffee, Franklin
Sanguisorba canadensis L.—Carter
Spiraea alba Du Roi—Johnson

RUBIACEAE
Houstonia montana (Chickering) Small—Carter

SCROPHULARIACEAE
Collinsia verna Nutt.—Clay, Sumner
Gerardia pseudaphylla (Pennell) Pennell—Coffee, Dickson, Warren
Tomanthera auriculata (Michx.) Raf.—Carroll

XYRIDACEAE
Xyris tennesseensis Kral—Lewis

THREATENED

ALISMATACEAE
Sagittaria graminea Michx. var. graminea—Coffee, Fentress
Sagittaria graminea Michx. var. platyphylla Engelm.—Cumberland, Lake, Obion

APIACEAE
Ammoselinum popei T. & G.—Davidson, Rutherford, Wilson
Hydrocotle americana L.—Scott, Van Buren
Polytaenia nuttalli DC.—Davidson, McNairy, Robertson

ARALIACEAE   Ginseng
Panax quinquefolium L.—Infrequent over the state and commercially exploited

ASTERACEAE
Aster ericoides L.—Loudon, McNairy
Aster sericeus Vent.—Coffee, Roane
Cacalia suaveolens L.—Dickson, Humphreys, Lewis, Williamson
Chrysogonum virginianum L.—Claiborne, Bradley
Eupatorium luciae-brauniae Fern.—Fentress, Pickett, Scott
Hieracium scabrum Michx.—Johnson, Pickett
Polymnia laevigata Beadle—Franklin, Marion, Rhea ?
Prenanthes roanensis (Chickering) Chickering—Carter
Silphium laciniatum L.—Carroll, Henry
Silphium pinnatifidum Ell.—Coffee

BETULACEAE
Alnus crispa (Ait.) Pursh—Carter

BRASSICACEAE
Cardamine clematitis Shuttlew.—Blount, Carter, Polk, Sevier
Cardamine flagellifera O. E. Schultz—Blount, Knox, Polk, Sevier
Cardamine rotundifolia Michx.—Grainger, Hawkins, Scott, Unicoi
Draba ramosissima Desv.—Blount, DeKalb, Putnam
Leavenworthia exigua Rollins var. exigua—Central Basin
Leavenworthia stylosa Gray—Central Basin
Leavenworthia torulosa Gray—Central Basin and southern Great Valley
Lesquerella densipila Rollins—Central Basin
Lesquerella globosa (Desv.) Wats.—Cheatham, Davidson, Maury, Montgomery, Trousdale
Lesquerella lescurii (Gray) S. Wats.—Central Basin and dissected Highland Rim

CAMPANULACEAE
Lobelia canbyi Gray—Coffee, Cumberland, Warren
Lobelia gattingeri Gray—Central Basin

CAPRIFOLIACEAE
Diervilla lonicera Miller—Anderson, Cheatham
Diervilla sessilifolia Buckley var. rivularis (Gattinger) Ahles—Marion, Polk, Unicoi

CARYOPHYLLACEAE
Arenaria groenlandica (Retz.) Sprengel var. groenlandica—Carter
Arenaria lanuginosa (Michx.) Rohrback—Giles
Silene caroliniana Walt. var. pensylvanica (Michx.) Fern.—Carter, Sullivan, Washington

CHENOPODIACEAE
Chenopodium standleyanum Aellen—Cumberland, Knox, Maury

CYPERACEAE
Carex misera Buckley—Blount, Sevier
Carex muricata L. var. ruthiii (Mackenzie) Gl.—Blount, Cocke, Sevier
Cymophyllus fraseri (Andr.) Mackenzie—Blue Ridge Province of East Tennessee
Fimbristylis puberula (Michx.) Vahl—Blount, Coffee, Rutherford
Rhynchospora perplexa Britt. ex Small—Coffee, Marion, Warren
Scirpus cespitosus L.—Carter, Sevier
Scirpus expansus Fern.—Carter, Unicoi

DROSERACEAE
Drosera capillaris Poir.—Bledsoe, Cumberland, McNairy, Van Buren
Drosera leucantha Shinners—Coffee, Franklin, Lewis, Warren

ERICACEAE
Gaylussacia dumosa (Andrz.) T. & G.—Coffee, Franklin
Gaylussacia ursina (M. A. Curtis) T. & G. ex Gray—Blount, Polk, Sevier
Leucothoe racemosa (L.) Gray—Morgan
Monotropsis odorata Schweinitz—Blount, Grundy, Sevier
Pieris floribunda (Pursh) B. & H.—Sevier
Vaccinium macrocarpon Ait.—Johnson, Polk

FABACEAE
Astragalus tennesseensis Gray—Central Basin
Lespedeza angustifolia (Pursh) Ell.—Coffee, Lincoln, Warren
Petalostemum candidum (Willd.) Michx.—Davidson, Rutherford, Wilson
Petalostemum foliosum Gray—Davidson, Rutherford, Williamson
Petalostemum gattingeri Heller—Central Basin and southern Great Valley
Psoralea subacaulis T. & G.—Central Basin

FUMARIACEAE
Adlumia fungosa (Ait.) Greene—Blount, Campbell, Cocke, Sevier, Washington

GENTIANACEAE
Gentiana austromontana Pringle & Sharp—Carter, Johnson, Unicoi, Washington
Gentiana linearis Froel.—Sevier

HAMAMELIDACEAE
Fothergilla major (Sims) Lodd—Grainger, Greene, Scott, Sevier

HYDROPHYLLACEAE
Hydrophyllum virginianum L.—Sevier, Sullivan
Phacelia ranunculacea (Nutt.) Constance—Montgomery, Obion

HYMENOPHYLLACEAE
Trichomanes boschianum Sterm. ex Bosch.—Franklin, Monroe, Scott

HYPERICACEAE
Hypericum adpressum Barton—Coffee, Warren
Hypericum graveolens Buckley—Johnson, Sevier
Hypericum mitchellianum Rydb.—Blount, Sevier, Unicoi

IRIADACEAE
Iris prismatica Pursh—Coffee, Warren

JUNCACEAE
Juncus gymnocarpus Coville—Blount, Sevier

LAMIACEAE
Meehania cordata (Nutt.) Britt.—Hancock
Pycnanthemum montanum Michx.—Cocke, Johnson, Sevier, Unicoi
Salvia reflexa Hornem.—Davidson
Scutellaria saxatilis Riddell—Cumberland, Johnson
Stachys clingmanii Small—Blount, Monroe, Sevier

LENTIBULARIACEAE
Utricularia subulata L.—Cumberland, Fentress, Morgan

LILIACEAE
Lilium canadense L.—Middle & East Tennessee; commercially exploited
Lilium michiganense Farw.—Cumberlands & Middle Tennessee; commercially exploited
Tofieldia racemosa (Walt.) BSP.—Coffee
Trillium lancifolium Raf.—Hamilton, Marion
Schoenolirion croceum (Michx.) Gray—Davidson, Rutherford, Wilson
Xerophyllum asphodeloides (L.) Nutt.—Blount, Greene
Zigadenus leimanthoides Gray—Coffee, Grundy

LYCOPODIACEAE
Lycopodium alopecuroides L.—Coffee, Fentress
Lycopodium selago L.—Carter, Sevier

LYTHRACEAE
Didiplis diandra (DC.) Wood—Hardin, Henry

NYCTAGINACEAE
Mirabilis albida (Walt.) Heimerl—Rutherford
Mirabilis linearis (Pursh) Heimerl—Maury

ONAGRACEAE
Ludwigia sphaerocarpa Ell.—Coffee, Franklin

ORCHIDACEAE
Platanthera psycodes (L.) Lindley—Carter, Johnson, Monroe, Sevier

POLEMONIACEAE
Phlox bifida Beck—Davidson, Rutherford, Wilson

POLYGONACEAE
Eriogonum longifolium Nutt. var. harperi (Goodm.) Reveal—DeKalb, Putnam, Smith
Polygonum arifolium L.—Henry, Johnson

PORTULACACEAE
Talinum calcaricum Ware—Central Basin
Talinum mengesii W. Wolf—Grundy, Marion, Morgan, Rhea ?

POTAMOGETONACEAE
Potamogeton tennesseensis Fern.—Cumberland, Fentress, Morgan, Polk

PRIMULACEAE
Lysimachia fraseri Duby—Hamilton, Polk, Sevier, Stewart

RANUNCULACEAE
Cimicifuga rubifolia Kearney—River bluffs of the Ridge and Valley Province and Montgomery Co.
Hydrastis canadensis L.—Infrequent over the state and commercially exploited
Ranunculus flabellaris Raf.—Lake, Obion, Robertson
Thalictrum coriaceum (Britt.) Small—Claiborne, Sevier

ROSACEAE
Amelanchier sanguinea (Pursh) DC.—Cumberland, Sevier

SANTALACEAE
Buckleya distichophylla (Nutt.) Torr.—Carter, Cocke, Greene, Sullivan, Unicoi, Washington

SAXIFRAGACEAE
Heuchera longiflora Rydb. var. aceroides (Rydb.) Rosend., Butt. & Lak.—Cocke, Greene, Sullivan

SCHISANDRACEAE
Schisandra glabra (Brickell) Rehder—Shelby, Tipton

SCROPHULARIACEAE
Pedicularis lanceolata Michx.—Coffee, Union, Warren

TAXACEAE
Taxus canadensis Marsh.—Pickett

VIOLACEAE
Viola egglestoni Brainerd—Central Basin

VITACEAE
Vitis rupestris Scheele—Davidson

XYRIDACEAE
Xyris fimbriata Ell.—Coffee

SPECIAL CONCERN

ACERACEAE
Acer saccharum Marsh. ssp. leucoderme (Small) Desmarais—Hamilton, Polk

APIACEAE
Heracleum maximum Bartram—Blount, Cocke, Unicoi

ARISTOLOCHIACEAE
Hexastylis virginica (L.) Small—Johnson

ASCLEPIADACEAE
Asclepias purpurascens L.—Stewart

ASPLENIACEAE
Dryopteris cristata (L.) Gray—Carter, Johnson
Thelypteris phegopteris (L.) Slosson—Sevier
Woodsia scopulina D. C. Eat.—Carter, Johnson, Unicoi
Woodwardia virginica (L.) Smith—Blount, Coffee, Marion, Van Buren

ASTERACEAE
Carduus spinosissimus Walt.—Bradley
Spilanthes americana (Mutis(Hieron. var. repens (Walt.) A. H. Moore—Lauderdale, Tipton

BRASSICACEAE
Arabis glabra (L.) Bernh.—Wayne
Arabis hirsuta (L.) Scop.—Wilson
Arabis patens Sullivant—Knox

CAMPANULACEAE
Campanula aparinoides Pursh—Blount, Carter, Johnson

CAPRIFOLIACEAE
Lonicera canadensis Marsh.—Sevier
Lonicera dioica L.—Johnson, Washington, Duncan (1967)
Lonicera flava Sims—Franklin, Hamilton, Duncan (1967)
Lonicera prolifera (Kirchner) Rehder—Davidson, Duncan (1967)

CARYOPHYLLACEAE
Stellaria alsine Grimm—Carter, Cocke
Stellaria longifolia Muhl.—Blount, Johnson

CELASTRACEAE
Euonymus obovatus Nutt.—Knox, Monroe, Sevier

CISTACEAE
Helianthemum bicknellii Fern.—Blount
Helianthemum canadense (L.) Michx.—Blount, Sevier
Helianthemum propinquum Bicknell—Blount, Coffee, Monroe
Lechea leggettii Britt. & Holl.—Coffee, White

CYPERACEAE
Bulbostylis ciliatifolia (Ell.) Fern. var. coarctata (Ell.) Kral—Fayette
Carex barrattii Schwein. & Torr.—Coffee
Carex buxbaumii Wahlenb.—Cumberland
Carex comosa Boott—Lewis, Montgomery
Carex crawei Dewey—Davidson, Wilson
Carex davisii Schwein. & Torr.—Davidson
Carex gravida Bailey—Knox, Meigs, Montgomery
Carex hirtifolia Mackenzie—Coffee
Carex howei Mackenzie—Chester, Hardeman
Carex lacustris Willd.—Montgomery
Carex lanuginosa Michx.—Madison, Underwood (1945)
Carex molesta Mackenzie—Davidson, Underwood (1945)
Carex muricata L. var. angustata Carey—Campbell, Grundy, Johnson
Carex muskingumensis Schwein.—Montgomery
Carex oxylepis Torr. & Hook var. pubescens J. K. Underwood—Cheatham, Lawrence, Rutherford
Carex pedunculata Muhl.—Marion
Carex reniformis (Bailey) Small—Gibson, Hancock, Humphreys
Carex rostrata Stokes—Johnson, Robertson, Sullivan
Carex schweinitzii Dewey—Knox, Underwood (1945)
Carex trisperma Dewey—Blount
Carex vestita Willd.—Franklin
Cyperus brevifolioides Thieret & Delahoussaye—Washington
Cyperus engelmannii Steudel—Obion
Cyperus dentatus Torr.—Polk
Cyperus haspan L.—Coffee, Warren
Cyperus plukenetii Fern.—McNairy, Sequatchie
Eleocharis equisetoides (Ell.) Torr.—Fentress, Putnam, Rutherford
Eleocharis intermedia (Muhl.) Schult.—Campbell, Union
Eriophorum virginicum L.—Cumberland, Fentress, Johnson, Marion, Polk
Fuirene squarrosa Michx.—Henderson, Polk
Rhynchospora alba (L.) Vahl.—Johnson
Rhynchospora macrostachya Torr.—Coffee
Scleria verticillata Muhl. ex Willd.—Coffee

ERICACEAE
Menziesia pilosa (Michx.) Juss.—Carter, Sevier
Vaccinium tenellum Ait.—Lincoln

FABACEAE
Rhynchosia latifolia Nutt.—Chester

FAGACEAE
Quercus nuttallii Palm.—Gibson, Obion, Shelby

HYDROCHARITACEAE
Elodea nuttallii (Planchon) St. John—Lake

IRIDACEAE
Iris fulva Ker—Dyer, Lake, Gibson

LAMIACEAE
Salvia azurea Lam. var. grandiflora Benth.—Henry

LILIACEAE
Clintonia borealis (Ait.) Raf.—Cocke, Sevier, Unicoi
Erythronium rostratum Wolf—Lawrence, Wayne
Smilax laurifolia L.—Polk
Streptopus roseus Michx.—Sevier
Trillium cernuum L.—Carter

LOGANIACEAE
Gelsemium sempervirens (L.) Ait. f.—Hamilton, Rhea

MAGNOLIACEAE
   Magnolia virginiana L.—McNairy, Polk

NAJADACEAE
   Najas quadalupensis (Spreng.)—Grundy, Lake
   Najas minor All. Magnus—Grundy

ONAGRACEAE
   Epilobium angustifolium L.—Unicoi
   Ludwigia leptocarpa (Nutt.) Hara—Lake, Shelby

OPHIOGLOSSACEAE
   Botrychium alabamense Maxon—Hamblen, Monroe, Putnam
   Botrychium matricariaefolium A. Br.—Sevier, Washington
   Ophioglossum crotalophoroides Walt.—Shelby

OROBANCHACEAE
   Orobanche ludoviciana Nutt.—Lauderdale

ORCHIDACEAE
   Spiranthes ovalis Linkley—Montgomery, Roane

PINACEAE
   Abies fraseri (Pursh) Poir.—Carter, Sevier
   Tsuga caroliniana Engelm.—Carter, Johnson, Sevier, Unicoi

POACEAE
   Agrostis borealis Hartm. var. americana (Scribn.) Fern.—
      Carter, Sharp *et al.* (1956)
   Bromus ciliatus L.—Sevier
   Elymus interruptus Buckley—Putnam
   Elymus svensonii Church—Davidson, Smith
   Festuca paradoxa Desv.—Coffee, Fayette
   Glyceria acutiflora Torr.—Grundy, Warren
   Glyceria grandis S. Wats.—Johnson
   Glyceria laxa Scrib.—Johnson
   Gymnopogon brevifolius Trinius—Coffee
   Muhlenbergia torreyana (Schult.) Hitchc.—Coffee, Svenson
      (1941)
   Panicum curtifolium Nash—Franklin
   Panicum ensifolium Baldwin ex Ell.—Fentress
   Panicum hemitomon Schult.—Coffee, Warren
   Panicum leucothrix Nash—Coffee, Warren
   Paspalum bifidum (Bertoloni) Nash—McNairy
   Poa palustris L.—Carter
   Poa saltuensis Fern. & Wieg.—Blount, Van Buren
   Sacciolepis striata (L.) Nash—Polk
   Tridens flavus (L.) Hitchc. var. chapmanii (Small) Shinners—
      McNairy

POLYGALACEAE
   Polygala boykinii Nutt.—Rutherford
   Polygala mariana Miller—McNairy
   Polygala nana (Michx.) DC.—Rhea
   Polygala nuttallii T. & G.—Coffee

POTAMOGETONACEAE
   Potamogeton epihydrus Raf.—Blount, Johnson, Polk
   Potamogeton foliosus Raf. var. macellus Fern.—Cumberland

PRIMULACEAE
   Lysimachia terrestris (L.) BSP.—Coffee, Unicoi, Warren
   Trientalis borealis Raf.—Sullivan

RANUNCULACEAE
   Ranunculus allegheniensis Britt.—Carter, Hawkins, Morgan

ROSACEAE
   Potentilla tridentata Ait.—Carter, Unicoi
   Prunus virginiana L.—Blount, Polk, Sevier

RUBIACEAE
   Galium asprellum Michx.—Johnson
   Galium mollugo L.—Carter, Davidson

SALICACEAE
   Populus grandidentata Michx.—Morgan, Sevier, Washington

SCROPHULARIACEAE
   Seymeria cassioides (J. F. Gmelin) Blake—Bradley, Pennell
      (1935)

SPARGANIACEAE
   Sparganium androcladum (Engelm.) Morong—Blount

SYMPLOCACEAE
   Symplocos tinctoria (L.) L'Her.—Carroll, Polk

ULMACEAE
   Ulmus crassifolia Nutt.—Shelby

VERBENACEAE
   Verbena stricta Vent.—Henry, Obion, Weakley

XYRIDACEAE
   Xyris iridifolia Chapman—Coffee, Grundy, Warren
   Xyris jupicai Richard—Putnam, Warren

LITERATURE CITED

Bailey, L. H. 1961. Manual of Cultivated Plants. The Macmillan Co., New York. 1116 pp.

Dennis, W. M. 1976. A biosystematic study of *Clematis* Section *Viorna* Subsection *Viornae*. Ph.D. Diss. The University of Tennessee, Knoxville, 176 pp.

Duncan, W. H. 1967. Woody vines of the Southeastern States. Sida 3: 1-76.

Farmer, J. A. and J. L. Thomas. 1969. Disjunction and endemism in *Croton alabamensis*. Rhodora 71: 94-103.

Gattinger, A. 1901. The Flora of Tennessee and a Philosophy of Botany. Gospel Advocate Publishing Co., Nashville. 296 pp.

Goff, F. G., R. L. Stephenson, and D. Lewis. 1975. Rare, endangered and endemic taxa and habitats of the East Tennessee Development District. ORNL EDFB-IBP-75-2.

Krochmal, A. 1968. Medicinal plants and Appalachia. Economic Botany 22: 332-337.

Lord, L. P. 1960. The genus *Saxifraga* in the southern Appalachians, Ph.D. Diss. The University of Tennessee, Knoxville, 106 pp. ,

Pennell, F. W. 1935. The Scrophulariaceae of eastern temperate North America. Monograph No. 1, Acad. Nat. Sci., Philadelphia, Pa.

Public Law 93-205 (1973). Endangered Species Act of 1973. 87 Statute 884-903. 93rd Congress, S. 1983, 21 pp.

Radford, A. E., H. E. Ahles, and C. R. Bell. 1968. Manual of the Vascular Flora of the Carolinas. University of North Carolina Press, Chapel Hill. 1183 pp.

Richards, E. L. 1968. A monograph of the genus *Ratibida*. Rhodora 70: 348-393.

Scribner, F. L. 1894. The grasses of Tennessee. Part II. Bull. Agr. Exp. Sta. Univ. of Tennessee, Vol. 7: 1-141.

Sharp, A. J. 1948. Wildflowers in need of protection in Tennessee. The Volunteer Gardner 2: 11.

Sharp, A. J., R. E. Shanks, J. K. Underwood, and E. McGilliard. 1956. A preliminary checklist of monocots in Tennessee. Unpubl. report. The University of Tennessee, Knoxville. 33 pp.

Sharp, A. J., R. E. Shanks, H. L. Sherman, and D. H. Norris. 1960. A preliminary checklist of dicots in Tennessee. Unpubl. report. The University of Tennessee, Knoxville. 114 pp.

Sharp, A. J. 1974. Rare plants of Tennessee. The Tennessee Conservationist 40: 20-21.

Shields, A. R. 1939. *Trichomanes petersii* A. Gray in North Carolina. Amer. Fern Jour. 29: 113-114.

Styron, C. E. 1976 (Ed.) Symposium on rare, endangered, and threatened biota of the Southeast The ASB Bulletin 23: 137-167.

Svenson, H. K. 1941. Notes on the Tennessee flora. Jour. Tenn. Acad. Sci. 16: 111-160.

Underwood, J. K. 1945. The genus *Carex* in Tennessee. Am. Midl. Nat. 33: 613-643.

U. S. Department of the Interior. Threatened or Endangered Fauna or Flora. Federal Register 40, No. 127, 1 July 1975, 27824-27924.

————. Endangered and Threatened Species. Federal Register 41, No. 117, 16 June 1976, 24524-24572.

Page 301

Scogin, R.

1977. Sperm Whale Oil and the Jojoba Shrub. <u>Oceans</u> 10(4): 65-66.

Page 302

Wilkes, G.

1977. Breeding Crisis for our Crops: Is the Gene Pool Drying Up?
<u>Horticulture</u> 55(4): 52-59.

The Role of Botanical Gardens in the Conservation of Endangered Plants

Page 305

Hardy, D. S.

1976. The Role Played by the Botanic Garden in the Conservation and
Preservation of Plants With Special Reference to the Succulent
Plants of South Africa. <u>Excelsa</u> 6: 26-28.

109722

DATE DUE